T0191916

Advances in Intelligent Systems and Computing

Volume 760

Series editor

Janusz Kacprzyk, Polish Academy of Sciences, Warsaw, Poland
e-mail: kacprzyk@ibspan.waw.pl

The series "Advances in Intelligent Systems and Computing" contains publications on theory, applications, and design methods of Intelligent Systems and Intelligent Computing. Virtually all disciplines such as engineering, natural sciences, computer and information science, ICT, economics, business, e-commerce, environment, healthcare, life science are covered. The list of topics spans all the areas of modern intelligent systems and computing such as: computational intelligence, soft computing including neural networks, fuzzy systems, evolutionary computing and the fusion of these paradigms, social intelligence, ambient intelligence, computational neuroscience, artificial life, virtual worlds and society, cognitive science and systems, Perception and Vision, DNA and immune based systems, self-organizing and adaptive systems, e-Learning and teaching, human-centered and human-centric computing, recommender systems, intelligent control, robotics and mechatronics including human-machine teaming, knowledge-based paradigms, learning paradigms, machine ethics, intelligent data analysis, knowledge management, intelligent agents, intelligent decision making and support, intelligent network security, trust management, interactive entertainment, Web intelligence and multimedia.

The publications within "Advances in Intelligent Systems and Computing" are primarily proceedings of important conferences, symposia and congresses. They cover significant recent developments in the field, both of a foundational and applicable character. An important characteristic feature of the series is the short publication time and world-wide distribution. This permits a rapid and broad dissemination of research results.

Advisory Board

Chairman

Nikhil R. Pal, Indian Statistical Institute, Kolkata, India
e-mail: nikhil@isical.ac.in

Members

Rafael Bello Perez, Universidad Central "Marta Abreu" de Las Villas, Santa Clara, Cuba
e-mail: rbellop@uclv.edu.cu

Emilio S. Corchado, University of Salamanca, Salamanca, Spain
e-mail: escorchado@usal.es

Hani Hagras, University of Essex, Colchester, UK
e-mail: hani@essex.ac.uk

László T. Kóczy, Széchenyi István University, Győr, Hungary
e-mail: koczy@sze.hu

Vladik Kreinovich, University of Texas at El Paso, El Paso, USA
e-mail: vladik@utep.edu

Chin-Teng Lin, National Chiao Tung University, Hsinchu, Taiwan
e-mail: ctlin@mail.nctu.edu.tw

Jie Lu, University of Technology, Sydney, Australia
e-mail: Jie.Lu@uts.edu.au

Patricia Melin, Tijuana Institute of Technology, Tijuana, Mexico
e-mail: epmelin@hafsamx.org

Nadia Nedjah, State University of Rio de Janeiro, Rio de Janeiro, Brazil
e-mail: nadia@eng.uerj.br

Ngoc Thanh Nguyen, Wroclaw University of Technology, Wroclaw, Poland
e-mail: Ngoc-Thanh.Nguyen@pwr.edu.pl

Jun Wang, The Chinese University of Hong Kong, Shatin, Hong Kong
e-mail: jwang@mae.cuhk.edu.hk

More information about this series at http://www.springer.com/series/11156

Sanjiv K. Bhatia · Shailesh Tiwari
Krishn K. Mishra · Munesh C. Trivedi
Editors

Advances in Computer Communication and Computational Sciences

Proceedings of IC4S 2017, Volume 2

 Springer

Editors
Sanjiv K. Bhatia
Department of Mathematics
 & Computer Science
University of Missouri
St. Louis, MO
USA

Shailesh Tiwari
Computer Science Engineering
 Department
ABES Engineering College
Ghaziabad
India

Krishn K. Mishra
Department of Computer Science
 & Engineering
Motilal Nehru National Institute
 of Technology
Allahabad, Uttar Pradesh
India

Munesh C. Trivedi
Department of Computer Science
 & Engineering
ABES Engineering College
Ghaziabad
India

ISSN 2194-5357   ISSN 2194-5365 (electronic)
Advances in Intelligent Systems and Computing
ISBN 978-981-13-0343-2   ISBN 978-981-13-0344-9 (eBook)
https://doi.org/10.1007/978-981-13-0344-9

Library of Congress Control Number: 2018940399

© Springer Nature Singapore Pte Ltd. 2019
This work is subject to copyright. All rights are reserved by the Publisher, whether the whole or part of the material is concerned, specifically the rights of translation, reprinting, reuse of illustrations, recitation, broadcasting, reproduction on microfilms or in any other physical way, and transmission or information storage and retrieval, electronic adaptation, computer software, or by similar or dissimilar methodology now known or hereafter developed.
The use of general descriptive names, registered names, trademarks, service marks, etc. in this publication does not imply, even in the absence of a specific statement, that such names are exempt from the relevant protective laws and regulations and therefore free for general use.
The publisher, the authors and the editors are safe to assume that the advice and information in this book are believed to be true and accurate at the date of publication. Neither the publisher nor the authors or the editors give a warranty, express or implied, with respect to the material contained herein or for any errors or omissions that may have been made. The publisher remains neutral with regard to jurisdictional claims in published maps and institutional affiliations.

Printed on acid-free paper

This Springer imprint is published by the registered company Springer Nature Singapore Pte Ltd.
The registered company address is: 152 Beach Road, #21-01/04 Gateway East, Singapore 189721, Singapore

Preface

The IC4S is a major multidisciplinary conference organized with the objective of bringing together researchers, developers, and practitioners from academia and industry working in all areas of computer and computational sciences. It is organized specifically to help computer industry to derive the advances in next-generation computer and communication technology. Researchers invited to speak will present the latest developments and technical solutions.

Technological developments all over the world are dependent upon globalization of various research activities. Exchange of information and innovative ideas is necessary to accelerate the development of technology. Keeping this ideology in preference, the International Conference on Computer, Communication and Computational Sciences (IC4S 2017) has been organized at Swissôtel Resort Phuket Patong Beach, Thailand, during October 11–12, 2017.

This is the second time the International Conference on Computer, Communication and Computational Sciences has been organized with a foreseen objective of enhancing the research activities at a large scale. Technical Program Committee and Advisory Board of IC4S include eminent academicians, researchers, and practitioners from abroad as well as from all over the nation.

In this book, selected manuscripts have been subdivided into various tracks named intelligent hardware and software design, advanced communications, intelligent computing techniques, Web and informatics, and intelligent image processing. A sincere effort has been made to make it an immense source of knowledge for all and includes 85 manuscripts. The selected manuscripts have gone through a rigorous review process and are revised by authors after incorporating the suggestions of the reviewers. These manuscripts have been presented at IC4S 2017 in six different technical sessions. A gift voucher for the value of EUR 150 (one hundred and fifty euros) for session-wise Best Paper Presentation of paper has been awarded by Springer Nature at IC4S 2017 through which authors can select and buy e-book from Springer Nature link.

IC4S 2017 received around 425 submissions from around 650 authors of 15 different countries such as USA, Iceland, China, Saudi Arabia, South Africa, Taiwan, and Malaysia. Each submission has gone through the plagiarism check. On the basis of plagiarism report, each submission was rigorously reviewed by at least two reviewers with an average of 2.3 per reviewer. Even some submissions have more than two reviews. On the basis of these reviews, 85 high-quality papers were selected for publication in this proceedings volume, with an acceptance rate of 20%.

We are thankful to the keynote speakers—Prof. Phalguni Gupta, IIT Kanpur, India; Prof. Jong-Myon Kim, University of Ulsan, Ulsan, Republic of Korea; Dr. Nitin Singh, MNNIT Allahabad, India; Mr. Aninda Bose, Senior Editor, Hard Sciences, Springer Nature, India; and Dr. Brajesh Kumar Singh, RBS College, Agra, India, for enlightening the participants with their knowledge and insights. We are also thankful to the delegates and the authors for their participation and interest in IC4S 2017 as a platform to share their ideas and innovation. We are also thankful to Prof. Dr. Janusz Kacprzyk, Series Editor, AISC, Springer Nature, for providing continuous guidance and support. Also, we extend our heartfelt gratitude to the reviewers and Technical Program Committee Members for showing their concern and efforts in the review process. We are indeed thankful to everyone directly or indirectly associated with the conference organizing for team leading it toward the success.

Although utmost care has been taken in compilation and editing, a few errors may still occur. We request the participants to bear with such errors and lapses (if any). We wish you all the best.

St. Louis, USA   Sanjiv K. Bhatia
Ghaziabad, India   Shailesh Tiwari
Allahabad, India   Krishn K. Mishra
Ghaziabad, India   Munesh C. Trivedi

Contents

Part II Intelligent Image Processing

About the Editors

Dr. Sanjiv K. Bhatia received his Ph.D. in Computer Science from the University of Nebraska, Lincoln, in 1991. He presently works as Professor and Graduate Director (Computer Science) at the University of Missouri, St. Louis. His primary areas of research include image databases, digital image processing, and computer vision. He has published over 40 articles on those areas. He has also consulted extensively with industry for commercial and military applications of computer vision. He is an expert in system programming and has worked on real-time and embedded applications. He serves on the organizing committee of a number of conferences and on the editorial board of international journals. He has taught a broad range of courses in computer science and was the recipient of Chancellor's Award for Excellence in Teaching in 2015. He is a senior member of ACM.

Dr. Shailesh Tiwari is currently working as Professor in Computer Science and Engineering Department, ABES Engineering College, Ghaziabad, India. He is also administratively heading the department. He is an alumnus of Motilal Nehru National Institute of Technology Allahabad, India. He has more than 15 years of experience in teaching, research, and academic administration. His primary areas of research are software testing, implementation of optimization algorithms, and machine learning techniques in software engineering. He has also published more than 40 publications in international journals and in proceedings of internal conferences of repute. He has served as a program committee member of several conferences and edited Scopus and E-SCI-indexed journals. He has also organized several international conferences under the banner of IEEE and Springer Nature. He is a Senior Member of IEEE, Member of IEEE Computer Society, and Executive Committee Member of IEEE Uttar Pradesh Section. He is a member of reviewer and editorial board of several international journals and conferences.

Dr. Krishn K. Mishra is currently working as Visiting Faculty, Department of Mathematics and Computer Science, University of Missouri, St. Louis, USA. He is an alumnus of Motilal Nehru National Institute of Technology Allahabad, India, which is also his base working institute. His primary area of research includes evolutionary algorithms, optimization techniques, and design and analysis of algorithms. He has also published more than 50 publications in international journals and in proceedings of international conferences of repute. He is serving as a program committee member of several conferences and also editing few Scopus and SCI-indexed journals. He has 15 years of teaching and research experience during which he made all his efforts to bridge the gaps between teaching and research.

Dr. Munesh C. Trivedi is currently working as Professor in Computer Science and Engineering Department, ABES Engineering College, Ghaziabad, India. He has rich experience in teaching the undergraduate and postgraduate classes. He has published 20 textbooks and 81 research papers in international journals and proceedings of international conferences. He has organized several international conferences technically sponsored by IEEE, ACM, and Springer Nature. He has also worked as member of organizing committee in several IEEE international conferences in India and abroad. Dr. Trivedi is on the review panel of IEEE Computer Society, International Journal of Network Security, Pattern Recognition Letters, and Computer & Education (Elsevier's Journal). He is also Member of Editorial Board for International Journal of Computer Application, Journal of Modeling and Simulation in Design and Manufacturing (JMSDM), and International Journal of Emerging Trends and Technology in Computer Science and Engineering. He has been appointed member of board of studies as well as in Syllabus Committee of different private Indian universities and member of organizing committee for various national and international seminars/workshops. He is Executive Committee Member of IEEE UP Section, IEEE Computer Society Chapter India Council, and also IEEE Asia Pacific Region 10. He is an active member of IEEE Computer Society, International Association of Computer Science and Information Technology, Computer Society of India, International Association of Engineers, and a life member of ISTE.

Part I
Web and Informatics

Internet of Thing and Smart City: State of the Art and Future Trends

Ako A. Jaafar, Karzan H. Sharif, Mazen I. Ghareb and Dayang N. A. Jawawi

Abstract Fast growing of cities, urban places, and population presents major challenges in our daily lives. Finding proper solution from different perspectives of cities became a concern by both researchers and industries concern. Smart city (SC) is the answer to overcome this issue. Internet of thing (IoT) is a crucial part of SC which has a tremendous impact on all the cities' sectors such as governance, health care, education, environment, and transportation. This paper provides a state of the art on Internet of thing (IoT)-based smart city (SC) from platform, architecture, application domain, and technology perspectives, also it concludes the relations between different technologies used in developing smart cities. Challenges and future trends of SC based on IoT will be discussed. As this research conducts, the SC solution based on IoT still confronts many challenges and problems which required further researches.

Keywords Smart City · Internet of thing · SC · IoTs · Cloud · RFID · WSN

A. A. Jaafar (✉)
Computer Science Department, University Technology Malaysia, Johor Bahru, Malaysia
e-mail: ako.abubakr@uhd.edu.iq; akodyar@yahoo.com

A. A. Jaafar · K. H. Sharif · M. I. Ghareb
Computer Science Department, University of Human Development, Sulaymaniyah, Iraq
e-mail: karzan.hussein@uhd.edu.iq

M. I. Ghareb
e-mail: mazen.ismaeel@uhd.edu.iq

M. I. Ghareb
Computer Science Department, University of Huddersfield, Huddersfield, UK

D. N. A. Jawawi
Software Engineering Department, University Technology Malaysia, Johor Bahru, Malaysia
e-mail: dayang@utm.my

© Springer Nature Singapore Pte Ltd. 2019
S. K. Bhatia et al. (eds.), *Advances in Computer Communication and Computational Sciences*, Advances in Intelligent Systems and Computing 760, https://doi.org/10.1007/978-981-13-0344-9_1

1   Introduction

Fast growing of cities, urban places, and population presents major challenges in our daily lives. The research prediction anticipates that the majority of the world's population will live in the cities in a future; according to Unisex by 2050, 70% of all population lives in urban cities [1]; currently, more than half of the world's population lives in the cities [2]. While population is increasing, the cities confront more and more demands in governance, health, transportation, energy, water, and education services. However, all of these services are integrated together in the cities. In that context footprints to come up with environmental solution has become research and industries concern [3–5].

The heterogeneous systems and devices participate in SC due to different types of service from multiple sources, and SC collects massive amount of data from each source and device to provide required service to its citizens and industries. The powerful SC must be able to integrate all encompassed domain together and finding smartness across collected data [6]. Therefore, advance technology is the answer to find the needed solution for SC. From that perspective, IoT techniques (IoT) are key technology paradigm to enable SC to handle a broad range of applications [7]. IoT is the idea of providing the global interconnection of devices and objects, sensors, mobile phones, actuators, computers, etc., based on unique addressing and geographical location around the world [8, 9]. In the IoT, data will be collected using sensor services and analyzed by an analyzing system for various domains that use IoT infrastructure [10].

This paper provides a state of the art on SC and IoT as well as the impact of IoT on enabling SC. The rest of this paper is organized as follows: Sect. 2: IoT in SC—this section demonstrates the state of the art of using IoT in SC from various points such as using cloud, different architectures, and technology for IoT in SC. Section 3 presents the analysis of previous works based on different aspects of IoT in SC. Beside that, it provides the current issue and challenges of IoT and SC.

2   IoT in SC

For the last decade, IoT has been one of the most focused research topics by researchers and industries. While the SC is enabled by available technologies in IoT [9], a lot of previous work concentrated on IoT technologies and its implementations in SC. IoT implementation in SC has been discussed from many perspectives such as architecture, platform, devices, technologies. Figure 1 shows the IoT perspective in SC.

Fig. 1 IoT perspective in SC

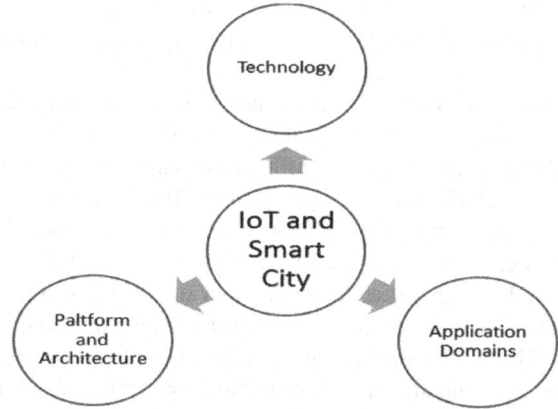

2.1 SC-IOT Platforms

Smart cities promise to provide better services and lifestyles to their citizens. Since that in SC various domains and heterogeneous resources are collaborating together such as transportation and traffic, electricity and water energy, health and education services, implementing such system requires large-scale, scalable, and open platforms to facilitate collaboration in order to integrate multiple systems in one holistic SC [11]. On the one hand, SC must be set up in accepting and overseeing gigantic measure of data. On the other hand, they have to be exceptionally accessible and give a uniform interface to all deployed technologies. Furthermore, the governments spend a lot of money on IoT technologies for modern city [12]. Therefore, adequate platform is required for SC based on IoT solution. In that context, many SCs have been deployed by researchers and industries.

SEN2SOCS platform in SmartSantander project is proposed to facilitate interaction between sensors and social networks and collect information to expose any irregular environmental situation in a city. This platform integrates different kinds of IoT technologies and application domains in large scale, while most current platforms are deployed for specific application domains. Furthermore, natural language generation (NLG) system is embedded into the platform that is able to collect data from sensors and convert into the understandable message by humans [13, 14].

CityPlus platform provides real-time processing and large-scale analysis for specific domains such as car parking availability, air pollution, and finding most efficient public transportation [15]. Civitas platform provides integration between multi-entities such as public institutions, companies, people, and householders through proposed devices called Civitas plug which allows different entities to consume/provide information to/from SC.

CHARIOT facilitates integration between different types of applications and computing group collection such as smart grid, smart home, public cloud, and micro-data center and cloudlets. In CHARIOT, each computing group collection consists of three categories of computing group (1) edge category, (2) cloudlet category, and (3) cloud category which promise integration of various existing applications and computational models which are related to SC [11].

A star-city platform integrates several heterogeneous sources based on semantic technologies to predict traffic congestions [16]. Barcelona Intelligent City offers a SC platform to collect information and building services from different deployed sensors; their goal is to develop applicable platform to both large and small cities [17].

Implementation of SC platform contains various technologies such as cloud computing, architecture, security and privacy, data analytics [18]. SmartSantander platform has been presented with a four-tiered architecture to provide sensor and social networking collaboration to deliver valuable information [2].

The SC platform categorized into [19] into four categories which are government-based, company-based, production-/enterprise-based, and pure business-oriented. They propose cloud-based parking middleware to solve car parking issues using cloud technologies which consists of four-layer architecture [19]. Another project of SC platform breaks down IoT platform into three different categories which are network-centric, cloud-centric, and data-centric [18]. European Platform for Intelligent Cities (EPIC) which combine different technologies of IoT, cloud, and semantic Web to implement SC in all European countries. SlapOS is another project that proposes an open SC platform which is a combination between IoT and cloud technologies [20]. The second large city in Denmark works toward being a SC, and in the process of this transformation, they propose open data platform to allow peoples and organizations to access city data to boost their innovation of services for citizens [21].

2.2   SC-IoT Architecture

The ambivalent definition of smart cities and their structure variety of SC architectures have been discussed in the literature. Here, a number of architectures will be presented to explore the fundamental components of a SC. SC architecture still suffers from lack of a standard architecture to integrate holistic SC functionalities. Consequently, the unpractical solution of the current presented work designs depends on a hypothetical approach and not on a practical one with genuine usage. Four perspectives of architecture have been recognized such as service-centric architecture, user-centric architecture, event-centric architecture, and data-centric architecture.

Service-Centric Architecture.

Architecture based on service flow and service-oriented architecture (SOA) in a SC widely have been discussed and implemented. In SOA, the communication between various services in SC services is implemented through information exchange among them and the ability of each service to go about overall movement for another service. From that point, [22] attempted to make a typical design for SC named enterprise architecture in which they proposed a five-layered generic SC architecture from logical and physical perspectives. As indicated by [23], the multilayered architecture is mandatory in order to implement IoT, which starts from bottom with an information obtaining layer, up to the top where there is an application layer.

The work in [24] gives a summary of smart city architectures, which concluded that a SC is made up of three distinctive layers. A layer where all types of digital content are stored is a first layer called information storage layer. Another layer has virtual content which is organized and presented as services for applications for users, known as the software layer. A user interface is needed as a third layer, and it is accomplished by ramifying Web applications with a way of using maps, three-dimensional images, textual content, charts, and other interface tools. Atzori et al. [25] proposes a framework as a straightforward model that is made up of three layers for IoT: first layer dedicated to node cooperation as well as information procurement known as the sensing layer which works in local and short-ranged networks, the second layer which deals with exchanging information crosswise over different networks known as the network layer, and lastly an application layer though which middleware functions and IoT applications are packed and sent; Generally [26] argued that many SC architectures consist of three layers: capillary network layer for recognizing the devices, service layers to receive data from network layer, and application layer—the data analyzer. While [27] implies that there are impediments in the practical application with three-layer architecture of IOT. So as to underline the level of IOT smart application, they present five-layer framework design which can better present the significance and components of the IOT including presentation layer, support layer for applications, a layer for network transmission followed by a layer for network access, and lastly a layer for perception.

In 2012, four layers had been proposed by [28], the physical layer is for defining the objects and devices, the second is the grid layer, and process management is the third layer operating on statistics data mining, while the control layer comprises the final layer where services are encompassed as well. Adding more, service-situated engineering is a software outline system in light of organized accumulations of discrete programming modules, known as services, which on the whole give the entire functionality of a large programming application [29].

User-Centric Architecture.

Most of current SC architectures are service-centric, and there is lack of user-centric approach with smart edge devices. For that reason, [30] proposed Aura Minora which is a novel bottom-up approach design for IOT which contains methodology to show how user-centric architecture would help end users in SC. In addition, [31] proposed five-level pyramid designs for smart urban communities where collaboration between humans and the system is taken into consideration.

Event-Centric Architecture.

This architecture deals with event identification, usage, and response. The sensors can detect any change of the events, while the system controls the output events. It is very challenging to implement this architecture in SC when heterogeneity is so high. From this point, [22] recommend a SC construction modeling in light of event-driven architecture (EDA) to which they separated smart urban areas under two parts: First part is knowledge processors (KPs), and the second part is semantic information brokers (SIBs). All the data saved in SIBs will act as a server for KPs. Once KPs would connect with SIBs, when KPs connect with SIBs, the Smart Space Access Protocol (SSAP) triggers the operations, and through it the sessions and transactions that took place between producers and consumers can be managed by KPs. In addition, [32] presented a construction modeling of event-driven SC. In these kinds of cities, the interoperability between IoTs, Internet of services (IoSs), and Internet of people (IoP) can be empowered through the use of digital artifacts so as to enable tenants with rapid respond on bigger mixture of events.

The work [33] presents an architecture that is event-based in light of SOFIA project which permits machine to machine (M2M) to collaborate with each other and set administrations as needed. In view of this architecture, they direct a contextual investigation in a vehicular setting.

Data-Centric Architectures.

From the perspective of data, [34] mentioned that conventional data organization processes cannot fulfill the prerequisites of a SC. Gathering information challenge finds an efficient approach to store data, later to be managed and analyzed in the same manner as well. In this manner, to help better in comprehending what a SC is, they propose general versatile multilayered SC architecture which contains the acquisition, transmission, vitalization, and storage of data. A secure IoT architecture for smart cities has been proposed by [35] which proposes a structure with four basic IoT architectural blocks: black network, trusted SDN controller, unified registry, and key management system. Together, those primary IoT-centric blocks allow a secure SC.

In addition, [36] provides an all-around structure with the common features after observing and combining all the current systems. At the base of the platform, there is a capillary network layer, within this layer are the actuators and the sensors, which holds the records, like records for warehouses where ancient is stored, along with metadata and real time. In this layer, data is controlled and manipulated as well, through database nodes and safety infrastructures. From here, the data is transferred to the service layer, where they integrated and simplified for use in all manners. It has the capability to receive and manage huge statistical data like open and streaming data, at the same time providing analytics services. At last, the data is manipulated further in the application layer, where useful data is produced and served to citizens via established interfaces.

The work in [37] proposes a consolidated IoT-based framework for SC advancement and urban arranging utilizing Big Data analytics. An architecture that is made up of four levels is suggested: (1) a level in charge of IoT sources, information era, and gathering called bottom level 1, (2) a second level in charge of a wide range of correspondence called intermediate level 1, an example would be between sensors, base stations, transfers, and then the Internet, (3) a third level that uses Hadoop structure to manage and process information called intermediate level 2, (4) lastly, the top level that is in charge of use and utilization of the information analysis and the results generated.

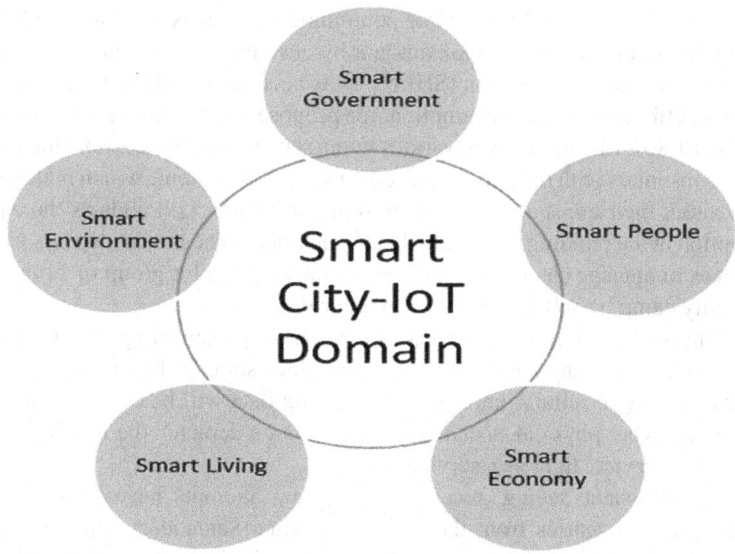

Fig. 2 IoT-based SC application domains

2.3 SC-IoT Application Domain

SC consists of many application domains that have focused on by researcher from many perspectives. The work in [38] categorized SC domain into seven categories which are resources, urban infrastructure, economy, government, living, transportation, and lastly coherency. Other vital domains in a SC are smart mobility, education, health care, and environment, all to be managed in a smart way in SC. Figure 2 shows the main application domains in SC.

As this research conducts, that SC domain mostly consists of smart people with smart living in a smart environment which has a smart economy and smart governance.

Smart Environment.

Smart environment aims to add smartness and intelligence into applications, devices, and knowledge surrounding us [39]. Many researchers concentrate on smart environment in SC like waste management, energy saving, water saving, green and healthy environment. Dynamic system scheduling has been proposed by [40] which uses IoT in SC as well as cloud-based system to provide services to multiple stakeholders in waste collection such as administrator, waste truck, smart bins, dump management.

Due to enormous increase in energy consuming and climate changes, many researchers are attempting to come up with a new smart module to save energy and provide smart environment. From that point, [41] propose a software architecture to minimize energy consumption by providing efficient energy management;

the system uses smart grid, IoT, and Fog communication and technologies, while the system collects energy consumption amount by customer using smart meter. Smart home energy management system (SHEMS) has been developed by [42] to manage and provide efficient energy consumption; the proposed system uses Zigbee technology and IEEE802.15.4 to integrate various home sensors and devices, and the system provides consumers with the real-time energy usage of their home which makes them concern about their consumption. Corotinschi and Găitan [29] address the energy consumption in the building, especially heating services. The project deploys various IoT devices to manage electricity and energy consumption for group of 11 building in university camp by self-learning action algorithm.

Enabling outdoor lighting in smarts city has been presented by [43] to guarantee energy efficiency and sustainability; the outdoor street light will be connected together by mesh or cellular network, and lighting data will be collected and controlled in one cyber physical system to enable light on demand; the lighting will be available as a service, light as a service (LaaS).

In terms of water saving, recently many new attempts have been done by researchers and industries from those attempts SmartSantander project in one of its services focused on water saving in the process of irrigation of plants and green areas in the city, and the proposed solution provides on-demand irrigations based on price schedule [14]. Lynggaard and Skouby [44] proposed four-layer models for smart home which multi-services can be deployed. The model has been applied in water waste energy which leads to save 50% to 75% of water energy in smart homes. The human behavior toward energy consumptions can be detected using smart home technology through IoT modules that integrate energy appliances including smart meters [45].

Smart Government.

Regarding the ongoing development of smart thing and smart objects for various domains and how the current domain's solutions become smart, smart government became controversial topic in both industry and academic sectors. The phrase smart government emerged in the late 90s and defined as comprehensive utilization of information technology to provide better live to its citizens [46], and lately smart government defined by [47] as a planning and managing multiple sectors and domains in a government based on integration of information, communication, and operational technologies with mentioning the great role of IoT trend in transforming government into smart governments. Smart government integrates soft domain and hard domain of SC. Also it consists of three main areas which are economic development, a vibrant political life, and supporting open innovations [48].

Smart Transportation and Mobility.

The unexpected growth of cities with huge increasing of vehicle has brought new challenges to road and traffic management like mitigation, congestions, accident, and air pollutions, and the previous available traffic management system was insufficient to build a reliable and secure traffic management system. Smart transportation, traffic, parking, and mobility are critical domains in SC; the traffic and transportation have been focused on since the end of twentieth century such as smart traffic management system based on artificial intelligence and decision support system to come up with solution of freeway and arterial traffics [49]. Lately many attempts have been done IoT-based transportation and mobility in a SC. IoT-based solution for traffic management for smart cities has been suggested by [50]. They proposed dynamic traffic signal controlling algorithm to put different situation to control traffic, in order to do that they use Raspberry Pi (RPi) as single-board computer (SBC). A new environment has been presented to manage traffic by using mobile, wireless sensor, M2M communication, social media feeds, and low-power embedded device to sense and collect traffic data, while the collected data will be used in a proposed modeling for a modular traffic where a road network can be constructed in its full resemblance by a user while still under investigation. IoT-based car parking has been addressed by many researchers to overcome with the solution of traffic conjunction. IoT–cloud-based parking services based on three-layer architecture with supporting cloud, Web, and mobile application have been proposed by [19] to solve poor car parking service. A management system for parking violations based on cloud and location-centric IoT has been proposed to enable SC to assist officers with mechanisms that make management of parking violations far more efficient, designing a PVM system with an architectural model that is efficient.

Dinh and Kim [51] focused on new business model to extend cloud architecture to allow parking provider to register their parking lots and parking seeker to find and book a vacant parking spot. IoT middleware was developed to achieve improbability among parking service and information. The A* algorithm used to find parking spot as well as the best path to go and exit to/from the parking spot based on collected information through IoT and cloud-based sources.

One of the main components of intelligent transportation system (ITS) is vehicle and their communication together. Mobility encompasses aspects such as economy and environment, which requires both good citizen behavior and high technologies and from the fact that the behavior of drivers is crucial to avoid serious problem and accident. Hernafi et al. [52] proposed a way to simulate how different types of drivers' attitude would affect the city through a smart mentality framework and also the proposed a VANET model to alert drivers when neighbors' vehicles are detected nearby in order to send and receive among neighbor's vehicles and in order to avoid or minimize vehicle accidents; the key enablers of ITS are vehicle-to-infrastructure (V2I) along with vehicle-to-vehicle (V2V) communication technologies, which both are IoT-dependent technologies, and therefore [53] proposed a mechanism to overcome with solution of congestion problem through V2I communication in a way that vehicle can communicate with installed nodes in order to find shortest path and making decisions to achieve lower congestion every moment due to its high dynamic behavior of traffics. Many researchers tried to analyze current mobility models in a SC; they address the importance of technologies and IoT in smart mobility and how open context awareness changes people's behavior toward sustainable transportation [54].

Smart Economy.

Smart economy is one of the most indicators of SC which is a key of city growth and its ability economically. Smart economy encompasses economic aspects in the city, both at a national stage with its integration into the global market and at a local stage including innovations, trademarks, entrepreneurship, and how flexible and productive the labor market can be. Smart economy categorized into three components such as entrepreneurship, productivity, and local and global interconnectedness [55]. The ability of a city to capitalize economy and innovate in the field is what defines it as intelligent. A SC must allow profits to be maximized, similar to all other models drawn for the development of economy. Since ICTs have a short life span, a constant growth of economy should be the aim in SC.

Smart People.

Transforming citizens to an aware citizen is focal dimension in SC to avoid project failing, since even advanced and costly IT solutions are subject to fail when they does not come with aim of increasing citizen awareness; therefore, many researchers addressed the smart people or smart citizens as one of the key indicators of SC [10, 39, 56]. Citizen engagement with SC with real participation via defining proper citizen engagement, accessing related information, well-training, and social engagement are the great factor toward smarter people [57]. The importance role of social media and people profile to take part in SC has been address by [58] Gamification solution has been used to change people's behavior toward SC [59].

Smart Living.

One of the most aspects of smart living is a smart home, from the point of finding comfortable live for citizens smart home challenges, and solution has begun from late 90s, while [60] propose an aware home infrastructure based on radio frequency and signaling in order to make home aware of what happened. Human tracking, individual interacting with home, and finding lost object were addressed. Recently many works have been done on smart home techniques and modules based on IoT, and [61] realize smart home based on IoT by proposing scalable architecture to integrate home devices. Through the proposed architecture, the current technology and device can be added to a new smart home via adding radio-frequency identification (RFID) tag for each home appliance and implementing RFID tags by its interface. Therefore, all devices will be controlled by intelligent devices in order to be availed to be accessed by user interface. Skouby and Lynggaard [62] propose an approach to integrate smart homes into one coherent SC based on IoT, cloud of thing, AI, and M2M communication via 5G in order to integrate smart home into a solid SC. The proposed infrastructure provides the intelligent things interconnecting and scalability to add new services and technology as well as centralizing data storage through cloud of thing (CoT). The combination of IoT, cloud, and Web service has been addressed by [62] to explore smart home. Zigbee technology, cloud technology, and JSON Web services have been used to integrate intelligence sensors, data storage, easy access, and efficient data exchanges.

Providing smart health is another aspect to facilitate smart health living; therefore, many researches and projects have been conducted in this area. Smart health can be highlighted as providing smart monitoring, health warning, and health support of patient through patient parameters such as heart condition and pulse, body temperature, and breathing to support patients in all ages especially elders and infants [63]. The attempts to establish better health service and smart live through technology and smart solution such as wearable device and hospital home concept began from late 90s [64]. Kahiwa-no-ha addressed environment friendliness, new industry creation, health, and secured living for all ages. Smart health is one of the subprojects that concern about citizen's health by collection of data using uses wristband data recorder which provides detailed health about citizens and their health conditions. The issue of rising elder's people number became global issue from the taking of them perspective [65]. Dubai SC project concentrates on health care, transportation, tourism, building, telecom, utility, and education, especially smart healthcare system [66]. In addition, [67] proposed a model to monitor elder's live by recognizing human behavior such as sitting which includes eat a meal, reading, or activities such as swimming. The model uses two type of sensors, ambient sensor to measure and monitor living environment, and the other is body sensor to monitor body's activity. A system to monitor the health live of elder's people is proposed by [41], and multiple IoT devices such as camera, microphone, and smart devices will be installed to capture patient condition using their facial expression and fixed phrase. It is important to mention that the collecting enormous amount data from patients and analyzing them in smart health care has a great impact such as providing effective public health care, population management, health cost reduction, and social network for patients [68].

2.4   IoT Technologies

In SC, technology is crucial, because with the application of ICT, considerable transformation can happen in the daily life of the citizens, including their working time [69].

Most of the studies available on the practice of the smart cities are mainly concerned with the infrastructure of technology and how to enable it, with their main focus being accessibility and availability of systems [70, 71].

We present a taxonomy from a high-level technology perspective in order to define the required components for SC and IoT that are aided by most researchers as shown in Fig. 3.

Network Technology.

RFID.

It is an outstanding technology, enabling wireless to have microchips that allow data communication through embedded communication paradigm. They act as an electronic barcode that can be attached to anything and help with identifying them

Fig. 3 IoT-based SC technology

automatically [72]. Various applications are possible through IoT, because it allows sensors to connect and communicate, also to deploy the received information. With RFID being a prerequisite for IoT, and the use of radio tags, it is thought possible to identify everything in our daily life; [73] has stated the creation of a European Platform for Intelligent Cities (EPIC) that is being intended for future implementation in all European cities. They also stated that geospatial position and 3D display can be enabled through IoT with the use of sensors and RFID.

Wireless sensor network (WSN).

Wireless sensor network allows valuable information that has been stored in a variety of environments to be collected, processed, analyzed, and disseminated. What allowed WSN to achieve its significant role in urban area's sensing applications was the new coming sensors; these sensors are more intelligent; at the same time, they are cheaper and smaller, allowing more widespread applications (an embedded camera for an example) [18]. Furthermore, [74] has claimed to have designed a smart parking system based on adaptive recommendation mechanism in a way that the system facilitates parking with less time, fuel consuming, and air pollution. Therefore, various technologies are needed for the implementation; more precisely, it uses WSNs to search the area around the parking space for vehicles.

Near-Field Communication (NFC).

Many useful application modes and new blueprint for the developing SC can be achieved by combining NFC and mobile communication technology as well as using a cloud architecture that uses NFC technology, which receives messages through unified interface and processes them in the server based on cloud computing technology [75].

In addition, a system that makes use of NFC tags was proposed by [76] in which unique ID was given to the clothes by an embedded NFC tag. Scanning the NFC tag and linking it to a stored database in the cloud by a mobile application is implemented resulting in an environment that allows manufactures to interact digitally with the end users or independently.

Addressing.

IoT success is dependent on uniquely identifying "things." Besides the obvious advantage of the unique identification of billions of devices, controlling them through the Internet also becomes a possibility. IPv4 is limited in its identification capability, working to some extent on a group of cohabiting sensor devices, but not individually. Some of the identification problems have been solved by IPV mobility attributes, but the problem is complicated at the wireless nodes because of their heterogeneous nature, and because of concurrent operations, production of variable types of data, and the data sent from all the devices at the same time leading to confluence [72, 77].

Thus, for the development of systems based on IoT, a system that uses Uniform Resource Naming (URN) is regarded. Moreover, into IPv6 ready sensor devices, [18] has proposed a method where uIP library can be ported to WSN devices.

3S.

3S technology is a requirement for geospatial information science and technology, which consists of global positioning system (GPS), geographic information system (GIS) as well as remote sensing technology (RS). The creation and design of SC require 3S when the geospatial information for accurate position is needed, and 3D visualization uses for city constructions [78].

Satellite communication system.

The satellite communication system becomes an integral part of Internet of remote things (IoRTs), according to [79]. What make IoRT feasible through satellite are interoperability with sensors/actuator and its support for IPv6. Needless to say, the efficiency of IoT over satellite can be enhanced by radio resource management algorithms.

Li-Fi.

As conducted by [78], Li-Fi networking paradigm is quite an appealing backup option for networking in IoT, since it offers performance enhancements; furthermore its indoor capability makes it more appealing for large-scale indoor communication in the next-generation wireless networking environments. Furthermore, [79] took a step further and presented a Li-Fi-enabled IoTs as an additional feature. To resolve the issue of the radio-frequency bandwidth shortage with Li-Fi and with the help of existing technology, perhaps new communication channels will be made. Since Li-Fi comes with high speed and security, makes it ideal for IoTs connectivity and to automate the real-time image processing system.

Data Technology.

Big Data and Data Analytic.

It is a common opinion among experts that the unchallenged commodity is data in twenty-first century, leading to rapid development of IoT and Big Data as the two main constitution of this age, which sets them on an inevitable collision with methodologies that represent conventional data processing and database systems [80, 81].

As stated by [82], the area of IoT will be crucial for data mining through the analysis of Big Data streams from sensors and devices. The big document summarization method is a valuable technique for IoT data management by extracting semantic feature and distributed parallel processing of NMF-based cloud technique of Hadoop [83].

One of the formidable tasks according to [84] is storing and analyzing smart building data coming in at real time with high-speed that are huge in size at the same time. Therefore, they present a framework named IoT Big Data Analytics (IBDA), in order to store the data produced at real time from smart buildings by the IoT sensors. In addition, [85] explains that the data layer can be defined as the "brain" of SC.

The inclusion of an intricate and comprehensive data-intensive computing and application system is an integral part of smart cities, as stated by [86], and sensing vast amounts of data presents a big challenge which requires a data accumulation that is extremely efficient and powerful in collecting vast amount of data in sensor layer. Schaumont [87] states that, in order to make data meaningful and useful, it has to be preprocessed. So, useful tools for this purpose could be used such as system on a chip (SoC) in which through a single chip all the parts of the electronic system are integrated into it using software embedding.

Semantic.

Reducing incompatibilities among data formats is a well-known advantage of semantic technologies, and also, it acts as an additional layer for new applications, to further reasoning data and extract new meaningful information [80]. Since we need a more interoperable and collaborate IoT, Web of things (WoTs) is the first step. WoT adds a common stack to IoT based on Web services; therefore, to obtain homogenous access through the web, semantic Web of things (SWoT) could be used to integrate semantic Web on WoT [88].

According to [79], in order to have interoperability between IoT applications that are semantic-based, a unified system is needed which requires assistance between SWoT generators and the developers of IoT projects: The templates of semantic-based IoT applications are needed to build applications that are interoperable.

In addition, [89] has suggested using already available data standards and read-ied communications to form semantic Webs and gateways which would allow an architecture for IoT that would fill the need for interoperable systems. The messages from different protocols such as MQTT, CoAP, and XMPP can be translated via multi-protocol proxy architecture using semantic gateway services (SGS). Gyrard et al. [90] stated that IoT can apply ontology methodologies, engineering along with best practices and tools.

Cloud.

To be able to provide and make available powerful decentralized services in the IoTs, and for innovative platforms of IoT to be integrated, efficient and fast performance, ubiquity, scalability, and lastly reliability are required. The demand for what is mentioned, in the current technological trends, leads to the fated meeting of IoT and cloud [91].

Most of smart city projects based on IoT works based on cloud platforms, or they use cloud in their platforms [2, 11, 12, 16–21]. In addition, [92] clarified how in the field of medicine, such as applications for disease prevention and beforehand alerting used current technologies, like Big Data processing, IoT systems and cloud. Innovative integrated fog cloud IoT (IFCIoT) architectural paradigm promised decreased latency, more efficient energy use with quicker response times, even increased local accuracy, and better scalability, all these while giving an improved performance [93]. Moreover, [94] demonstrated a model for the interaction between IoTs and cloud (IoT–cloud) that is location-based.

The location and interest of the mobile user dictates what type of sensing service is provided by the IoT–cloud. It is the role of the cloud to make schedules on demand between the networks of the physical sensors, acting as a virtual sink responding to the demands based on the mobiles user's locations. Compared to the conventional WSNs, this model shows outstanding progress as seen through the extensive analysis results.

Visualization.

A user-friendly visualization of the data represented is crucial for a successful SC application, and easy comprehension by users; in that context, [18] improve visualization schemes by plugging into GIS platforms and integrating geo-related information. SmartSantander allows the visualization and easy monitoring of historic sensed and real-time data by the support of its sensor network which uses Web application [95]. The work in [34] stated that the main process of utilizing massive data once it is compiled in a SC is the vitalization of data, in which the cleaning, maintenance, evolution, and association are emphasized.

Security and privacy.

Despite the undeniable prized benefits of IoT, it does not come without its downsides, as we are exposed on a daily basis to security threats of varied types. Furthermore, the privacy concern is ever so high in IoT environment since many sensitive and personal data is shared and delivered among connected things. As such, protective mechanisms are needed in place, to secure data, while having its flow monitored as it moves to cloud from things [96].

Therefore, [97] has presented a system named Lamina whose sole purpose is to provide users that are in a public IoT space with the privacy and security they desire. However, users are needed to share limited amount of data with the public

IoT spaces about their interests and habits. To provide protection and privacy for users, lamina applies a unique rotation mechanism for MAC addresses as well as an encryption which is CryptoCoP-based. Nonetheless, sufficient data is allowed for IoT space to provide targeted service accurately. A context-aware, purpose-aware, dynamic, adaptive and composite privacy and security mechanism is presented by [98], founded on architectural principles. IoT architectures have being the source of heated discussions in [99], with security, privacy, and user-trust perspective as a main concern.

Middleware.

Despite variation of devices and service with the different implementation logic in SC, they have to integrated together in order to achieve data and application heterogeneity, real-time response, and manage the high volume data. This could be a challenging, expensive, and time-consuming task, because the integration must be based on adapting different data representations into a single one that is able to offer easily accessible services at the application layer. This is the moment when a middleware layer has to be considered, for its most important functionalities to deal with problems such as scalability, interoperability, or data heterogeneity, aiming to shield the underlying complexity of a system, typical of different operating systems, computer architectures, or network protocols. An essential role needs to be fulfilled by the middleware, if smart city's future is to be a success. Middleware is a fundamental technology which is often explained as a software system with the function to act as the intermediary connecting IoT devices with applications [100].

To address this challenges, a new information-centric network (ICN) was proposed in [101]; as an IoT middleware, it would leverage many promising features of ICN, like naming. The efficiency of the service discovery and demonstrated dependability are proposed in ICN-based IoT middleware. What [102] states that IoT needs the support of middleware that makes the interaction of consumers and developers of IoT an easy experience, regardless of how different each side perceives the IoT system. And because of that, bridging the gap between consumers and developers of IoT is a main focus of their software.

A middleware framework [103] has proposed for context-aware applications that generates intermediate, reusable context extracted from input when context engines or functional unit sets are obtained from applications when they are broken down. The middleware that was suggested in [104] for IoT devices makes real-time processing in a distributed manner of data streams a possibility. This middleware adds functionality such as distributing issued task by the application software into subtasks and then allows the subtasks to be distributed to numerous IoT devices for execution, and the data streams are distributed to the IoT devices meant, as well as allowing the data streams to be analyzed at real time, and lastly actuators and sensors can be integrated seamlessly. In addition, the actuator network middleware presented in [105], to transmit and control the behavior of the environment or physical system.

3   The Challenges and the Recommendations

Despite there are many attempts to implement SC, but this process is still in the early stage. It confronts a lot of challenges and obstacle in terms of technology, security, heterogeneity, and interoperability. Table 1 shows the analysis of previous attempts in terms of SC and IOT.

Table 1 SC and IoT works

Works	Domain	Platform	Architecture	Data Collection Technique	Homogenization Technique	Ongoing Issue
2	Smart parking	SmartSantander	Tiered architecture—four tiered	Regular sensor, NFC tag/QR, smart phone	Semantic, information modeling (metadata)	Unclear security policy
13	Smart environment	SmartSantander	Component-based design	Sensors for environmental purposes, smart phones	Data mining techniques	Testing the proposed platform design and its functionality
14	Smart City test bed	SmartSantander	Tiered architecture—three-tiered	NFC tag, mobile phones	Device tier in multi-tier architecture	Lack of high homogenization to be able to add more variety devices
15	Real-time data stream processing	CityPlus	Layered architecture—three-layered	Sensors, mobile phones	Semantic annotations, adaptive processing, aggregation and federation	–
17	Smart city	Barcelona Intelligent City	Layered architecture-multilayer	Sensors such as acoustic sensors, temperature, humidity, gas	Standard Web service such as RSTFUL and JSON used to transport data the data will be exposed based on common XML format wit sensorML encoding	Unclear security policy, Poor interoperability solution
18	IoT platform based on cloud in Melbourne City	SmartSantander	Smart Santander architecture—three- tiered	wireless sensor network (WSN), mobile phones	Using standard cloud service and advance QoS-based interpretation tool	Having interoperable backbone to have efficient plug and play object in SC

(continued)

Table 1 (continued)

Works	Domain	Platform	Architecture	Data Collection Technique	Homogenization Technique	Ongoing Issue
19	Intelligent car parking system	IoT platform based on Cloud (Platform as a Service)	Layered architecture-three-layered	RFID, laser, passive infrared, microwave radar, ultrasonic, passive acoustic array sensors, CCTV	IoT integrated services portal to allocate different car parking services	Poor interoperability solution with other smart systems in SC
20	Smart city based on IoT and cloud computing	Cloud open-platform (SlapOS)	Slave and Master nodes	RFID, EPC	Cloud open-platform integration of existing M2M approaches.	–
29	Smart building	Smart Platform based on cloud and IoT	Layered architecture	RFID, WSN	REST API	Poor interoperability solution
32	Smart city	Event-driven Smart City software platform	Event-driven architecture	Various types of sensors	Semantic Web	–
33	Smart city-vehicular context—vehicle maintenance services	M2M cloud computing platform	Event-driven architecture	WSN	semantic information brokers (SIB), and knowledge-based processors (KBP)	Unclear security policy
40	Waste management	Cloud-based decision support system	Multi-tier	RFID, WSN, camera, and actuators	XML-based data and task description	Poor interoperability solution
41	Smart health	IoT platform based on Cloud	Multi-tier	Video camera, microphone, and smart devices	Web service and cloud	Poor interoperability solution
42	Smart home	Smart Home platform based on (ZigBee and IEEE802.15.4) technologies	Multi-tier	Sensor (ZigBee technologies)	Context awareness, Context broker, Data representation based on Resource Description Framework (RDF)	Poor interoperability solution, Unclear security policy

(continued)

Table 1 (continued)

Works	Domain	Platform	Architecture	Data Collection Technique	Homogenization Technique	Ongoing Issue
43	Smart energy	Citywide wireless Outdoor Lighting Networks (OLNs)	Multi-tier	WSN	Infrastructure discovery called (Discovery)	Poor interoperability solution, Unclear security policy
44	Smart energy	IoT platform based on Cloud (Cloud of Thing(CoT))	Multi-tier	Regular sensor	Using Cloud of Thing in layered architecture	–
50	Smart road management	Raspberry Pi (RPi)	Multi-tier	RPi, smart phones	WebIOPi which is Representational State Transfer (REST) API	Poor interoperability solution, Unclear security policy
63	Smart home	Smart home platform based on CoT and 5G Technology	Multilayer—four-layered	WSN and embedded WSN	Cloud platform based on M2M approaches.	Poor interoperability solution, Unclear security policy
66	Smart city	Smart phone platform, e-platform	Multi-tier	RFID, smart phone, sensor	virtualization of sensors data (Virtual Sensor), Middleware with reasoning data from multiple object	–
67	Smart living	Smart home platform based on IoT	Multi-tier	Wearable device	–	–
95	Smart city	SEN2SOC-SmartSantander	Multi-tier enterprise architecture	Sensor and social network	Semantic and Sensor data monitoring	–

3.1 Heterogeneity and Interoperability Challenges

Due to the increase in IoT technologies and interaction different technologies, data, and application in a SC projects, heterogeneity and interoperability became the highest challenges in SC. Although, a lot of activity happens to implement interoperable platform, architecture, or middleware to integrate heterogeneous device, data, and application, still further study is required to promote more sophisticated architecture, platform, and middleware.

3.2 Technology Challenges

IoT has been recognized as an essential part of our future lives. Things in daily life that are identified uniquely and addressed will have the ability to collect information about their actual environment and themselves, and then store, process, and then communicate this information. Also to the regards of different aspects of life of the citizens, such as entertainment, energy, their transportation, the environment they live in, and lastly the healthcare system. For this interconnected network of devices to become economically feasible, scientists and engineers still have a long way to go. From the literature review, several technological challenges can be recognized that IoT developers and users currently faced: the problems associated with device capabilities, power consumption, sensing, unification of connectivity standards, integration of multiple cloud services, data management, intelligent analysis and actions, compatibility, interoperability standard issues, security, privacy; to advance through this new age and trend of technology, truly many manufacturers and entire enterprises have to work and cooperate together.

3.3 Data Challenges

Other challenges in SC based on IoT are data-related problems in terms of data representation, collection, analyzing, standardization, and visualization. Data in SC provided by multiple stakeholders, providers, IoT devices, and domains needs to be integrated, and validated by multiple sources and clients. Different applications and sensor should understand different data format by translating available data into a standard information, while data and information standardization make the applications interoperability in SC. Lack of proper standardization as well as poor data sharing by data source owners and companies leads to poor interoperable SC application. Furthermore, the gigantic amount of data in SC required advance Big Data prediction algorithms.

3.4 Security Challenges

The security challenge in SC is the real-time monitoring of the physical world which has near to billions if not trillions of interconnections between smart things, while providing smart services to them. The SC with its ubiquitous sensing and the control and processing systems for intelligent information derived from an infrastructure of a heterogeneous network has truly become a new paradigm, recently emerged. However, security and privacy concerns arise, and there are many security and privacy challenges in SC based on IoT that lead to open many ways for researchers to find solutions for fixing those challenges: A potential risk can arise any time in the IoT networks from any sensor or device. So SC entities have a trillion points of vulnerability and preserve the collected data in a confidential and secure way, and as we send the data, the integrity must be maintained (trust and data integrity),; these are matters of collecting data properly, then its protection in a secure and private way. Physical attacks to the numerous physical interfaces and many end-point devices also come into play, and there are vulnerabilities that can be used, as local networks can be spoofed, authentication and identification issues, and a type due deployment leading to improper updates of security. One security issue that is standing out these days is the tablets and cell phones which are quite abundant on the Internet.

4 Conclusion

SC is a new phenomenon, which is as yet new in writing. Numerous different sciences investigate the SC area from different perspectives, either from the industry point of view or the scholarly community, as their interest is clearly evident from the numerous researches, schools, and journals that has been done or is presently underway. The SC domain indeed encompasses all the sciences, and each one is approaching its domain from an alternate point of view. Researchers and schools over the world are being or have been examined this wonder, and a characteristic "picture" is given. This research showed a state of the art of current technologies, platforms, architecture of smart cities in different projects such as parking system, health care, factories' energy consumption. Furthermore, it categorizes SC-IoT domain in different categories, for instance, smart people with a smart life using a smart economy in a smart environment, and governed by a smart governance system. Moreover, it showed that most smart cities' project platform was cloud and common architecture was combined architecture. The research explained current IoT technologies that have been used such as RFID, NFC, addressing, 3S, satellite communication system, and Li-Fi. The big challenges of smart cities are the data technology for processing the data and security issues. Finally, the scholars, industrial companies, and the smart cities project owners find that SC-IoT faces many challenging in terms of technology, security, heterogeneity, and interoperability.

Acknowledgements We would like to thank University of Human Development and University of Technology, Malaysia, for their grant and financial support to complete this work.

References

1. Unicef (ed.): The State of the World's Children 2012: Children in an Urban World. eSocialSciences (2012)
2. Lanza, J., Sánchez, L., Gutiérrez, V., Galache, J.A., Santana, J.R., Sotres, P., Muñoz, L.: SMART CITY services over a future internet platform based on IoTs and cloud: the smart parking case. Energies 9(9), 719 (2016)
3. Rees, W., Wackernagel, M.: Urban ecological footprints: why cities cannot be sustainable—and why they are a key to sustainability. Environ. Impact Assess. Rev. 16(4–6), 223–248 (1996)
4. Wackernagel, M., Kitzes, J., Moran, D., Goldfinger, S., Thomas, M.: The ecological footprint of cities and regions: comparing resource availability with resource demand. Environ. Urban. 18(1), 103–112 (2006)
5. Satterthwaite, D.: Sustainable cities or cities that contribute to sustainable development? Urban Stud. 34(10), 1667–1691 (1997)
6. Nam, T., Pardo, T.A. (eds.): Conceptualizing SMART CITY with Dimensions of Technology, People, and Institutions, pp. 282–291. ACM (2011)
7. Petrolo, R., Loscrì, V., Mitton, N.: Towards a SMART CITY based on cloud of things, a survey on the SMART CITY vision and paradigms. Trans. Emerg. Telecommun. Technol. (2015)
8. Duquennoy, S., Grimaud, G., Vandewalle, J.-J. (eds.): The Web of Things: Interconnecting Devices with High Usability and Performance, pp. 323–330. IEEE (2009)
9. Atzori, L., Iera, A., Morabito, G.: The IoTs: a survey. Comput. Netw. 54(15), 2787–2805 (2010)
10. Arasteh, H., Hosseinnezhad, V., Loia, V., Tommasetti, A., Troisi, O., Shafie-khah, M., Siano, P. (eds.): IoT-Based Smart Cities: A Survey, pp. 1–6. IEEE (2016)
11. Pradhan, S., Dubey, A., Neema, S., Gokhale, A. (eds.): Towards a Generic Computation Model for SMART CITY Platforms, pp. 1–6. IEEE (2016)
12. Boulos, M.N.K., Al-Shorbaji, N.M.: On the IoTs, smart cities and the WHO Healthy Cities. Int. J. Health Geogr. 13(1), 10 (2014)
13. Samaras, C., Vakali, A., Giatsoglou, M., Chatzakou, D., Angelis, L. (eds.): Requirements and Architecture Design Principles for a SMART CITY Experiment with Sensor and Social Networks Integration, pp. 327–334. ACM (2013)
14. Sanchez, L., Muñoz, L., Galache, J.A., Sotres, P., Santana, J.R., Gutierrez, V., Ramdhany, R., Gluhak, A., Krco, S., Theodoridis, E.: SmartSantander: IoT experimentation over a SMART CITY testbed. Comput. Netw. 61, 217–238 (2014)
15. Barnaghi, P., Tönjes, R., Höller, J., Hauswirth, M., Sheth, A., Anantharam, P. (eds.): Citypulse: Real-Time IoT Stream Processing and Large-Scale Data Analytics for SMART CITY Applications (2014)
16. Lécué, F., Tallevi-Diotallevi, S., Hayes, J., Tucker, R., Bicer, V., Sbodio, M.L., Tommasi, P. (eds.): Star-City: Semantic Traffic Analytics and Reasoning for City, pp. 179–188. ACM (2014)
17. Gea, T., Paradells, J., Lamarca, M., Roldan, D. (eds.): Smart Cities as an Application of IoTs: Experiences and Lessons Learnt in Barcelona, pp. 552–557. IEEE (2013)
18. Jin, J., Gubbi, J., Marusic, S., Palaniswami, M.: An information framework for creating a SMART CITY through IoTs. IEEE IoTs J. 1(2), 112–121 (2014)
19. Ji, Z., Ganchev, I., O'Droma, M., Zhao, L., Zhang, X.: A cloud-based car parking middleware for IoT-based smart cities: design and implementation. Sensors 14(12), 22372–22393 (2014)
20. Suciu, G., Vulpe, A., Todoran, G., Gropotova, J., Suciu, V. (eds.): Cloud Computing and IoT for SMART CITY Deployments, pp. 1409–1416 (2013)
21. Tönjes, R., Barnaghi, P., Ali, M., Mileo, A., Hauswirth, M., Ganz, F., Ganea, S., Kjærgaard, B., Kuemper, D., Nechifor, S. (eds.): Real Time IoT Stream Processing and Large-Scale Data Analytics for SMART CITY Applications (2014)

22. Anthopoulos, L., Fitsilis, P. (eds.): From Digital to Ubiquitous Cities: Defining a Common Architecture for Urban Development, pp. 301–306. IEEE (2010)
23. Bandyopadhyay, D., Sen, J.: IoTs: applications and challenges in technology and standardization. Wirel. Pers. Commun. **58**(1), 49–69 (2011)
24. Komninos, N. (ed.): The Architecture of Intelligent Clities: Integrating Human, Collective and Artificial Intelligence to Enhance Knowledge and Innovation, pp. 13–20. IET (2006)
25. Atzori, L., Iera, A., Morabito, G., Nitti, M.: The social IoTs (sIoT)–when social networks meet the IoTs: concept, architecture and network characterization. Comput. Netw. **56**(16), 3594–3608 (2012)
26. Vilajosana, I., Llosa, J., Martinez, B., Domingo-Prieto, M., Angles, A., Vilajosana, X.: Bootstrapping smart cities through a self-sustainable model based on big data flows. IEEE Commun. Mag. **51**(6), 128–134 (2013)
27. Zhong, C.-L., Zhu, Z., Huang, R.-G. (eds.): Study on the IOT Architecture and Gateway Technology, pp. 196–199. IEEE (2015)
28. Carretero, J.: ADAPCITY: a self-adaptive, reliable architecture for heterogeneous devices in Smart Cities. In: European Commissions-ICT Proposers (2012)
29. Corotinschi, G., Găitan, V.G. (eds.): Smart Cities Become Possible Thanks to the IoTs, pp. 291–296. IEEE (2015)
30. Shaikh, T., Ismail, S., Stevens, J.D. (eds.): Aura Minora: A User Centric IOT Architecture for SMART CITY, p. 59. ACM (2016)
31. Al-Hader, M., Rodzi, A., Sharif, A.R., Ahmad, N. (eds.): SMART CITY Components Architecture, pp. 93–97. IEEE (2009)
32. Cretu, L.-G.: Smart cities design using event-driven paradigm and semantic web. Informatica Economica **16**(4), 57 (2012)
33. Wan, J., Li, D., Zou, C., Zhou, K. (eds.): M2M Communications for SMART CITY: An Event-Based Architecture, pp. 895–900. IEEE (2012)
34. Wenge, R., Zhang, X., Dave, C., Chao, L., Hao, S.: SMART CITY architecture: a technology guide for implementation and design challenges. China Commun. **11**(3), 56–69 (2014)
35. Chakrabarty, S., Engels, D.W. (eds.): A Secure IoT Architecture for Smart Cities, pp. 812–813. IEEE (2016)
36. Palattella, M.R., Accettura, N., Vilajosana, X., Watteyne, T., Grieco, L.A., Boggia, G., Dohler, M.: Standardized protocol stack for the internet of (important) things. IEEE Commun. Surv. Tutor. **15**(3), 1389–1406 (2013)
37. Rathore, M.M., Ahmad, A., Paul, A., Rho, S.: Urban planning and building smart cities based on the IoTs using big data analytics. Comput. Netw. **101**, 63–80 (2016)
38. Anthopoulos, L.G.: Understanding the SMART CITY domain: a literature review. In: Transforming City Governments for Successful Smart Cities, pp. 9–21. Springer (2015)
39. Arroub, A., Zahi, B., Sabir, E., Sadik, M. (eds.): A Literature Review on Smart Cities: Paradigms, Opportunities and Open Problems, pp. 180–186. IEEE (2016)
40. Medvedev, A., Fedchenkov, P., Zaslavsky, A., Anagnostopoulos, T., Khoruzhnikov, S. (eds.): Waste Management as an IoT-Enabled Service in Smart Cities, pp. 104–115. Springer (2015)
41. Hossain, M.S. (ed.): Patient Status Monitoring for Smart Home Healthcare, pp. 1–6. IEEE (2016)
42. Han, D.-M., Lim, J.-H.: Design and implementation of smart home energy management systems based on ZigBee. IEEE Trans. Consum. Electron. **56**(3) (2010)
43. Murthy, A., Han, D., Jiang, D., Oliveira, T. (eds.): Lighting-Enabled SMART CITY Applications and Ecosystems Based on the IoT, pp. 757–763. IEEE (2015)
44. Lynggaard, P., Skouby, K.E.: Complex IoT systems as enablers for smart homes in a SMART CITY vision. Sensors **16**(11), 1840 (2016)
45. Bhati, A., Hansen, M., Chan, C.M.: Energy conservation through smart homes in a SMART CITY: a lesson for Singapore households. Energy Policy **104**, 230–239 (2017)

46. Sudan, R.: Towards SMART government: the Andhra Pradesh experience. Indian J. Public Adm. **46**(3), 401–410 (2000)
47. http://www.gartner.com/newsroom/id/2707617. Accessed 18 Apr 2017
48. Rochet, C., Correa, J.D.P.: Urban lifecycle management: a research program for smart government of smart cities (2016)
49. Ritchie, S.G.: A knowledge-based decision support architecture for advanced traffic management. Transp. Res. Part A: Gen. **24**(1), 27–37 (1990)
50. Misbahuddin, S., Zubairi, J.A., Saggaf, A., Basuni, J., Sulaiman, A., Al-Sofi, A. (eds.): IoT Based Dynamic Road Traffic Management for Smart Cities, pp. 1–5. IEEE (2015)
51. Dinh, T., Kim, Y.: A novel location-centric IoT-cloud based on-street car parking violation management system in smart cities. Sensors **16**(6), 810 (2016)
52. Hernafi, Y., Ahmed, M.B., Bouhorma, M. (eds.): An Approaches' Based on Intelligent Transportation Systems to Dissect Driver Behavior and Smart Mobility in SMART CITY, pp. 886–895. IEEE (2016)
53. Parrado, N., Donoso, Y.: Congestion based mechanism for route discovery in a V2I-V2V system applying smart devices and IoT. Sensors **15**(4), 7768–7806 (2015)
54. Poslad, S., Ma, A., Wang, Z., Mei, H.: Using a SMART CITY IoT to incentivise and target shifts in mobility behaviour—is it a piece of pie? Sensors **15**(6), 13069–13096 (2015)
55. Yau, K.-L.A., Lau, S.L., Chua, H.N., Ling, M.H., Iranmanesh, V., Kwan, S.C.C. (eds.): Greater Kuala Lumpur as a SMART CITY: A Case Study on Technology Opportunities, pp. 96–101. IEEE (2016)
56. Letaifa, S.B.: How to strategize smart cities: revealing the SMART model. J. Bus. Res. **68**(7), 1414–1419 (2015)
57. Bull, R., Azenoud, M.: Smart citizens for smart cities: participating in the future (2016)
58. Crowley, D., Curry, E., Breslin, J.G.: Citizen actuation for smart environments: evaluating how humans can play a part in smart environments. IEEE Consum. Electron. Mag. **5**(3), 90–94 (2016)
59. Kazhamiakin, R., Marconi, A., Martinelli, A., Pistore, M., Valetto, G. (eds.): A Gamification Framework for the Long-Term Engagement of Smart Citizens, pp. 1–7. IEEE (2016)
60. Kidd, C.D., Orr, R., Abowd, G.D., Atkeson, C.G., Essa, I.A., MacIntyre, B., Mynatt, E., Starner, T.E., Newstetter, W. (eds.): The Aware Home: A Living Laboratory for Ubiquitous Computing Research, pp. 191–198. Springer (1999)
61. Jie, Y., Pei, J.Y., Jun, L., Yun, G., Wei, X. (eds.): Smart Home System Based on IoT Technologies, pp. 1789–1791. IEEE (2013)
62. Skouby, K.E., Lynggaard, P. (eds.): Smart Home and SMART CITY Solutions Enabled by 5G, IoT, AAI and CoT Services, pp. 874–878. IEEE (2014)
63. Sarin, G. (ed.): Developing Smart Cities Using IoTs: An Empirical Study, pp. 315–320. IEEE (2016)
64. Lymberis, A. (ed.): Smart Wearables for Remote Health Monitoring, from Prevention to Rehabilitation: Current R&D, Future Challenges, pp. 272–275 (2003)
65. Ma, J. (ed.): Internet-of-Things: Technology Evolution and Challenges, pp. 1–4. IEEE (2014)
66. Kaur, M.J., Maheshwari, P. (eds.): Building Smart Cities Applications Using IoT and Cloud-Based Architectures, pp. 1–5. IEEE (2016)
67. Chen, Y.-H., Tsai, M.-J., Fu, L.-C., Chen, C.-H., Wu, C.-L., Zeng, Y.-C. (eds.): Monitoring Elder's Living Activity Using Ambient and Body Sensor Network in Smart Home, pp. 2962–2967. IEEE (2015)
68. Sakr, S., Elgammal, A.: Towards a comprehensive data analytics framework for smart healthcare services. Big Data Res. **4**, 44–58 (2016)
69. Hollands, R.G.: Will the real SMART CITY please stand up? Intelligent, progressive or entrepreneurial? City **12**(3), 303–320 (2008)
70. Housing, U., OTB, M.S.: Smart cities Ranking of European medium-sized cities

71. Giffinger, R., Gudrun, H.: Smart cities ranking: an effective instrument for the positioning of the cities? ACE: Archit. City Environ. **4**(12), 7–26 (2010)
72. Gubbi, J., Buyya, R., Marusic, S., Palaniswami, M.: IoTs (IoT): a vision, architectural elements, and future directions. Future Gener. Comput. Syst. **29**(7), 1645–1660 (2013)
73. Ballon, P., Glidden, J., Kranas, P., Menychtas, A., Ruston, S., Van Der Graaf, S.: Is There a Need for a Cloud Platform for European Smart Cities, pp. 1–7 (2011)
74. Horng, G.-J.: The adaptive recommendation mechanism for distributed parking service in SMART CITY. Wirel. Pers. Commun. **80**(1), 395–413 (2015)
75. Wang, Y., Zhou, Y. (eds.): Cloud Architecture Based on Near Field Communication in the SMART CITY, pp. 231–234. IEEE (2012)
76. NG, K.K.R., Rajeshwari, K. (eds.): Interactive Clothes Based on IOT using NFC and Mobile Application, pp. 1–4. IEEE (2017)
77. Zorzi, M., Gluhak, A., Lange, S., Bassi, A.: From today's intranet of things to a future IoTs: a wireless-and mobility-related view. IEEE Wirel. Commun. **17**(6) (2010)
78. Hu, M., Li, C.: Design SMART CITY Based on 3S, IoTs, Grid Computing and Cloud Computing Technology. In: IoTs, pp. 466–472. Springer (2012)
79. Gyrard, A., Bonnet, C., Boudaoud, K., Serrano, M. (eds.): Assisting IoT Projects and Developers in Designing Interoperable Semantic Web of Things Applications, pp. 659–666. IEEE (2015)
80. di Martino, B., Cretella, G., Esposito, A.: Big data, IoT and semantics. In: Handbook of Big Data Technologies, pp. 655–690. Springer (2017)
81. Sezer, O.B., Dogdu, E., Ozbayoglu, M., Onal, A. (eds.): An Extended IoT Framework with Semantics, Big Data, and Analytics, pp. 1849–1856. IEEE (2016)
82. De Francisci Morales, G., Bifet, A., Khan, L., Gama, J., Fan, W. (eds.): IoT Big Data Stream Mining, pp. 2119–2120. ACM (2016)
83. Ji, Y.-K., Kim, Y.-I., Park, S.: Big data summarization using semantic feature for IoT on cloud. Contemp. Eng. Sci. **7**(21–24), 1095–1103 (2014)
84. Bashir, M.R., Gill, A.Q. (eds.): Towards an IoT Big Data Analytics Framework: Smart Buildings Systems, pp. 1325–1332. IEEE (2016)
85. Jara, A.J., Sun, Y., Song, H., Bie, R., Genooud, D., Bocchi, Y. (eds.): IoTs for Cultural Heritage of Smart Cities and Smart Regions, pp. 668–675. IEEE (2015)
86. Deshpande, A., Guestrin, C., Madden, S.R., Hellerstein, J.M., Hong, W. (eds.): Model-Driven Data Acquisition in Sensor Networks, pp. 588–599. VLDB Endowment (2004)
87. Schaumont, P.: A Practical Introduction to Hardware/Software Codesign. Springer Science & Business Media (2012)
88. Jara, A.J., Olivieri, A.C., Bocchi, Y., Jung, M., Kastner, W., Skarmeta, A.F.: Semantic web of things: an analysis of the application semantics for the IoT moving towards the IoT convergence. Int. J. Web Grid Serv. **10**(2–3), 244–272 (2014)
89. Desai, P., Sheth, A., Anantharam, P. (eds.): Semantic Gateway as a Service Architecture for IoT Interoperability, pp. 313–319. IEEE (2015)
90. Gyrard, A., Serrano, M., Atemezing, G.A. (eds.): Semantic Web Methodologies, Best Practices and Ontology Engineering Applied to IoTs, pp. 412–417. IEEE (2015)
91. Kelaidonis, D., Rouskas, A., Stavroulaki, V., Demestichas, P., Vlacheas, P. (eds.): A Federated Edge Cloud-IoT Architecture, pp. 230–234. IEEE (2016)
92. Suciu, G., Vulpe, A., Fratu, O., Suciu, V. (eds.): M2M Remote Telemetry and Cloud IoT Big Data Processing in Viticulture, pp. 1117–1121. IEEE (2015)
93. Munir, A., Kansakar, P., Khan, S.U.: IFCIoT: Integrated Fog Cloud IoT Architectural Paradigm for Future IoTs. arXiv:1701.08474 (2017)
94. Dinh, T., Kim, Y., Lee, H. (eds.): A Location-Based Interactive Model for IoTs and Cloud (IoT-Cloud), pp. 444–447. IEEE (2016)
95. Vakali, A., Anthopoulos, L., Krco, S. (eds.): Smart Cities Data Streams Integration: Experimenting with IoTs and Social Data Flows, p. 60. ACM (2014)
96. Hwang, Y.H. (ed.): IoT Security & Privacy: Threats and Challenges, pp. 1–1. ACM (2015)

97. Harris, A.F., Sundaram, H., Kravets, R. (eds.): Security and Privacy in Public IoT Spaces. IEEE, pp. 1–8 (2016)
98. Seetharaman, S., Ghosh, S.: Adaptive and composite privacy and security mechanism for IoT communication
99. Kozlov, D., Veijalainen, J., Ali, Y. (eds.): Security and Privacy Threats in IoT Architectures, pp. 256–262. ICST Institute for Computer Sciences, Social-Informatics and Telecommunications Engineering (2012)
100. Ngu, A.H., Gutierrez, M., Metsis, V., Nepal, S., Sheng, M.Z.: IoT middleware: a survey on issues and enabling technologies. IEEE IoTs J. (2016)
101. Li, S., Zhang, Y., Raychaudhuri, D., Ravindran, R., Zheng, Q., Dong, L., Wang, G. (eds.): IoT Middleware Architecture over Information-Centric Network, pp. 1–7. IEEE (2015)
102. Boman, J., Taylor, J., Ngu, A.H. (eds.): Flexible IoT Middleware for Integration of Things and Applications, pp. 481–488. IEEE (2014)
103. Venkatesh, J., Chan, C., Akyurek, A.S., Rosing, T.S. (eds.): A Context-Driven IoT Middleware Architecture (2015)
104. Nakamura, Y., Suwa, H., Arakawa, Y., Yamaguchi, H., Yasumoto, K. (eds.): Design and Implementation of Middleware for IoT Devices Toward Real-Time Flow Processing, pp. 162–167. IEEE (2016)
105. Nan, C., Lee, Y., Tila, F., Lee, S., Kim, D.H. (eds.): A Study of Actuator Network Middleware Based on ID for IoT System, pp. 17–19. IEEE (2015)

End-to-End Monitoring of Cloud Resources Using Trust

D. Usha and K. Chandrasekaran

Abstract Cloud computing is the fast-growing technology in the current era. With the growth, it brings with it multiple issues like privacy, security. Trusting the service provider and his resources is the major drawback for the user. In this paper, we have proposed a trust model which builds an end-to-end trust between the service provider resources and user. Our model calculates trust based on four parameters, namely utilization, saturation, failure, and availability. Our model builds a strong trust decision which enables the user to trust the resources of the service provider.

Keywords Trust · Trust management · Resource monitoring · Trust parameters
Trust decision

1 Introduction

Nowadays, customers of any computing solutions expect results instantaneously without waiting for even a minute. Because of this requirement, it is imperative that all businesses that use Web-based solutions are expected to keep up with customer demands. Almost all present-day Web-based applications use the cloud computing; for example, online banking, shopping, email, and more make use of cloud computing to support billions of customers every day. Therefore, in this context, it becomes inevitable to upkeep the cloud computing resources, which essentially leads to the path of establishing what we call as "cloud resource monitoring" in this paper. We propose to use the concepts related to trust as one of the security measures for monitoring computing resources of the cloud.

D. Usha (✉) · K. Chandrasekaran
National Institute of Technology Karnataka Surathkal, Mangalore, Karnataka, India
e-mail: ushachavali@gmail.com

K. Chandrasekaran
e-mail: kchnitk@ieee.org

© Springer Nature Singapore Pte Ltd. 2019
S. K. Bhatia et al. (eds.), *Advances in Computer Communication
and Computational Sciences*, Advances in Intelligent Systems
and Computing 760, https://doi.org/10.1007/978-981-13-0344-9_2

29

End-to-end monitoring implies the monitoring of computer resources in a cloud to meet the end-user requirement or satisfaction in conjunction with that of the resource provider. The two end points in the cloud ecosystem (user and the provider) are the important entities in supporting the proposed cloud monitoring, and this can be achieved by using the trust between the two. In those properly deployed cloud computing applications, such monitoring can remove problems between the two entities before they become an issue, thus reducing down time of resource and facilitating continuous improvement.

Resource monitoring in a cloud computing ecosystem is the meaning of healthy functioning and actual consumption of cloud resources. These are the most vital to keep an eye on in order to enforce the Service-Level Agreements (SLA), make sure that the resources scaling appropriately, and upkeep the applications' availability always to all customers.

Some of the parameters of the cloud computing resource need to be measured include utilization, saturation, failure rate, and availability. Utilization refers to the percentage of the resource that is currently being used. Saturation is the amount of work waiting to be completed by the resource—usually, this will only occur when utilization is at or near 100%. Failure rate is related to the non-functioning of resources. Availability is the percentage of time that the resource has been responding to the monitoring mechanism or tool or other requests. It is customary that in a cloud computing environment, resource providers want to monitor all computing resources. Most important resources in a given cloud computing environment are storage, CPU, and memory. In our work, the end-to-end cloud monitoring of resources, we consider storage as the resource for our study. For a storage disk, utilization would measure the amount of time the device was working, saturation would measure the length of the wait queue to write or read, errors would report any disk problems, and availability would be the percent of the time the device has been available to write.

1.1 Existing Trust Models

Privacy and security are the basic hurdles pertaining to implementation of the cloud computing model. Trust is recognized as an important aspect for decision making in distributed and auto-organized applications [1].

Authors [2] have proposed a trust model that calculates trust value from direct trust and also suggested formal representations and rules to calculate trust value.

Authors [3] have proposed trust on a three-factor basis namely *subjectivity*, *expected probability,* and *relevance*. Subjective trust is the belief of uncertainty.

Authors [4] have proposed a method for calculating trust in cloud platform based on a Trusted Computing Platform (TCP) which in turn is used to provide authentication, confidentiality, and integrity of the system.

Authors [5] have proposed a trust-based cloud architecture that takes into account trust delegation and reputation system for cloud resource in cloud environment.

Authors [6] have proposed a trust model for firewall in cloud computing which uses various security policies, transaction contexts, historic data of entities, and their influence in the dynamic evaluation of the trust value.

Authors [7] have proposed a trust model based on graph abstraction that calculates three types of trust, namely individual service trust, session trust, and composite trust which in turn helps in evaluating the services in service-oriented architecture applications.

Authors [8] have proposed a quantitative assessment metrics method to handle the security in private cloud. The metrics used for assessment are vulnerability density, vulnerability severity, and patching behavior when implemented proved that there is considerable improvement in security aspects.

Authors [9] have proposed a trust-based evidence collection to ensure security assurance of the services by the service provider to the cloud user. A trust computing platform module is used to calculate the system performance as evidence in cloud environment.

Authors [10] have highlighted on several trust-based reputation models that are used in Web services where the single Web service model helps the user in selecting the appropriate Web service, composite Web service model integrates set of services to perform composite functions.

Authors [11] have proposed a workflow-oriented trust definition based on the observation and consideration from existing works. Trust is classified for better comprehension. A broad range of trust properties and workflow characteristics are presented which will enhance security applied to service workflows.

Authors [12] have given a detailed description of the different security threats that could occur in the cloud environment and the different metrics to measure these threats so as to minimize the threats.

Authors [13] have proposed a self assessment-based trust worthy resource allocation which consists of job clustering, trust computation, monitoring, load balanced scheduler, and local resource manager component which in turn reduce cost and also improvise the job success rate for service provider.

Even though multiple researchers have proposed multiple models, still trust is the major issue of cloud environment.

2   End-to-End Resource Monitoring Model Using Trust

The proposed model, which is depicted below, has the users, (resource) service providers and the resources themselves; and, in our end-to-end working model, cloud customers—users and cloud service providers who are the custodians of the resources too—are expected to have involved in the monitoring process of the model. Also, this model assumes that although the entities may be in different domains of ownership, sufficient data will be made available, by both parties involved, in order that end-to-end monitoring can be quantified.

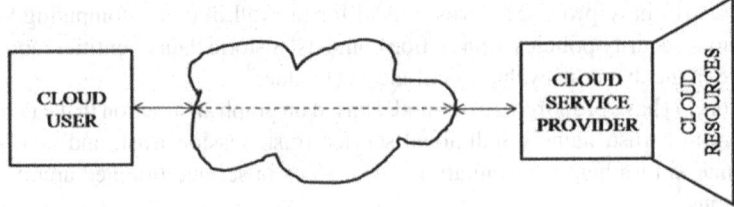

Fig. 1 Cloud customers and cloud service providers

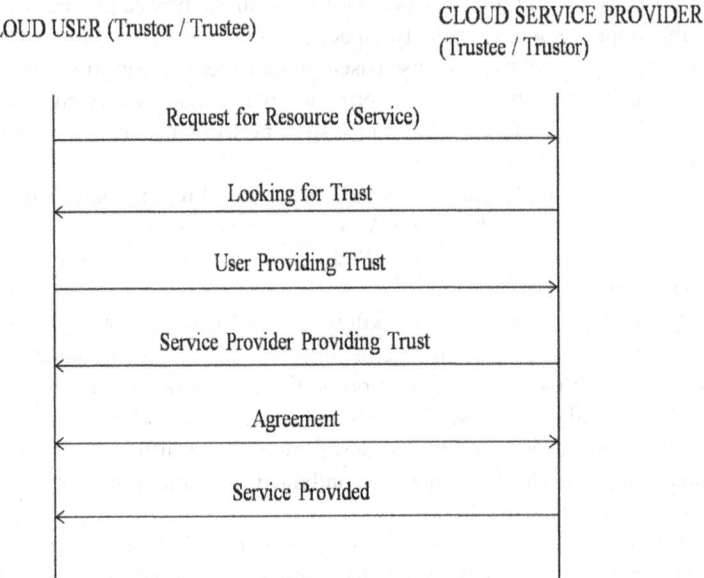

Fig. 2 Sequence diagram

In Fig. 1 below, there are cloud customers and cloud service providers who are connected via Internet as known to all. The mutual trust verification—monitoring by both parties on each other—is the essence of this model. This process is illustrated in the sequence diagram shown in Fig. 2.

Main advantages of this end-to-end model are as follows: (a) each party, i.e., cloud consumer as well as cloud service provider monitor each other's trust through a bilateral mechanism (Figs. 1 and 2), (b) cloud consumer maintains its trust as its profile-based trust (Fig. 3), and (c) cloud service provider maintains his resources' trust-related values (Tables 1 and 2) that are based on the four parameters as explained earlier in this chapter.

Trust in cloud consumers is maintained as per the profile-based trust procedure explained as per the Fig. 3 below. Here, we assume that the users–customers have already completed the procedures related to access control or registration in the

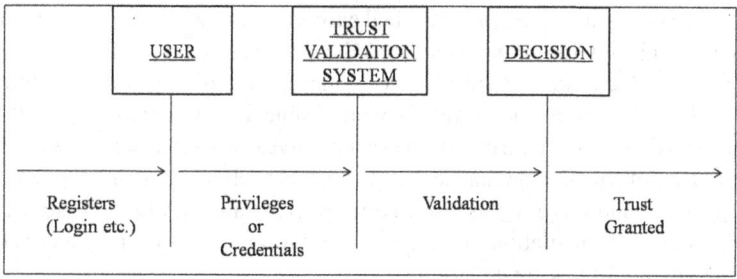

Fig. 3 Trust for cloud consumer–user

Table 1 Trust for the quantitative evaluation of parameter

Trust value	Description	Decision
Zero	No trust	Not accepted
0–0.4	Low trust	Not accepted
0.41–0.6	Average trust	Accept but verify
0.61–0.89	High trust	Accept
0.9–1.0	Very high trust	Accept

Table 2 Factors affecting trust decision

Utilization $W_1 = 35\%$	Saturation $W_2 = 15\%$	Failure $W_3 = 15\%$	Availability $W_4 = 35\%$	Trust decision
High	High	High	High	High
High	Low	Low	High	High
High	High	Low	High	High
High	Low	High	High	High
Low	High	High	Low	Low
Low	Low	High	Low	Low
Low	High	Low	Low	Low
Low	Low	Low	Low	Low
High	High	Low	Low	Low
Low	Low	High	High	Low

system such as login, in the given cloud environment. The proposed profile-based trust for users in our model is an entity that has some pre-defined privileges to access cloud services. When a user accesses the cloud and provides the credentials or the privileges, the system in place validates the user credentials and verifies the user profile. Once the user profile is validated, the trust is granted to the user for all the services that are asked for.

Trust in cloud service provider's side is maintained by keeping the consolidated value of trust based on the trust weight that is given for the four parameters considered in our work. These parameters, as described earlier, are utilization, saturation, failure rate, and availability. Cloud consumer should be able to consider whether a trust value related to a parameter of a cloud resource is acceptable or not, based on the base value which we have considered in our work. Table 1 represents values that were established to determine the trust for the quantitative evaluation of a parameter.

As per the information given in the above table, cloud consumer trusts cloud service provider and thereby the resource from trust value T which is greater than or equal to 0.6. The calculation of trust of a cloud resource service provider is being represented by the following equation.

$$Tv_{(CSP,CU)} = \sum W_1 * U + W_2 * S + W_3 * F + W_4 * A \qquad (1)$$

We know that utilization and availability are more necessary and are of equal importance of any transaction. We also assume that failure is possible only when the resource is not available or not utilized properly, so in the Eq. 1, W_1 is the weight factor for resource utilization, and it is 35% in our model, W_2 is the weight factor for saturation, and it is 15%, W_3 is the weight factor for failure rate of the resource and it is 15%, and, W_4 is the weight factor for availability of the resource, and it is 35% in our assumption and trust calculation. The trust decisions based on these weight factors are tabulated here—Table 2.

Four parameters have greater impact on evaluation of trust value of a cloud provider on cloud consumer about the resources, as shown in Table 2. Better utilization and availability and capacity have more weightage in selection of a resource on which consumer is more reliable, as these features are more responsible to ensure the reliability and handling of that resource. The trust values range from [0 to 1], and these values are variable over time. Based on trust value, a provider can have his resource consumption increased or decreased.

3 Simulation and Results

Based on the weights being assigned, we have evaluated the trust of a cloud provider for a cloud consumer. The simulation was initiated with the cloud consumer trusting cloud provider by assigning the initial value 1 to all metrics. Simulation was performed using the standard Monte Carlo method to generate random values for the four metrics. The values for each of the metrics are assigned randomly varying between 0 and 1, as shown in Table 3.

From Table 3, we can see that the values of the metrics in each simulation directly influence the decision to trust or not. This procedure enables the cloud consumer to make a decision based on the trust to invoke and monitor the services of the provider and resources. In return, the cloud provider is given the trust of the cloud consumer through his profile-based approach. This end-to-end trust model has the benefits of mutually trusting each other in the given cloud environment.

Table 3 Simulation and trust decision

Iteration	Utilization	Saturation	Failure	Availability	Trust decision
1	1	1	1	1	Trust
2	0.82	0.43	0.51	0.76	Trust
3	0.06	0.11	0.30	0.89	Not trust
4	0.68	0.01	0.50	0.32	Not trust
5	0.70	0.34	0.61	0.74	Not trust
6	0.25	0.67	0.32	0.54	Not trust
7	0.76	0.98	0.64	0.69	Not trust
8	0.77	0.24	0.87	0.87	Not trust
9	0.98	0.42	0.33	0.89	Trust
10	0.56	0.31	0.25	0.71	Trust

4 Conclusion

Cloud computing is currently the focus of several researchers, which makes evident the significance and need of trust model that ensures reliable and secure handling of resources. In this paper, we proposed a trust model, known as end-to-end trust model to monitor resources that are based on four important parameters which govern the reliability of the resource. This model is to ensure trust values being exchanged between consumers and providers are of great importance, due to the fact that mutual trust values are used for the resource's utilization. In this model, the trust value of a given resource is obtained from a set of four important parameters for monitoring resource-related operations. Cloud providers and thereby resources with greater trust values are subsequently chosen by the cloud consumers.

References

1. Marsh, S.P.: Formalising trust as a computational concept. Ph.D. Thesis, University of Stirling (1994)
2. Beth, T., Borcherding, M., Klein, B.: Valuation of trust in open networks. In: ESORICS 94, Brighton, UK, Nov 1994
3. Abdul-Rahman, A., Hailes, S.: A distributed trust model. In: Proceedings of the 1997 New Security Paradigms Workshop, pp. 48–60 (1998)
4. Shen, Z., Li, L., Yan, F., Wu, X.: Cloud computing system based on trusted computing platform. In: IEEE International Conference on Intelligent Computation Technology and Automation (ICICTA), vol. 1, pp. 942–945, China (2010)
5. Dillon, T., Wu, C., Chang, E.: Cloud computing: issues and challenges. In: 24th IEEE International Conference on Advanced Information Networking and Applications (AINA), pp. 27–33, Australia (2010)
6. Yang, Z., Qiao, L., Liu, C., Yang, C., Wan, G.: A collaborative trust model of firewall-through based on Cloud Computing. In: Proceedings of the 2010 14th International Conference on Computer Supported Cooperative Work in Design, pp. 329–334, 14–16, Shanghai, China (2010)

7. Azarmi, M., Bhargava, B.: An end-to-end dynamic trust framework for service-oriented architecture. In: 10th International Conference on Cloud Computing. IEEE (2017)
8. Torkura, K.A., Cheng, F., Meinel, C.: Application of quantitative security metrics in cloud
   computing. In: The 10th International Conference for Internet Technology and Secured Transactions (ICITST-2015)
9. Anisetti, M., et al.: Towards transparent and trustworthy cloud. IEEE Cloud Comput. 4(3),
   40–48 (2017)
10. Wahab, O.A., Bentahar, J., Otrok, H., Mourad, A.: A survey on trust and reputation models for
    web services: single, composite, and communities. Decis. Support Syst. 74, 121–134 (2015)
11. Viriyasitavat, W., Martin, A.: A survey of trust in workflows and relevant contexts. IEEE
    Commun. Surv. Tutor. 14(3), 911–940 (2012)
12. Tariq, M.I.: Towards information security metrics framework for cloud computing. Int. J. Cloud
    Comput. Serv. Sci. (IJ-CLOSER) 1(4), 209–217 (2012)
13. Varalakshmi, P., Judgi, T., Fareen Hafsa, M.: Local trust based resource allocation in cloud. In:
    2013 Fifth International Conference on Advanced Computing (ICoAC). IEEE (2013)

A Big Step for Prediction of HIV/AIDS with Big Data Tools

Nivedita Das, Manjusha Pandey and Siddharth Swarup Rautaray

Abstract Human activity is quickly transforming an ecosystem. How this transformation collision health of humanity, whose health is at risk, and the extent of the associated disease burden are relatively new subjects within the area of environmental health. HIV/AIDS, tubercular, and malaria are three major global public health an expression of intent to injure and cause substantial morbidness, undeadliness, not positive socioeconomic impact and nonstandard. The HIV/AIDS is actual of an occurrence disease to proceed with to be an important global challenge of health, and actual, to bear HIV-1 virus burden testing is more and more needed at the point of care (POC). The presence of Big Data is everywhere. It is not actually data, but is a concept which actually explains about the gathering of data, organizing the data, analyzing the data and getting information out of the data. More applications are created everyday to extract the value from it which is professional and practical. The use of Big Data technologies in enterprise data warehouse and business intelligence results in better business insights and decisions. Now, Big Data analytics recently used in the point-of-care delivery and disease penetration. Big Data analytics tools are essential and useful tools, which gives strength to companies to analyze entire data related to their customers and the flea market in which they perform. As this data holds a large amount of information concerning the specific type, commodity, client service satisfaction, and client sentiment, many companies have taken the use of Big Data analytics tool.

Keywords Predictive analysis · Symptoms analyzer · Big Data tools

N. Das (✉) · M. Pandey · S. S. Rautaray
School of Computer Engineering, KIIT University, Bhubaneswar, Orissa, India
e-mail: niveditads26@gmail.com

M. Pandey
e-mail: manjushafcs@kiit.ac.in

S. S. Rautaray
e-mail: siddharthfcs@kiit.ac.in

© Springer Nature Singapore Pte Ltd. 2019
S. K. Bhatia et al. (eds.), *Advances in Computer Communication
and Computational Sciences*, Advances in Intelligent Systems
and Computing 760, https://doi.org/10.1007/978-981-13-0344-9_3

1   Introduction

The huge amount of data is the actual Big Data is a concept and that represents, which both are structured and unstructured. Big Data means really a Big Data, and it is not merely a data, rather it has become a complete subject, which includes different tools, frameworks, and techniques. But it does not describe the amount of data. Actually Big Data describes it as a gathering of data elements whose capacity, velocity, category, and elaboration help to a person to find, accept, and bear a new hardware and software system. Big Data has successfully stored, resolving problems, and the creation of the data concept [1]. Big Data refers a data which is large and composite in their nature. Actually traditional data processing application software is insufficient to share with them [2]. Big Data analytics spreads a role in the handling data which is formed by various resources to provide distinctive characteristics of information [3]. Healthcare transformation has many challenges such as access to health care, consumer engagement, level and quality of care, and cost efficiency. We can take advantage of Big Data to gain a broad idea about the health of the patient and to create a unique customer background. Big Data has six challenges. These are described here. The **Data Storage** is distributed in nature. That storage is a framework that is designed exactly for storing, managing, and retrieving of the large amount of data [4]. That data enables to store and sort the Big Data and we can easily access, use, and process it by using applications and services which are using Big Data. In the **Data Processing**, Big Data is used as a data locality. This data locality where all data is present and they are in running stage as shown in Fig. 1. In **Very Large File**, large data sets can be in the form of massive files that do not suitable into the vacant memory and for processing the files are taking a long time to complete. So, we use MATLAB for this problem.

The **High-End Hardware** is cheaper and low-end hardware. It uses high-end RAM and high-end processor. In the **Backup and Recovery**, if the entire node having same mirror data gets crashed, then Hadoop provides a solution for this problem. The **Disaster Management (DR Activity)**, suppose a data center is crashed, that's why put the same copy of data in different data center. This is called DR Activity. In the field of health care, Big Data hits an important role. Big Data has some characteristics like volume, velocity, variety, veracity, and value. Here, Big Data introduces a concept of 5V's as shown in Fig. 2. Since the information is spreading immensely nowadays, Big Data defines both size and vision from unstructured, composite, noisy, mixed,

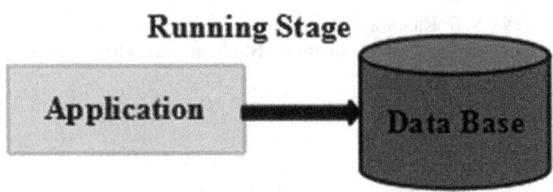

Fig. 1   Concept of data processing

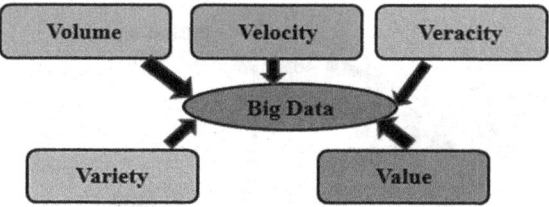

Fig. 2 Characteristics of Big Data

representation, and the volume of data [5]. These 5 V's are discussed in details as: **Volume**, Big Data suggests a large weight of data. It used to be an individual's created data. Nowadays, data is generated by digitally on systems such as social media, and the volume of data to be analyzed is humongous. **Velocity**, speed at which data is being created is called Data Velocity which flows of data in the form of sources like business processes, machines, networks, and interaction of human with things like social media, mobile devices. The data is very large and continuous in nature. **Veracity**: the veracity concept in Big Data deals with bias, noise, and unstructured. Big Data feels veracity in data analysis is the major issue when it compares to volume and velocity. **Variety**, different types of data being created are called Data Variety. This concept is to direct the attention to a lot of sources and different types of data which are structured and unstructured. We have accustomed to store data from sources like databases, file system, and spreadsheets. Nowadays, data comes in the form of emails, photographs, videos, pdf, audio, etc. **Value**, importance of data or the value of information includes data is called data value. The word value in Big Data plays an important role. It includes a massive volume and different varieties of data which are easy to access and delivers quality analytics that helps in making decisions. It provides the actual technology.

The HIV virus is a type of organism. Once someone infected with HIV, it stays in the body for the whole life. This virus damages the immune system. The immune system defects the body against the diseases. HIV virus is the source of AIDS. The full form of AIDS is acquired immune deficiency syndrome, which means deficient immune system. The difference between Big Data and relational databases is that the traditional table-and-column figure does not have by Big Data but that relational databases have. In classical relational databases, a schema for the data is required [6]. Each portion of data is in its precise place. Data is extracted from source systems in its original form, storing in a disorganized distributed file system. The storing of data across several data nodes in a simple hierarchical form of directories of files is known as Hadoop Distributed File System (HDFS) [7]. Commonly, data is accumulated in 0.064 GB files in the data nodes with a high rank of compression.

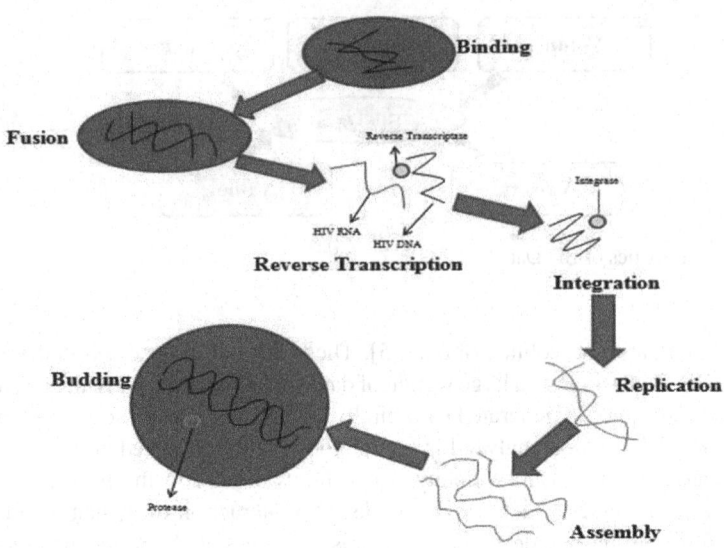

Fig. 3 HIV life cycle

1.1 Life Cycle of HIV

The HIV life cycle follows these seven stages as shown in Fig. 3. These stages are Binding, Fusion, Reverse Transcription, Integration, Replication, Assembly and Budding. In the life cycle of HIV, another name of **Binding concept** is attachment. HIV virus binds to cells on the surface of CD4 cell. In the **Fusion**, envelope of HIV and membrane fuse of CD4 allows entering the HIV to the CD4 cell. The **Reverse Transcription** combines infected DNA from infected RNA and then formed a new complementary DNA strand. On the **Integration of HIV Life Cycle**, HIV releases an HIV enzyme in the inner part of CD4 nucleus. HIV adopts proteins to integrate its infected DNA into the CD4 cell of the DNA. In the concept of the **Assembly,** all the proteins come together and form a new HIV enzyme named as HIV RNA moves toward the surface of the cell and assemble into nontoxic HIV. In **Budding**, the newly formed nontoxic HIV itself applies forces without the owner CD4 cell. HIV releases the HIV protease. This protease is acting as to disintegrate the lean protein interconnected rings that form the inexperienced virus [8, 9].

The presented research paper was followed organized manner. Section 1 is the introduction of the predictive analysis system which also briefs the characteristics of Big Data. Section 2 discusses the literature about the predictive analysis system, its categories and gives an overview of tools and technology framework. Section 3 is the symptom analyzer of HIV/AIDS. Section 4 is the discussion of predictive analysis for HIV/AIDS using Big Data tools; finally, the paper ends with references.

2 Literature Survey

The primary objection of Big Data analytics in healthcare domain and it includes secure, store, search, share, and analyze the healthcare data. Actually, Big Data refers to structured, semi-structured, unstructured, and the complex data generated by an automatic process. Health care is an example of acceleration, variation, and quantity, which are typical aspects of the data what it produces [10]. This data is spread among different healthcare systems, health guarantee, investigator. For adding these three V's, the veracity of healthcare data is also critical for its meaningful use toward developing untraditional research. Actually, predictive analysis is used to predict the present conditions and future events. A person may get AIDS after becoming infected with the HIV virus. Some of them may get infected with the virus by unsafe sex, contact with blood of an infected person, or by using a corrupted injection needle [11]. During pregnancy from mother to child, childbirth, or breastfeeding, the virus is spreading via: blood, sperm, vaginal fluids, pre-seminal fluids, breast milk. There is no risk of becoming contaminated with the virus by froth or shaking hands. When a person has HIV virus, but there are no symptoms yet, the person may be "Seropositive". This means there are HIV antibodies in the person's blood. A person does not need to have any indication of disease yet. An HIV infection transmitted through different stages. In the first stage, the body can show signs of disease like continuous fever and swelling in the glands, while some people who are infected remain asymptomatic in the first stage. In the second stage, recurring infections through air, skin, mouth, and genital injury often occur. In the third stage, a person may have complaints like continued diarrhea, excess loss in weight, tuberculosis in the lungs and other serious infections like meningitis. Finally, besides serious infections, the nervous system may be harmed in the fourth stage, which can result in motor loss or AIDS-related dementia. It may take five to fifteen years before a person knows that he/she has got AIDS. This is because sometimes it takes longer for symptoms to occur. HIV and AIDS can be handled with proper treatment, but cannot be cured. Antiretroviral slows down the multiplication of the virus, but it does not wipe out [12]. To support the medication, AIDS patients often get treatment to boost the immune system and fight against diseases. A person can prevent AIDS by having safe sex and using clean needles. To make this entire paper, take the help of these following papers (see Table 1).

In these, all paper in table representation is about data space for HIV vaccine, HIV infection model, and classification of HIV data by constructing a social network, prediction of HIV Drug Resistance and Predictive analytics in health care. With the help of other different papers, the entire paper was completed and the paper which was help is given below.

Javier Andrew et al. published a paper which provides an overall growth in Big Data on the basis of health care and biomedical. Moreover, in this paper the author will discuss the value of Big Data for medical and health informatics, transnational bioinformatics, sensor informatics, and imaging informatics. Big Data can serve to boost the relevancy of medical scrutiny studies into the related literary work about

Table 1 State of art

Year	Author	Paper title	Description
2016 [1]	David McColgin et al.	The Data Space for HIV Vaccine Study	To design the better understand activity and to explore the new idea in the entire paper
2016 [2]	Saroj Kumar Sahani et al.	Analyze of a Delayed HIV Infection Model	Uses of different types of infection models to burst of the infected cells
2016 [3]	Yunus can Koc at et al.	Classification of HIV data By Building A Social Network with Frequent Item sets	HIV-1 has used protease intermammary sulcus dataset from the UCI machine learning repository and to demonstrate our purpose method by comparing it with a decision tree, Naive-Bayes, and k-nearest neighbor methods
2016 [4]	Syed Mohd Ali et al.	Big Data Representation: Tools and Challenges	Now, all the work is done digitally. Here, the discussion is about Big Data visualization tools and challenges
2016 [5]	Zoya Khalid et al.	Combining Sequence and Structural Properties forecast by HIV Drug Resistance	Data mining and expert systems are used for prediction of HIV Drug Resistance
2016 [6]	A. Rishika Reddy et al.	Predicts the Big Data Analytics in the area of Healthcare	Analysis of Big Data hits an important role in the healthcare area. The most discussion is helpful for facing challenges and gave the actual solution

real world, where population diversity is an impediment. It specifically provides the chance to empower effective and precision medicine by performing patient stratification. This is definitely an important task toward personalized health care.

V. Kavitha et al. in this paper actually maintained and stored in a huge number of data in one system is taking a long time to access to minimize this kind of problem it used replicas of that data at a large number of disks. It is helpful for consuming time for accessing the data. The healthcare data is growing quickly and is probable to go up significantly in arrival timing. In the future, we are reliable to perceive extensive usage of Big Data analytics across the different area in the healthcare sector. This paper provides the mechanism to increase the value of Big Data analytics by making a better level of excellence of MapReduce, whose appropriate selection can give

Table 2 Big Data tools and its application

Sl. No	Big Data components	Application
1	MapReduce	It appears as a programming framework used for parallel computing application which proposed by Google. It provides a flexible and scalable foundation for analytics
2	Hadoop	It is an open-source framework. It is a technology which helps in the treatment of cancer and genomics. It is used to monitor the patient vital
3	HDFS	It provides high throughput access to application data
4	Hive	Hive programming is similar to database programming. Hive jobs are optimized for scalability
5	Pig	Pig is a platform for constructing data flows for extraction, transformation and load (ETL) processing and analytics of large data sets
6	Sqoop	It is a way to transfer data from relational databases and HDFS. It stores structure data in relational databases

promising results. Big Data aids to support the all types of analysis of health care and provides the best output to the user.

The tools and their application are presented in this below in Table 2.

Hadoop is an open-source framework. Framework is a collection of predefined classes. We can extend the framework functionality and change the existing functionality. It is a ready-made one. It has two methods: one is Hadoop Distributed File System (HDFS), and another is MapReduce. Hadoop plays a dominant role in health care. Its capability to store lots and lots of data and we always used MapReduce to analysis its text, pattern of the chances of fraud detection in the healthcare sector. **Cloudera** can help us in business to form an enterprise data hub to assign people in our group to better accessibility to those data we are storing. Generally, Cloudera is an enterprise service solution to help business manage to their Hadoop Ecosystem.

3 Symptom Analyzer

This symptom analyzer, that is my proposed framework, analyzes the HIV symptom using Big Data tools. Here, all the features of HIV symptoms most impact on body parts. In this Fig. 4, we discuss some of the body parts which are affected by HIV. Through the contact with the bodily fluids such as blood, breast milk, ovule and marginal fluid, the HIV is transmitted. When the impact of HIV is on a brain, the nerve of the brain gets damaged and a very great pain in the head [13, 14]. When on an eye, it causes the blindness which can lead to bleed in the retina. Impact on the mouth causes a small, white, gray, or pinkish swollen area forms that look like

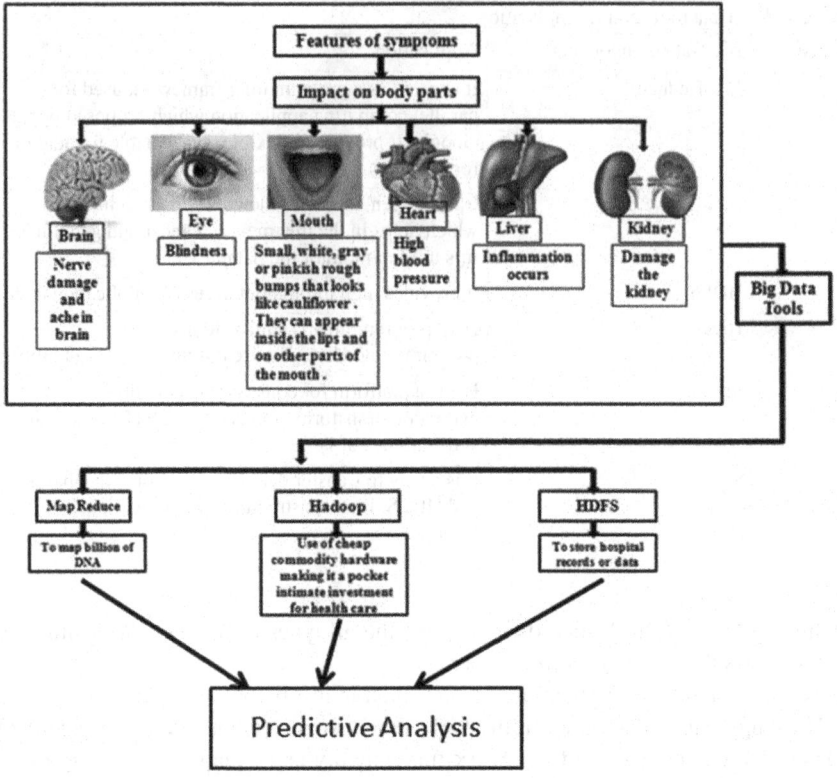

Fig. 4 Proposed framework

cauliflower. They can appear in the lips and on another portion of the mouth. When on heart, the human gets diabetes and cholesterol. When impact on kidney causes high blood pressure and diabetes, the regular diet and healthy exercise may hold to control diabetes and high blood pressure under control. The HIV affects the liver, and it may get hepatitis and a kindling of the liver.

The whole features of symptom come in the contact of Big Data tools like MapReduce, Hadoop, and HDFS; it maps numbers of the billion of DNA and Hadoop makes the cheap commodity which is used the purpose of health care as a pocket investment. The HDFS is to store hospital records or data. Then, the whole process is going to predictive analysis to give the result of the public awareness.

4 Discussion

In using this above-proposed framework, we can also design a mobile application which means "mobile apps" for the public. Through these mobile apps, HIV disease can be detected and hence cured at an early stage. So, it will combat the spread of HIV [15]. The Government decides to provide free treatment for HIV positive patients. That whole treatment is under the "Test and Treat policy for HIV".

To improve the healthcare facility in future also, the research community must provide clinical delivery innovation, patient dialogue, and collaboration, securing optimum operations of the healthcare system. The successful implementation of healthcare solutions is in high expectation. To achieve the said healthcare facility that provides the coordination between patients and coordinators for gaining higher self-satisfactions is mandatory. As the clinical innovation is extending decision making and delivery of the patient service to be also beneficial for improving their health has become inevitable. Ability to, access to, analyze and document the patients' record anywhere, thus providing health care in any location, have become the challenge of the day.

A new future can be created for health care by using Big Data technologies. Today medical costs are continually bruising worldwide. Using this Big Data technology, we can lower the increasing cost in medical examinations and differed results through disease prevention, early discovery of disease, thus improving the quality of life, predicting with a high degree of accuracy, analyze the interrelationship of examination value and diseases would help in the prevention of many fatal diseases similar to AIDS like various forms of cancer, infections, and other physical ailments.

References

1. Arora, Y., Goyal, D.: Big data: a review of analytics methods & techniques. In: 2016 2nd International Conference on Contemporary Computing and Informatics (IC3I). IEEE (2016)
2. Yadav, K., Pandey, M., Rautaray, S.S.: Feedback analysis using big data tools. In: 2016 International Conference on ICT in Business Industry & Government (ICTBIG), Indore, pp. 1–5 (2016)
3. Vijayaraj, J., et al.: A comprehensive survey of big data analytics tools. In: 2016 Online International Conference on Green Engineering and Technologies (IC-GET). IEEE (2016)
4. Ali, S.M., et al.: Big data visualization: tools and challenges. In: 2016 2nd International Conference on Contemporary Computing and Informatics (IC3I). IEEE (2016)
5. Thara, D.K., et al.: Impact of big data in healthcare: a survey. In: 2016 2nd International Conference on Contemporary Computing and Informatics (IC3I). IEEE (2016)
6. McColgin, D., Hoover, P., Igra, M.: The DataSpace for HIV vaccine studies. In: 2016 IEEE Conference on Visual Analytics Science and Technology (VAST), 23–28 Oct, Baltimore, Maryland, USA 978-1-5090-5661-3/16/$31.00 ©2016. IEEE (2016)
7. Sahani, S.K.: Analysis of a delayed HIV infection model. In: International Workshop on Computational Intelligence (IWCI). IEEE (2016)
8. Wang, J., et al.: Regularity of herbal formulae for HIV/AIDS patients with syndromes based on complex networks. In: 2014 IEEE Symposium on Computational Intelligence in Big Data (CIBD). IEEE (2014)

 9. Afzal, S., Maciejewski, R., Ebert, D.S.: Visual analytics decision support environment for epidemic modeling and response evaluation. In: 2011 IEEE Conference on Visual Analytics Science and Technology (VAST). IEEE (2011)
10. Khalid, Z., Sezerman, O.U.: Prediction of HIV drug resistance by combining sequence and structural properties. In: IEEE/ACM Transactions on Computational Biology and Bioinformatics (2016)
11. Tiwari, T., Sampath, P., Lavanya, S.: Predictive methodology for diabetic data analysis in big data. Procedia Comput. Sci. **50**, 203–208 (2015)
12. Zhang, X., Wang, J., Liang, B., Qi, H., Zhao, Y.: Mining the prescription-symptom regularity of TCM for HIV/AIDS based on complex network
13. Rao, A.S., Thomas, K., Sudhakar, K., Maini, P.K.: HIV/AIDS epidemic in India and predicting the impact of the national response: mathematical modeling and analysis (2009)
14. Lin, G., et al.: Fast supervised hashing with decision trees for high-dimensional data. In: Proceedings of the IEEE Conference on Computer Vision and Pattern Recognition (2014)
15. Narasimhan, Ravi, Bhuvaneshwari, T.: Big data—a brief study. Int. J. Sci. Eng. Res. **5**, 1–4 (2014)

Linguistic Rule-Based Ontology-Driven Chatbot System

Anuj Saini, Aayushi Verma, Anuja Arora and Chetna Gupta

Abstract Chatbots are designed to change the way system and users interact. As the IT-support system interaction behaves like developer asks a query regarding a project from system, then system searches and results the best suitable answer. Chatbots usually roam around two components of a conversational system—intent and entity. In order to build a generic chatbot, it is important to employ natural language understanding system and process machine to understand language as does we human interpret. In this research work, linguistic rules, ontology, and similarity indexing are performed in a uniform manner to build a generic chatbot system. Initially, linguistic rules have been implied to understand context in correspondence to detect intent and entity in user query. Then, ontology is used to map intent and entity in varying parts of question–answering system. Even, ontology helps in finding similarity among relations used in query and relations in its ontological structure while applying syntactic ambiguity resolution. Syntactic ambiguity resolution is used to edit distance, cosine similarity, N-gram matching, and semantic feature similarity as a set of text preprocessing, information retrieval, and similarity algorithms. System provides a confidence score to consume resolved entity, and one can safely uses resolved entity if its score is above than defined threshold value.

Keywords Intent · Entity · Question–Answering · Linguistic rules · Ontology

A. Saini (✉)
Sapient Global Markets, Gurugram, India
e-mail: asaini13@sapient.com

A. Verma
Hays Business Solutions, Noida, India
e-mail: aayushi291091@gmail.com

A. Arora · C. Gupta
Jaypee Institute of Information Technology, Noida, India
e-mail: anuja.arora29@gmail.com

C. Gupta
e-mail: chetnagupta04@gmail.com

© Springer Nature Singapore Pte Ltd. 2019
S. K. Bhatia et al. (eds.), *Advances in Computer Communication and Computational Sciences*, Advances in Intelligent Systems and Computing 760, https://doi.org/10.1007/978-981-13-0344-9_4

1   Introduction

An automatic question–answering system and its market need to have evolved it as an active research area for several years. This has become necessity for varying domains such as law domain and IT-support domain and has enhanced frequently asked question section of numerous Web applications. Law domain in such a way that if lawyer asks a query relevant to law terminology, the system is capable enough to retrieve the relevant subsection from the law knowledge base along with the order of document relevance. Researchers have used both structured and unstructured datasets for this purpose. Efficiency of question–answering system depends on the suitability and coherence of system generated answer from the knowledge base for the asked question by the user. Varying QA systems have been developed for varying domains obtaining varying paradigm and solution approaches. Classification is detailed below.

- Database querying methods are usually used as domain-dependent methods which search results corresponding to query for a particular domain [1]. These methods are useful for finding similarity questions from frequently asked question set. In database querying method, content is arranged as lists of answers associated with metadata like users' preference for respective answers and weightage to best preferred answer [2]. Some research work merged database query method with information retrieval approach topic modeling [3] such as latent Dirichlet allocation (LDA) [4] and probabilistic latent factor techniques [2] are applied to improve efficiency of QA systems.
- Information retrieval methods: Usually, questions are written in natural language and information retrieval techniques in hybrid manner are recently applied to achieve better results. Verma and Arora [5] incorporated features using both information retrieval and natural language processing techniques in their projected reflexive hybrid approach to provide precise answers. They have integrated term frequency/inverse document frequency, POS tagging, and Word2Vec approach to retrieve relevant document according to user queries.
- Knowledge base methods: Knowledge-based QA systems seem to be most influential approach in this aspect and help in developing generic QA system. It is known as generic because this can work in domain-independent manner and can handle any query in spite of having its associated content. Knowledge-based question–answering systems are developed in two directions—Factoid QA system and non-Factoid QA systems. Factoid QA systems are about concise/objective facts, for example—"What is population of India?" In contrast, non-factoid is an umbrellas term which covers all questions beyond factoid QA and might have some overlap with factoid QA. For example, can you update the task "ABC" with a new date of completion? Google's knowledgebase and freebase are well-known commercial QA systems available in market. Most knowledge-based systems use resource description framework (RDF) to generate standardized chatbot systems.

Scenario: User: Create New Project	Intent: Create; Entity: Project
Bot: in Which tool you want to create Project?	Intent: Create; Entity: Project, Tool
User: "Github"	
Bot: What is the Project Name and Description?	Entity: Project Name; Description
User: Name: New, Description: NLP POC	
Bot: Ok. I am going to create "New" project in Github with description "NLP POC"	

Fig. 1 Example QA query scenario, chatbot results, and enlisted intent–entity

In this paper, we focus on building the generic chatbot for factoid QA system using knowledge-based methods. Factoid question depends on existence of intent and entity in the questions; question responds according to intent and entity. Figure 1 shows a scenario corresponding to extracted intent, entity, and bot's reply. Query in this scenario is "Create New Project".

This chatbot exploits the query information as mentioned in mechanism and present our findings:

- We start out by automatically detecting intent and entity of Web query without any conscious control of manual intent and entity list for specific domain. To do this, we propose linguistic rules to exploit context along with linguistic phrases to extract intent and entity.
- We then build an ontological mapping to annotate relationship among intent and entities for question–answering system, and these annotations help in reasoning system to provide accurate answer to user.
- Finally, we apply syntactic ambiguity resolution employing set of similarity, information retrieval, and transliteration methods.

The paper is organized as follows: Sect. 2 details about the pertinent and significant research work of automatic question–answering system/chatbots system. Section 3 depicts proposed linguistic rule-based ontology-driven chatbot system sketch. Sections 4, 5, and 6 provide detailed description of three phases—linguistic rule-based intent–entity extraction, ontological mapping, and syntactic resolution, respectively, of proposed approach. Experimental results and evaluation are described in Sect. 7 and finally conclude in Sect. 8.

2 Related Work

Several researches have been done in the field of question–answering system earlier; however, it has significantly progressed in recent years [1]. Relevant literature has been reviewed to analyze the related work as summarized in Table 1.

Table 1 Techniques and dataset used in question–answering system

Reference	Techniques used	Dataset used
[6]	POS tag, shallow parsing Semantic parsing NE tagger Extracting entity associations	100 questions from Text Retrieval Conference (TREC-8)
[1]	TF-IDF, Word2Vec, POS tag	General sample question set
[7]	Semantic Web Semantic text mining Agent ontology	1000 articles of People's Dairy corpus out of which 50 questions were prepared
[8]	Subject–predicate extraction, Topic entities, Predicate scoring, Answer patterns Ranking	Training data and testing data of 14609 and 9870 question–answer pairs, respectively, published by NLP&CC2016
[9]	Classification, reformulation NLP parser, Wordnet, dice coefficient, similarity algorithms	Advanced Knowledge Technologies Ontology for academic domain
[10]	RDF-based ontology answer ranking using noun phrases	Snippets of Web documents using available ontology

3 Linguistic Rule-Based Ontology-Driven Chatbot System Sketch

Most of the knowledge-based QA systems maintain a manual formed intent and entity gazetteer list corresponding to a domain. Certainly, these systems are accurate and robust for specific domain chatbot systems because while developing QA system users' expectations toward chatbot, well-known question sets are limited, for example, chatbot for food-ordering Web site "Swiggy". List of intent and entities are preknown and cannot be changed. Few entities for Swiggy Application are dish, restaurant, city, ingredients, etc., and similarly intents are order, find, refund, complaints, etc. Even for such intent–entity combination, specific action/dialog is to be performed. These systems are easy and quick to build. Whereas to design a generic chatbot system which returns highly accurate and relevant results by understanding the searcher's intent and contextual meaning of the terms as they appear in the search query string is a complex issue. System is complex to design if intent–entity list is unknown or contains substantial number of intents and entities which keep on expanding by time or varying variants of intent for same entity such as create, remove, edit for same entity "cart". Therefore, in our proposed approach, to build a generic chatbot, we employ natural language understanding to extract intent–entity. As shown in Fig. 2, our process is divided into three phases—linguistic rule generation and intent–entity extraction; ontological mapping; syntactic ambiguity resolution.

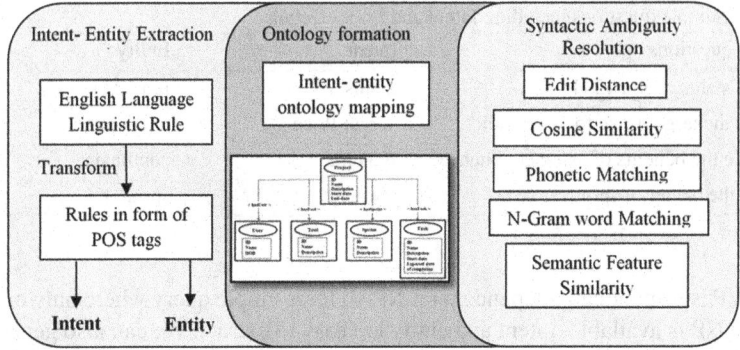

Fig. 2 Linguistic rule-based ontology-driven chatbot system sketch

4 Linguistic Rules and Intent–Entity Extraction Phase

To form linguistic rules, we have observed that entities are always noun phrases (NP) and intents are corresponding actions means verb phrases (VP). On the other end, we noticed various linguistic patterns such as numbers of NPs or VPs exist in a sentence; there could be more than one intent/entity in a given question, etc. Therefore, for appropriate identification of intent and entity understanding of context is required. The proposed system uses linguistics of English language in the form of POS tagging and apply shallow parsing to extract phrases. Then on top of that apply linguistic rules (see some POS tags in Table 2) to exploit context along with linguistic phrases to extract intent/entity.

Some sample proposed linguistic rules of English language that we have used are as follows:

- Rule 1: "<VP><NP>", "<VP: Intent><NP: Entity>"
  Whenever there is a verb phrase (verb + noun or adverb + verb) followed by a noun phrase (NP), we are going to treat verb phrase as intent and noun phrase as entity.
- Rule 2: "<VP><PP><NP>", "<VP: Intent><PP><NP: Entity>"
  Whenever there is a sequence of VP, PP, and NP that is verb phrase then proposition and then noun phrase, we are going to treat verb phrase as intent and noun phrase as entity.

Table 2 POS tags used to describe linguistic rules

Tag	Description	Tag	Description
NP	Noun phrase	ADVP	Adverb phrase
VP	Verb phrase	ADJP	Adjective phrase
PP	Proposition phrase	O	Splitter

Table 3 Sample questions including Intent and Entity Details

Sample questions	Intent	Entity
What is status of jira 1234?	Status	jira 1234
What is the target price for Microsoft?	Target Price	microsoft
What are the benefits of using confluence?	Benefits	confluence
Who is the owner of sprint 3?	Owner	sprint 3

As VP is semantically dependent on NP, so for a simple query where only one VP and one NP is available, intent and entity are easy to extract. We can also generalize that intents are actions to be performed such as find, create, delete and entities are proper noun or generic nouns such as person, location, computer, jira, ticket. Some sample questions consisting of intent and entity details are depicted in Table 3.

This rule can help us to solve most of what/who/where/when kind of simple questions such as—What is the capital of India?; What is the name of prime minister of india?; Who is the CEO of Google?; When is the anniversary of the Vietnam war?

Some other sample rules based on adverb, adjective phrase are as follows:

- Rule 3: "<ADVP><ADJP><PP><NP>", "<ADVP: Intent><ADJP><PP><NP: Entity>"
  If there is a sequence of adverb, adjective, preposition, and noun phrase, we are going to use adverb as "Intent" and NP as an "Entity".
- Rule 4: "<ADVP><VP><VP>", "<ADVP><VP: Entity><VP: Intent>"
  An adverb phrase followed by verb phrase and further verb phrase, then use first verb phrase as "Entity" and second verb phrase as "Intent".
- Rule 5: "<ADVP><O><NP><VP>", "<ADVP><O><NP: Entity><VP: Intent>")"
  An adverb phrase followed by a splitter (O), noun phrase (NP), and verb phrase (VP), then use noun phrase as "Entity" and verb phrase as "Intent".

5 Ontological Mapping

Domain ontology provides the conceptual structure of the formal specification of related terminologies, properties, attributes, and interrelationships of the entities among them [11]. If ontology mapping fails to recognize an important entity–intent relationship, the whole QA system is altered. Similarly, if the context in a question is not interpreted perfectly, then whole response of the question is lost. Building of ontology model helped us in aggregating and extracting information to answer user queries, as described in the following example.

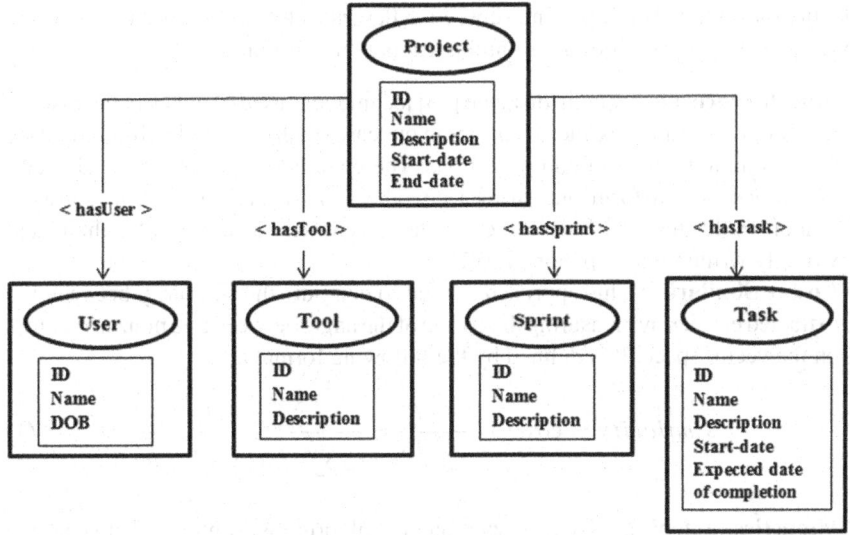

Fig. 3 Sample pictorial ontology representation for project entity

User query: How many tasks are there in project "ABC" current sprint 5 which assigned to user A?

The extracted intent and entities from this query are mapped to our IT-Support domain ontology to fetch the exact answer of user query.

A sample pictorial representation of the ontology for "project" entity is shown in the Fig. 3.

It shows the entity "project" possesses some properties such as Id, Name, Description and is connected to four entities which are user, tool, sprint, and task by given relationships. These entities are further linked to others with their respective properties and relationships in the complete ontology.

For the above query asked by the user, initially the ontology model searches for project entity with name: ABC which is mapped to sprint entity with id: 5, user entity with name: user A, task entity and then compute the number of tasks to extract the most appropriate answer.

6 Syntactic Ambiguity Resolution

In chatbot systems, syntactic ambiguity is a serious and considerable challenge. Many words in a sentence can belong to more than one syntactic category [12]. The process of resolving such syntactic behavior using different approaches is known as Syntactic Resolution. In this process, focus is toward the structural syntactic resolution ambiguity for two main constraints—grammatical constraint and word identity constraint. The following subsection describes few techniques that are used to resolve extracted entities after ontological mapping is done.

- **Edit Distance**: Levenshtein distance [13] technique is used to quantify the dissimilarity of two strings (extracted entities in our case) with one another by computing the minimum number of edit operations. There are basically three allowable edit operations to transform one string to another—insertion, removal, or substitution of single character [14]. For example, Levenshtein distance of "python" and wrongly written word "pyhon" is 0.1.
- **Cosine Similarity**: This approach is used to compute the similarity between the extracted entities by measuring the cosine of the angle between two nonzero vectors on the vector space. It is defined by the following formula:

$$similarity = \cos \theta = \frac{A \cdot B}{|A||B|} = \frac{\sum_{i=1}^{n} A_i B_i}{\sqrt{\sum_{i=1}^{n} A_i^2} \sqrt{\sum_{i=1}^{n} B_i^2}} \tag{1}$$

- **Phonetics matching**: Using fuzzy matching algorithms, similarity between two strings can be calculated in combination with phonetic algorithms. For example, brandname, brand-name, name brand, branding name, brand na can be resolved into brand_name.
- **N-Gram**: N-gram has also been used to resolve syntactic ambiguity problem. It has been considered as a feature to match question term with entity. For example, the query "What is the stock price for microsoft in NSDAQ" consists of 2-gram term set "stock price".
- **Semantic feature similarity**: Semantic matching [5] is added to similarity matching pipeline to capture matching beyond the exact word match and into the meanings at the conceptual and contextual level. For example, if we search for a specific skill such as "big data", it is also matched with related terms such as "Hadoop", "Hive", and "Pig", because of the combined usage of such keywords together in the corpus.

7 Experimental Results

Intent–entity annotation is performed in such a way that entity references and relationships with intent are connected based on their ontology mapping status. The ontological relationship mapping of the number of entities corresponding to their intent is shown in Fig. 4. The proposed approach generated "getIntentEntities" outcome snapshot is shown in Fig. 5, which shows entities, intent, requirement of relevant mapping for a specific service request "create task in JIRA".

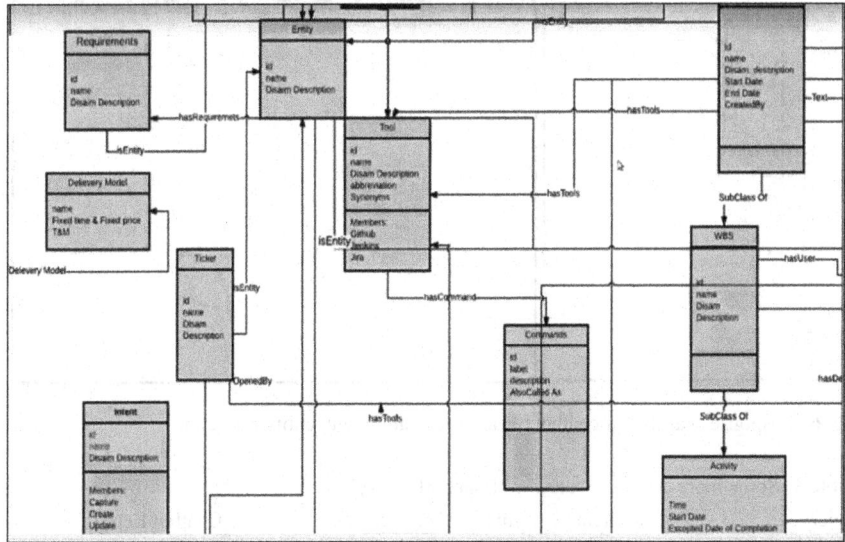

Fig. 4 Ontology representation of project

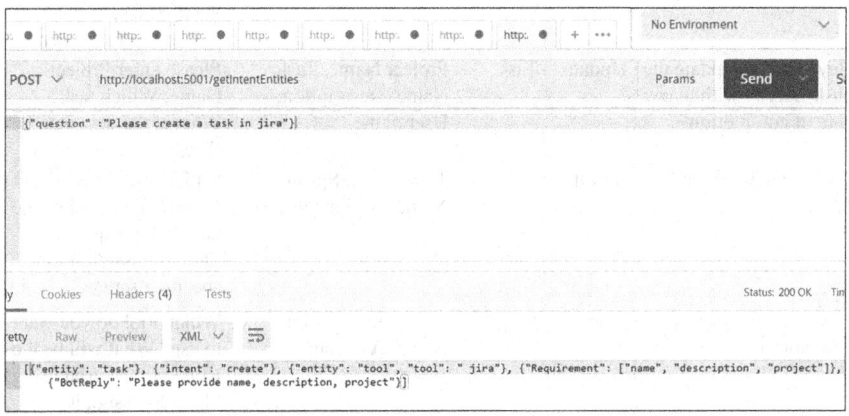

Fig. 5 Response snapshot of intent–entity extraction

Figure 6 shows results of a NL query "show all tasks across all branches", question parsing query shows "show" as intent, "task" as entity and mapped requirements are "tool, branch". Corresponding response is shown in Fig. 6. Here we compute confidence score of outcome corresponding to asked NL query. Table 4 shows all chatbot results having confidence score threshold more than 80%.

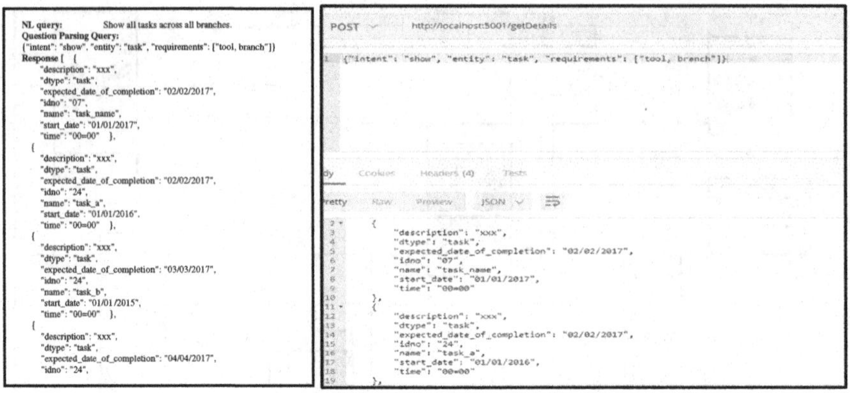

Fig. 6 Response snapshot of chatbot result of relevant intent–entity extraction

Table 4 Some more chatbot results for IT-support Query

Query	Intent	Entity	Requirement	Chatbot Reply
Please Create a Task in JIRA	Create	Task	Project Name, Task Name, Description, User name	What is the Project Name? Please enter task name, What will be the task description? Please enter user name
Hey can you Update the Task "ABC" with a new date of completion	Update	Task	Project Name, Task Name, Description, User name	Please enter Project Name, Which task? What is the new date? Please enter user name
Create a new Sprint in JIRA	Sprint	Sprint	Tool Name, Sprint Name, Assignee	Which tool do you want to use? What will be the name of the sprint? Should I assign it to a specific user?
Hey, can you Start a new Sprint	Start	Sprint	Tool Name, Sprint Name, Assignee	Which tool do you want to use? What will be the name of the sprint? Should I assign it to a specific user?

8 Conclusion

Automatic question–answering systems using chatbots are becoming necessity in current market situations. Through this paper, we focus on building the generic chatbots for factoid QA system using knowledge-based methods. Our QA approach successfully distinguishes itself on following finding through the presented approach by (1) automatically detecting intent and entity of Web query using linguistic rules to exploit context along with linguistic phrases to extract intent and entity, (2) building an ontological mapping to annotate relationship among intent and entities to support reasoning system for providing accurate answer to user, (3) applying syntactic ambiguity resolution employing set of similarity, information retrieval, and transliteration methods to produce correct responses. The experimental results are promising in terms of effectiveness, efficiency, and precision.

References

1. Sneiders, E.: Automated question answering using question templates that cover the conceptual model of the database. In: NLDB, vol. 2, pp. 235–239, June 2002
2. Komiya, K., Abe, Y., Morita, H., Kotani, Y.: Question answering system using Q & A site corpus Query expansion and answer candidate evaluation. SpringerPlus 2(1), 396 (2013)
3. Mcauliffe, J.D., Blei, D.M.: Supervised topic models. In: NIPS'07, pp. 121–128 (2007)
4. Zhang, K., Wu, W., Wu, H., Li, Z., Zhou, M.: Question retrieval with high quality answers in community question answering. In: Proceedings of the 23rd ACM International Conference on Information and Knowledge Management, pp. 371–380. ACM, Nov 2014
5. Verma, A., Arora, A.: Reflexive hybrid approach to provide precise answer of user desired frequently asked question. In: 2017 7th International Conference on Cloud Computing, Data Science & Engineering-Confluence, pp. 159–163. IEEE, Jan 2017
6. Li, W., Srihari, R.K., Li, X., Srikanth, M., Zhang, X., Niu, C.: Extracting Exact Answers to Questions Based on Structural Links. Association for Computational Linguistics Stroudsburg, PA, USA (2002)
7. Guo, Q.-l.: A novel approach for agent ontology and its application in question answering. J. Cent. South Univ. Technol. (2009) 16: 0781–0788
8. Lai, Y., Lin, Y., Chen, J., Feng, Y., Zhao, D.: Open domain question answering system based on knowledge base. In: Lin, C.-Y., et al. (eds.) NLPCC-ICCPOL 2016, LNAI 10102, pp. 722–733 (2016)
9. Vargas-Vera, M., Motta, E.: AQUA—ontology-based question answering system. In: Monroy, R., et al. (eds.) MICAI 2004, LNAI 2972, pp. 468–477 (2004)
10. Tartir, S., McKnight, B., Budak Arpinar, I.: SemanticQA: web-based ontology-driven question answering. In: SAC '09 Proceedings of the 2009 ACM symposium on Applied Computing, pp. 1275–1276, Honolulu, Hawaii, U.S.A. (2009)
11. Gupta, K., Arora, A.: Web search personalization using ontological user profiles. In: Proceedings of the Second International Conference on Soft Computing for Problem Solving (SocProS 2012), 28–30 Dec 2012, pp. 849–855. Springer, New Delhi (2014)
12. Franz, A.: Automatic Ambiguity Resolution in Natural Language Processing: An Empirical Approach, Volume 1171 of Lecture Notes in Artificial Intelligence. Springer Science & Business Media, 13-Nov-1996—Computers, 3540620044, 9783540620044
13. Yujian, L., Bo, L.: A normalized Levenshtein distance metric. IEEE Trans. Pattern Anal. Mach. Intell. 29(6), 1091–1095 (2007)
14. https://nlp.stanford.edu/IR-book/html/htmledition/edit-distance-1.html

Fuzzy Based Steganalysis Pattern Discovery for High Accuracy

Mohammed Abdullah Salim Al Husaini and Shaik Asif Hussain

Abstract Steganalysis pattern is used to identify and detect the stego from cover images. Embedded Mechanism is used to combine cover image with stego pattern for hiding significant information. In order to classify stego from cover image a pattern need to be constructed for steganography algorithm. In the proposed work high detection accuracy is expected, using evolutionary algorithm where significant features are classified for stego image via fuzzy rules. These rules are based on linguistic knowledge and numerical data. The fuzzy classifier system will automatically generate if then rules from training patterns and fitness value. This multi pattern classification is compact and high performance for accuracy. Through the pattern discovered steganography method is predicted by involving different image patterns used for stego images. Based on the discovered knowledge suitable models are trained and saved in databases. Image databases are validated using the methods investigated using Steganalysis and steganography techniques. The steganography method classifies the image into multiple classes and each class has feature pattern for stego and cover images. These features extracted are stored in databases by stego pattern classification (SPD). The experimental results obtained define the stego pattern extracted from training samples. Steganalysis techniques used are based on the text, Image formats like JPEG, BMP, GIF and PNG and audio. Steganography algorithms are used to hide information in cover media to obtain stego media. Steganography techniques protects and safeguards personal information of individuals from criminals. The performance results are evaluated using Anova tool and statistical calculations are done through Matlab for improved results.

M. A. S. A. Husaini · S. A. Hussain (✉)
Middle East College, Muscat, Sultanate of Oman
e-mail: sah.ssk@gmail.com; shussain@mec.edu.com

M. A. S. A. Husaini
e-mail: PG15F1572@mec.edu.om

© Springer Nature Singapore Pte Ltd. 2019
S. K. Bhatia et al. (eds.), *Advances in Computer Communication and Computational Sciences*, Advances in Intelligent Systems and Computing 760, https://doi.org/10.1007/978-981-13-0344-9_5

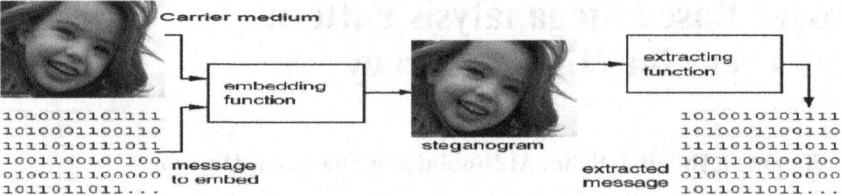

Fig. 1 shows the steganography process (slideshare.net [1])

Table 1 shows the basic differences between steganography and cryptography (scialert.net)

S. no	Steganography	Cryptography
1.	Hides message within another message as a cover and looks like normal file or graphics	Encrypted message is like meaningless jumble of characters
2.	Collection of image files, graphics, sound files do not leave any suspicion	Collection of random characters on disk always leave suspicion
3.	Smart eavesdropper easily detect something from sudden change of message format	Smart eavesdropper detect secret communication from encoded data
4.	Requires caution when reusing pictures or sound files	Requires caution when reusing keys

1 Introduction

Steganography term is usually derived from a word 'Steganos' and 'graphia' where Steganos mean covered and graphia mean writing. The main idea behind this term is to transmit secret data. Steganalysis method is to find the presence of hidden information. Embedding of data into digital pictures as covert refers to image steganography (Fig. 1).

The most popular media used in steganography is image as it is widely used as carrier in internet. One of the best attribute with image in Human visual system is it finds difficulty in differentiating large conceal data. Hence images find their significant place in steganography (Table 1).

There are various image formats like JPEG, BMP, TIFF are used for cover objects. There are compressed and uncompressed data used in image formats. PNG format provides best compression and color support and it can be alternative to GIF format for presenting web images. There are different types of images that can be used in steganography such as Binary, gray scale and RGB or true color image.

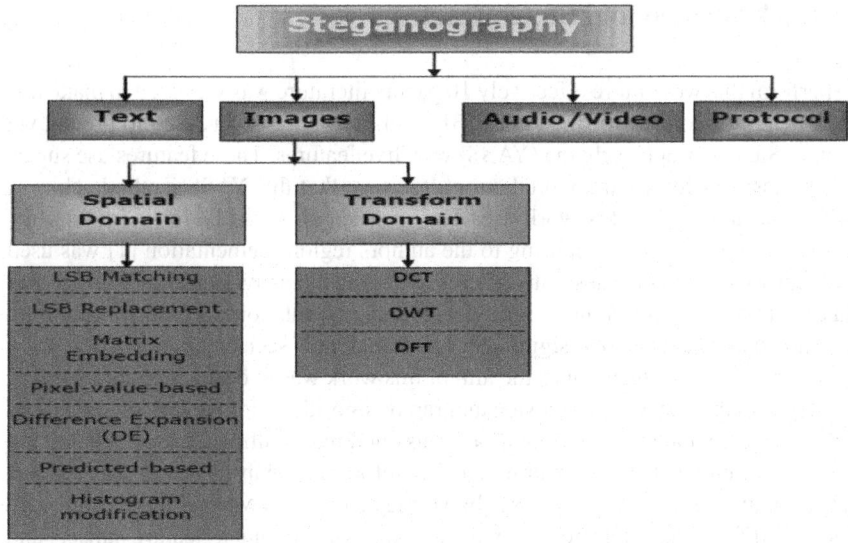

Fig. 2 Shows the Image steganography classification (sciencedirect.com)

1.1 Image Steganography Classification

Image Steganography is based on file formats for text, image, audio or video and protocol [2]. Usually these images are classified into spatial and transform domain for different methods. Its classification is shown below Fig. 2.

1.2 Image Steganalysis

This method is used to identify the artifacts present in the suspicious file and form that as an embedded message. This method goal is for exploiting illicit hidden information. The methods for detecting the presence of hidden message are divided as Statistical and Feature based Steganalysis. Where statistical again classified as spatial and transform domain in which spatial refers to difference in the pixels and transform domain for frequency coefficients are calculated. Feature based method is employed in this work as it is used to extract the selected and relevant information for identifying the covert message in an image.

2   Back Ground and State of Art

To perform this work more effectively 10 papers literature was surveyed to make this work better. According to Bhatt et al. [3] blind Steganalysis was used to predict yet another Steganography scheme (YASS) with five features. These features use supervised classifiers for accuracy prediction. It is seen that this YASS method achieves 94% as accuracy [3]. This work uses Huffman length statistics for predicting the steganography method. According to the authors region segmentation [1] was used for preprocessing to give absolute difference of gray intensity. Here entire image was classified into three different regions where higher correlation was related according to region characteristics and significant improvement is seen in Accuracy. Based on various Steganalysis techniques the aim of this work was to design a multi classifier for stego images depending on steganography algorithms for separating cover and stego images. Smaller Image blocks are classified into multiple classes [4] and find the class for either stego or cover image. Therefore Steganalysis detect the presence of information from multiple classes by voting process. SVM classifiers [5] are also best suited for image steganography as they easily differentiate feature subsets and detection through multi classifiers by training and fusing classifier. The Steganalysis method used achieves better detection performance when compared with voting and other Bayesian methods. This work analyzed much on audio and image stego media for embedding the process into cover image and detection through Steganalysis algorithms. The OLSB technique used to hide secret data in cover image with minimum distortion and space used [6]. The preprocessing step proposed by authors is neural networks for classification system which can reduce complexity by Bhattacharyya distance. The relevant features are extracted from spatial and transform domain and this removal of redundant features is done by steghide [7], PQ an nsF5 and uses SVM, NN for classification with 20% improvement in feature selection.

3   Proposed Methodology

The proposed work is carried in four different phases as

(1)   Pattern discovery and Image Database.
(2)   Feature space selection.
(3)   Detection methods and classification.
(4)   Matching similarity and accuracy.

The above phases are defined in different steps which define cover image means which do not have any secret information inside and similarly stego means message is hidden inside.

(1)   Image data base creation: Different image formats like JPEG, BMP, TIFF and PNG are collected and they are trained to form stego and cover images. The image steganography process depends on size, histogram and pixel range.

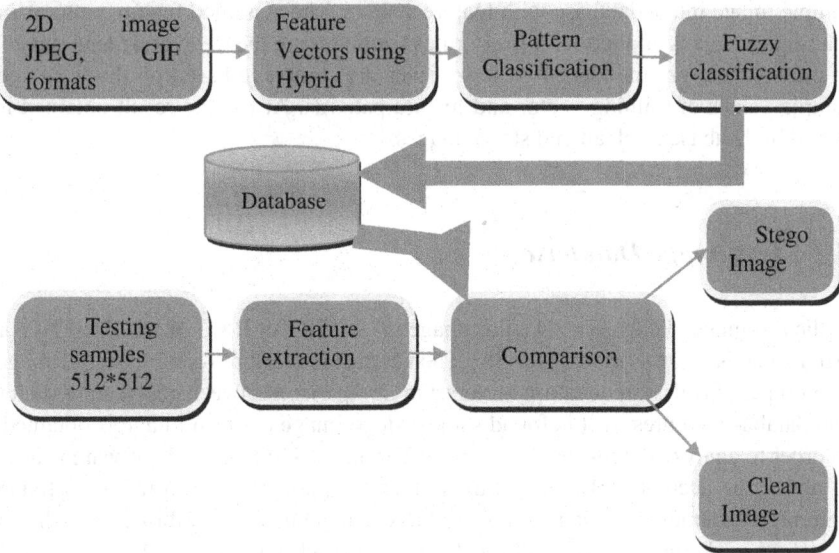

Fig. 3 Steganography block diagram

(2) Pair sampling: Embedding mechanism takes place to hide significant information to form stego image. This sampling mechanism depends on features.

(3) Feature selection and extraction: Image features are classified based on multi classifier SVM and embedded into cover image to form stego with appropriate matching.

(4) Classification of patterns: Stego pattern discovery is done by using support vector machine which is a non-discriminant function for images.

(5) Training and Testing samples: Samples of images are trained with the features selected and stored in database and retrieved when it is required.

(6) Fuzzy rule classification: if-then rules are used to classify the fuzzification based on numerical data and linguistic knowledge for extraction and detection (Fig. 3).

The block diagram section is divided into two phases. The first phase defined the training sample of second phase represents the decoding and testing image database sample.

3.1 Training Samples

In this sample a 2D color image (512 * 512) of JPEG or BMP or PNG or GIF format is used. Then this image is converted into gray scale image. The obtained image is added with three different steganography methods Model Base (MB), Perturbed Quantization (PQ), and yet another steganography Scheme (YASS) for training the sample image into a database with feature vectors. The classified feature patterns of the stego image is applied with fuzzy rules for obtaining the stego and clean image. In order to analyze the algorithm fuzzy rules are compared based on the weighted sample values of training vector and feature patterns. At last the result obtained is stored in database as clean and stego images.

3.2   Test Image Database

In the test image database a 2D color image (512 * 512) of JPEG or BMP or PNG or GIF format is given as input to test [8]. Then extract the feature vector from the image are compared with stored feature vectors. On Comparison between new feature vector and database features. If it is found same a stego image or clean image is obtained. In order to analyze the results Analysis Of Variance (ANOVA) tool software is used to detect the accuracy. By using Fuzzy rules the pattern of stego image is found with high accuracy. The first step is used to calculate the compatibility of each test samples and define certainty factor CF_j for clean and stego image [9].

3.3   Selection of Feature Vectors

In order to define the similarity of feature vector Bhattacharya Coefficient is used to decide the matching between two samples of stego and clean image

$$\text{Bh} = \left(\frac{1}{8}\right)(m_i - m_j^T)\left(\frac{C_i - C_j}{2}\right)^{-1}(m_i - m_j) + \left(\frac{1}{2}\right)\ln\left(\frac{\frac{C_i - C_j}{2}}{\sqrt{C_i C_j}}\right) \qquad (1)$$

3.4   Conversion of Binary Classifier

Steganalysis technique used is to verify either the data transmitted from one user to another user contains secret message or not. This classification helps for taking accurate decision for unknown image of stego or clean image [10]. To determine the performance the accuracy is detected by using

$$A_{\text{detect}} = 1 - P_{\text{Error}} \qquad (2)$$

From the above P_{Error} is known as average error Probability. To calculate the errors the statistics process uses false positives and false negatives. Actually false positives defines when secret message is detected from a cover image and false negatives represents when secret message is not detected from a cover image.

3.5 Fuzzy Logic

The fuzzy set is used for the feature vectors defined for MB, PQ and YASS and fuzzy logic used is with linguistic variables as large, medium, small, very small, medium large, don't care conditions. Three steps signify the fuzzy rules for stego and cover
Step 1: Each sample is calculated for training values
Where Xp = (Xp1,Xp2,.....,Xpn) uses if-then rule for knowledge based for below rule

$$\mu_j(X_p) = \mu_{j1}(X_{p1})x \ldots \ldots x \mu_{jn}(X_{pn}) \text{ Where } P = 1, 2, \ldots M \qquad (3)$$

where $\mu_j(X_{pi})$ represents feature membership function and sample of ith and Pth is the region and M it describes the number of samples
Step 2: Relative sum difference for training samples is calculated using below formula for clean and stego images

$$\beta_{clean}(R_j) = \frac{\sum_{X_p \varepsilon\, clean} \mu_j(X_p)}{N_{clean}} \qquad (4)$$

$$\beta_{Stego}(R_j) = \frac{\sum_{X_p \varepsilon\, Stego} \mu_j(X_p)}{N_{Stego}} \qquad (5)$$

where the $\beta_{Stego}(R_j)$ and $\beta_{Clean}(R_j)$ represents relative sum difference for stego and clean images respectively. N_{Stego} and N_{Clean} training samples of clean and stego.
Step 3: Certainty factor depends of capability of fuzzy rules and weights of the samples

$$CF_j = \frac{(\beta_{clean}(R_j) - \beta_{Stego}(R_j))}{(\beta_{clean}(R_j) + \beta_{Stego}(R_j))} \qquad (6)$$

All the above rules classify stego and clean image for the feature classification from Eqs. (3) (4) and (5).

3.6 Steganography Methods

3.6.1 YASS Algorithm

The basic fundamental operations in YASS algorithm is defined with image size and in the next step the secret information is encoded and after the image is divided into consecutive display as B and H blocks. Next step is to apply DCT block for quantization with the help of RA coded data [11]. These quantized coefficients to inverse DCT to round off the values by applying iterative embedded process.

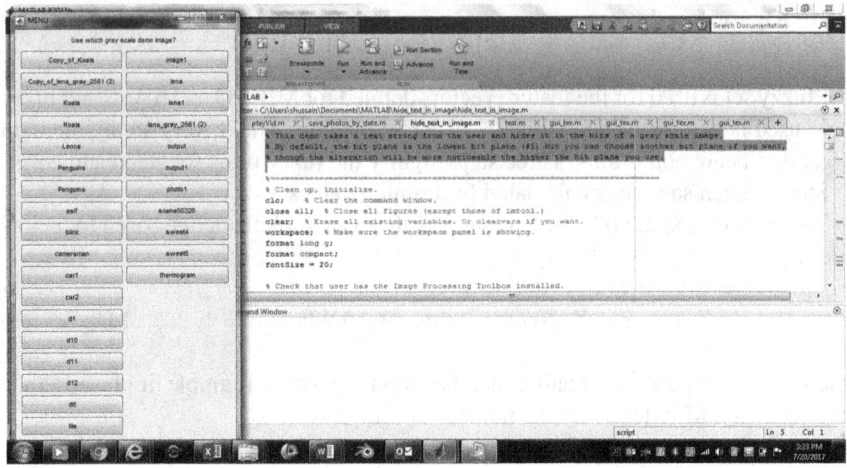

Fig. 4 shows the basic GUI for hiding string in the cover image

3.6.2 MB and PQ Algorithms

To work with image data bases variety of payloads for each classifier [12] built is carried out in blind Steganalysis with certain size and range of observed and stego image [13].

4 Experimental Result and Analysis

In order to understand and evaluate the detection accuracy of proposed method a total number of 1300 images are used as source images. All the images used were JPEG and BMP formats of 512×512 size. All codes of MB, PQ and YASS are written using MATLAB Software. In order to understand the above methods the given Fig. 4 describes the color image is converted in gray scale and its feature vectors are classified for pattern classification and the given input string from the user is hidden in LSB and it is decoded at destination (Fig. 5).

The above figure shows how a text is hidden in the image by using LSB bit and string is recovered from the image. Model based method achieves higher embedded efficiency based on the length of the message hidden with JPEG images.

The below figure shows how stego and cover images are separated for image databases. The classifier code used has produced separate images of 700 stego and 600 clean images with image databases for clean and stego images (Figs. 6, 7 and 8).

The above figure shows steganography model with YASS algorithm for audio signal hidden inside

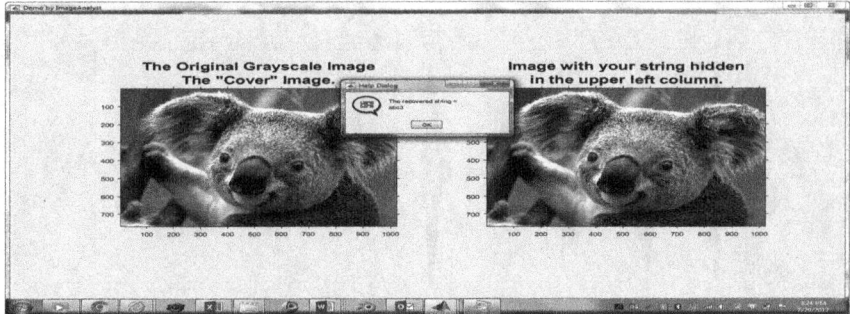

Fig. 5 Is a basic gray scale image with a input string hidden using Model Based method

Fig. 6 Shows how cover Image and stego image is separated from Image database

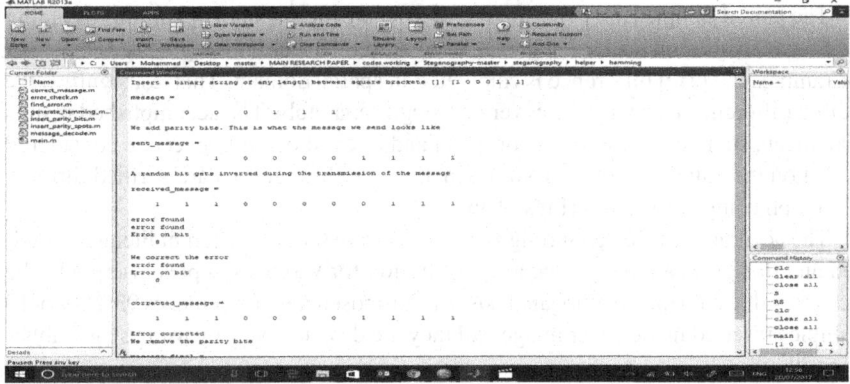

Fig. 7 Cyclic Redundancy checking with parity bits for the hidden message

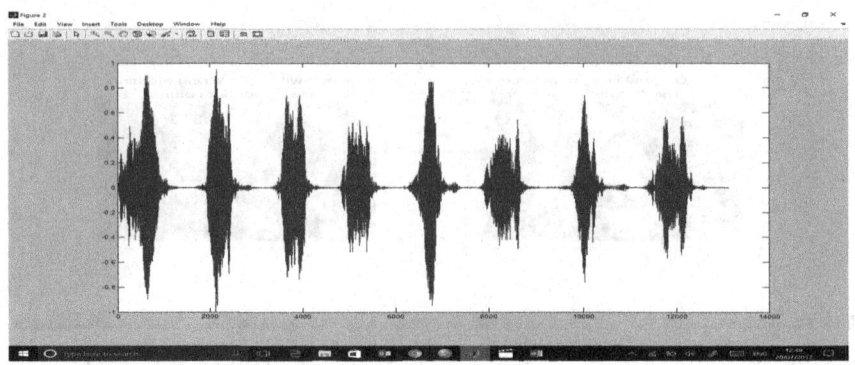

Fig. 8 describes audio steganography with chirp set audio signal

4.1 Computational Efficiency

The values in the images appear to be the same but on fact there are different values in each picture. By using Matlab different values for clean image and stego image are observed. The comparison between them is graphical shown in ANOVA tool. Cover1 is first line values in clean image and steg8 is first line values in stego image as shown in Fig. 7. Statistical values of mean, standard deviation, interval of confidence and correlation is shown through Anova tool (Table 2).

The standard deviation defines the data point of mean and variance is mean of all values and deviation is square root of variance as $S = \sqrt{\frac{\sum (x-x')^2}{n-1}}$.

The paired samples test is performed to evaluate the standard conditions of hiding by Steganalysis tools such as stegdetect, stegsecret, JP seek and stegbreak. In this pair of samples the pixel difference is calculated in spatial domain. This test signifies the paired difference normally for cover and stego separately. The accomplished results like mean, standard deviation, error mean and interval of confidence are considered based on the values obtained from the matlab and these results are verified through the graph using Anova tool (Figs. 9 and 10).

The accuracy is detected through anova software tool for different literature studied and this gives a good glance and illustration for various samples inserted to the accuracy detected through steganalysis. The proposed method achieves 93% for 5-10 samples inserted in the cover image and they are detected back by using steganalysis algorithms.

Table 2 Samples test of each pair with stego and Cover Images (Anova tool)

Paired samples test

		Paired differences					t	df	Sig. (2-tailed)
		Mean	Std. deviation	Std. error mean	95% confidence interval of the difference				
					Lower	Upper			
Pair 1	cover1 steg8	−7.75000	25.34430	6.33607	−21.25502	5.75502	−1.223	15	0.240
Pair 2	cover2 steg9	−8.87500	26.66802	6.66701	−23.08539	5.33539	−1.331	15	0.203
Pair 3	cover3 steg10	−11.18750	27.74700	6.93675	−25.97283	3.59783	−1.613	15	0.128
Pair 4	cover4 steg11	−14.06250	29.10777	7.27694	−29.57294	1.44794	−1.932	15	0.072
Pair 5	cover5 steg12	−19.06250	29.72422	7.43105	−34.90142	−3.22358	−2.565	15	0.022
Pair 6	cover6 steg13	−22.37500	30.59602	7.64901	−38.67847	−6.07153	−2.925	15	0.010
Pair 7	cover7 steg14	−25.75000	30.29081	7.57270	−41.89083	−9.60917	−3.400	15	0.004

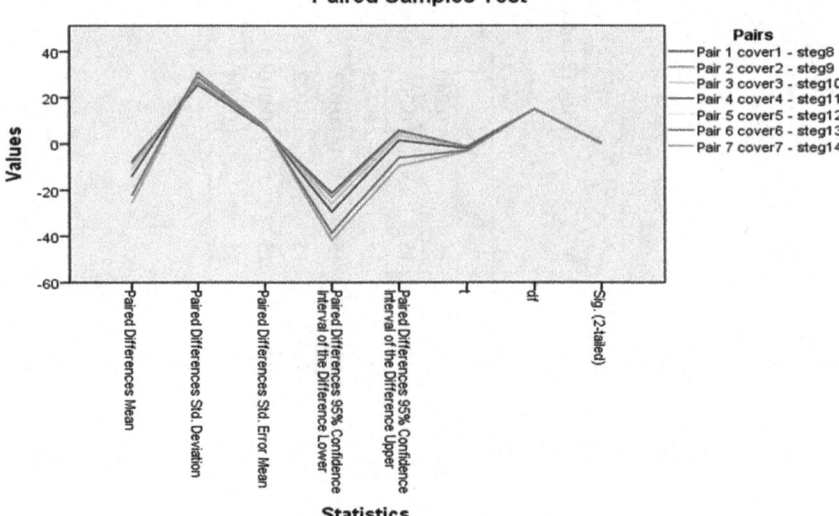

Fig. 9 Sample testing for detecting computational efficiency

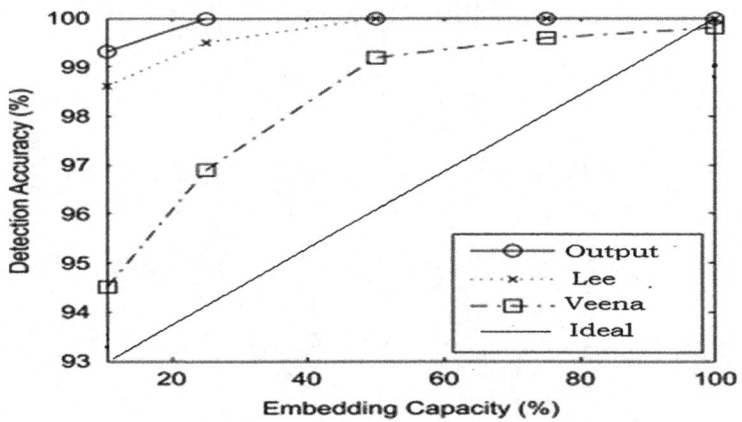

Fig. 10 Accuracy detection through Anova tool

5 Conclusion

The proposed work achieves greater detection accuracy for stego and clean image. To automatically generate feature vectors fuzzy if-then rules are used for different patterns. The learning procedure is evaluated with image databases separately for stego and clean image. This multidimensional approach with three different methods of MB, PQ and YASS has yielded better performance time when compared with other conventional methods. On the whole 13 features were used for pattern classification during comparison. JPEG images used were more robust and efficient in steganalysis rather than PNG and GIF images. The algorithms used in steganography were tested with different samples. This algorithms proved better security in encryption and decryption. The experimental results showed significant improvement in detection accuracy using Image based approach. It is evident that the feature selection algorithms used have removed redundant features during pattern classification. The proposed approach defined in two stages as training and testing samples. In first stage the database images are discovered as stego and clean image seperately. The image statistics for paired samples were tested and graphical analysed in Anova tool. The performance accuracy is tested for the paired samples of cover and stego image and it is compared with other methods studied literature to say that the desired accuracy of 94% is achieved.

References

1. Jingxin Hong, J.H.H.Z.: A Study on Preprocessing Method for Steganalysis Feature Extraction. IEEE (2016)
2. Badr, S.M.G.I.S.G.M.IS.A.H.K.: A review on steganalysis techniques: from image format point of view. Int. J. Comput. Appl. 102 Sept 2014
3. Bhat, V.H., Krishna, S., Deepa, S.P., Venugopal, K.R., Patnaik, L.M.: Steganalysis of YASS using Huffman Length statistics. Int. J. Hybrid Inf. Technol. 4(3) (2011)
4. Jay Kuo, C.C., Cha, B.H., Cho, S., Wang J.: Block Based Image Steganalysis for a Multi Classifier. IEEE Explore (2010)
5. Peiqing Liu, F.L.C.Y.: Improving Steganalysis by Fusing SVM Classifiers for JPEG Images, pp. 185–190. IEEE (2015)
6. Fatnassi, A.H.G.S.B.: A new hybrid steganalysis based approach for embedding image in audio and image cover media. Sci. Direct (2016)
7. Rita Rana Chhikara, M.K.: Significance of feature selection for image steganalysis. IEEE (2016)
8. Ghareh Mohammadi, F.M.S.A.: Image steganalysis using abee colony based feature. Sci. Direct 35–43 (2014)
9. DR.N.Kamaraj, S.A.: Optimized image steganalysis through feature selection using MBEGA. Int. J. Comput. Netw. Commun. (IJCNC) 2 (2010)
10. Ishibuchi, H.T.N.T.M.: A fuzzy classifier system that generates fuzzy if-then rules for pattern classification problems, vol. 2, pp. 759–764. IEEE (1995)
11. Teddy Furon, F.C.G.D.P.B.: Information Haiding, 9th edn. Springer, s.l. (2007)
12. Seongho Cho, B.-H.C.C.-C.K.: Block-based Image Steganalysis: Algorithm and Performance Evaluation, vol. 24(7), pp. 846–856. Elsevier (2013)
13. Soukal, J.F.M.G.D.: Perturbed Quantization Steganography. Springer, Oct 2005

Nearest Neighbor-Based Clustering Algorithm for Large Data Sets

Yadav Pankaj Kumar, Sriniwas Pandey, Mamata Samal and Mohanty Sraban Kumar

Abstract Clustering is an unsupervised learning technique in which data or objects are grouped into sets based on some similarity measure. Most of the clustering algorithms assume that the main memory is infinite and can accommodate the complete set of patterns. In reality, many applications give rise to a large set of patterns which does not fit in the main memory. When the data set is too large, much of the data are stored in the secondary memory. Input/outputs (I/O) from the disk are the major bottlenecks in designing efficient clustering algorithms for large data sets. Different designing techniques have been used to design clustering algorithms for large data sets. External memory algorithm is one class of algorithms which can be used for large data sets. These algorithms exploit the hierarchical memory structure of the computers by incorporating locality of reference directly in the algorithm. This paper contributes towards designing clustering algorithms in the external memory model (proposed by Aggarwal and Vitter) to make the algorithms scalable. In this paper, it is shown that the *Shared near neighbors* algorithm is not I/O efficient since the computational complexity and the I/O complexity both are same and high. The algorithm is redesigned in the external memory model reducing its I/O complexity without any change in its computational complexity. We substantiate the theoretical analysis by showing the performance of the algorithms with their traditional counterpart by implementing in STXXL library.

Y. Pankaj Kumar · S. Pandey (✉) · M. Sraban Kumar (✉)
PDPM Indian Institute of Information Technology, Design and Manufacturing, Jabalpur, Jabalpur, India
e-mail: snp.kecian@gmail.com

M. Sraban Kumar
e-mail: sraban@iiitdmj.ac.in

Y. Pankaj Kumar
e-mail: pankaj.yadav@iiitdmj.ac.in

M. Samal
Indian Institute of Technology Guwahati, Guwahati, India
e-mail: msamal@gmail.com

© Springer Nature Singapore Pte Ltd. 2019
S. K. Bhatia et al. (eds.), *Advances in Computer Communication and Computational Sciences*, Advances in Intelligent Systems and Computing 760, https://doi.org/10.1007/978-981-13-0344-9_6

Keywords Clustering of large data sets · Nearest neighbor clustering · Shared near neighbors clustering · External memory clustering algorithms

1 Introduction

Clustering is an unsupervised learning technique in which data or objects are grouped into sets based on some similarity measure. A universal criteria for clustering is that the data points in a group must be more similar than the points across the groups. There are few typical requirements for a good clustering technique in data mining [10, 24]. Versatility, ability to discover clusters with different shapes, minimum number of input parameters, robustness with regard to noise, insensitivity to the data input order, scalability to high-dimensional and large data sets are few of them.

1.1 Clustering of Large Data Sets

The performance of the algorithm should not decrease with the increase in the data size. Most of the clustering algorithms are designed for small data sets, and they fail to fulfill this important requirement. Many scientific, engineering, and business applications frequently produce very large data sets [1]. Though the definition of "Large" varies with the change in technology, specifically with the change in the memory size and the computational speed of the computers, it is a common consensus that the data size is increasing at much faster than the technologies to handle it. Majority of the clustering algorithms are not designed to handle large data sets.

Few approaches like decomposition [20], incremental [6], parallel implementation [19], summarization [5], approximation [13], distribution [22] are proposed in the literature to handle large data sets. Though these approaches make algorithms suitable for large data, but there are some limitations and challenges attached with each approach. For example, incremental approach introduces dynamic nature in clustering and complete data set is not required in advance, and clusters are evolved with the incoming data points. However, the order of input points effects the clustering process and cost increases if initial points do not represent the distribution properly [15, 20]. Approximation techniques are used when reaching to a perfect optimum is intractable; hence, we settle for an approximate solution. But approximation techniques are complicated in terms of implementation and data structure usage [21]. Parallelization is another technique used to handle large data sets. However, it is not always feasible and easy to parallelize every algorithm. Parallelization overhead and over dependency of different parallel modules are the constraints on parallelization [14]. Distributed clustering is another way of handling large data; local clustering is done at different sites, and global clustering is done at a central site using the local representatives. The technique is fast and reasonable in real-world scenario when data are being collected from different geographical sites, but if dependencies exist between data items in different locations, results may be inappropriate [11].

Balanced iterative reducing and clustering using hierarchies (BIRCH) [25], CLARANS [18], clustering using representatives (CURE) [9], scalable K-means ++ [4] are few algorithms designed for large data. Support for spherical and fixed-size clusters, dependency on initial parameters, parameter tuning, no guaranteed convergence and main memory restriction, etc., are important issues associated with these algorithms. Besides these issues, these algorithms are designed to work well in any particular scenario and are not universally applicable.

In the traditional algorithm design, it is assumed that the main memory is infinite and it allows uniform and random access to all its locations. But in reality, the present-day computers have multiple levels of memory and accessing data from each level has its own cost and performance characteristics. Most of the clustering algorithms assume that the main memory is large enough to accommodate the data set. However for large data sets, this is not a realistic assumption. If the data are too large to fit in the main memory, then it has to be stored in the disk of the machine and disk access is millions times slower than the main memory access. Hence in case of large data, the usual computational cost does not prove to be an appropriate performance, metric but the number of input/outputs (I/O cost) can be more apt measure for performance. External memory algorithm is one such class of algorithms which exploits the hierarchical memory structure of the computers by incorporating locality of reference directly in the algorithm [2]. The external memory model was introduced by Aggarwal and Vitter in 1988. The input/output model (I/O-model) views the computer consisting of a processor, internal memory (M), and external memory (disk). The external memory is considered unlimited in size and is divided into blocks of B consecutive data items. Transfer of a block of data between disk and RAM is called an I/O.

1.2 Contribution of the Paper

Shared near neighbor *(SNN)* [12] is a technique, in which similarity of two points is defined based on the number of neighbors, the two points share. The main advantage of the shared near neighbor-based clustering algorithms is that the number of clusters is not required as input to the algorithm and it is autogenerated by the algorithm. Document clustering, temporal time series clustering are few examples where SNN clustering technique is used. In this paper, *SNN*-based clustering algorithm is designed in external memory model to make it scalable. The computational as well as the I/O complexity of the SNN algorithm is $O(N^2 k^2)$ (where N is data set size and k is number of neighbors considered for each data point); hence, the SNN algorithm is not I/O efficient and not suitable for large data sets. We show that the I/O complexity of the proposed algorithm is $O(N^2 k^2 / BM)$ which is a BM factor improvement over the traditional SNN algorithm while the computational complexity remains same. Both traditional as well as proposed algorithm are implemented, and the performance of the proposed algorithm is compared with its traditional counterpart. The proposed algorithm outperforms the in-core algorithm, as expected from the theoretical results. A previous version of our paper exists in arxiv [23].

1.3 Organization of the Paper

This paper is organized as follows: In Sect. 2, the proposed scalable shared near neighbor-based clustering algorithm and its I/O analysis is described. Section 3 contains the experimental results and observations. The concluding remarks and future works are given in Sect. 4.

2 Proposed Scalable Clustering Algorithm Based on SNN

Shared near neighbor (SNN) is a technique in which similarity of two points is defined based on the number of neighbors, the two points share. It can efficiently generate clusters of different sizes and shapes.

The inputs of the SNN algorithm are two parameters: k (size of the nearest neighbors list) and θ (similarity threshold). The performance of the algorithm depends upon these parameters. An analytic process exists to find the most appropriate values of the input parameters [16].

2.1 Traditional Shared Near Neighbors Algorithm

In the two-step SNN algorithm, the first step is to calculate the k-nearest neighbors for all data points. The k-nearest neighbors of a point are arranged in ascending order. As each point is its own zeroth neighbor, so first point of each neighborhood row indicates the data point itself. In the second step, shared near neighbor of each data point pair is calculated. If ith and jth point ($i < j$) are having at least θ (similarity threshold) common neighbors and both points belong to each other's neighborhood, then the larger index j is replaced by the smaller index i or in other words j point is labeled as label of ith point. I/O complexity of this algorithm is as follows: For the first step, I/O complexity is $O(N^2 D)$, and for second step, number of I/Os is $O(N^2 k^2)$. Hence, overall complexity of the traditional algorithms is $O(N^2 k^2)$ which is same as computational complexity.

2.2 Proposed Scalable Shared Near Neighbors Algorithm

The in-core method is not I/O efficient making it unsuitable for massive data sets. In this section, we design an external memory algorithm to make it I/O efficient and hence scalable. The outputs of the algorithm are exactly same as the in-core method since the computational steps of both the algorithms are same. The data structure and access pattern are modified to reduce the amount of data transfer.

Algorithm 1 Proposed Algorithm For Generating the K-nearest neighbors matrix.

Input: Set of data points $S \in R^{N \times D}$, and M, the main memory size.
Output: knn[N][k] // k-Nearest Neighbors Matrix.

$t = M/2(D + k)$
for $i=0$ to $N/t - 1$ **do**
  for $j=0$ to $N/t - 1$ **do**
    Read S_i and S_j and knn_i
    Do the following computations in main memory.
    for $l = (i)t$ to $(i + 1)t - 1$ **do**
      for $m = (j)t$ to $(i + 1)t - 1$ **do**
        sum = distance(l, m)
        if dist[l][k] > sum **then** dist[l][k] = sum
        for $d = k$ to 1 & dist[l][d] < dist[l][d-1] **do**
          Find the appropriate position in $dist_{l, d}$
          matrix block and also swap the index
          value of that point into the different
          matrix block $knn_{l, d}$
    Write the matrix block $knn_{l, d}$ into disk.

2.2.1 Computation of K-Nearest Neighbor Matrix

The k-nearest neighbor matrix is generation in the first step of the algorithm. Assume that the data set of size $N \times D$ is partitioned into N/t number of $t \times D$ blocks. Here t is a parameter which is dependant on the available main memory. Any two blocks S_i and S_j are brought into the main memory, and the distance between each pair of points of these blocks is calculated in the main memory using an incore method. The distance and corresponding points index are stored in a temporary vector of size $t \times k$ called "dist" and in knn matrix block of size $t \times k$ respectively. The k-nearest neighbors of the block S_i is computed and written into the disk. The process is repeated N/t times. This procedure to generated knn matrix is described in Algorithm 1. The main memory M contains 2-blocks of size $t \times D$ and 2-blocks of size $t \times k$.

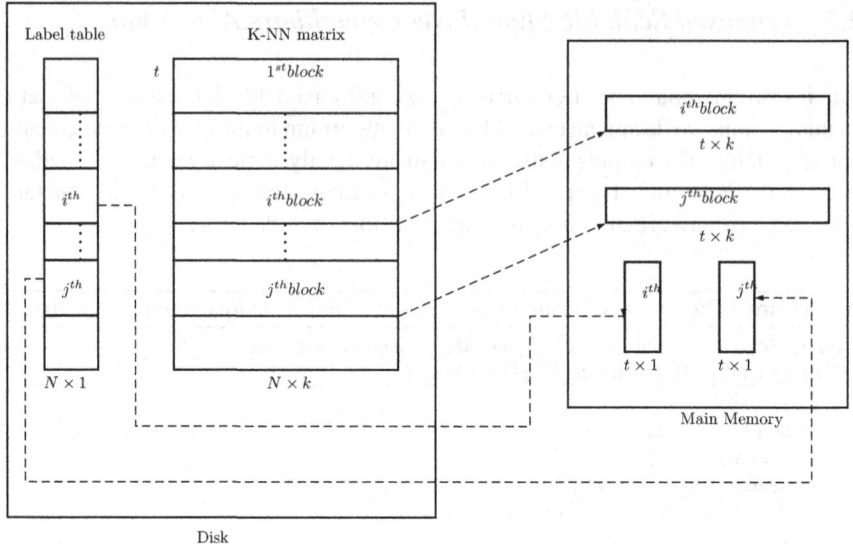

Fig. 1 Snapshot of the transfer of blocks between disk and main memory

2.2.2 The Clustering Step

In the first step of the algorithm, knn matrix of size $N \times k$ is generated which is the input to the next phase of the algorithm. Assume that the matrix is divided into N/t blocks of size $t \times k$ each. Also assume that label table of size N is divided into N/t blocks of size t each. Read any two blocks knn_i and knn_j where $i \leq j$ and also read two blocks of label table, $label_i$ and $label_j$ into the main memory. Then find all possible point-pairs satisfying the SNN similarity criteria in knn_i and knn_j blocks. In this way, the label of all the points of the knn_j block is calculated. Repeating the process N/t times will generate cluster labels for all points. The procedure is described in Algorithm 2. The main memory contains 2-blocks of size $t \times k$ and 2-blocks of size $t \times 1$. The transfer of blocks between main memory and disk is shown Fig. 1.

2.3 I/O Analysis

Traditional algorithm takes $O(N^2 k^2)$ number of I/Os. I/O complexity of the proposed algorithm is described here.

Algorithm 2 Proposed Shared Near Neighbors Algorithm

Input: θ //Similarity threshold.
        $knn[N][k]$ //k-nearest neighbor matrix
        $label[N]$ //cluster label of each point which is initialized as $label[i] = i$
Output: label[N]   //Cluster labels.

$t = M/k$
for $i = 0$ to $N/t - 1$ **do**
  for $j = i$ to $N/t - 1$ **do**
    Read matrix block knn_i and knn_j
    Do the following computations in main memory
    for $r = (i)t$ to $(i + 1)t - 1$ **do**
      for $l = (i)t$ to $(i + 1)t - 1$ **do**
        for $m = 0$ to k **do**
          if $knn[r][0]==knn[l][m]$ **then** count=1
        for $m = 0$ to k **do**
          if $knn[l][0]==knn[r][m]$ **then** count++
        if $count==2$ **then**
          for $p = 1$ to k **do**
            for $q = 1$ to k **do**
              if $knn[r][p] == knn[l][q]$ **then** n++
        if $n > \theta$ **then**
          if $i == j$ **then**
            Read the block $label_i$ of size $1 \times t$ into main memory
            if $label[r] > label[l]$ **then**
              $label[r] = label[l]$
            else
              $label[l] = label[r]$
          else
            Read the block $label_i$ of size $1 \times t$ into main memory
            if $label[r] > label[l]$ **then**
              $label[r] = label[l]$
            else
              $label[l] = label[r]$
      $count = 0, n = 0$

2.3.1   I/O Complexity of Algorithm 1

The main memory contains 2-blocks of size $t \times D$ and 2-blocks of size $t \times k$. Hence, $M = 2tD + 2tk$; i.e., $t = \Theta(M/(D + k))$. Total number of input/outputs required to generate the knn matrix is $O((ND/B)N/t) = O((N^2D/B)((D + k)/M)) = O((N^2Dk + N^2D^2)/BM)$.

2.3.2   I/O Complexity of Algorithm 2

While execution of second algorithm, we know $t = \Theta(M/k)$. Total number of I/Os required by the algorithm is $O(((Nk + N)/B)N/t) = O((N^2k + N^2)/tB) = O(N^2k^2/BM)$.

So the total number of I/Os incurred by two phases of the algorithm is $O((N^2Dk + N^2D^2 + N^2k^2)/BM)$. The dimension D is a constant, so ignoring the constant term, the I/O complexity of the algorithm is $O(N^2k^2)/BM)$ which is a BM factor improvement over the traditional algorithm.

3   Experimental Results

3.1   Performance of the Proposed Algorithm

Many software libraries like STXXL [8], LEDA-SM [7], TPIE [3] exist for external memory. STXXL is used in our work which is an implementation or adaptation of C++ STL (standard template library) for external memory computations [17]. Both the traditional and the proposed algorithm are implemented in STXXL. Since both the algorithms follow exactly same computational steps, computational complexity remains same. Both of the algorithms generate same set of clusters; hence, the quality analysis is omitted. Our main focus is on analyzing and reducing the I/O complexity of algorithms.

Our experiments are run in a restricted environment to demonstrate the improvement in I/O complexity. Though 8 GB/16 GB, main memory is very common in most of the systems nowadays; using it to full capacity is not possible. If we use whole memory or even half of the main memory to run our experiment, program will be out of CPU before going out of memory. It can be explained as: suppose 16 GB main memory is available and we are using half (8 GB) of it to calculate k-neighborhood. Then number of points, having dimension 64 and required space for each dimension 4 bytes, that can accommodate in this memory is $\approx 31 million$ ($8 \times 10^9 \div (64 \times 4)$). To calculate the pair-wise distance, total number of floating point operations required is $\approx 10^{17}$, i.e., 100 $PetaFlops$. This number is out of range for normal desktop computers. If the CPUs with such capabilities are available, then the experiment can be run for very large size input data using the total available main memory.

Effectiveness of the proposed algorithm can be understood in following way. Suppose a system without any restriction on processing capability is available and data are so large that it can not be accommodated in main memory and we have to go for out of core implementation. As the usual block size (B) in normal computers is in KBs and main memory (M) is in GBs, number of I/Os will get improved by an order of 2^{40} $(BM = 2^{30} \times 2^{10})$. If the size of data set is same as example in the previous paragraph, number of I/Os will be reduced from Pis to Kis (Peta Vs. Kilo).[1]

The data sets are generated randomly. The dimension of the data is 64, and the size of the data set varies from 10000 to 200000. The main memory size is restricted

---

[1] https://en.wikipedia.org/?title=Binary_prefix.

to 1 MB only, and the disk size is 150 GB. The algorithm is implemented on Ubuntu 12.04 system with a 2.0 GHz CPU (Intel Core 2 Duo) and 2 GB main memory.

For ease of implementation the algorithmic block size (t) is set same as the disk block size (B). When the block size is set to 8 KB and available main memory is restricted to 1 MB, the total number of reads or writes goes beyond 500×10^6 for 10×10^4 data points in case of traditional algorithm. While in proposed algorithm, the number of read or writes is less than 100×10^6 even for 20×10^4 data points. The similar results were obtained for total number of I/Os and total data read and written. Total number of I/Os for traditional algorithm exceeds 900×10^6 for 10×10^4 data points, while it is less than 150×10^6 for 20×10^4 data points in case of proposed algorithm. In-core algorithm fails to give result after 5 days for 15×10^4 points. Figure 2a–d illustrates number of reads, writes, total read plus writes, and input–outputs, respectively.

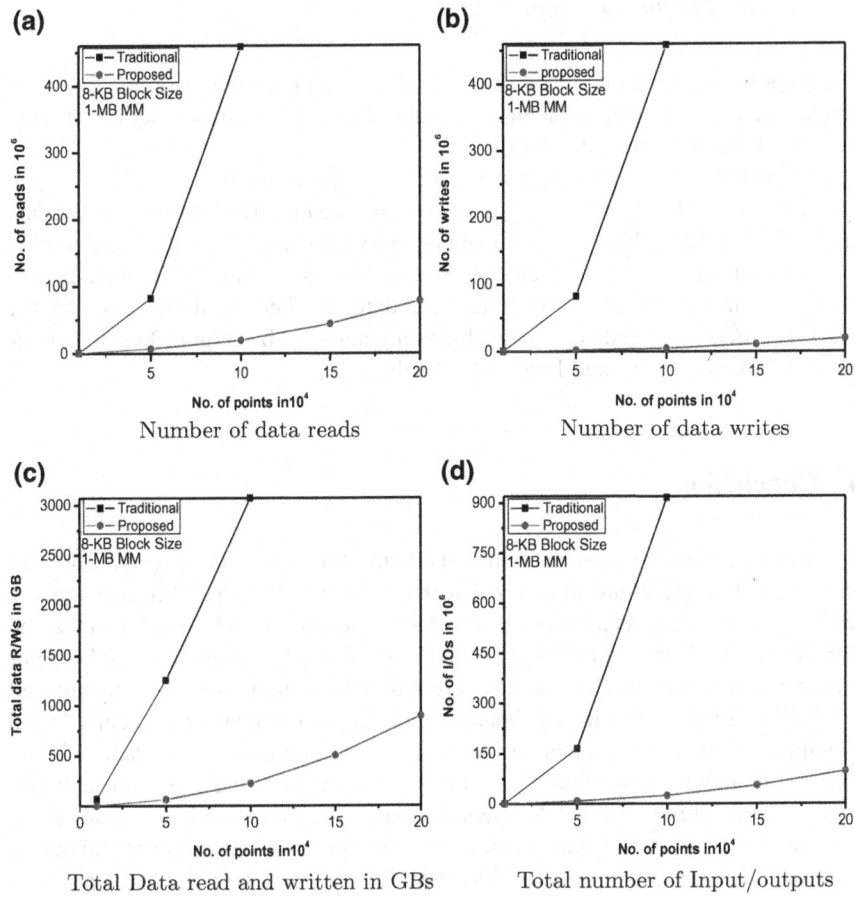

Fig. 2 Comparison of proposed and traditional algorithm

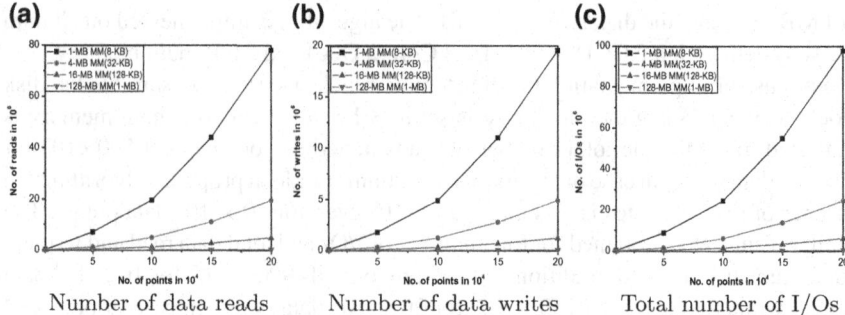

(a)   **(b)**   **(c)**

Number of data reads   Number of data writes   Total number of I/Os

Fig. 3 Effect of main memory size on algorithm performance

3.2 Effect of Main Memory Size on the Performance of the Proposed Algorithm

The proposed algorithm is run on different sizes of main memory to study the effect of main memory on the performance of the algorithm. The main memory is restricted to 1 MB, 4 MB, 16 MB, and 128 MB.

It is clear from the graph that the I/O count reduces as the main memory size increases. If we closely observe a graph, we can see that when the number of data points is $20 \times 10^4 (\approx 100 \text{ MB})$ and main memory size is 128 MB, the line denoting the total number of I/O is very close to x-axis. Similar effect of main memory size can be seen in other graphs as well. It substantiates the theoretical I/O claim that the number of I/Os is dependent on available main memory size. Figure 3a–c illustrates number of reads, writes, and I/Os, respectively.

4 Conclusion

This paper makes some contribution in the field of big data clustering by redesigning the existing algorithm in external memory model. The shared near neighbors (SNN) algorithm has been designed on external memory model. It is shown that the I/O complexity of the proposed algorithm is $O(N^2 k^2 / BM)$ which is a BM factor improvement over the traditional SNN algorithm. Both algorithms are implemented in STXXL to compare their performance. Our goal is to demonstrate that this design technique can be used to adapt various existing algorithms for large data. Without theoretical analysis, it is often difficult to say which clustering algorithm will perform better for different sized data sets. So one of our future work is to analyze the I/O complexity of the best-known clustering algorithms of the literature and design them on the external memory model to make them suitable for massive data sets.

References

1. Abello, J., Pardalos, P.M., Resende, M.G.: Handbook of Massive Data Sets. Springer (2002)
2. Aggarwal, A., Vitter, J.: The input/output complexity of sorting and related problems. Commun. ACM **31**(9), 1116–1127 (1988)
3. Arge, L., Procopiuc, O., Vitter, J.: Implementing I/O-efficient data structures using TPIE. In: Algorithms ESA 2002, pp. 88–100. Springer (2002)
4. Bahmani, B., Moseley, B., Vattani, A., Kumar, R., Vassilvitskii, S.: Scalable k-means++. Proc. VLDB Endow. **5**(7), 622–633 (2012)
5. Ball, G.H., Hall, D.J.: A clustering technique for summarizing multivariate data. Behav. Sci. **12**(2), 153–155 (1967)
6. Charikar, M., Chekuri, C., Feder, T., Motwani, R.: Incremental clustering and dynamic information retrieval. In: Proceedings of the Twenty-Ninth Annual ACM Symposium on Theory of Computing. pp. 626–635. ACM (1997)
7. Crauser, A., Mehlhorn, K.: LEDA-SM: Extending LEDA to secondary memory. In: Algorithm Engineering, pp. 228–242. Springer (1999)
8. Dementiev, R., Kettner, L., Sanders, P.: STXXL: standard template library for XXL data sets. Softw. Pract. Exp. **38**(6), 589–637 (2008)
9. Guha, S., Rastogi, R., Shim, K.: CURE: an efficient clustering algorithm for large databases. In: ACM SIGMOD Record, vol. 27, pp. 73–84. ACM (1998)
10. Han, J., Kamber, M.: Data Mining. Concepts and Techniques. Morgan kaufmann, Southeast Asia Edition (2006)
11. Januzaj, E., Kriegel, H.P., Pfeifle, M.: Dbdc: Density based distributed clustering. In: Advances in Database Technology—EDBT 2004, Lecture Notes in Computer Science, vol. 2992, pp. 88–105 (2004)
12. Jarvis, R.A., Patrick, E.A.: Clustering using a similarity measure based on shared near neighbors. IEEE Trans. Comput. **100**(11), 1025–1034 (1973)
13. Kanungo, T., Mount, D.M., Netanyahu, N.S., Piatko, C.D., Silverman, R., Wu, A.Y.: A local search approximation algorithm for k-means clustering. In: Proceedings of the Eighteenth Annual Symposium on Computational Geometry, pp. 10–18. ACM (2002)
14. Kim, W.: Parallel clustering algorithms: survey (2009). http://www.cs.gsu.edu/~wkim/indexfiles/SurveyParallelClustering.pdf
15. Liu, Y., Guo, Q., Yang, L., Li, Y.: Research on incremental clustering. In: 2nd International Conference on Consumer Electronics, Communications and Networks (CECNet), 2012, pp. 2803–2806, April 2012
16. Moreira, G., Santos, M.Y., Moura-Pires, J.: SNN Input Parameters: how are they related? In: International Conference on Parallel and Distributed Systems (ICPADS), pp. 492–497. IEEE (2013)
17. Musser, D.R., Derge, G.J., Saini, A.: STL Tutorial and Reference Guide: C++ Programming with the Standard Template Library. Addison-Wesley Professional (2009)
18. Ng, R.T., Jiawei, H.: CLARANS: a method for clustering objects for spatial data mining. IEEE Trans. Knowl. Data Eng. **14**(5), 1003–1016 (2002)
19. Olson, C.F.: Parallel algorithms for hierarchical clustering. Parallel Comput. **21**(8), 1313–1325 (1995)
20. Rokach, L., Maimon, O.: Clustering methods. In: Data Mining and Knowledge Discovery Handbook, pp. 321–352. Springer (2005)
21. Wikipedia: Approximation algorithm (2015). Accessed June 2015
22. Xu, X., Ester, M., Kriegel, H.P., Sander, J.: A distribution-based clustering algorithm for mining in large spatial databases. In: Proceedings of 14th International Conference on Data Engineering, 1998, pp. 324–331. IEEE (1998)
23. Yadav, P.K., Pandey, S., Mohanty, S.K.: Nearest neighbor based clustering algorithm for large data sets. arXiv:1505.05962

24. Zaïane, O.R., Foss, A., Lee, C.H., Wang, W.: On data clustering analysis: Scalability, constraints, and validation. In: Advances in Knowledge Discovery and Data Mining, pp. 28–39. Springer (2002)
25. Zhang, T., Ramakrishnan, R., Livny, M.: BIRCH: an efficient data clustering method for very large databases. In: ACM SIGMOD Record, vol. 25, pp. 103–114. ACM (1996)

CWordle: A Visual Analytics System for Extracting the Topics of Speech

Xiang Tang, Xiaoju Dong and Hengjia Zhang

Abstract Speech information is a large part of the Internet data type, which represents many valuable resources in the real world. However, extracting topics of speech is much challenging because there is not only a requirement for an analyzation of multi-level speech data, but also a requirement for valid methods for filtering and displaying extraction results. Meanwhile, the time consumption is also an important factor. Our work provides a visualization system for analyzation and extraction of speech. By means of case studies of real-world speech data, the practicability and effectiveness of the method are demonstrated.

Keywords Speech · Topic extraction · Time consumption · Tag cloud · Analysis

1 Introduction

With the rapid pace of social information and network technology, people are communicating with each other through the Internet instead of writing letters. As a result, different types of data are generated and the amount of data is increasing. Therefore, it is of great significance to deal with these mass data and conduct data mining. Most of the data generated by the network is in the form of text and speech. However, compared to text data, the mining of speech data has not been paid much attention to. There are few papers about the topic extraction of speech. The research of speech data can extract the subject and get the central and auxiliary ideas expressed by the

X. Tang · X. Dong (✉)
Department of Computer Science, Shanghai Jiao Tong University, Shanghai, China
e-mail: xjdong@sjtu.edu.cn

X. Tang
e-mail: tangxiang@sjtu.edu.cn

H. Zhang
University of Michigan-Shanghai Jiao Tong University Joint Institute, Shanghai Jiao Tong University, Shanghai, China
e-mail: hengjia@umich.edu

© Springer Nature Singapore Pte Ltd. 2019
S. K. Bhatia et al. (eds.), *Advances in Computer Communication and Computational Sciences*, Advances in Intelligent Systems and Computing 760, https://doi.org/10.1007/978-981-13-0344-9_7

speaker. Therefore, there exists an urgent research request for developing speech topic analytics visualization to make senses out of a large amount of speech data.

However, given the structural complexity and rich information of speech data, most current speech analytics researches try to obtain the whole text of speech merely. Although there are few approaches that aggregate and visualize the topics of speech, none of them provides a time-efficient method with interactive abilities. Therefore, it is difficult to tell the topic of speech with no need of listening to the speech. Moreover, there are few visualization systems displaying the topics of speech by current speech analytics methods. It is difficult to well understand the importance of topics. Such functionalities are not supported by existing approaches.

In our work, a visual analytics method is proposed for extracting the topics of speech at low time consumption. We are particularly concerned with the visualization of topics and time cost of speech analysis, where the goal is to make the extraction highly efficient and visually intuitional. Our method for dealing with this challenge utilizes speech recognition tools and word segmentation methods to extract the possible topics of speech. We have exploited an adequately operational system and applied it to the real-world speech data. Moreover, we demonstrate the effectiveness and practicability of the approach by case studies.

2 Related Work

2.1 Speech Analytics

Recently, topic analysis in speech data has abstracted much attention in the field of data mining. Most existing approaches focus on summarizing speech by accuracy and time-consuming analysis such as the dynamic programming technique based on a target compression ratio [5] and based on word significance [3, 9], model analysis, and full-text machine learning [11]. We make use of existing speech analysis techniques while designing our visualization system.

However, few existing approaches have been developed for visualizing topics of speech and leverage visualization to further analyze the speech data at low time consumption. Therefore, our research devotes to extracting topics of speech with the support of the visualization method called Tag Cloud.

2.2 Tag Cloud

Tag cloud, a visual design that changes font size based on word frequency to provide users with efficient information extraction, was first proposed in 1997 [4]. The tremendously huge popularity of Wordle attracts people to use the tag cloud visualization [12, 15]. It is easy for people to create their own visually attractive tag clouds, not just consumers of an art [10, 16]. Afterward, developments of tag cloud techniques such as TagCrowd and Tagul [13, 14] give the tag cloud the ability on Chinese coding and add the outline for the tag cloud, improving the expressive power and visual effects [1]. ManiWordle develops based on Wordle, using Greedy algorithm like Wordle and creatively using scanline algorithm while iterating every word in logical order and drawing path on canvas, which makes logically related words put closely [8]. WordlePlus adds some functions like zooming and merging words [6]. Geo Word Clouds redefines the meaning of positions where words are put [2].

The existing research proves the fact that Tag Cloud enables to display the valuable information according to its importance point. However, the existing approaches do not give the tag cloud the capability on extraction of various data, especially speech data. There lacks an integrated approach that combines the extraction and display of data in a single visualization and enables efficient summarization of speech data.

3 Data and Task Abstraction

The selected data and task abstraction are introduced in this section for a better understanding of the problem domain of visual speech analysis.

3.1 Speech Data

Speech data can be news speeches obtained by network or a talk by a celebrity. The common features of speech data are the structural complexity and invisibility. As people are not able to preview the content of speech, it is of great value to find a method helping users to be aware of characteristic data of speech. Meanwhile, the time length of speech data can be very long, so it is not efficient enough to spend much time analyzing the speech and to obtain topics after few minutes later.

3.2 Text Data

For the high efficiency of the topic extraction, we first transform speech data to text data. Text data is mainly composed of Chinese text and English text. Our work can support the data in both languages. However, differences between two languages increase the difficulty in dealing with text data, because English words are separated by spaces or punctuation marks while Chinese words often come together. On the other hand, words consist of nouns, verbs, adjectives, function words, and so on, so we should filter the function words of text data because they always make no sense.

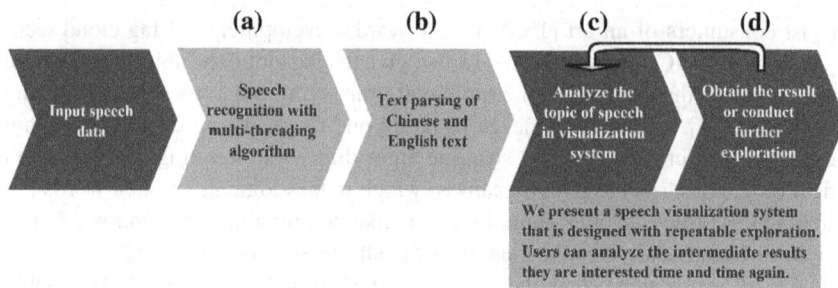

Fig. 1 Approach overview: **a** multithreading speech parsing; **b** Chinese-and-English text parsing; **c** speech topic analysis visualization system; **d** obtain the result

3.3 Task Abstraction

General tasks of speech topic analysis involve summarizing key topics and calculating their importance in speech. We obtain a task list for topic analysis of speech data using a method which combines tasks of two research domains:

- **Topic extraction**: What are the main topics involved in a speech? What is the significance difference between two of these topics?
- **Low time consumption**: Is it possible for users to extract the invisible speech data with low time consumption and with no need of listening to the speech?
- **Result Visualization**: The existing approaches do not enable visualization of extraction results, which make it difficult to tell the difference of importance between different topics. Hence, results should be displayed through visualization technology.

4 System Overview

The visual analysis proposed in the paper is consisted of three modules: a multithreading speech parser, a Chinese-and-English text parser, and a visualization system (Fig. 1). The multithreading speech parser uses multithreading algorithm to parse the source speech data in order to accelerate parsing speed. The Chinese-and-English text parser uses Chinese and English word divided syncopation technology to analyze the words in the text and calculate the words and their importance based on statistics. Next, the visualization system makes it easy to understand the difference between different topics.

5 Speech Topic Analysis

Methods for speech parsing and text parsing are introduced in this section.

5.1 Speech Parsing

We conduct our speech parsing work with the approach of multithreading and concurrency because there exists a task to provide users with the result of topic extraction in short time.

- **Multithreading algorithm**: Our work uses multithreading algorithm for the reason that it is unacceptable to get the text result after few minutes later. The principle of the algorithm is that the speech parsing costs CPU time while writing the text result into files costs IO time. It takes much time to analyze the speech data and little time to output the result, so we are supposed to utilize most of the available threads to do CPU jobs while the remaining threads to do IO jobs. We conduct our work on the computer with an 8GB memory and a 64-bit processor, so the maximum number of threads on this computer is about two thousand. The relationship between the amount of threads used and the efficiency changes as the application environment changes. In the application environment of our research, there exists a parameter k which means the amount of CPU threads is k times the amount of IO threads.
- **Synchronization mechanism**: To ensure that the parsed text stream is consistent with the source speech data, thread synchronization mechanism is used to output the text stream. By numbering threads, divided speeches with smaller numbers are placed before those with larger numbers, so the parsed text stream corresponding to smaller number must also be output in front of that corresponding to larger number. As the result of synchronization mechanism, we can achieve correct results while increasing efficiency.

5.2 Text Parsing

For the high efficiency of extracting topics from speech, we first transform speech data to text data. Text data is mainly composed of Chinese text and English text. Our work can support the data in both languages. We adopt a Chinese word segmentation approach [7] to analyze Chinese words in the generated text, which is able to tell the property of every word. We extract Chinese words from the text and calculate the occurrence frequency of notional words, including nouns, verbs. Meanwhile, we adopt an English word segmentation approach while analyzing the English words. After parsing the text, notional words and their corresponding weight are calculated, which is the source of visualization.

6 Speech Topic Analysis Visualization

The task of providing a visual representation of the complex structured and unstructured information posts a challenge in the consideration of the speech analysis process. To design a splendid visualization, out system should meet the following requirements which are based on the analytic tasks in Sect. 3.3:

- **Obtain the speech summarization**: The visualization should provide a summarization of what this speech is concerned about and help users to quickly identify specific topics of speech data and make the judgment.
- **Enable intuitive expression of topic importance**: It is important to visually tell the different weight of different topics because users cannot easily understand the value of one topic by observing the result of a key-value map.
- **Support further analysis**: The visualization should not only provide users with the display of results, but also the further analysis of the output. Combining the analysis of visualization and speech parsing, we can obtain more accurate topics of the speech data.

6.1 Visualization Design and Interface

Figure 2 indicates the speech analysis visualization based on the requirements above. Our visualization system uses a whole-and-part framework to settle structural complexity. Speech visualization is the central part that coordinates topic views and distributed views. It is encoded in coherent visual sense, but at different detail levels. Specifically, the visualization employs overall-preview plus partial filter layout to describe the speech because of the advantages of the layout exhibited in visualizing the overall summarization in short time and the relationship between whole and part as well.

Specific visual encodings are presented below:

- **Topic view**: The topic view indicates overall topic polarities of speech (Fig. 2a). The popularity can be indicated from the size of word visualization, e.g., the number of this topic appearing in the speech. A text visualization approach called Tag Cloud is used to convey the weight of every topic in the speech. When a topic is of great importance, its font size appears big and its position is close to the central. Moreover, to help users identify the most important topics of the speech, we adopt drawing algorithm in descending order of topic weight in spiral way so that the most significant topic could appear close to the center.
- **Word distribution view**: The word distribution view shows the relationship between the weight and class of words in the speech (Fig. 2b). For example, we can visualize the proportion of verbs in different weight level numbered from zero, as well as the proportion of important words with different classes of words such as verbs, adjectives, adverbs, nouns.

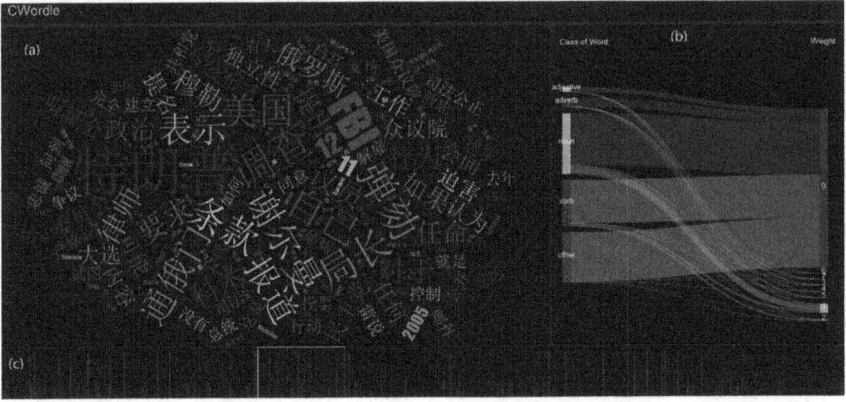

Fig. 2 Visual interface; **a** the topic view indicates the speech is concerned about American politics; **b** the word distribution view shows the relationship between the weight and the class of words; **c** the timeline view indicates the significance distribution

- **Timeline view**: The timeline view provides a summarization of the topic weight of speech with the visual encoding of timeline (Fig. 2c). The horizontal axis represents the time process of speech while the vertical axis represents the weight of words in corresponding positions of speech. Through the timeline view, it is intuitive to judge which part of speech is of great importance, which helps users conduct targeted extraction of speech data. For example, from Fig. 2c we could tell that the end of the speech is less important so that we can ignore the beginning, while we should pay much attention to the range of 35–45% because the height of rectangles is high in average in the middle.

6.2 Interactive Exploration

The speech visualization includes two major interactive functions for supporting the location of a specific topic and analysis of a specific part of speech.

- **Location of the specific topic**: Fig. 3 shows that users can pick any topic they are interested into overview the occurrence locations of this topic, which enables them to analyze the specific part of the speech.
- **Analysis of the specific part of speech**: We design a brush in timeline view to select a range of speech, as a result of which the topic view and the word distribution view will change in response. This design is used for the deep analyzation of speech data because there should be a function of selecting and analyzing the specific part of speech where users are interested in. Therefore, when the brush tool is used (Fig. 2c), users are able to extract the topics and the word distribution of this part.

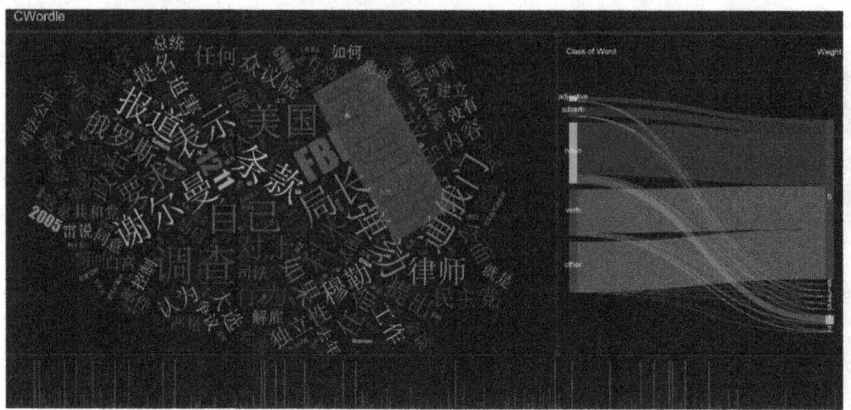

Fig. 3 Locations of the specific topic in the timeline view

7 Case Study

Our system was applied to the speech data extracted from a news portal for the evaluation of the approach raised in the paper. Here, two interesting examples are highlighted, where we explored speech data about topics of "American politics" and "Internet science and technology". Below, we present the exploration of these two speech data with the focus on the topic extraction which is summarized in Sect. 3.3.

7.1 American Politics

Observing the tag cloud from the topic view allows us to understand the main theme of the speech: The speech is about American politics, including the diplomatic relation with Russia and the fairness of the law enforcement.

We can easily identify that the weight of "Trump" is a little bigger than "FBI" and much bigger than "The White House". Therefore, this speech is mainly concerned about "Trump". As mentioned above in Sect. 6.2, when we select the topic "Trump", the locations of "Trump" are almost all over the speech (Fig. 3). From the timeline view Fig. 2c, it shows that the middle and the end of the speech are less important while the range between 5 and 25% as well as the range between 35 and 45% are of great significance. According to the aforementioned interactive exploration, we can brush on the timeline with the filter between 5 and 25%. Through the child tag cloud is in the specific range in Fig. 4, it is obvious for users to find that the weight of the topic "FBI" in this specific part is much smaller than that of the whole speech while the whole weight of "Trump" is almost same as that in part.

Fig. 4 Child tag cloud in speech of American politics

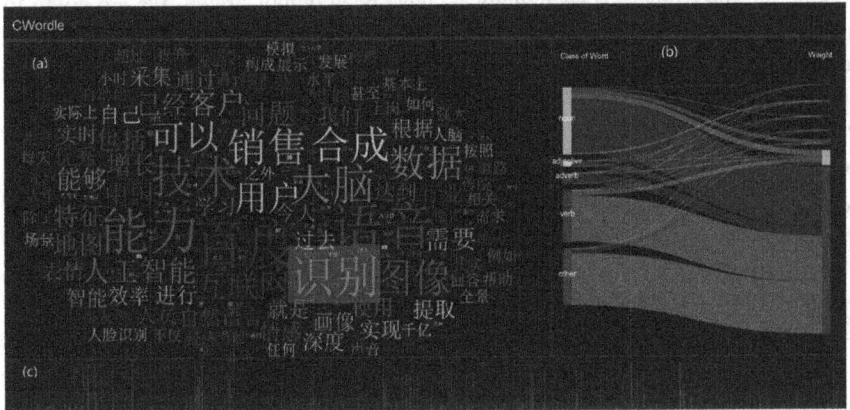

Fig. 5 Analyze the speech of Internet Science and Technology; **a** the topic view; **b** the word distribution view; **c** the timeline view

7.2 Internet Science and Technology

Taking "Robin Li's keynote speech in 2016: The next scene in the Internet is artificial intelligence" as an example, the topic view in Fig. 5a shows that Baidu is focusing on the latest technology such as "Artificial intelligence" and "Speech recognition", which he thinks are the core competitiveness of future. From view Fig. 5a, it indicates that "Baidu", "Speech", and "Recognition" are the most important topics of the speech because of their font size and central location. As mentioned above in Sect. 6.2, when we select the topic "Recognition", the locations of "Recognition" are mainly in the range between 25 and 50%, as well as the end of the speech (Fig. 5b). From the timeline view Fig. 5c, it shows that the very beginning and the range between 40 and 45% is less important, while the range between 25 and 32% and the range between 50 and 55% are of great significance. According to the aforementioned interactive exploration, we can brush on the timeline with the filter between 50 and 80%. Through the child tag cloud in the specific range, it is obvious to find that this range of the speech is about speech synthesis.

8 Conclusion

In this paper, a method of visual analysis of speech data is proposed for the first time. The main feature of the method is that it is capable of obtaining results of topic extraction in short time with the multithreading algorithm and visually observing the summarization as well as weight of topics in speech. By means of case studies, the practicability and effectiveness in analyzing complicated speech data, particularly in conducting the Chinese speech analysis, have been demonstrated. We are going to use the visual speech topic analysis technique to analyze more complex speech data from different sources and apply to different languages of speech data such as French and Japanese. In future work, we will devote our effort to improve the accuracy and practicality of the approach.

Permission
No permission is required for the images used in the proposed work or study as all the images are generated by the authors themselves.

Acknowledgements The work is supported by the National Nature Science Foundation of China (61100053) and CCF-Venustech Hongyan Research Initiative (2016-013).

References

1. Albers, P., Hernández, O.: Tagxedo: using digital technologies to support the writing of students in an alternative education program, pp. 51–79 (2014)
2. Buchin, K., Creemers, D., Lazzarotto, A.: Geo word clouds. In: IEEE Pacific Visualization Symposium, pp. 144–151 (2016)
3. Furui, S., Kikuchi, T., Shinnaka, Y., et al.: Speech-to-text and speech-to-speech summarization of spontaneous speech. IEEE Trans. Speech Audio Process. **12**(4), 401–408 (2004)
4. Gas, F.B., Wattenberg, M.: TIMELINES: tag clouds and the case for vernacular visualization. Interactions **15**(4), 49–52 (2008)
5. Hori, C., Furui, S.: A new approach to automatic speech summarization. EURASIP J. Adv. Signal Process. **2003**(2), 1–12 (2003)
6. Jo, J., Lee, B., Seo, J.: WordlePlus: expanding wordle's use through natural interaction and animation. IEEE Comput. Graph. Appl. **35**(6), 1–1 (2015)
7. Jieba Chinese word segmentation. https://github.com/fxsjy/jieba
8. Koh, K., Lee, B., Kim, B., et al.: ManiWordle: providing flexible control over Wordle. IEEE Trans. Visual. Comput. Graph. **16**(6), 1190–1197 (2010)

9. Mckeown, K., Hirschberg, J., Galley, M., et al.: From text to speech summarization. In: A Wittgenstein Symposium, Girona, 1989, pp. 399–408. Rodopi (2005)
10. Mcnaught, C., Lam, P.: Using Wordle as a supplementary research tool. Qual. Rep. **15**(3), 630–643 (2010)
11. Murray, G., Renals, S.: Towards online speech summarization. In: INTERSPEECH 2007, Conference of the International Speech Communication Association, Antwerp, Belgium, August. DBLP, pp. 2785–2788 (2011)
12. Seifert, C., Kump, B., Kienreich, W., et al.: On the beauty and usability of tag clouds, pp. 17–25 (2008)
13. TagCrowd. http://tagcrowd.com
14. Tagul. https://tagul.com/
15. Viégas, F.B., Wattenberg, M., Feinberg, J.: Participatory visualization with Wordle. IEEE Trans. Visual Comput. Graph. **15**(6), 1137–1144 (2009)
16. Viegas, F.B., Wattenberg, M., Ham, F.V., et al.: ManyEyes: a site for visualization at internet scale. IEEE Trans. Visual Comput. Graph. **13**(13), 1121–1128 (2007)

Frequently Used Devanagari Words in Marathi and Pali Language Documents

Milind Bhalerao, Sanjiv Bonde and Madhav Vaidya

Abstract Optical character recognition (OCR) deals with the recognition of printed or handwritten characters. India being the multilingual country, and possessing the historical information in some of the old languages being practiced in India since ancient time, it is obvious that important information is still to be discovered from these ancient available documents. In this paper, we have devised a method which will identify the language of the script under observation. Character recognition itself is a challenging problem because of the variation in the font and size of the characters. In this paper, a scheme is developed for complete OCR for Marathi and Pali languages. The proposed system successfully segments out the lines and words of the Marathi and Pali documents. The proposed system is evaluated on ten Marathi and ten Pali documents comprised of 552 text lines and 6430 words. We obtained the promising results on the line segmentation with an accuracy of 99.25% and 98.6% and for word segmentation 97.6% and 96.5%, respectively, on Marathi and Pali language documents. Using K-NN classifier, the most frequently used words in Marathi and Pali documents are identified.

Keywords Feature extraction · Marathi · Pali · Line segmentation
K-means classifier

1 Introduction

Optical character recognition deals with the printed or handwritten documents and converts it to a form suitable for machine interpretation. It can also be classified as online and offline character recognition. The applications like signature verification, bank check reading, postal address interpretation drive the interest of the researcher in optical character recognition. It is observed that much more research has taken place in printed Chinese, Japanese, and Korean scripts [1]. There are many research groups

M. Bhalerao (✉) · S. Bonde · M. Vaidya
Shri Guru Gobind Singhji Institute of Engineering and Technology, Nanded, India
e-mail: mvbhalerao@sggs.ac.in

© Springer Nature Singapore Pte Ltd. 2019
S. K. Bhatia et al. (eds.), *Advances in Computer Communication and Computational Sciences*, Advances in Intelligent Systems and Computing 760, https://doi.org/10.1007/978-981-13-0344-9_8

working in Devanagari, Bangla, Oriya, Gurumukhi, Urdu, and other scripts [2]. Substantial work has been reported for Devanagari script which is adopted for different languages like Hindi, Sanskrit, Gujarati. [3–5]. According to Jayadevan et al. in [6], an ideal combination of classifiers was to be developed for the purpose of recognition. Similar shaped characters affect the performance of OCR system as reported by Pal et al. in [7]. Kaur and Gujral [8] divided each character into zones and then structural and region-based features are extracted from each zone. Sushama Shelke et al. proposed a multi-feature multi-classifier scheme for handwritten Devanagari character recognition [9]. S. Arora et al. [5] used shadow feature and intersection junction feature extraction techniques. It is observed that improved feature extraction methods, multi-feature approach, multi-classifiers approach are used to improve the recognition rate. A substantial amount of work has been reported for Devanagari in the field of OCR but after going through the previous work done in OCR, it is observed that no work is done for the ancient Pali language. This paper deals with the Pali and Marathi documents to find out frequently used characters in both the languages.

1.1 About Marathi and Pali Language

Marathi language is very much popular in the state of Maharashtra (India). Nearly, 71 million people speak the Marathi language. Maharashtri was one of the *Prakrit* languages. Marathi is thought to be originated from Maharashtri language. Many inscriptions were found on the stones and the copper plates in the eleventh century. The *Modi* alphabets were used to write in Marathi till twentieth century. Thereafter, Devanagari script is being used to write the Marathi language [10]. Marathi language is having 13 vowels and 36 consonants as shown in Fig. 1. Figure 1a shows the 13 distinct vowels, and Fig. 1b shows the 36 distinct consonants used in Marathi language. Marathi language is written using the mixture of these basic characters composed of vowels and consonants.

Since 1870 optical recognition of characters has been a popular research area. The researchers have dealt with the challenges in English, Chinese, Latin, Arabic [11], Japanese [12], Thai [13], and Devanagari scripts [14]. There are many scripts and languages in India but not much work has been done for the recognition of Marathi characters.

1.2 About Pali Language

In ancient India, it was the fourth century B.C., *Panini* (a well-known *Sanskrit* linguist, grammarian, and a revered scholar) was giving his best to the *Sanskrit* language, in the same era *Aadi-Prakrut*—the language of common man was also taking a new

Fig. 1 Basic Marathi characters

(a)

अ आ इ ई उ ऊ ऋ ए ऐ ओ औ अं अः

Vowels

(b)

क ख ग घ ङ
च छ ज झ ञ
ट ठ ड ढ ण
त थ द ध न
प फ ब भ म
य र ल व श
ष स ह ळ क्ष
ज्ञ

Consonants

Fig. 2 Basic Pali characters

(a)

अ आ इ ई उ ऊ ए ओ

Vowels

(b)

"*Aghosh*" characters   "*Ghosh*" characters

क ख ग घ ङ
च छ ज झ ञ
ट ठ ड ढ ण
त थ द ध न
प फ ब भ म
य र ल व स
ह ळ अं

Consonants

form as *Magadhi* which is also referred to as *Magadhi* of *Lord Buddha* era. The same *Magadhi* language further became popular as *Pali, Tanti, Ardh Magadhi* [15].

There are 41 distinct characters in Pali language. It includes 8 vowels and 33 consonants as shown in Fig. 2. The 8 distinct vowels are shown in Fig. 2a, whereas the 33 consonants of Pali language are shown in Fig. 2b. Unlike in Devanagari or Marathi basic characters, the letter 'अं' which is treated as vowel is considered as consonant in Pali. The letter 'अं' in Pali is referred to as "*Niggahit*." The consonants are divided in five categories viz क, च, ट, त, प categories. The consonants are further divided into two types like "*Ghosh*" and "*Aghosh*." The first two consonants in every category are "*Aghosh*," and the remaining three consonants are treated as "*Ghosh*." Apart from these five categories 'य', 'र', 'ल', 'व','ह','ळ' are treated as "*Ghosh*" and the consonant 'स' is treated as "*Aghosh*" [16].

The writing style of both of these languages is from left to right direction. Also, there are compound characters, modifiers, numerals, and also there exit similar look-

Fig. 3 A Pali word showing
three parts and Shirolekha

ing characters in Pali and Marathi languages. The script currently used in Marathi is called "Balbodh" which is a modified version of Devanagari script [17, 18].

The letters in Marathi and Pali languages take a unique identity consisting of specific lines, curves, and circles written below or above and before or after the vowels and consonants. Also, the combination of vowels and consonants leads to new shapes which form the compound characters. The distinctive feature of the Pali and Marathi language is the presence of a horizontal line on the top of all characters, called as "Shirolekha." These words can be divided into three different parts like upper, middle, and lower parts. Figure 3 shows the three parts of a Pali word, viz upper, middle, and lower part.

In this paper, we propose a system toward the recognition of frequently used Devanagari characters in Marathi and Pali languages.

2 Proposed Method

For sorting out the Pali and Marathi documents, we need to follow the workflow as mentioned in the flowchart given in Fig. 4.

The samples consisting of Pali and Marathi documents are scanned with the flatbed scanner, namely HP LaserJet M1005 MFP with 150 dpi. OCR system by Chaudhuri and Pal [19] proposed the same set of algorithm for the two scripts. This algorithm can be used for the document digitization and the skew detection. The line, word, and character segmentation are carried out in accordance with the same algorithm.

2.1 Preprocessing

This process composes of five sub-processes, i.e., binarization, inversion, labeling connected components, skeletonization, and normalization.

Binarization

The scanned input gray image is binarized. The OSTU [20] method is employed to convert the grayscale image into binary image. The advantage of using this method is that it chooses threshold automatically to reduce the intraclass difference between black and white pixels (Fig. 5).

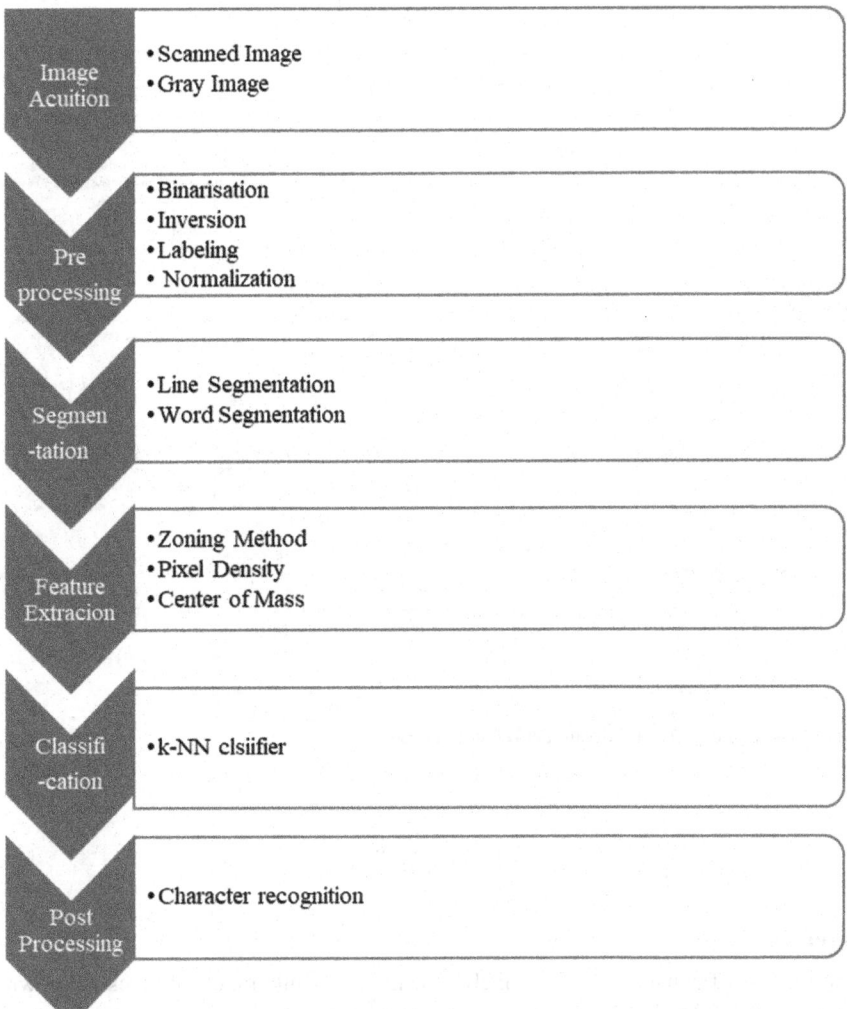

Fig. 4 Proposed method

$$\sigma_w^2(t) = w_0(t)\sigma_0^2(t) + w_1(t)\sigma_1^2(t) \tag{1}$$

w_0 and w_1 are the probabilities of the two classes separated by a threshold t and σ_0^2 and σ_1^2 are the variances of these two classes.

The class probability w_0 and w_1 are computed from the L histogram

$$w_0(t) = \sum_{i=o}^{t-1} p(i) \tag{2}$$

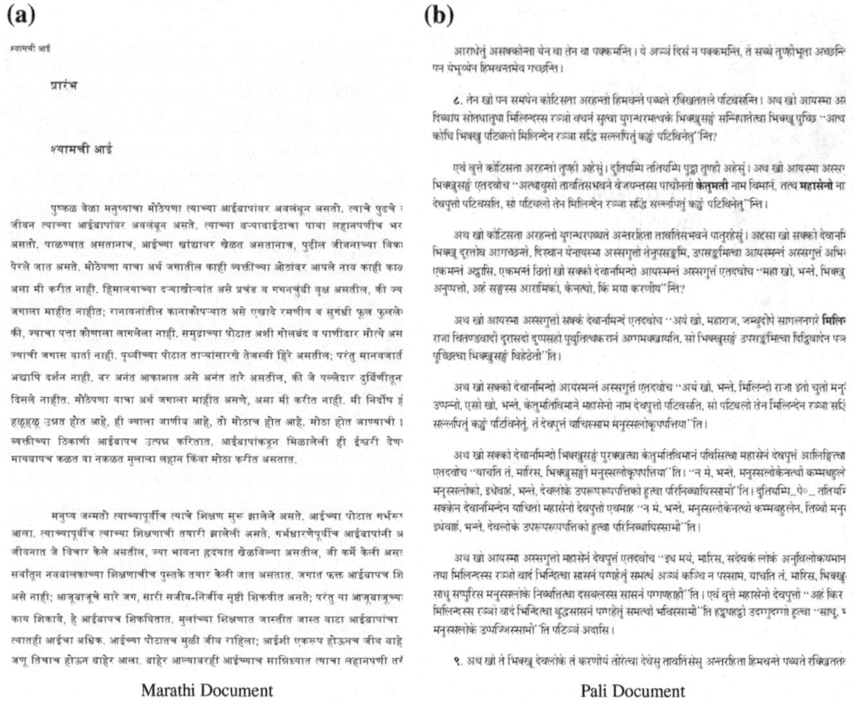

(a)

(b)

Marathi Document

Pali Document

Fig. 5 Binarized **a** Marathi document **b** Pali document

$$w_1(t) = \sum_{i=t}^{L-1} p(i) \tag{3}$$

Inversion

The process of converting black pixels to white and white pixels to black is known as inversion of the document image. After converting document into binary format, the image consists of foreground black pixels in white background. So the number of pixels which are having value 1 is more as compared to the number of pixels having value 0. As all the methods are applicable for pixel value 1, inversion of the image is necessary to minimize the calculation (Fig. 6).

2.2 Segmentation

Segmentation is carried out at two different levels that is at line and word. The input scanned document is first binarized using Ostu's method, and lines are segmented.

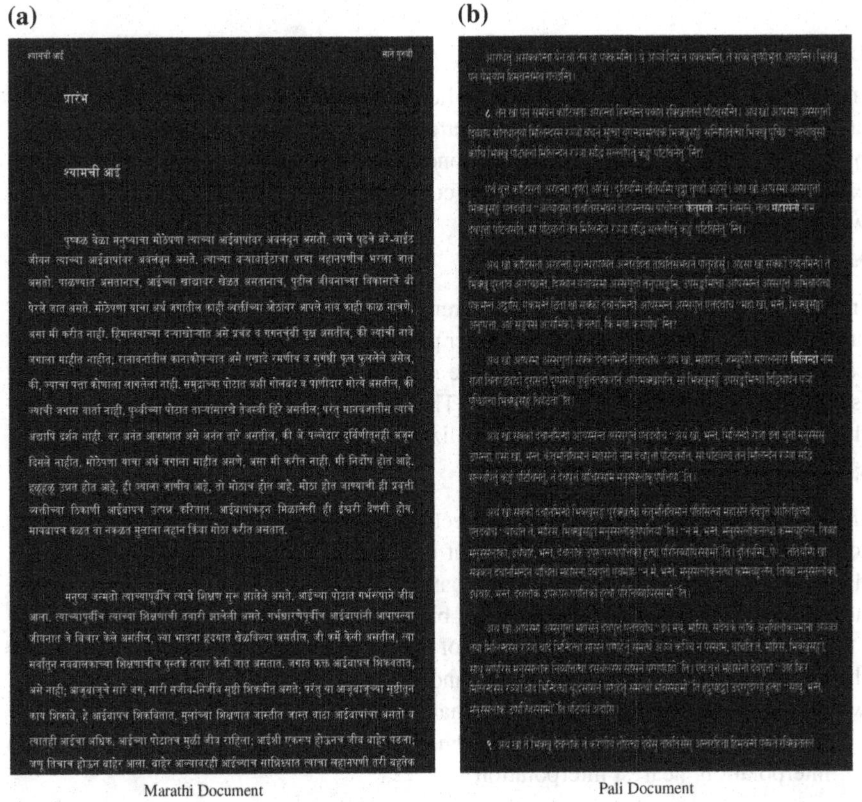

(a) **(b)**

Marathi Document Pali Document

Fig. 6 Inverted **a** Marathi document **b** Pali document

Fig. 7 Segmented **a** Marathi word **b** Pali word

(a) **(b)**

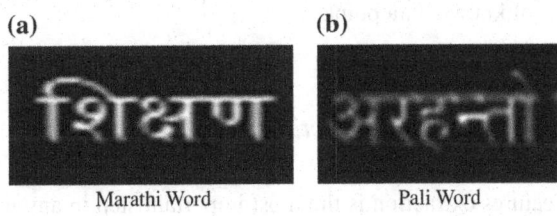

Marathi Word Pali Word

The text lines are extracted by computing the horizontal projection profile of input image along X and Y directions. From each of the segmented line, the words are then extracted by computing the vertical projection profile of input image.

Analysis of geometrical correlation between the horizontal and vertical projection profiles leads to the orientation of the text lines. The Y-cut algorithm [21] is used to extract the horizontal text lines from these projection profiles. The isolated words are extracted employing the X-cut algorithm. The isolated horizontal lines and the isolated words are shown in Fig. 7.

Labeling Connected Components

In this step, we are applying the labels to the connected pixel components in the image and retaining those components which are having maximum number of connected pixels. This labeling of connected components will help in feature extraction step for extracting the features of the particular connected object that is word in the image which is unique in that pattern image.

Normalization

In the process of normalization, the different image parameters are being transformed and the new appropriate values for further process are obtained. Image normalization can be achieved in different ways. If the intensity of all the images is equalized, it is called the "Intensity Normalization." The width, the slant, and the aspect ratio of the characters can be adjusted by normalizing the image.

Size Normalization

In size normalization, we keep the aspect ratio of the images and normalize them to fixed sizes. First, we binarize and cut the original images into a rectangle with the same height and width of original patterns. There are various methods of size normalization like bi-cubic interpolation, bilinear interpolation, nearest interpolation.

In size normalization, the aspect ratio of the images is kept constant and normalizes them to fixed sizes. First, we binarize and cut the original images into a rectangle with the same height and width of original patterns.

There are various methods of size normalization like bi-cubic interpolation, bilinear interpolation, nearest interpolation.

Bi-cubic Interpolation

In Mathematics, a method by which a new data point is being constructed, given the set of known data points.

2.3 Feature Extraction

Features extraction is the most important step in any recognition system because it helps to improve the recognition rate. Features are generally classified into two types, which are global features and local features. Global features are extracted globally form the entire pattern image like histograms, transforms, and the local features are extracted from the different regions of image.

In this paper, we have extracted the features from the statistical distribution of points using zoning, moments, and histogram of oriented gradients.

Fig. 8 Zoning with 8 × 8 grids

(A) Zoning

The given image is divided into the different part called as zones. The local characteristics of the input pattern image can be extracted using zoning.

For a zoning method Z_M, the given image is divided into M different zones where M is an integer value. The M zones are represented as $Z_M = \{z_1, z_2,..., z_M\}$, each of the partition contains the local information. The proposed technique uses 8 × 8 regular grids, which creates uniform partitions of the pattern image having regions of identical shape.

In case where the number of pixels in x or y direction does not fulfill 8 pixels, the zeros will be appended to get equal size zones (Fig. 8).

Extraction of Pixel Density, Center of Mass Using Static Uniform Zoning Method

The segmented word is now divided into $N \times M$ zones. From each zone, features are extracted to form the feature vector. The goal of zoning is to obtain the local characteristics instead of global characteristics.

(a) Pixel Density

It is the density of foreground pixels in each zone which is calculated as the ratio of the average pixel count of foreground pixels, representing a single word, to the total number of pixels in the block. It is simple feature calculated for each block. The pixel density is calculated as

$$Pixel\ Density = \frac{No\ of\ foreground\ pixels\ in\ a\ zone}{Total\ number\ of\ pixels\ in\ that\ zone} \tag{4}$$

(b) Center of Mass:

The feature vector to be calculated also consists of a feature as the center of mass of each zone.

2D moment of order $r = p + q$ of discrete image function $f(x, y)$ of size $N \times M$ is given by

$$m_{pq} = \sum_{x=0}^{M-1} \sum_{y=0}^{N-1} x^p y^q f(x, y) \tag{5}$$

where $p.q \geq 0$

$m_{10}/m_{00}, m_{01}/m_{00}$ give the x, y coordinates of the center of gravity.

(B) Histogram of Oriented Gradients

The histogram of oriented gradient (HOG) is a method to represent the shape and local portions of objects in the image which can be described by using oriented gradients of intensity or by the edge directions.

Two features mainly strength and angle of gradient are calculated. The direction of gradient is calculated as given in Eq. 6. These directions are quantized into 32 directions, and each direction is accumulated with strength of gradient.

$$\text{Strength of gradient:} g(u, v) = \sqrt{(\Delta x)^2 + (\Delta v)^2} \tag{6}$$

Angle of gradient is

$$\theta(u, v) = \tan^{-1} \frac{\Delta y}{\Delta x} \tag{7}$$

where

$$\Delta x = f(x + 1, y + 1) - f(x, y)$$
$$\Delta y = f(x + 1, y) - f(x, y + 1) \tag{8}$$

$f(x, y)$ is intensity value at point (x, y).

After the calculation of these two features, a separate feature called as histogram of gradient (HOG) is obtained. HOG is a feature descriptor which counts total number of occurrences of gradient, i.e., depending upon the direction of pixel intensity, it counts intensity value. A feature vector of 1×392 is obtained. The above features which calculated using zoning technique are combined with HOG features to form a final feature vector of dimension 394. These features are fed to K-NN classifier for grouping the similar words in the document.

2.4 Classification

Classification is the important step when we have to classify the objects based on the features extracted in the previous step.

K-Nearest Neighbor Classifier

K-nearest neighbor is the simplest of all classification algorithms which stores feature vectors of all available input samples and classifies new samples according to the similarity measure.

Steps in K-NN

i. Store input training sample. Assign a label to every training sample.
ii. Compute the distance between test and training samples.

Table 1 Percentage segmentation of Marathi words

Input image	Line segmentation accuracy (%)	Word segmentation accuracy (%)
Doc1	99.4	97.45
Doc2	99.52	97.36
Doc3	98.99	97.66
Doc4	99.22	97.21
Doc5	99.21	97.58
Doc6	99.23	96.99
Doc7	99.18	97.4
Doc8	99.2	97.58
Doc9	99.15	97.48
Doc10	99.25	97.6

iii. Store the training samples of which the distance from the test sample is minimum.

iv. Go for the most common label from the step 3.

To calculate the minimum distance between the test sample and the training sample, we have used Euclidean distance as depicted in Eq. (9).

$$D_{st} = \|x_s - y_t\| = \sqrt{(x_s - y_t)C^{-1}(x_s - y_t)'} = \sum_{i=1}^{m} \left((x_s - y_t)^2\right)^{1/2} \qquad (9)$$

3   Results

The proposed scheme was evaluated on ten Marathi and ten Pali document images. The ten Marathi documents comprised a total of 262 lines. Out of which 260 lines were correctly segmented using the proposed technique. The accuracy of line segmentation for Marathi documents obtained is found to be 99.23% and the accuracy of the word segmentation is 97.43%. The ten Pali documents comprised a total of 290 lines. Out of which 288 lines were correctly segmented. The accuracy of line segmentation for Pali documents obtained is found to be 99.31% and the accuracy of the word segmentation is 96.50% (Tables 1 and 2).

$$\% Line\ segmentation = \frac{Number\ lines\ segmented}{Total\ number\ of\ lines} \times 100 \qquad (10)$$

$$\% Word\ segmentation = \frac{Number\ words\ segmented}{Total\ number\ of\ words} \times 100 \qquad (11)$$

After segmentation, using K-NN classifier, the most frequently used words are found out in Marathi as well as the Pali documents. Table 3 shows the top ten

Table 2 Percentage segmentation of printed Pali words

Input image	Line segmentation accuracy (%)	Word segmentation accuracy (%)
Doc1	99.5	97
Doc2	99.4	97.2
Doc3	99.35	96.1
Doc4	99.85	96.6
Doc5	99.6	96.67
Doc6	99	95.9
Doc7	99.4	96.75
Doc8	99.31	95
Doc9	99.1	98
Doc10	98.6	96.5

Table 3 Top ten words used most frequently in Marathi

Words	Frequency
नाही	8
असतील	7
असते	5
जात	4
असे	4
की	4
आईच्या	3
असतो	3
मोठेपणा	3
त्याच्या	3

frequently used words in a single Marathi document while Table 4 depicts the top ten frequently used words in a single Pali document.

Database used:

For the evaluation of the proposed method, 20 documents images are obtained out of which 10 documents belong to Marathi and the other 10 documents belong to the Pali literature. The document images contain only text lines. These document pages are scanned with the flatbed scanner, namely Hp Scanjet Enterprise flow 7000s2 with 150 dpi.

Table 4 Top ten words used most frequently in Pali

Words	Frequency
खो	16
भिखूसंघ	9
अथ	9
एतदवोच	6
अस्सगुत्तो	5
देवपुत्तो	4
महासेनो	4
तं	4
आयस्मा	4
भिक्खू	4

4   Conclusion

There are different words used in different languages and there are certain words in a particular language which are used most frequently. Depending upon the frequency of these words, the language of the tested document can be predicted.

In this paper, we have presented a feature extraction method of which the features are based on zoning and histogram of oriented gradients. Also, the K-NN classifier is implemented for the recognition of the characters. This feature extraction method and the classifier are commonly used for extracting the features as well as for the recognition of Marathi and Pali language characters.

The experiments carried out on 20 scanned documents comprising of 10 Marathi and 10 Pali language documents each, it is observed that the character "खो" is used most frequently in 10 of documents and the character "नाही" is used most frequently in the remaining 10 documents. Since the character "खो" is used the most in Pali language and the character "नाही" is used the most in Marathi language, the 20 scanned documents can easily be divided in Pali and Marathi languages documents.

In future, the implemented work may be extended considering the different features and classification techniques for the recognition of handwritten Pali language documents and also for understanding the ancient sculptured in Pali language.

References

1. Govindaraju, V., Srirangaraj.: Guide to OCR for Indic scripts. Springer-Varlag London Limited (2009)
2. Bhattacharya, U., Chaudhuri, B.B.: Handwritten numeral databases of indian scripts and multistage recognition of mixed numerals. IEEE Trans. Pattern Anal. Mach. Intell. **31**(3), 444–457 (2009)

3. Arora, S., Bhattacharjee, D., Nasipuri, M., Malik, L., Basu, D.K., Kundu, M.: Performance comparison of SVM and ANN for handwritten Devnagari character recognition. IJCSI Int. J. Comput. Sci. Issues **7**(3), No. 6 (2010)
4. Dongre, V.J., Mankar, V.H.: A review of research on Devnagari character recognition. Int. J. Comput. Appl. **12**(2) (2010)
5. Arora, S., Bhattacharjee, D., Nasipuri, M., Basu, D.K., Kundu, M.: Combining multiple feature extraction techniques for handwritten Devnagari character recognition. In: IEEE Region 10 Colloquium and the Third ICIIS, IIT Kharagpur, India, Dec 2008
6. Jayadevan, R., Kolhe, S.R., Patil, P.M., Pal, U.: Offline recognition of Devnagari script: a survey. IEEE Trans. Syst. Man Cybern.- Part C: Appl. Rev. **41**(6) (2011)
7. Pal, U., Chanda, S., Wakabayashi, T., Kimura, F.: Accuracy improvement of Devnagari character recognition combining SVM and MQDF. In: Proceeding of 11th ICFHR, Kolkata, India, pp. 367–372 (2008)
8. Kaur, R., Gujral, S.: Recognition of similar shaped isolated handwritten gurumukhi characters using machine learning. In: IEEE 5th International Conference- Confluence the Next Generation Information Technology Summit, pp. 251–256 (2014)
9. Shelke, S., Apte, S.: A novel multi-feature multi-classifier scheme for unconstrained handwritten Devnagari character recognition. In: 12th International Conference on Frontiers in Handwritten Character Recognition, Kolkata, India, pp. 215–21 (2010)
10. Ajmire, P., Warkhede, S.E.: Handwritten Marathi character (vowel) recognition. Lap Lambert Academic Publ (2012)
11. Annoud, I.A.: Am. J. Appl. Sci. **4**(11), 857–864 (2007)
12. Wakahara, T., Kimura, Y., Mutsuo.: Handwritten Japanees character recognition using adaptive normalization by global affine transformation. In: Proceeding of 6th ICDAR Vol., Issue (2001)
13. Mitrpanont, J.L., Limkonglap, U.: Using countour analysis to improve feature extraction in Thai handwritten character recognition systems. In: Proceeding of 7th IEEE ICCIT (2007)
14. Pal, U., Sharma, N., Wakabayashi, T., Kimura, F.: Off-line handwritten character recognition of Devnagari script. In: 9th ICDAR (2007)
15. Dharmarakshit, B.: Pali Sahityaka Itihas, 1st edn. Gyanmandal Limited, Varanasi
16. Kamble, J.: Pali Bhasha Parichay, 1st edn. Dr. Babasaheb Ambedkar Research and Training Institute, Pune
17. Kamble, P.M., Hegadi, R.S.: Handwritten Marathi character recognition using R-HOG Feature. Procedia Comput. Sci. **45**, 266–274 (2015)
18. Dongare, V.J., Mankar, V.H.: A review of research on Devanagari character recognition. Int. J. Comput. Appl. **2**(14), 8–15
19. Chaudhuri, B.B., Pal, U.: An OCR system to read two Indian language scripts: Bangla and Devanagari (Hindi). Doc. Anal. Recognit.
20. Otsu, N.: A threshold selection method from gray-level histograms. IEEE Trans. Syst. Man Cybern. **9**(1), 62–66 (1979)
21. Wang, A.B., Fan, K.C., Wu, W.H.: Recursive hierarchical radical extraction for handwritten Chinese characters. Pattern Recognit. **30**(7), 1213–1227 (1997)

Novel Concept of Query-Similarity and Meta-Processor for Semantic Search

Shilpa S. Laddha(ID) **and Pradip M. Jawandhiya**

Abstract This paper is the extension of the work which explores the overall working of the proposed semantic information retrieval (IR) system and the query prototype technique used for matching the user-entered query with the prototypes defined in the system. The objective of the proposed design is to improve the performance of the results provided by keyword-based retrieval systems using semantic approach. This paper explores the concept of query-similarity, meta-processor, and keyword manager which are the promising paradigm of the proposed design for matching the query, identifying the service, providing the meta-information with reference to the results retrieved for the user-entered query and thereby providing the direct, precise, and relevant information along with title, time, and brief information of the resulting URLs.

Keywords Information retrieval · Semantic search engine · Query prototype
Tourism

1 Introduction

IR is concerned with the representation, storage, organization, and access to information items. Earlier target of the IR area include indexing text and searching for useful documents in a collection. Nowadays, research in IR includes modeling, Web search, text classification, systems architecture, user interfaces, data visualization, filtering, and languages, etc. In the beginning, IR was the topic of interest mostly

S. S. Laddha (✉)
Government College of Engineering, Aurangabad 431001, India
e-mail: kabrageca@gmail.com

P. M. Jawandhiya
PL Institute of Technology and Management, Buldana 443001, India
e-mail: pmjawandhiya@rediffmail.com

© Springer Nature Singapore Pte Ltd. 2019
S. K. Bhatia et al. (eds.), *Advances in Computer Communication and Computational Sciences*, Advances in Intelligent Systems and Computing 760, https://doi.org/10.1007/978-981-13-0344-9_9

limited to librarians and information experts. A single fact, the introduction of the Web, changed these insight. The Web has become the largest storage area of knowledge in human history. Due to its huge size, discovering meaningful information on the Web usually requires rigorous search, and searching on the Web is all about IR and its technologies. This is the significance of IR. Thus, suddenly overnight, IR has acquired a place with other technologies at the center of the stage [1]. Users of modern IR systems, such as search engine users, have information needs of varying complexity. The full description of the user information need is not necessarily understandable by the IR system. It is required to submit a good query to the IR system. So the first step is translation of the user information need into a query. This translation process yields a set of keywords, or index terms, which summarize the user information need [2, 3]. Given the user query, the key goal of the IR system is to retrieve information that is useful or relevant to the user. That is, the IR system must rank the information items according to a degree of relevance to the user query [4, 5]. Finding useful information on the Web is normally a tiresome and tricky task. For instance, to satisfy the information need, the user might navigate the space of Web links (i.e., the hyperspace) searching for information of interest. However, since the hyperspace is huge and almost unknown, such a navigation task is usually inefficient. For inexperienced users, the problem becomes more difficult, which might completely discourage all their efforts [6]. The major difficulty is the absence of a well-defined fundamental data model for the Web, which implies that information definition and structure is frequently of low quality. These difficulties have attracted renewed interest in IR and its techniques as promising solutions. Because without stating the appropriate keywords, expected result is not possible. The extension of the current Web is the semantic Web where the information is given well-defined meaning, thus playing a crucial role in enhancing current keyword-based Web search, as the objective is to create machine-readable data. Semantic search engines have gained popularity in the last decade [7]. The main intension of semantic Web is to make Web content understandable by human as well as machine. To enhance the IR on the Web, an alternative approach of semantic web searching specifically for the tourism domain is proposed in this paper. The search reported in this paper explores the important concept of query-similarity, meta-processor, and keyword manager which are the promising paradigm of the proposed design for matching the query, identifying the service, and thereby providing the direct, precise and relevant information along with meta-information in terms of title, time and in brief of the resulting URLs. The remainder of this paper is structured as follows. Section 2 briefs the problem by means of challenges. Sections 3, 4 and 5 are the core of the paper presenting the system architecture, query similarity mapper implementation, ontological synonym set generation of semantic search. Section 6 describes the meta-processor module. Section 7 briefs the template manager and URL generator module of the semantic search. Section 8 concludes the paper and outlines future work.

2 Challenges

The key challenge for the IR system is to retrieve all the items that are relevant to a user query, with as few irrelevant items as possible. Therefore, relevance is a subjective notion. Several experiments and researches have been done to assess relevance. The notion of relevance is of central importance in IR [8, 9]. The several techniques used in semantic search engines are artificial intelligence, natural language processing, and machine learning. Kngine utilizes the efficiency of knowledge-based approach and the power of the statistical approach [10], whereas the popular search engine Google uses its own search technology which called Hummingbird algorithm [11]. The architecture methodology and algorithm with results of query processing system for Urdu is described in [12]. The challenge is to develop a technique to allow retrieval of higher quality. Another challenge that needs to be addressed is to implement techniques which will yield faster indexes with smaller query response time [13, 14]. The proposed semantic IR system will significantly improve the performance over the traditional keyword-based information retrieval systems in terms of recall and precision by providing smart, intelligent, and even spontaneous relevant results with meta-information as dreamed by father of WWW Tim Berners-Lee [15] along with user-friendly user interaction with the system.

3 System Architecture

The overall working of the Search Engine [16] is as shown in Fig. 1. Query controller takes user-entered query as input and generates the relevant output along with the time required for execution and meta-information. The query controller gives the input query to the semantic query mapper and query similarity module which first checks the validity of query. It checks whether the user entered the (i) help query, (ii) empty query, (iii) non-tourism domain query, (iv) tourism domain query. If query is specific to tourism domain, then it first identifies the specific service using the novel concept of query prototype and query similarity and accordingly displays the result. Meta-processor is also designed which will generate the meta-information of the result in the background. This result is again given to the query controller for the proper display along with meta-information and required time for execution in the user-understandable form. This is in short the overall working of the Tourism Search Engine.

4 Query-Similarity Mapper

The system is implemented for tourism domain. The domain is fixed. Another issue is to identify the subdomain or service about which the user has entered the query as shown in Fig. 2. As discussed in [16], for different services/subdomains with respect

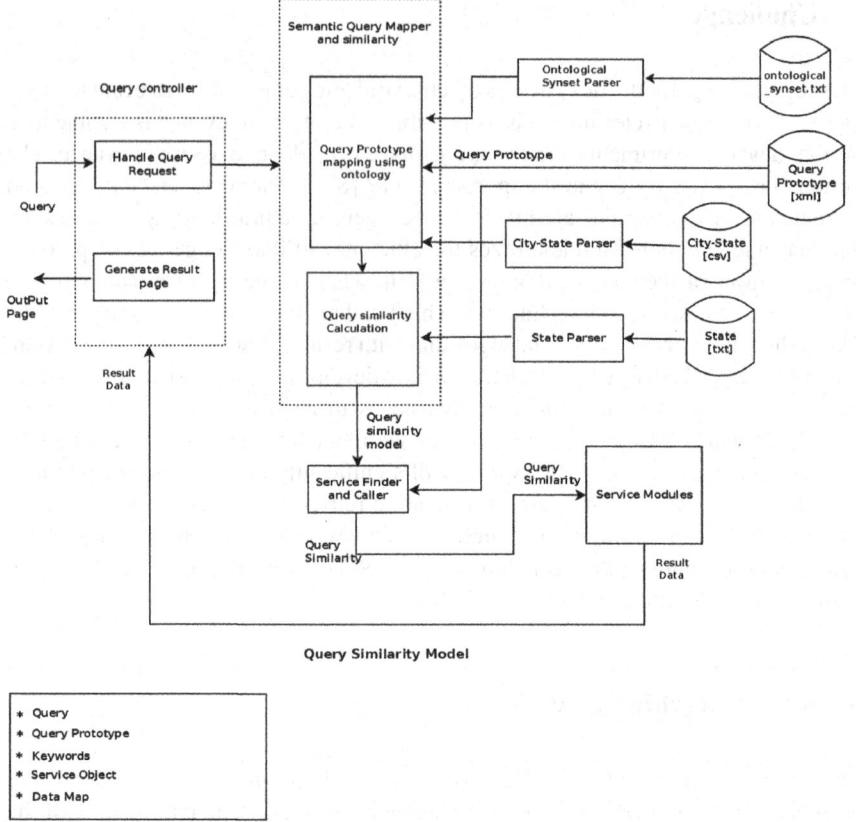

Fig. 1 System architecture

to tourism domain, expected query patterns called as query prototypes are defined and stored in query-prototype.xml using spring injection technology. Next issue is to match the user-entered query with the defined query prototype. To achieve this, the novel concept of query similarity is designed. There is the possibility that user-entered query may match with more than one-defined query prototype, to address such scenario, query similarity manager is implemented which matches the user-entered input query against all the query prototypes stored in query-prototype.xml file using uniquely designed query-prototype parsers. The query-prototype parser parses the query-prototype.xml. It then maps user-entered query with each defined query prototype, calculates similarity using the uniquely derived query similarity formula as given below. Query similarity mapper calculates the similarity within range 0–1. Zero (0) indicates NO similarity and One (1) indicates 100% similarity. The query similarity iterates tokens of query prototype and checks each token of query prototype in query. The algorithm is designed which works as follows:

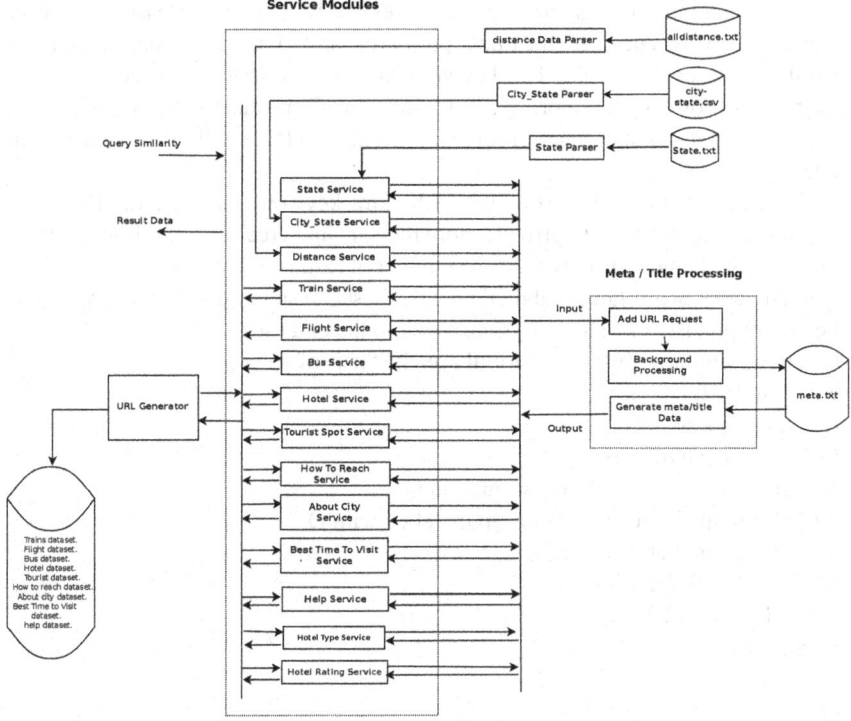

Fig. 2 Service finder

- Query prototype parser parse all query prototypes from queryprototype.xml file
- Match query with defined query prototype lists, return respective service
- Compare query-tokens length with query-protocol-tokens length
- Find the state and city from given input query

  - Check whether the token at index is template token or not, e.g., [state], [city], [from-city]
  - Check whether the token contain state word or not using the state parser and put state name in hash map, if matched else null.
    All the states and territories of India are found and stored in separate file which is used by the state parser to parse the states.
  - Check whether token contain city word or not using the city parser and then put city name in hash map
  - All the state-wise city names are found and stored in separate file which is used by the city–state parser to parse the city names

- Check for simple tokens in query and match them with query prototype
- Check for ontological tokens, e.g., (distance), (train)

- Collect remaining tokens from query as keywords. This is the case when user entered query matches with the query prototype 100% but also contain some extra words called as keywords. The keyword manager is designed to address such queries for providing more precise, relevant result. For example, train from Mumbai to Pune Janshatabdi, here Train from Mumbai to Pune will 100% match with query prototype
(train) from [from-city] to [to-city] with extra keyword Janshatabdi. The Keyword(s) are further used to provide more relevant and precise result. If after 100% matching with query prototype, if the number of results is 7, out of which 3 results contain the keyword Janshatabdi, then only these 3 results were shown to the user; i.e., more precise relevant results are given to the end user.
- Calculate similarity by using formula on the basis of
K, T1 and T2;
Q = Input query
Qp = Prototype of query
K = Total number of keywords found in Q
T1 = Total number of tokens of Qp matched with Q
T2 = Total number of tokens of Qp
S1 = (T1 * 1.0)/(K + T1);
S2 = (T1 * 1.0)/T2;
if (S2 == 1)
similarity = S1 * S2;
else
similarity = 0;

This way the query-similarity manager found the query-prototype with maximum similarity value and thereby identify the service/subdomain of user interest. On identification of service and the respective query prototype, the prototype tokens are replaced by actual values in the user-entered query; i.e., the URL is not fixed. If the user is entering the query train from Pune to Mumbai then query-prototype mapper matches the query with (train) from [from-city] to [to-city] prototype and then locate the user at the respective Web page on the Web site. Again if the user is entering train from Pune to Bangalore, then query will match with the same prototype (train) from [from-city] to [to-city] prototype and will place the user at the respective Web page on the Web site. So even if the prototype match in both the cases is same, the Web page fetched or shown to the user is different in both case and thus query-similarity is the novel approach. In other words, URLs are not fixed, but out of hundreds or thousands of pages on that particular URL, which Web page is of user interest is decided at run time and accordingly the Web URL is created dynamically. So novel approach of query-similarity introduced intelligence in the proposed system which will increase as it grows older.

5   Ontological Synonym Set

In specific domain for one keyword, multiple similar meaning words (synonyms) are available. For example, for distance domain, one token may be Find. But some may use get or list. So for Find, Get/list are not synonyms but with respect to specific domain one word may have several other words which also gives the same meaning. Ontology refers to abstract sense, but for domain specific, their is requirement of domain-specific ontological words, i.e., beyond ontology. While defining the query prototype, the round brackets ( ) are used to define the ontological token which at run time will refer the ontological synonym set created in the following way. The World-Net is used to find the synonym of all possible words used in the tourism domain. For this, each word related to tourism domain is given as input to the WorldNet which in turns gives all possible synonym sets as output. But for the tourism domain, the output provided by WorldNet is not perfect, the words which fit for tourism domain are kept and unfit are removed from the output list manually. Also some words which are not given by WorldNet but are required and matching with the word given as input for tourism domain are added manually. This process is done for all possible words for the tourism domain, and a new ontological synonym set for tourism domain is derived. This concept drastically improves the performance of the proposed system by reducing the number of query-prototypes which were defined before using the ontological synonym set. Also it covers maximum all possible queries which different user may enter in different ways, and the proposed system interprets them correctly and provides the relevant, precise result. In other words, this ontological sense increases the proposed semantic IR system scope drastically.

Ontological synonym set parser is designed which parse the ontological synonym set file where the lines are stored in following fashion.

Base word Synonym-1, Synonym-2, …, Synonym-n

with new base word per line and is stored in ontological synonym set file.

For example, season flavor, flavor, harden, temper, mollify, time of year, seasons.

Base word and respective synonyms are stored on one line. This novel format of storing ontological synonym set is also uniquely designed for reducing the number of query prototypes and handling maximum queries.

Ontological Synonym set Parser works as follows: To Read all the words from ontological synonym set file and store in dataMap.

- Get the ontological base word
- Get all synom words of base ontological word
- Return array list of base word
- Repeat Step 1 to 4 to parse all lines in ontologicalsynset.txt

For e.g.: If user wants to know the information about some city then some possible ways to fire the query are as follows: about [city]

- abt [city]
- ABOUT [city]

- About [city]
- information [city]
- info [city]
- detail [city]
- details [city]
- information of [city]
- information about [city]

So here [ ] is used for template token, city, which will give information about any city provided at run time but the use of ( ) for ontological token, reduced the above number of queries as follows

- (about) info [city]
- (about) [city]
- (about) of [city]
- (about)[city]
- (about)[city]
- (about) [city] information

So when the ontological token appears in the query, the user-entered ontological token is replaced by base word and hence handle n number of user queries using single prototype. Remaining tokens are treated as simple tokens. This way the novel concept of query prototype is derived. All the query prototypes with respect to all different services are stored in prototype.xml file. The use of [ ] and ( ) brackets and also how to use that ontological synonym set on query prototype concept is the new paradigm introduced in the proposed system.

6 Meta Processor

Google in its initial stage starts crawling from four Web sites but now the number of Web sites it crawls, increase drastically. The information gathered during crawling is processed in the background and the page ranking algorithm is used for displaying the result to the end user. The proposed system is implemented for tourism domain. The meta-processor is designed to provide meta-information like title, time and brief information with respect to relevant URLs of the user-requested information. Whenever the user enters any query first time, only the Web links are displayed to the user, but at the same time a thread is spawn, which in the background fetches the meta-information and dump it on the server. Because processing these URLs for meta and title at run time, will take more time, because it require connection with multiple servers for this information. But the users time is precious, need to provide instant results. But immediately from the second run of the same query, user gets the relevant information along with the meta-information. So processing the meta-information is a background process, and meta-processor is doing this job. Google faces the same problem of meta-information processing at the initial phase but now it has huge metadata available with it. While processing the meta-information in the background, in some cases there is the possibility that meta- processor could not be able to fetch the meta-information of some web links due to issues like server slow, server down or data not accessible. In some cases, the link is same but the server is moved permanently, but in such cases also the proposed metaprocessor is working properly and giving the required information by processing the thread in next runs. The novel design of meta-processor is responsible for making the proposed system more intelligent. URLPrototype.xml with respect to each service store all the relevant URLs for respective service. New URLs can be added, existing URLs can be updated or deleted without recompiling/redeploying the system.

7 Template Manager

Some services in the tourism domain require result in one line or paragraph instead of link like about city service, best time to visit service. For this, the novel template manager is designed which works in background. Templates are Web site specific. For adding new URL, template of that Web site need to be added. Because different Web sites use different structure/template to display/store the information. This novel approach of template manager helps in getting this information. Based on the result of query similarity mapper, the service finder find the service and then invoke the appropriate service. On invocation of the service module the results for the user-entered query are fetched and provided to the user using meta-processor and template manager. URL generator generates the actual URL, and the result is provided to the user along with the time required for execution and meta-information as shown in Fig. 2.

8 Conclusion

This paper proposed the novel and promising concepts of query similarity, meta-processor and keyword manager for semantic information retrieval system for the tourism domain. The novel concept of query similarity matches the user-entered query with the defined query prototype thereby provides the service using the query prototype with which it matches the maximum using uniquely derived query simi-larity formula. The keyword manager is designed and implemented to handle extra keywords in the query after matching with the query prototype and providing more precise result. The ontological synonym set for the tourism domain is generated. The format of the file where the ontological similar words are stored is uniquely designed and then used for prototype matching. Some services like distance, about city, best time to visit with reference to tourism domain expects direct answer. Tem-plate manager is designed and implemented for this where the templates are created for different URLs as the structure for different URL is different. The meta-processor is the new paradigm of the proposed system to provide time required for execution, title, meta-information of relevant links as part of result.

References

1. Andrews, K.: Visualizing cyberspace: information visualization in the harmony internet browser. In: Proceedings '95 Information Visualization, Oct 1995, USA, Atlanta, pp. 97–104
2. Baeza-Yates, R., Ribeiro-Neto, B.: Modern Information Retrieval. ACM Press, New York, USA (1999)
3. Chowdhury, G.: Introduction to Modern Information Retrieval, 3rd edn. Facet Publishing (2010)
4. Information Retrieval. www.wikipedia.org. Accessed 22 June 2011
5. Amit, S.: Modern information retrieval: a brief overview (2001)
6. Alis J. Technologies: Web languages hit parade. http://babel.alis.com:8080/palmares.html (1997)
7. Humm, B.G., Ossanloo, H.: A semantic search engine for software components. In: Isaas, P. (ed.) Proceedings of the International Conference WWW/Internet 2016, pp. 127–135. IADIS Press, Mannheim, Germany (2016). ISBN 978-989-8533-57-9
8. Meadow, C.T., Boyce, B.R., Kraft, D.H., Barry, C.L.: Text Information Retrieval Systems. Academic Press (2007)
9. Van Risjbergen, C.J.: The Geometry of Information Retrieval. Cambridge UP (2004)
10. http://www.kngine.com/Technology.html
11. Sullivan, D.: FAQ: All About the New Google Hummingbird Algorithm | Why is it called Hummingbird? (2013)
12. Thaker, R., Goel, A.: Article: domain specific ontology based query processing system for Urdu language. Int. J. Comput. Appl. 121(13), 20–23 (2015)
13. Laddha, S.S., Laddha, A.R., Jawandhiya, P.M.: New paradigm to keyword search: a survey. In: 2015 International Conference on Green Computing and Internet of Things (ICGCIoT), Noida, pp. 920-923 (2015). https://doi.org/10.1109/ICGCIoT.2015.7380594
14. Laddha, S.S., Jawandhiya, P.M.: An exploratory study of keyword based search results. Indian J. Sci. Res. 14(2), 39–45 (2017)
15. http://searchmicroservices.techtarget.com/definition/Semantic-Web
16. Laddha, S.S., Jawandhiya, P.M.: Semantic search engine. Indian J. Sci. Technol. 10(23) (2017). https://doi.org/10.17485/ijst/2017/v10i23/115568

Design and Implementation of Trigger on OceanBase

Liang-yu Xu, Xiao-fang Zhang, Li-jun Zhang and Jin-tao Gao

Abstract Trigger technology is an important part of database design and development. It is an important function module of database management system. It is a special type of stored procedure. The implicit implementation of the trigger makes it an effective way to ensure data integrity in the database. OceanBase is a new type of mass distributed database system, but the existing OceanBase open-source version does not support the trigger function, affecting the system to promote and use. Based on the deep analysis of the working principle of traditional database triggers and the overall architecture of OceanBase, this paper designs and implements the trigger function of OceanBase.

Keywords Trigger · Distributed database · OceanBase · SQL

1 Introduction

With the advent of the digital era, the explosive growth of information has made it more difficult to save and process data. Traditional databases can't handle such a large amount of data, although the database can be modified according to the business characteristics of each enterprise, the cost will be greatly increased. In recent years, due to the mature of distributed database, many enterprises gradually began to choose the distributed database, using its good expansibility and fault tolerance for their business needs in the mass data storage and processing.

Database for the data processing ability includes online analytical processing (OLAP) and online transaction processing (OLTP) [1], OLAP has the characteristics of less statement execution and processing a large amount of data, the characteristics of the current more mature distributed database have a strong ability of online analytical processing, such as Hive systems designed to build on the basis of the Hadoop AMPLab by Facebook, Berkeley university laboratory constructed a system

L. Xu (✉) · X. Zhang · L. Zhang · J. Gao
School of Computer Science, Northwest Polytechnical University, Xi'an 710129, China
e-mail: xuliangyu1991@163.com

© Springer Nature Singapore Pte Ltd. 2019
S. K. Bhatia et al. (eds.), *Advances in Computer Communication and Computational Sciences*, Advances in Intelligent Systems and Computing 760, https://doi.org/10.1007/978-981-13-0344-9_10

of the Shark on the Spark system, and the Dremel system, BigTable System [2], and Spanner System [3] designed and developed by Google. OLTP has the characteristics of high transactional and fast response, generally high availability online system, which has higher requirements for the execution efficiency of statements. Such as Alibaba, independent research and development of OceanBase [4] in Taobao shopping when processing data of high efficiency, as well as the VoltDB [5] database's high efficiency in transaction processing.

These distributed databases have an outstanding ability to store and process massive amounts of data. However, compared with the traditional relational database, there is a certain lack of function, such as the lack of SQL function, which leads to the problem of data integrity. This allows more businesses to focus on enhancing the capabilities of the database to meet their business needs.

OceanBase is a relational distributed database developed by Alibaba, which supports basic SQL operation and satisfies the basic characteristics of the relational database. However, compared with the current mainstream relational database, there are some shortcomings, such as not supporting the storage procedure, cursor, trigger, and other functions. Triggers play a very important role in ensuring the integrity of data, and the integrity of data is self-evident in software systems.

This paper analyzes the realization of the function of mainstream open-source database Mysql, PostgreSQL triggers, and studies the key technologies to realize the trigger, and combining the OceanBase architecture and its characteristics, propose the design and implementation scheme of database architecture which is suitable for OceanBase triggers, which provides an effective way to ensure data integrity in OceanBase, making the OceanBase function more complete and adapting to the functional requirements of the enterprise.

This paper is arranged as follows: Sect. 2 introduces the OceanBase structure; Sect. 3 introduces the relevant knowledge points of triggers; Sect. 4 introduces the design of triggers on Oceanbase; Sect. 5 summarizes the full text.

2 Architectural Description of OceanBase

As shown in Fig. 1, OceanBase system architecture, divided into four types of servers: RootServer (RS), the master server, responsible for managing all other servers in the cluster. UpdateServer (UPS), incremental server, responsible for providing Ocean-Base system write service. ChunkServer (CS), baseline data server, responsible for storing the OceanBase system baseline data. MergeServer (MS), responsible for handling SQL requests. The figure shows a single cluster configuration, OceanBase also supports multi-cluster configuration [6].

OceanBase clients communicate with MS, by means of the Mysql client, Java client, or C client to complete. The latter two communication mode need to first request server for MS address list to get a MS to communicate, after get in connection with MS, send a SQL statement to MS. The MS on the SQL statement do lexical, grammar analysis, generate the corresponding syntax tree and then the syntax tree

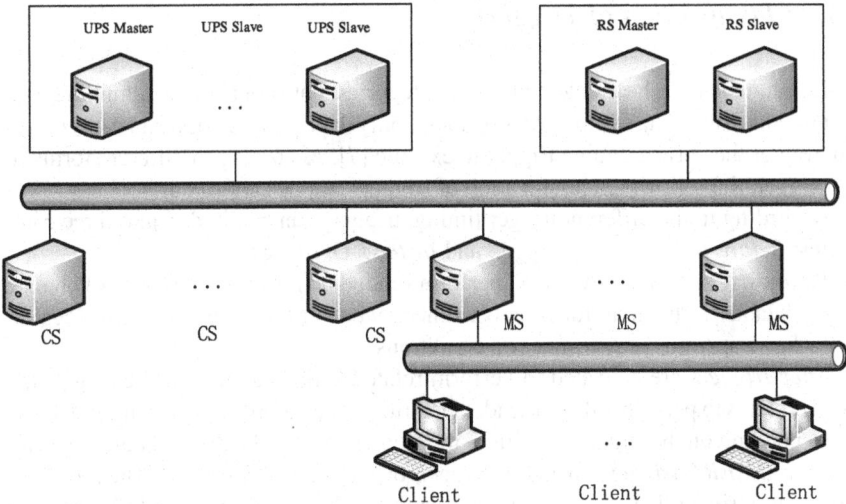

Fig. 1 Basic architecture Of OceanBase

generate logic and physical plans accordingly. The physical structure of the plan is to form a variety of different operators of MS, and then forwarded to the corresponding UPS or execution of CS for operator and pull data such as operation to complete the execution of SQL statements. In the process, UPS is responsible for the storage and processing relative to the baseline data of incremental data, CS is responsible for the storage and query have merged the baseline data, RS is responsible for the management to coordinate the implementation of all servers.

3 Introduction of Trigger

Trigger is a programmer or a method used to guarantee data integrity, data analyst, which is a special stored procedure, can be used to form the integrity of the implementation of a complicated constraints, when modifying the tables, the trigger is called automatically, which can prevent the table data is not correctly revised [7]. The need for complex data integrity in the business can be met with its characteristics. As described above, the execution of a trigger is neither a program call nor a manual start, it is triggered by an event.

3.1 Classification of Triggers

SQL statements can be divided into four categories: data query, data definition, data manipulation and data control. Among them, only data manipulation and data control can trigger the corresponding trigger to execute [7]. According to different forms of the trigger, the following classification of triggers can be carried out.

According to the different trigger timing, triggers can be divided into three categories: *Before-trigger*, *After-trigger*, and *Instead-Of-trigger* [8].

Before-triggers are activated to perform related actions prior to a record change in the database. The main function is to check whether the record change is legal or to further constrain the modification conditions.

After-triggers are activated to perform related actions after record changes; the main effect is to perform other cascade operations on a record change in the database.

Depending on the trigger condition, the trigger can also be divided into three categories: *INSERT-trigger*, *DELETE-trigger*, or *UPDATE-trigger* [8]. The *INSERT-trigger* is a trigger that is triggered when an INSERT is performed, and likewise, the *DELETE-triggers* and *UPDATE-triggers* are the triggers that are triggered when the delete and update actions are performed.

Depending on the trigger granularity, triggers can be divided into row granularity triggers and statement granularity triggers [8].

On OceanBase, trigger of data manipulation language realizes the trigger when the data in the database insert, delete, update operations, to detect whether there will be corresponding trigger definition, and then based on the trigger and trigger the granularity of the trigger to complete for the execution of the trigger.

3.2 Definition of Triggers

The definition of a trigger needs to meet certain rules, where the most basic form of the rule consists of three parts: events, conditions, and actions. When a set of rules are defined, the database will perform the conditional evaluation when the corresponding event is executed and perform the corresponding action when the condition is set up [9].

Event detection is the basis for triggers implementation and is used to describe why a trigger is triggered. Trigger events from the internal operation of the database, such as the database table to insert, update or delete operations. This kind of event can be a simple operation, and it can be a combination of several kinds of operations.

Condition is based on the occurrence of the incident, the implementation of the corresponding action additional conditions. The result of the condition is expressed by a Boolean expression. Where the conditions include database predicates, restrictive predicates, database queries, and so on. The database also supports default conditions, and the condition is constant. Because the conditions may need to determine the corresponding database query, it will affect the efficiency of the database, the need for the corresponding optimization.

Action refers to the operation of the database when the events and conditions are met including data modification operations, data storage operations, and other database commands. The specific execution time is also related to the definition of the trigger, the processing granularity. As action can operates the database, it can also cause other triggers to perform as cascading triggers.

On OceanBase, when triggers are defined, the events, conditions, and actions of the trigger are detected. The trigger is stored when the correct definition is met.

3.3   Execution Semantics of Triggers

3.3.1   Trigger Granularity

The trigger granularity is the frequency at which the trigger is detected and processed; the finest granularity is the row level that is related to each operation of a row data. The implementation of a trigger can perform insert a row of data or delete a row of data or update a row of data. A bit more granularity is the level of the statement, that is, the execution of a SQL statement, regardless of the number of rows in the SQL statement operating data, only the implementation of a trigger action. On OceanBase, the *after-trigger* supports two trigger granularities, and the *before-trigger* only supports row-level trigger granularity due to its characteristics.

3.3.2   Predicate

Since events can be combinations of several complex operations, the trigger predicate can do this when a different response is required depending on the operation. According to the different results of the predicate, define different branches to complete different responses. On OceanBase, predicates act as a form of trigger conditions that respond to different actions for different events.

3.3.3   Transactional

Trigger execution may success, also may fail, in order to keep the trigger's execution on the atomic properties of the database operation, the trigger is usually handled as a transaction, and when a certain action in the execution of the trigger fails, all the actions rollback to keep the database consistent. On OceanBase, for a data manipulation statement that triggers a trigger, the statement needs to be executed in a transaction with all subsequent triggers. If all statements are executed normally, the transaction is committed, or the transaction is rolled back.

3.3.4   Serial and Recursive Execution

Since multiple triggers can be executed for the same event, the execution order of the triggers is another aspect that needs to be considered. For multiple triggers that have been defined in the database, when the events and conditions are met, the action of the trigger is executed serially.

Triggers are triggered by INSERT, UPDATE, DELETE operations. Similarly, for *after-triggers*, you can also define INSERT, UPDATE, DELETE operations, these operations will also trigger a new trigger, this is called the trigger cascade [10].

Trigger cascade provides an effective way to achieve data integrity, and the rational use of cascading can make data processing more convenient. But the use of cascading also need to be cautious, if the cascade of too many triggers, may lead to death cycle or the entire database system data caused irreversible damage. For most database management systems, the number of cascading layers is added to prevent this error from occurring.

As shown in Fig. 2, an INSERT operation, triggering five triggers, including two *before-trigger* and three *after-trigger*. These five triggers are considered as the trigger level one, if one of them triggers the other triggers, as shown in below Fig. 2. In this, 'After11' trigger at level one triggers all other trigged called the next layer. At this point, interrupt the implementation of the first layer, that is, the first implementation of the second layer of the trigger After21 and After22. When the implementation of the second layer is completed, then the implementation of the first layer of the remaining trigger is After12 and After13. It is possible to generate a trigger on the third layer or even the nth layer, and the execution is triggered in the same manner as described above.

When the trigger defines a loop structure, such as the trigger T1 trigger trigger T2, and trigger T2 triggers the trigger T1, then the trigger will be trapped in an infinite loop to perform, thus undermining the structure of the database, So there is a need for some way to stop the trigger loop; most of the database defines the maximum number of cascaded triggers, such as the MySQL database defines the maximum number of cascading layers 16 to ensure data logic and security.

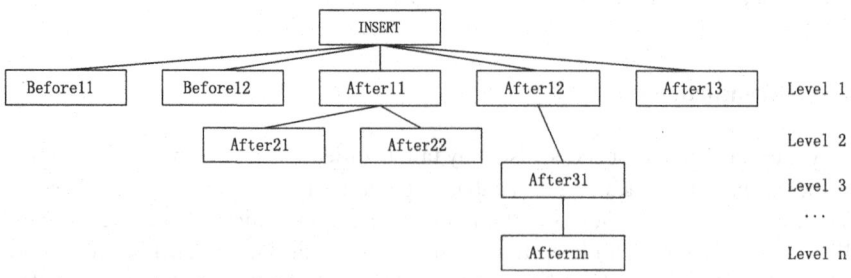

Fig. 2   Trigger cascade representation

On OceanBase, the triggering mode is implemented as described above, and the sequential execution mode is used in the same layer, while the different layers adopt the recursive execution mode. For cascading restrictions, the same way as MySQL is used, with the maximum number of cascading layers 16 to avoid the occurrence of an infinite loop.

4 OceanBase Trigger Design

On OceanBase, when the statement is executed, a check on the trigger is added, and if the condition is satisfied, the trigger is triggered. Since the function of the *before-trigger* is to check and modify the data, so when it has *before-trigger*, the changes will be applied directly to the data. When the statement is executed, the modified data is used directly. And the *after-trigger* is executed after the completion of the statement, so first save the trigger, when the implementation of the statement is completed, and then trigger the implementation of the trigger.

The complete execution of a trigger consists of three complete parts, including definition, storage, and execution. The following describes the implementation of the trigger from these aspects.

4.1 Creation of Triggers

4.1.1 The Syntax of the Trigger

The syntax for defining a trigger is shown in Fig. 3, where the OR REPLACE keyword is an optional keyword that indicates whether the original trigger in the database is overwritten. Trigger_name is the trigger name created, Trigger_time is the triggering time of the trigger, table_name is the trigger associated with the table, reference_define is the name of the NEW, OLD keyword associated with the trigger, and trigger_gran is the trigger granularity of the trigger. Trigger_action is the action that needs to be performed after the trigger is triggered; that is, the execution body of the trigger.

Trigger execution, in fact, is the executive body of the execution, because the need for triggers in different constraints to make different operations, so choose to use *PL/SQL* [11, 12] as the trigger of the host language.

```
CREATE [OR REPLACE] TRIGGER trigger_name ON table_name REFERENCING
          reference_define FOR EACH trigger_gran Trigger_action
```

Fig. 3 Defines the syntax of the trigger

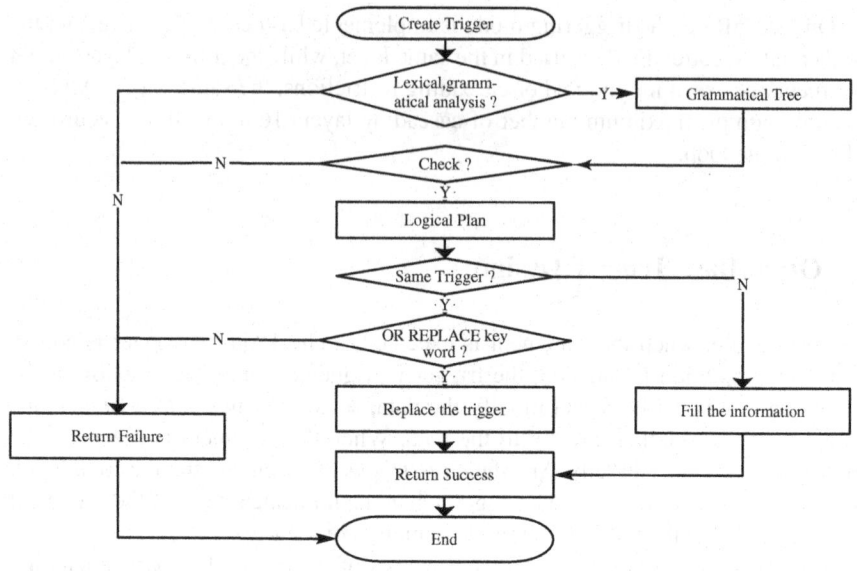

Fig. 4 Trigger creation process

4.1.2   Compilation of Triggers

In the trigger compilation phase, you need to check the trigger definition statement because the trigger is not directly invoked, but when an event triggered, the trigger stored in the database system will be compiled if meets certain conditions.

In the compilation phase of the creation of the trigger need to go through the lexical analysis, parsing, logic program generation, physical planning to generate four steps [13]. The specific steps to create are shown in Fig. 4.

When the MS receives the client to send the SQL statement used to create the trigger, first through the lexical analysis of each node, and then parse, analyze whether there is a grammatical error, if there is, return the error message, if not, then generate a syntax tree into the logic plan generation phase, in the logical planning stage, according to whether there are OR REPLACE keywords generate different logical plan, at the same time, the trigger associated with the table, as well as trigger conditions, variables, fields, etc., to check, If an error occurs during the period, an error message is returned.

On the contrary, the logical plan is generated. In the next step, the physical plan generation phase will first determine whether there is a trigger message with the same name in the system table (__all_trigger) that stores the trigger information, and if so, whether it is based on a logical plan of whether or not there are different plans generated by the OR REPLACE keyword, If there is no keyword, it will return an error message; if so, replace the previous created trigger. Finally, the information of

Table 1 Structure of _all_trigger

Fields	Descriptions	Fields	Descriptions
tg_name	Trigger name	tg_is_row	trigger granularity
table_id	Table id	tg_ref_new	new keyword
tg_type	Trigger type	tg_ref_old	old keyword
tg_cascade	trigger cascade	tg_ref_newtable	newtable keyword
tg_events	Trigger Events	tg_ref_oldtable	newtable keyword
tg_attribute	trigger column	tg_source	SQL statement

the trigger is stored in the system table (__all_trigger) where the trigger information is stored, and the success information is returned.

4.1.3 Storage of Triggers

The trigger information is stored by a system table (__all_trigger) whose table structure is shown in Table 1. Where tg_name represents the name of the trigger that was created, which is unique in the database and is not allowed to repeat. Table_id represents the id of the table corresponding to the currently created trigger event.

In Table 1, tg_name represents the name of the trigger that was created, which is unique in the database and is not allowed to repeat. Table_id represents the id of the table corresponding to the currently created trigger event. Tg_type indicates whether the type of the trigger is a front trigger or a post trigger. Tg_cascade indicates whether the triggers are cascaded. If the triggering is not allowed, the current trigger will not trigger the execution of other triggers when it is triggered. The default is to allow cascading. Tg_events represents any of the specified types of triggers, insert, update, delete, or any combination of them. Tg_attribute represents the corresponding column of the corresponding UPDATE column level trigger. Tg_is_row represents the trigger granularity of the current trigger, and if the value is true, the trigger granularity is the row.

Otherwise, the trigger granularity is the statement. Tg_ref_new represents the name of the new keyword corresponding to the current trigger, tg_ref_old represents the name of the old keyword corresponding to the trigger, and tg_ref_newtable represents the name of the new_table keyword corresponding to the trigger. Tg_ref_oldtable represents the name of the old_table keyword corresponding to the trigger. Tg_source represents the complete SQL statement defined by the trigger.

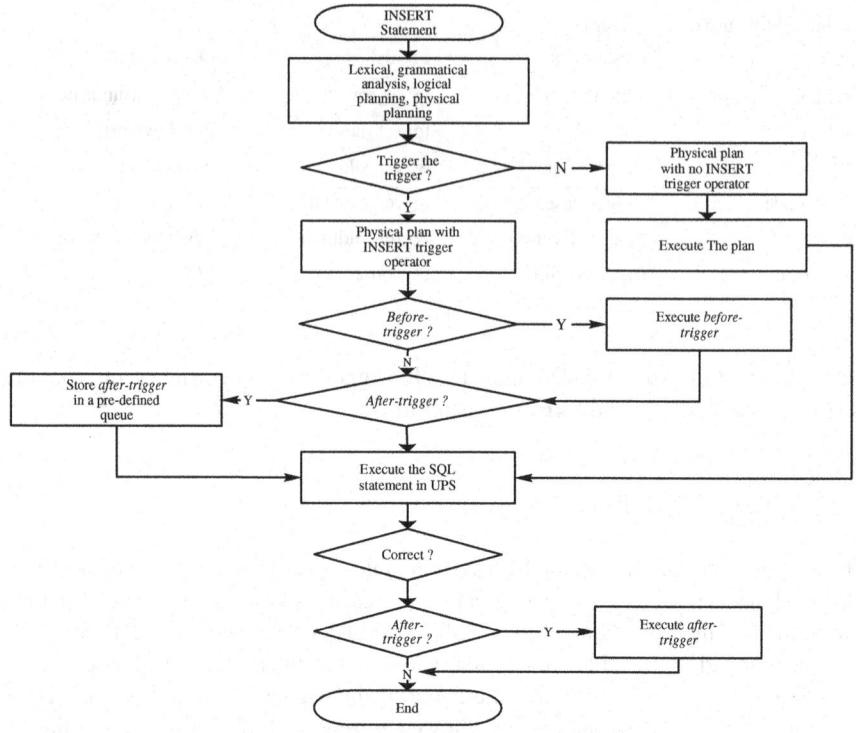

Fig. 5 Execution of *INSERT-trigger*

4.1.4   The Execution of the Trigger

The *INSERT-trigger*, which refers to the execution of an INSERT statement, triggers the execution of the defined statement in the trigger, which is called an INSERT-trigger, and its concrete execution process is shown in Fig. 5.

An INSERT statement, first through lexical syntax parsing, and then generate INSERT statement logical plan. When the physical plan of the INSERT statement is formed, it finds whether the INSERT statement triggers the trigger based on the inserted information into the system table (__all_trigger) that stores the trigger information. If not, executes the original INSERT process. If there is a trigger, the INSERT-trigger physical operator is built to populate the trigger information.

In the implementation of the INSERT-trigger physical operator, if there is a *before-trigger*, execute it and then stored *after-trigger* in a pre-defined queue. When a trigger in the queue needs to be executed, maintain a transaction that contains the current SQL statement and the triggers in the queue. After SQL statement executed correctly the triggers in the queue are executed sequentially, during execution, if the trigger generates a new trigger, the depth-first traversal trigger operation is performed in the same way until all triggers are executed or an error has failed.

In any step of the implementation of the trigger, as long as an error occurs, the database needs to roll back to the previous state. Otherwise, submit the transaction. The same is true for UPDATE and DELETE operations.

UPDATE-trigger and *DELETE-trigger* have the same execution process as *INSERT-trigger* described above.

5   Concludes

Triggers can quickly achieve the data integrity of the database system, consistency [14], but also control the data to meet the statement. In the database to deal with the logical relationship between the complex table and the table when the trigger can make the database design becomes simple and efficient [15].

As the open-source version of OceanBase lack of support for the trigger, affecting the system in the enterprise to promote the use of. Based on the architecture of OceanBase itself, this paper combines the implementation of other database triggers. It designs and realizes the function of OceanBase triggers, and supports the basic functions of triggers, and in the OceanBase can be used normally, in the testing process, the trigger function of the implementation of the efficiency is not expected high. In the future work, we need to add other relatively high-level trigger function, to support more trigger statements. And the relevant functions of the trigger to do some optimization, in order to improve its efficiency.

References

1. Plattner, H.: A common database approach for OLTP and OLAP using an in-memory column database. In: Proceedings of the 2009 ACM SIGMOD International Conference on Management of Data, pp. 1–2. ACM (2009)
2. OceanBase [EB/OL]. https://github.com/alibaba/oceanbase. Accessed 05 July 2016, 03 Jan 2017
3. Stonebraker, M., Weisberg, A.: The VoltDB main memory DBMS. IEEE Data Eng. Bull. **36**(2), 21–27 (2013)
4. Chang, F., Dean, J., Ghemawat, S., et al.: Bigtable: a distributed storage system for structured data. ACM Trans. Comput. Syst. (TOCS) **26**(2), 4 (2008)
5. Corbett, J.C., Dean, J., Epstein, M., et al.: Spanner: Google's globally distributed database. ACM Trans. Comput. Syst. (TOCS) **31**(3), 8 (2013)
6. Yang, Z.: The architecture of OceanBase relational database system. J. East China Normal Univ. (Nat. Sci.) (5), 141–148 (2014)
7. Wei-xue, L.I.U.: Further study on trigger of SQL server database. Comput. Technol. Dev. **23**(10), 48–51 (2013)
8. Silberschatz, A., Korth, H.F., Sudarshan, S.: Database System Concepts. McGraw-Hill, New York (1997)
9. Zuo, W.: Research on Active Database Theory. Jilin University (2005)
10. IBM [EB/OL]. https://www.ibm.com/support/knowledgecenter/SSEPEK_11.0.0/apsg/src/tpc/db2z_triggercascade.html. Accessed 03 Jan 2017

11. Zhu, T., Zhou, M., Zhang, Z.: Study on stored procedure implementation oriented to OceanBase. J. East China Normal Univ. (Nat. Sci.) (5), 281–289 (2014)
12. Zhu, J., Liu, B., Yu, S., Gong, X., Zhou, M.: Design and implementations of stored procedure in OceanBase. J. East China Normal Univ. (Nat. Sci.) (5), 144–152 (2016)
13. Aho, A.V., Sethi, R., Ullman, J.D.: Compilers, Principles, Techniques. Addison Wesley (1986)
14. Shen, Y.J., Yang, H.Q.: The realization of teaching case design database integrity by using triggers in SQLSever2005. In: Computer CD Software and Applications, vol. 5, no. 19, pp. 240–242 (2012)
15. Hu, H.N.: Application of SQL server triggers in database design. In: Computer Programming Skills and Maintenance, vol. 19, no. 8, pp. 37–38 (2012)

Text Clustering on Short Message by Using Deep Semantic Representation

Songze Wu, Huaping Zhang, Chengcheng Xu and Tao Guo

Abstract Text clustering is a big challenge in the text mining field; traditional algorithms are powerless when dealing with short texts. Short messages are a much more flexible form of data in social media, containing not only textual information, but also comment, time and regional information. We propose an algorithm to extract semantic and multidimensional feature representation from such texts. In particular, by using the fact that comments are semantically related to the short message, we can get the supervised information and train the text representation, with which we transform the problem into a semi-supervised problem. We use a convolutional-pooling structure that aims at mapping the text into a semantic representation. What's more, we expand the semantic representation with time- and region-related features, leading to a much more flexible and strong representation for short messages. Our approach shows great advantages in labelled data over traditional feature representation methods and performs better than other clustering methods via deep neural network representation.

Keywords Text clustering · Short message · Deep semantic representation
Multidimensional feature · Semi-supervised

S. Wu (✉) · H. Zhang · C. Xu
School of Computer Science, Beijing Institute of Technology, Beijing 100081, China
e-mail: wusongze2015@nlpir.org

H. Zhang
e-mail: kevinzhang@bit.edu.cn

C. Xu
e-mail: xuchengcheng2015@nlpir.org

T. Guo
Institute of Information Engineering, China Academy of Sciences, Beijing 100081, China
e-mail: Guotao@iie.ac.cn

© Springer Nature Singapore Pte Ltd. 2019
S. K. Bhatia et al. (eds.), *Advances in Computer Communication
and Computational Sciences*, Advances in Intelligent Systems
and Computing 760, https://doi.org/10.1007/978-981-13-0344-9_11

133

1 Introduction

With the rapid development of the Internet and digital technology, a variety of new media which interaction between users has sprung up. At the same time, social media on the Internet has become a preponderant channel for the public to express their emotions and share their opinions. We can collect text, media and voice information from users' Weiblogs, Tweets and so on. Users' friend relationships, open personal information and interests can be viewed by anyone. We can even get location information once the user has activated the locate function, and the messages are related during a period, which can be used to extract time information. This special information is called short messages in social media, which contains not only text information but location and time information. Text clustering is difficult in text mining and topic detection, especially for short texts, which lack enough content from which we can learn. However, traditional short text is less abundant and most of the information is short texts, which are very difficult to understand by machine. For now, traditional clustering algorithms perform poorly on sparse and noisy data. It is very difficult to use such less information to get a dense and flexible semantic representation for short text. In this paper, instead of building short messages with the traditional text representation and explicit information [1–3], we obtain a much more complicated and flexible representation of short messages [4].

Some deep learning models are aimed at learning a representation for short texts. Short text clustering via convolutional neural networks (STCC) [5] uses a convolutional neural network (CNN) to get a deep representation for short texts. Firstly, they pre-train a vector based on keyword features with a locality-preserving constraint, and choose Laplacian affinity loss, then use the difference between the vector and CNN output as the network loss to train the representation. Recently, by adding constraint information such as must link (ML) and cannot link (CL) [6], Semi-supervised clustering for short text via deep representation learning [7, 8] has been proposed, in which texts are represented as distributed vectors with neural networks, and a small amount of labelled data is used to specify the representation for clustering. The representation learning process and the K-Means clustering process are combined, and the objective is optimized with both labelled data and unlabelled data iteratively.

However, the STCC model just trains the representation of short text via unsupervised information, which will lead to bad performance once the dataset has much noise. For a semi-supervised deep presentation model, it is very hard to determine how many constraints need be added to each dataset, and labelling the dataset also wastes a lot of human labour.

Unlike traditional social media data, the short message data are much more flexible. Although a short message has less text information, a great deal of external information such as time-related information, location information, and even comments and users' relationship in the social network is available.

In our approach, we project the short text and its comments in a low-density representation via CNN network firstly. Then, we calculate the similarity between each short text and other comments via the representation. Next, we consider the short message to be strongly semantically related to its comments, so we take the similarity difference between related comments and unrelated comments as loss to train the CNN network. Finally, the representation is expanded with multidimensional features extracted from location, time and other information. Using such a strong representation to go through the text clustering is superior to the traditional representation, as well as other deep representation methods: STCC and the approach of Wang et al. [7].

The contributions of our paper are as follows:

(1) Making good use of the advantages of short messages, we use the external information rather than text only. It helps a lot when going using an unsupervised clustering algorithm.
(2) Unlike other deep representation models, with the help of a strong relationship between comments and the original short text, we make good use of comment information to train our CNN model.
(3) By expanding the short message features with multidimensional information, we get a much more flexible representation for clustering.

2 Related Work

Traditional clustering algorithms such as K-Means [9, 10], hierarchical clustering [11] and DBSCAN [12] perform well in many text mining tasks [13, 14]. However, such simple models cannot provide good representation of the complicated and deep semantic information contained in short messages. Some works [11] aim to get the semantic information for short sentences, instead of using a BOWs model to represent a sentence; the new approach uses distribution over the entire set of documents. However, these approaches are trying to find the explicit semantic representation rather than the latent semantic information. Some breakthroughs have resulted from adding some a priori knowledge into the clustering algorithm, which changes this task into one of semi-supervised clustering. A graph-based method for incorporating prior knowledge has been proposed [15], in which all texts are modelled as a graph, where the nodes represent the text and the edges mean similarity among them. By adding a new edge which means constrictions between the two texts, we can run a clustering algorithm on the graph. All of these approaches improve the text clustering results. Unfortunately, these methods have certain limitations:

- Lack of flexible information to represent short texts, with the result that it is hard to get the semantic information with explicit words.
- Even though prior knowledge does improve clustering results, in many situations we cannot get the proper constraints and information.

So getting the deep semantic representation using unsupervised learning rather than the explicit meanings of words is much more important for short message representation. The latent semantic analysis (LSA) is the most popular algorithm to get latent semantic information. Besides, generative topic models, such as probabilistic latent semantic analysis (PLSA) and latent Dirichlet allocation (LDA) [14], are also used. However, these methods are aimed at obtaining shallow semantic information. With the development of deep learning, many methods have been proposed to learn a representation of text by end-to-end method. We can use a convolutional-pooling structure to get the contextual representation of a text [16]. In terms of representation-based clustering algorithms, although we can get a representation with deep neural network end-to-end, this unsupervised representation is less helpful than we expect. Adding information such as constriction and semantic relationships will lead to a good representation. DSSM [17] and CLSM [18] are good architectures in the information retrieval (IR) field, which use supervised information to get a good representation. Similar to these architectures, we can use the external data of short messages as supervised information with which to train a good representation, and then feed the vector into a clustering algorithm.

3   Deep Semantic and Multidimensional Representation for Clustering

This section describes how to build our model and get the good representation of short messages. There are two main parts of our model; Sect. 3.1 describes how to use a deep semantic model with a convolution-pooling structure to get a semantic representation of short messages, and Sect. 3.2 describes how to train and configure the semantic extract model. Then, Sect. 3.3 is a feature expanding by using multidimensional information such as time and location. The model is aimed at getting much more flexible representation using multidimensional information, and furthermore, the representation contains deep semantic information which is a great and strong information for the short message. Finally, Sect. 4 describes how to achieve short message clustering by using such a deep semantic and flexible representation.

3.1   Semantic Representation with Convolutional-Pooling Structure

For text information, a short message is similar to a short text. We cannot get the semantic information contained in the message by using only the few words, because the text is non-structured and much more complicated, with deep semantic meaning. Although with the help of a complicated knowledge base, we can get the semantic information, there is still a gap between the deep semantic meaning and the meaning shown in the text words. In order to get the deep semantic representation, given that a simple BOWs model cannot represent deep semantic information, we turn to the convolutional-pooling neural network (CNN). Additionally, by considering that

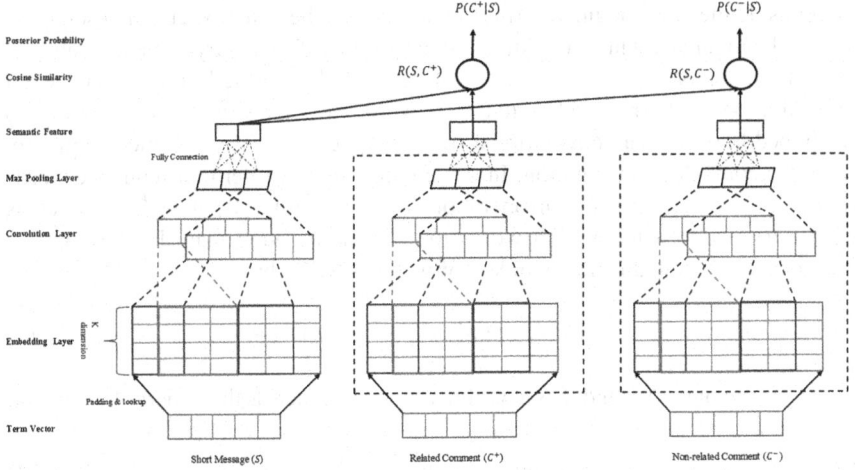

Fig. 1 Illustration of the convolutional-pooling semantic network

short messages and their comment messages are semantically related, we can train the CNN model output as a supervised problem.

By using a deep neural network with a convolutional-pooling structure, we can obtain content which contains semantic information. Besides, mapping the text features into the feature vectors in a semantic space, and we can use the relationship between the short message and the comments to get a good representation with strong semantic relationship. Some comments are closely related to the short message itself, so we can think of the two as being positively related. Other comments are less closely related, so we set them as the negative relations. The deep semantic model to get the representation can be described as follows:

(1) Using a convolutional-pooling structure neural network, we can map a term vector to its semantic representation for short messages and other comments, respectively.

(2) The relevance score between a short message and its comment is computed as the cosine similarity of their corresponding semantic representation vectors.

The architecture of the model is shown in Fig. 1. For any textual information of short messages (the short message and its comments), we can get a term vector by using a hash mapping. Firstly, the input is a term vector and is transformed the term vector into a fixed shape matrix by using pre-trained word embedding. Next, with the different sizes of a slides window, the convolutional layer can extract the contextual features for each word with its neighbours. Third, a max-pooling layer that discovers and combines salient word-n-gram features is used to form a fixed-length sentence-level feature vector. Finally, a semantic layer extracts a high-level semantic feature vector for the input word terms by using a fully connection. More details are shown in Fig. 1.

Let us define X as the input term vector, and y as the output vector in a semantic space. The input is a sentence with length n (padding if necessary) which is mapped into a term vector, $X = \{x_1, x_2, \ldots, x_n\}$, let $x_i \in \mathbb{R}^k$ be the k-dimensional word embedding vector corresponding to the ith word. After mapping the word term into the embedding layer, a convolutional over word sequence is performed implicitly to get the contextual information. In general, let the $X_{i:j}$ refer to a sequence words $x_i, x_{i+1}, \ldots, x_j$. A convolution operation involves a filter $W_c \in \mathbb{R}^{h \times k}$, which is applied to a slide window of h words to produce a feature map. For example, a feature c_h is generated from a window of words $X_{i:i+h-1}$ by

$$c_h = f(W_c \cdot X_{i:i+h-1} + b) \tag{1}$$

where b is the bias term and f is a nonlinear function such as the hyperbolic tangent:

$$\tanh(x) = \frac{1 - e^{-2x}}{1 + e^{-2x}} \tag{2}$$

A filter with window size h is applied to each possible context of words in the sentence $\{x_{1:h}, x_{2:h+1}, \ldots, x_{i:i+h-1}, \ldots, x_{n-h+1:n}\}$ to produce a feature mapping $C = \{c_1, c_2, \ldots, c_{n-h+1}\}$. Then, a max-pooling operation over the feature map takes the maximum value $\hat{c} = \max(C)$ as the feature corresponding to this particular filter. The idea is to extract the most important feature (the one with highest value) for each feature map. This pooling scheme naturally deals with variable sentence lengths.

For each filter with different window sizes and different weights for W_c, we repeat the convolution-pooling structure described above, and we can get the global feature after the max-pooling $v = \{\hat{c_1}, \hat{c_2}, \ldots, \hat{c_m}\}$, one more nonlinear transformation layer is applied to extract the high-level semantic representation,

$$y = \tanh(W_s \cdot v) \tag{3}$$

where W_s is the semantic projection matrix, and y is the representation of the input sentence in the latent semantic space, with a fixed dimensionality of L. The projection from the max-pooling layer to the semantic layer uses a fully connection.

3.2   Training the CNN Semantic Network

After the semantic representation for each text input is obtained, a deep semantic representation is formed. However, the use of this semantic vector to run a clustering algorithm constitutes completely unsupervised learning. By using the comment information of short messages, we can build a supervised training framework for their semantic representation.

Considering that in a short message, the comments are strongly semantically related to the short message, we can use this as a form of supervised information to train our representation. For instance, using a framework such as DSSM and CLSM for information retrieval, we calculate similarity between a short message and its comments, where the representation of short message and its comments on the semantic space, by using the cosine similarity,

$$R(A, B) = \cos(y_A, y_B) = \frac{y_A^T \cdot y_B}{\|y_A\| \|y_B\|} \tag{4}$$

The relevance between text A and text B is $R(A, B)$, where y_A, y_B is the semantic vector for text A and text B, respectively. In general, let the S corresponding the short message, and C^+ denotes the comment. We can transform the relevance score between the text S and its' comment C^+ to the posterior probability through a softmax function:

$$P(C^+|S) = \frac{\exp(\gamma R(C^+, S)}{\sum_{C' \in C} \exp(\gamma R(C', S))} \tag{5}$$

where γ is a smoothing factor in the softmax function, which is set empirically on a held-out dataset in our experiment, C denotes the comments set for all short texts, and we do not need to use all comments to calculate posterior probability. The hyper-parameter J is used to choose exactly J comments for calculation. For each short message S, we choose one comment $C^+ = C_1$ as a semantic positive sample, and $C^- = \{C_2, C_1, \ldots, C_J\}$ as the semantic negative samples, which are randomly chosen from all the other comments by the number $J - 1$. In this work, there is no research on how to choose the number of samples, and we treat J as the parameter for all the experiments.

Semantic relationships between short messages and their comments are well-suited to the use of supervised learning to train an accurate representation. In training the model to extract semantic vectors, model parameters are learned to maximize the likelihood of the comments by given the short message across the training set. In this situation, we minimize the loss function:

$$Loss = -\log \prod_{(S, C^+)} P(C^+|S) \tag{6}$$

In practice, all the weights in the network are initialized randomly. We choose the RMSProp algorithm as an optimizer to apply the gradient. The experiments were finished using a Nvidia Tesla K20 GPU with Tensorflow 1.0 framework.

3.3 Expansion Representation with Multidimensional Features

Being a kind of social media information, short messages are is time sensitive and regionally related. Definitions of these terms are given below:

- Time sensitive: two short messages are much more topic related if they have both been sent within a short time interval.
- Regionally related: two short messages are much more topic related if they are in the same region.

For example, Table 1 provides some examples of short messages with time and regional relationships. As the example shows, sentence 1 and sentence 2 differ greatly in terms of the explicit words used in each; we cannot learn the subject of these two sentences. Learning representation from the textual information is useless, but the location information can help us reach the subject and learn the relationship between this two short messages. In this situation, we use a hash map to encode and decode the location information. Actually, if we use the one-hot vector to represent the location feature, a large sparse vector will be obtained. So we use the binary code to encode and decode a location.

In terms of time sensitivity, deep semantic information cannot be learned from textual information, even by using a deep neural network. In the example of sentence 3 and sentence 4, it is very hard to learn the semantic relationship between "连续三年" (consecutive three years) and "第三季"(season 3), but by considering that the time

Table 1 Time and regional relationships in short messages

	SID	Short message example
Good regional related	1	Although it is less beautiful without the rape flower, more demure is shown; location at Wanfenglin; time at 20170503
	2	River is just like a silk, and the mountain is like a jade hairpin; location at Wanfenglin; time at 20170503
Good time sensitive	3	Golden State of Warriors (GSW) fight with Cleveland Cavaliers (CLE) season 3; location at Beijing; time at 20170518
	4	It is the third time that the NBA final is the GSW and CLE; location at Beijing; time at 20170518
Bad regional related	5	Take a walk to the Tiananmen Square; location at Tiananmen Square; time @ 20170510
	6	It's very excited to see the flag-raising; location at Tiananmen Square; time @ 20170510
Bad time sensitive	7	It's a good day to celebrate with your lover; location at Shanghai; time at 20170520
	8	I'm very sad for bad results about the examination; location at Shanghai; time at 20170520

interval between the two short messages is short and also that they are in the same field, we can reach the fact of their being semantically related. So considering the time information associated with short messages as a strong feature is reasonable.

However, even multidimensional information is represented as feature vector, such as the time- sensitive and regional-related information, which may also lead to bad clustering performance. For example, sentence 5 and sentence 6 show a bad case with the location feature in clustering. Sentence 7 and sentence 8 in Table 1 show bad performance when using the time feature for clustering. In these situations, it is implied that if getting the time interval and location information as feature representation, respectively, we cannot get the entire information for the short text, and we must use the semantic and textual information as the main feature representation.

To avoid these bad cases, we use the location and time information to expand the representation of short message. By using this external information of short messages, we can obtain a more flexible and complete representation.

4　Experiments

In this section, we describe our experiment about short message clustering by using deep semantic and multidimensional representation. Section 4.1 introduces the dataset and the evaluation method used. Additionally, the way in which the data were split into the training part (used to train the deep semantic network) and the clustering part is also shown in this section. Then, Sect. 4.2 shows the clustering result, where we compare our deep semantic representation of the three baseline methods: BOW, TF-IDF and average word embedding vector. Finally, Sect. 4.3 shows the experiment on the dataset with and without the multidimensional feature expanding.

4.1　Dataset and Evaluation Method

The short messages in Weiblog have a characteristic: if the short messages contains the format such as "#xxx#" in its text information, which implies that the short message is discussing the topic "xxx". More details are given in Table 2. In particular, we can use this characteristic to extract the labelled information from Weiblog. A set of short messages with the same topic information are in the same cluster in some aspects. In this way, we select ten hot topics in the social media and crawl for data related to these ten topics. The ten topics are as follows: "NBA总决赛" (NBA Final), "北京马拉松" (Beijing Marathon), "万峰林" (Wanfenglin) and so on. These topics were carefully selected to ensure there were many short messages for each topic, and that the short messages related to one topic are of high cohesion. In this way, we can get prior knowledge of whether the K value in K-Means algorithm is 10.

We crawl the data in Weiblog dated between May 1, 2016, and May 21, 2017. A short message filter was used to go through the dataset, and simply keeping short

Table 2 Example short messages' topics in Weiblog

Topic	Short message
NBA Final	I can't wait for the battle on June 2nd #NBA Final#; location at Beijing; time @ 20170518
NBA Final	I'm very excited about the battle on June 2nd #NBA Final#; location at Taiyuan; time @ 20170520
Beijing Marathon	You never know how happy I am when 30 thousand people across the Tiananmen Square #Beijing Marathon#; location at Beijing; time at 20160923
Beijing Marathon	It's hard for me to finish the whole distance for Marathon, so I just finish a half distance #Beijing Marathon#; location at Beijing; time at 20160923

messages with "#xxx#" topic labels corresponding to one of the chosen ten topics. Besides, we also extracted the time and location information from the original dataset, and each short message was kept if and only if it contained textual information, comments (keeping the most popular comment if there were more than one), time, and location information. We discarded the video information and emotion icons, to ensure semantic representation could be extracted from each short message. After the special filtering, we collected 18,736 short messages which satisfied all of the conditions previously outlined, each of which could be represented as a five-tuple: (topic id, short text, comment, time, location).

In order to get good representation in semantic space, we split the dataset into two parts, the first of which was used to train the supervised convolutional-pooling structure network to get the proper semantic representation (see Sect. 3.2). The second was used for clustering. We used 10,736 short messages for training semantic representation and 8,000 for clustering.

The cluster performance was evaluated with adjusted mutual information (AMI). As the K-Means algorithm is much influenced by the initial centres of clustering, so we use the average of 10 trials as the final score.

4.2 Deep Semantic Representation and NLP Feature Performance

In a training set with 10,736 samples, we split 20% of the dataset off to be used as the validation set to avoid issues with overfitting. When training the deep semantic representation (see Sects. 3.1 and 3.2), we used the embedding layer to lookup the word vector for each word in the text sequence. In all experiments, we used two pre-trained word embedding vector by word2vec toolkit (genism in python) and Fast-Text (open source) on the external corpus (all the articles in People's Daily over a four-month period) with dimension equal to 300. Because the dataset is in Chinese, we need to tokenize the data first. In our experiments, ICTCLAS is used to preprocess the data.

Table 3 Fixed parameters of the deep semantic network

Parameter	Value
N, the padding length of a input sentence	20
Filter size	[5, 7, 17]
Filter number	32
Semantic representation vector dimension	64
γ, the smooth factor in the softmax function	0.05

Table 4 Performance of all representation methods

Representation	AMI	Acc. (%)
BOW	0.022	13.45
TF-IDF	0.030	14.34
Average embedding (word2vec)	0.135	14.98
Average embedding (fast text)	0.138	15.03
Deep semantic (J = 4)	**0.420**	**49.54**
Deep semantic (J = 10)	0.408	48.44

The parameters in the semantic neural network architecture are shown in Table 3, and we can get a good semantic representation for the short message.

The baseline methods are also based on the K-Means algorithm, but they represent the short texts differently:

- **BOW**: represents a short text as a bag of words vector.
- **TF-IDF**: represents a short text as a TF-IDF vector.
- **Average word vector**: represents short text by calculating the average value in each dim of the word-embedding vector.

Deep semantic representation is much better than these traditional NLP representations, as shown in Table 4.

4.3 Comparison with Other Deep Representation Models

Unlike other deep learning representation models for short text clustering, by using the relationship between comments and the original short text, we use the loss Eq. 6 to train our model. This leads to stronger semantic representation that is shown in other deep models. We take two deep neural models as the baselines:

STCC: Short Text Clustering via Convolutional Neural Networks [5]. They embed the original features into compact binary code with a locality-preserving constraint and train the CNN output vector to fit the binary code.

Semi-supervised STC: Semi-supervised Clustering for Short Text via Deep Representation Learning (semi-supervised STC) [7]. Trains a good representation through

Table 5 Compare with other deep representation methods

Representation	AMI	Acc. (%)
STCC	0.407	48.38
Semi-supervised STC: add constraints (only ML)	0.401	48.25
Semi-supervised STC: add constraints (only CL)	0.381	47.71
Semi-supervised STC: add constraints (both)	0.411	48.89
Deep semantic	**0.420**	**49.54**

Table 6 Performance with/without multidimensional features

Representation	AMI	Acc. (%)
Deep semantic	0.420	49.54
Expand time	0.454	51.10
Expand location	0.460	51.00
Expand time and location	**0.487**	**52.34**

combining the CNN model and K-Means clustering, together with labelled data (Table 5).

4.4 Multidimensional Feature Expansion Performance

Using the binary code to encode a location into a vector, we can expand the representation with regional information. Similarly, we can use the time interval to encode a time into a vector. We performed some experiments to show how these additional dimensions influence our final clustering results. The final result is shown in Table 6.

By analysing the results, we reach the conclusion that expanding the multidimensional feature does improve the clustering result, although it will bring some noise into the representation.

5 Conclusion

In this paper, we use the relationship between comments and short texts, and treat comments as a positive document for short texts, using a CNN to extract the deep semantic representation from short texts. Firstly, we get the deep representation of short texts and their comment. Second, we calculate the similarity between the representation vector of short texts and its comment, and we use the margin maximizing loss function to train the CNN. In this way, we feed the representation of short messages into a K-Means clustering algorithm, which are training in a supervised fashion. Furthermore, our approach uses location and time information of short messages to expand the feature representation of our model.

References

1. Collobert, R., Weston, J., Bottou, L., Karlen, M., Kavukcuglu, K., Kuksa, P.: Natural language processing (Almost) from scratch. J. Mach. Learn. Res. **12**, 2493–2537 (2011)
2. Xing, E.P., Jordan, M.I., Russell, S.J., Ng, A.Y.: Distance metric learning with application to clustering with side-information. In: Proceedings of Advances in Neural Information Processing Systems, pp. 521–528 (2003)
3. Ying, Y., Li, P.: Distance metric learning with eigenvalue optimization. J. Mach. Learn. Res. 1–26 (2012)
4. Gabrilovich, E., Markovitch, S.: Computing semantic relatedness of words and texts in Wikipedia-derived semantic space. In: Proceedings of IJCAI, vol. 7, pp. 1606–1611 (2006)
5. Xu, J., Wang, P., Tian, G., et al.: Short text clustering via convolutional neural networks. In: Proceedings of NAACL-HLT, pp. 62–69 (2015)
6. Bair, E.: Semi-supervised clustering methods. Wiley Interdiscip. Rev. Comput. Stat. **5**(5), 349–361 (2013)
7. Wang, Z., Mi, H., Ittycheriah, A.: Semi-supervised clustering for short text via deep representation learning. CoNLL **2016**, 31 (2016)
8. Wang, Z., Mi, H., Ittycheriah, A.: Semi-supervised clustering for short text via deep representation learning (2016). arXiv:1602.06797
9. Hartigan, J.A., Wong, M.A.: Algorithm AS 136: a k-means clustering algorithm. J. Roy. Stat. Soc. Ser. C (Appl. Stat.) **28**(1), 100–108 (1979)
10. Pelleg, D., Baras, D.: K-means with large and noisy constraint sets. In: European Conference on Machine Learning, pp. 674–682. Springer, Berlin, Heidelberg (2007)
11. Johnson, S.C.: Hierarchical clustering schemes. Psychometrika **32**(3), 241–254 (1967)
12. Ester, M., Kriegel, H.P., Sander, J., Xu, X.: Density-based spatial clustering of applications with noise. In: International Conference on Knowledge Discovery and Data Mining, vol. 240 (1996)
13. Bilenko, M., Basu, S., Mooney, R.J.: Integrating constraints and metric learning in semi-supervised clustering. In: Proceedings of the Twenty-First International Conference on Machine Learning, p. 11. ACM (2004)
14. Vinh, N.X., Epps, J., Bailey, J.: Information theoretic measures for clusterings comparison: is a correction for chance necessary? In: Proceedings of the 26th Annual International Conference on Machine Learning, pp. 1073–1080. ACM (2009)
15. Ji, X., Xu, W.: Document clustering with prior knowledge. In: Proceedings of the 29th Annual International ACM SIGIR Conference on Research and Development in Information Retrieval, pp. 405–412. ACM (2006)
16. Kim, Y.: Convolutional neural networks for sentence classification (2014). arXiv:1408.5882
17. Huang, P.S., He, X., Gao, J., Deng, L., Acero, A., Heck, L.: Learning deep structured semantic models for web search using clickthrough data. In: Proceedings of the 22nd ACM International Conference on Information & Knowledge Management, pp. 2333–2338. ACM. (2013)
18. Shen, Y., He, X., Gao, J., Deng, L., Mesnil, G.: A latent semantic model with convolutional-pooling structure for information retrieval. In: Proceedings of the 23rd ACM International Conference on Conference on Information and Knowledge Management, pp. 101–110. ACM (2014)

The Use of Kernel-Based Extreme Learning Machine and Well-Known Classification Algorithms for Fall Detection

Osman Altay and Mustafa Ulas

Abstract Fall is a common occurrence in older people and causes dangerous consequences. Therefore, the fall detection needs to be done quickly and efficiently. In our developed expert system, three different algorithms have been tried to find fall detection by classification method. Two of these algorithms are the widely used naïve Bayesian and k-nearest neighbor algorithms for fall detection. The other used algorithm is kernel-based extreme learning machine algorithm. The main purpose of this study is to compare these three algorithms in detecting falls. As a result of the work, the kernel-based extreme learning machine algorithm has been shown to find the fall detection with better results than the k-nearest neighbor and naive Bayesian algorithms. The kernel-based extreme learning machine algorithm achieves a better F-measure value of 3.5% according to the k-nearest neighbor algorithm ($k=1$), 1.8% according to the k-nearest neighbor algorithm ($k=5$) and 6.6% according to the naïve Bayesian algorithm.

Keywords Fall detection · Classification · Kernel-based extreme learning machine · Machine learning

1 Introduction

Fall is a condition that can be frequent in older people and can lead to dangerous consequences. More than one in three people over 65 years of age are exposed to injuries that are not fatal due to falls at least once a year [1]. Apart from older people, people with visual impairment, gait disorders due to orthopedic reasons, balance problems due to various reasons, neurological and psychological disorders are faced with the problem of falling. Factors that can cause fall can be divided into two:

O. Altay (✉) · M. Ulas
Department of Software Engineering, Firat University, Elazig, Turkey
e-mail: oaltay@firat.edu.tr

M. Ulas
e-mail: mustafaulas@firat.edu.tr

© Springer Nature Singapore Pte Ltd. 2019
S. K. Bhatia et al. (eds.), *Advances in Computer Communication and Computational Sciences*, Advances in Intelligent Systems and Computing 760, https://doi.org/10.1007/978-981-13-0344-9_12

Internal factors and external factors. Among the internal factors associated with falls are aging, mental weakness, neurological and orthopedic disturbances, visual and balance disorders. External factors include multiple drug use, slippery floors, poor ground, unstable carpets, electrical and power cabling, ladders, and obstacles in the outdoor environment [2].

Wireless networks and low-power technologies have shown great improvement in recent years. Thus, the relevant circumstances to human health are being monitored by wearable technology products. With all these developments, falls detections have become important to prevent people from suffering from falls. The accelerometer has been used in a variety of research and applications. The accelerometer has been used for a variety of purposes in human movements such as metabolic energy expenditure, levels of physical activity, balance and posture sway, gait measurement and falls detection [3]. The biggest reason for the fall is sudden changes in the mass center of the human body [2]. By the World Health Organization, fall is defined as an event that causes a person to tumble down or wobble as a result of unwanted, unexpected, and uncontrollable events [4].

There are many studies related to fall detection in the literature. These studies can be divided into using wearable technology and camera systems. When working with wearable sensor, A. K. Bourke and G. M. Lyons used a threshold-based algorithm in their work [5]. In another study, support vector machine algorithm was used [6]. In a study by Jay Chen et al., they proposed a new wearable sensor for fall detection [7]. While others aimed to detect falls with camera systems. A study by C. Zhang et al. proposed a new fall detection system using RGB cameras [8]. In a study by C. Roguier et al., the shape analysis method has been applied to real data sets and obtain better results than other image processing methods [9]. A voxel person and a fuzzy logic system have been used to obtain a flexible system [10]. Semantic and event-based transcoding algorithms were used to detect falls in a multiple environment [11]. In Vinay Vishwakarma et al., using camera system and uses a set of extracted features to analyze, detect and confirm a fall [12]. Using the Omni camera, with the image processing method has achieved more than 75% success for fall detection [13]. Machine learning algorithms have been used in almost all of the fall detection related studies. Among these algorithms, algorithms widely used in classification studies such as support vector machines, naïve Bayesian (NB) algorithm, k-nearest neighbor (KNN) algorithm, C4.5 decision trees, and dynamic time bending algorithm have been preferred.

In this study, fall detection has been tried to be found using classification algorithms. The situation to be detected is divided into two classes as the daily movements of the person and the falling state. Three different classification algorithms have been used for fall detection in this study. First and second algorithms are the most commonly used algorithms for fall detection, KNN and NB algorithms, and third algorithm is the kernel extreme learning machine (KELM) algorithm. The results obtained from the three algorithms have been discussed.

2 Algorithms and Methods

Three different algorithms have been chosen to find fall detection using classification method. NB and KNN algorithms, which are widely used for fall detection, and KELM algorithm have been chosen. To our knowledge, KELM algorithm has not been used previously for fall detection. These algorithms and other methods used in the study are described below.

2.1 K-Nearest Neighbors Algorithm

KNN algorithm was first described at the beginning of the 1950s [14]. KNN algorithm was originally given to large training sets but did not gain popularity because there was not enough computing power on the computers. KNN algorithm has gained importance after the 1960s with the increase of computing power on the computers [15].

KNN algorithm is one of the most easily understood and applicable algorithms within data mining algorithms. Since the distance calculation is based on the KNN algorithm, it is easier to implement the training algorithm on numerical data sets than on the training sets that contain categorical data.

The steps of the KNN algorithm are given below;

1. For new data, determine the distance of each line in the training set,
2. Sort the determined distances,
3. Get the smallest k values between the sorted distances,
4. Find the class that repeats the most in k values and assign new data to this class.

Euclidean.
There are different methods for calculating the distances in the KNN algorithm. Euclid has been used in this study for calculating the distance between two objects. Euclidean distance is calculated by Eq. (1).

$$Euclidean_{i,j} = \sqrt{\sum_{k=1}^{n} \left(x_{ik} - x_{jk}\right)^2} \qquad (1)$$

Z-Score Normalization.
Z-score normalization, average and standard deviation of data are calculated and normalized. Equation (2) is used to normalize the data.

$$x_i' = \frac{x_i - \bar{x}}{\sigma_x} \qquad (2)$$

where σ_x denotes the standard deviation of x values, and $\bar{x}$ denotes the arithmetic mean.

2.2   Naive Bayesian Algorithm

The Bayes theorem emerged from studies of the probability and decision hypotheses of Thomas Bayes, a clergyman who did not obey the rules of society in the eighteenth century [15]. NB algorithm uses statistical methods when performing the classification process. Assuming that, we do not know $X\{x_1, x_2, x_3, \ldots, x_n\}$ belong to which class and we have a training set named C and that have m classes $\{C_1, C_2, C_3, \ldots, C_n\}$ in this training set.

According to the Bayes theorem, the probability of occurrence of the C_i when the X occurs, is given in Eq. (3) [16].

$$P(C_i\backslash X) = \frac{P(X\backslash C_i)P(C_i)}{P(X)} \tag{3}$$

To get faster results, simplification can be done for the probability of $P(X)$. If the values of X are considered to be independent of each other, equation can be used after simplification. If, during this process, the properties are all independent, then Eq. (4) used [16];

$$P(X\backslash C_i) = \prod_{k=1}^{n} P(x_k\backslash C_i) = P(x_1\backslash C_i) \times P(x_2\backslash C_i) \cdots \times P(x_n\backslash C_i) \tag{4}$$

Equation (5) is used to find the X learn to belong to which class [16].

$$\arg max = \{P(X\backslash C_i)P(C_i)\} \tag{5}$$

2.3   Extreme Learning Machine Algorithm

ELM is due to the lack of faster learning algorithms in artificial neural networks. Principle of studying the ELM algorithm given a training set $\aleph = \{(x_i, t_i)\|x_i \in R^n, t_i \in R^m, i = 1, \ldots, N\}$, activation function $g(.)$ and D is taken as the number of hidden neurons; The function $g(x)$ is expressed by Eq. (6) [17].

$$f_D(x_j) = \sum_{i}^{D} \beta_i g(w_i \times x_j + b_i), \quad j = 1, \ldots, N \tag{6}$$

The Eq. (6) can be written $f(x) = h(x)\beta$ where w_i are the input weights and b_i is the bias value. The value of β_i is the output weight that connects the hidden node in the i'th iteration to the output node [17]. The matrix form of the equation is given Eq. (7) $H\beta = T$;

$$H = \begin{bmatrix} g_{(w_1 \times x_1 + b_1)} \cdots g_{(w_D \times x_1 + b_D)} \\ \cdot \\ \cdot \\ \cdot \\ g_{(w_1 \times x_D + b_1)} \cdots g_{(w_D \times x_N + b_D)} \end{bmatrix}$$

$$\beta = [\beta_1 \ldots \beta_N]' \ and \ [t_1 \ldots t_N]' \tag{7}$$

In order to remove the disadvantages of the recursive learning algorithm, in the ELM algorithm is proposed use of Moore–Penrose generalized inversed. It is shown in Eq. (8) [17].

$$\check{\beta} = \mathbf{H}^* \mathbf{T} \tag{8}$$

where $\mathbf{H}^*$, H Moore–Penrose has been applied and $\check{\beta}$ expresses the output weights.

Kernel-Based Extreme Learning Machine.
In the KELM algorithm, if the hidden layer property mapping $h(x)$ is not known by the user, the kernel matrix can be defined for the ELM by the following Eq. (9) [18];

$$\delta_{ELM} = HH^T : \delta_{ELM_{i,j}} = h(x_i) \times h(x_j) = K_{(x_i, x_j)} \tag{9}$$

Then, the output function of the ELM can be written as in the following Eq. (10).

$$f(x) = h(x)H^T \left(\frac{1}{c} + HH^T \right)^{-1} T = \begin{bmatrix} K(x, x_1) \\ \cdot \\ \cdot \\ \cdot \\ K(x, x_N) \end{bmatrix}^T \left(\frac{1}{c} + \delta_{ELM} \right)^{-1} T \tag{10}$$

We used to here radial basis function (RBF) for to compute K [18].

2.4 Performance Evaluation

There are many different methods in the literature to measure the performance of the classification model [19]. In this publication, we used evaluation measures and calculated accuracy (11), sensitivity (12), specificity (13), precision (14), and F-measure (15) for three different algorithms. Since the value of k is taken as two different values, it is calculated twice for KNN.

Accuracy;

$$accuracy = \frac{TP + TN}{TP + FN + FP + TN} \tag{11}$$

Sensitivity;

$$sensitivity = \frac{TP}{TP + FN} \qquad (12)$$

Specificity;

$$specificity = \frac{TN}{TN + FP} \qquad (13)$$

Precision;

$$precision = \frac{TP}{TP + FP} \qquad (14)$$

F-measure;

$$fmeasure = 2 \times \frac{Precision \times sensitivity}{precision + sensitivity} \qquad (15)$$

3 Expert System for Fall Detection

Expert system, a branch of artificial intelligence, was developed by artificial intelligence community in the mid-1960s. The simplest explanation of expert system is to transfer a job to a computer that an expert person can do.

In order for expert systems to work, the information obtained by experts in the field is stored in the computer environment according to certain rules. The expert systems allow the computer to deduce from stored information using various algorithms or to achieve a targeted result. The expert system designed for fall detection is shown in Fig. 1.

Fig. 1 Expert system for fall detection

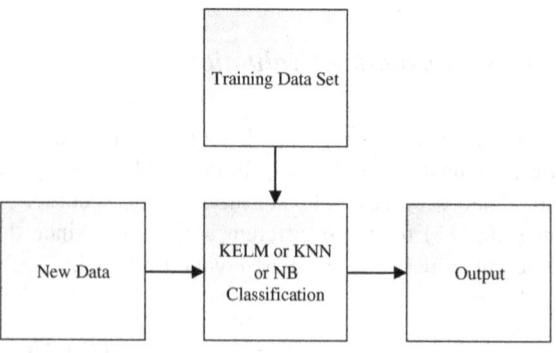

Fig. 2 Distribution of data within classes

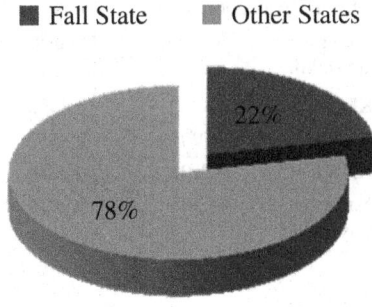

■ Fall State   ■ Other States

22%

78%

3.1 Data Acquisition

All data have been generated during an earlier experimental study. The six MTw sensor units that are part of the MTw Software Development Kit have been used [2]. The distribution of data within classes has been shown in Fig. 2. As you can see in Fig. 2, while 22% of the exits are fall state, the remaining 78% are other states.

3.2 Results and Discussion

KELM, KNN, and NB classifications have been used to realization the expert system. The first applied KELM is the design of the ELM algorithm. RBF kernel method is used in KELM algorithm. In the KNN algorithm, the distance between the data is calculated by Euclidean. For the second method, the KNN algorithm, z-score normalization was applied to the all data except for target data. Then k values were taken as 1 and 5 and different results were found in two values. Z-score normalization is applied to the data in the last method NB algorithm.

Four different results have been obtained at the end of the applications. The results are shown in Table 1. In the testing phase of the application, 70% (11,468) of all data have been set as training data set and the remaining 30% (4914) as test data set. Selected data for training and test data sets have been selected randomly. Then, accuracy, sensitivity, specificity, precision, and F-measure values are calculated for all the methods applied. It has been observed that the KELM classification better results than the KNN and NB classification.

When k value is taken as 1 in the KNN algorithm, sensitivity and specificity values are calculated as 0.8552 and 0.5640, respectively, while accuracy is 0.7904. F-measure value has been calculated as 0.8638. When the k value is taken as 5, the accuracy is 0.8101 and sensitivity and specificity values are calculated as 0.8840 and 0.5521, respectively. F-measure value has been calculated as 0.8786. In the last method NB classification, the sensitivity and specificity values have been calculated as 0.9016 and 0.1344, respectively, while the accuracy is 0.7308. F-measure value has been calculated as 0.8389. When the results are evaluated, it is observed that the

Table 1 Evaluation measures results

Using algorithms	Accuracy	Sensitivity	Specificity	Precision	F-measure
KELM	0.8260	0.8490	0.6729	0.9453	0.8946
KNN k = 1	0.7904	0.8552	0.5640	0.8726	0.8638
KNN k = 5	0.8101	0.8840	0.5521	0.8733	0.8786
Naïve Bayesian	0.7308	0.9016	0.1344	0.7843	0.8389

KELM algorithm relatively better results than the other two algorithms. Especially, sensitivity and specificity values can be considered better than others. These values of NB algorithm have achieved the worst results.

4 Conclusions

In this study, an expert system has been designed for fall detection. Three different classification algorithms have been used (KELM, KNN, and NB). One of these, KNN has been applied in two different ways ($k=1$ and $k=5$). F-measure value for the best performing KELM algorithm has been calculated as 0.8946. In the other two algorithms, F-measure has been calculated as 0.8638 for KNN algorithm ($k=1$), 0.8786 for KNN algorithm ($k=5$) and 0.8389 for NB algorithm. The KELM algorithm achieves a better F-measure value of 3.5% according to the KNN algorithm ($k=1$), 1.8% according to the KNN algorithm ($k=5$) and 6.6% according to the NB algorithm.

Fall in real life is instantaneous, so the detections must be so fast. Fall detections have to adapt to the real world and work on micro-controllers. Therefore, the algorithm to be used must have a low computational cost. Within the three tested algorithms, the KELM algorithm appears to be the best algorithm for detecting falls. The KELM algorithm increased the accuracy by 4.5% for KNN ($k=1$), 1.9% for KNN ($k=5$), and 13% for the NB algorithm.

References

1. Chen, J., et al.: Wearable sensors for reliable fall detection. In: 27th Annual International Conference of the Engineering in Medicine and Biology Society, IEEE-EMBS 2005 (2006)
2. Özdemir, A.T., Barshan, B.: Detecting falls with wearable sensors using machine learning techniques. Sensors **14**(6), 10691–10708 (2014)
3. Kochera, A.: Falls among older persons and the role of the home: an analysis of cost, incidence, and potential savings from home modification. Issue Brief (Public Policy Institute (American Association of Retired Persons)), (IB56), 1 (2002)
4. World Health Organization. http://www.who.int/en/. Accessed 23 July 2017

5. Bourke, A.K., Lyons, G.M.: A threshold-based fall-detection algorithm using a bi-axial gyroscope sensor. Med. Eng. Phys. **30**(1), 84–90 (2008)
6. Zhang, T., Wang, J., Xu, L., Liu, P.: Fall detection by wearable sensor and one-class SVM algorithm. Intelligent Computing in Signal Processing and Pattern Recognition, pp. 858–863 (2006)
7. Nyan, M.N., Tay, F.E., Murugasu, E.: A wearable system for pre-impact fall detection. J. Biomech. **41**(16), 3475–3481 (2008)
8. Zhang, C., Tian, Y., Capezuti, E.: Privacy preserving automatic fall detection for elderly using RGBD cameras. In: Computers Helping People with Special Needs, pp. 625–633 (2012)
9. Rougier, C., Meunier, J., St-Arnaud, A., Rousseau, J.: Robust video surveillance for fall detection based on human shape deformation. IEEE Trans. Circuits Syst. Video Technol. **21**(5), 611–622 (2011)
10. Anderson, D., Luke, R.H., Keller, J.M., Skubic, M., Rantz, M., Aud, M.: Linguistic summarization of video for fall detection using voxel person and fuzzy logic. Comput. Vis. Image Underst. **113**(1), 80–89 (2009)
11. Cucchiara, R., Prati, A., Vezzani, R.: A multi-camera vision system for fall detection and alarm generation. Expert Syst. **24**(5), 334–345 (2007)
12. Vishwakarma, V., Mandal, C., Sural, S.: Automatic detection of human fall in video. Pattern Recognit. Mach. Intell. 616–623 (2007)
13. Miaou, S.G., Sung, P.H., Huang, C.Y.: A customized human fall detection system using omnicamera images and personal information. In: 1st Transdisciplinary Conference on Distributed Diagnosis and Home Healthcare, D2H2, pp. 39–42. IEEE (2006)
14. Sillverman, B.W., Jones, M.C., Fix, E., Hodges, J.L.: An important contribution to nonparametric discriminant analysis and density estimation. Int. Stat. Rev. **57**(3), 233–247 (1951)
15. Han, J., Pei, J., Kamber, M.: Data Mining: Concepts and Techniques. Elsevier (2011)
16. Özkan, Y.: Veri Madenciliği Yöntemleri. Papatya, İstanbul (2013)
17. Huang, G.B., Zhu, Q.Y., Siew, C.K.: Extreme learning machine: theory and applications. Neurocomputing **70**(1), 489–501 (2006)
18. Huang, G.B., Wang, D.H., Lan, Y.: Extreme learning machines: a survey. Int. J. Mach. Learn. Cybernet. **2**(2), 107–122 (2011)
19. Sun, Y., Kamel, M.S., Wong, A.K., Wang, Y.: Cost-sensitive boosting for classification of imbalanced data. Pattern Recogn. **40**(12), 3358–3378 (2007)

Performance Test for Big Data Workloads on Various Emerging Memories

Cheng Qian

Abstract Mismatch of performance improvement between CPU and main memory (also called memory wall) is making main memory become an increasingly significant factor contributing to overall system performance. In recent years, the rising of big data applications even aggravates the memory wall problem. Thus, it is necessary to analyze the memory performance for different memory systems while running big data applications. In this paper, we test and analyze five different memory systems by running both conventional workloads and big data workloads. Through comparison and analysis, we make several conclusions to guide the design of main memory system for big data workloads.

Keywords Big data · Emerging memories · Performance test · Simulation

1 Introduction

Big data era has come, which also changes the processing mode of the current computer. As the data size to be processed increases quite rapidly, the memory system in a computer becomes the bottleneck, which is so-called memory wall. There are lots of researches focusing on relieving the memory wall problem. For instance, Hybrid Memory Cube, Micron (HMC) is such a 3D device to improve memory performance. Recently, new version of HMC (HMC 2.0) even integrates PIM logic unit [12], which can support part of atomic instructions. Multiple storage unit stacks on a logic unit linked by thousands of trough silicon via (TSV) making latency shorter and capacity larger. This novel technology relieves the memory wall problem in some ways.

C. Qian (✉)
State Key Laboratory of High Performance Computing,
National University of Defense Technology, Changsha 410073, China
e-mail: qiancheng@nudt.edu.cn

© Springer Nature Singapore Pte Ltd. 2019
S. K. Bhatia et al. (eds.), *Advances in Computer Communication and Computational Sciences*, Advances in Intelligent Systems and Computing 760, https://doi.org/10.1007/978-981-13-0344-9_13

157

Another promising method is the usage of non-volatile memory. Having attractive data-retention ability and not bad data access latency, non-volatile memory materials provide strong support to the current widely used DRAM material. Currently non-volatile memories are hotspots to research. Non-volatile memories helps to improve overall performance by acting as secondary storage or integrating with DRAM as memory hierarchy system.

There are several works evaluating the performance that conventional workload running on 3D DRAM or non-volatile memory integrated memory systems [9, 11], and these works definitely verity that novel 3D architecture and non-volatile memory do benefit to improve memory system performance. However, there are quite few works concentrating on evaluating memory performance. In big data era, memory system should have different features compared with the conventional system. Therefore, it is quite necessary to have a comprehensive evaluation on big data workloads running on novel memory architecture.

In this paper, we present our evaluation about running big data workloads on five different memory systems, namely 2D DRAM, 3D DRAM, PCM, RRAM, and STT-RAM. Our evaluations focus on several parameters to address the difference between conventional workloads and big data workloads; afterward we have a discussion, the trend of usage of 3D stacked and non-volatile technology on big data processing. From the experiment results, we obtain the following conclusion:

1. L2, L1D, L2I cache miss rate only have tiny variation for different memory systems when the capacities of memory systems are same.
2. As capacity of memory changes, the change of cache miss rates shows difference for conventional workloads and big data workloads. For conventional workloads, there is barely difference. For big data workloads, except using STT-RAM, there is inconspicuous reduction of cache miss rate.
3. Given the memory system has the same capacity, the value of IPC is closely related to the total bandwidth to/from main memory.
4. For big data workloads, given the capacity of memory system is limited, the impact that bandwidth has on IPC is different with conventional workloads.
5. The enlargement of capacity has limited effect on improving bandwidth and IPC. For conventional workloads and big data workloads, the reason is different.

The paper is organized as follows: Sect. 2 describes the background and related work; Sect. 3 presents the evaluation method and experimental environment; Sect. 4 illustrates the discussion based on the experiment result; Sect. 5 draws a conclusion.

2 Background and Related Works

2.1 Background

3D stacked memory is a promising technology that can efficiently relieve memory wall problem [11]. In the current commercial memory system, off-chip DDR3 memory channel is widely used. It is connected to the memory controller on the processor chip via PCI-e bus. As the processor pin counts near the scaling limits [6], it will lead to higher power consumption and limit per-pin memory capacity if just continuing boost processor pin bandwidth and also the memory capacity is limited. Naturally, expending the memory to the 3D level is a good way to improve not only capacity but also bandwidth via high-speed link (TSV). Micron cooperation has released a novel 3D memory device called Hybrid Memory Cube (HMC). In recent version HMC 2.0, even some atomic instructions are supported, although it is quite limited [12]. However, it shows that PIM is a promising way to save the bandwidth and power consumption.

Non-volatile memory is also a promising technology which has totally a different property compared with DRAM. DRAM has been used as the material of main memory for decades. Currently, DRAM is facing the scaling problem, while non-volatile memory can provide good support as its data-retention ability and not bad read/write latency. The kind the existing non-volatile memory is diverse. Several kinds of non-volatile memories are attractive to research, such as Flash, PCM, STT-RAM, RRAM and so on. PCM is the abbreviation of phase change memory. It is a non-volatile memory material comprised of chalcogenide glass. Though a heating element made of TiN either heat or quench the glass, it switches its condition as crystalline or amorphous. The unique cell programming method leads to relatively high write latency and power consumption. RRAM is a type of non-volatile random access computer memory that works by changing the resistance across a dielectric solid-state material often referred to as a memristor. RRAM has relatively low power consumption and quite good read/write latency. However, it is still under development currently and not largely commercialized [7]. STT-RAM is an advanced type of MRAM devices which uses a magnetic tunnel junction (MTJ) to store binary data. STT-RAM enables higher densities, low power consumption; the main advantage of STT-MRAM is the ability to scale the STT-MRAM chips to achieve higher densities at a lower cost [9].

As Table 1 shows the basic factors of these materials illustrate some different usage trend for different materials. For example, flash is now the most mature material for commerce. But in consideration of its large read/write latency, it is suitable for being used as secondary storage. Except flash, other materials show sweet properties, involving read/write latency, and so on. Therefore, in our experiment, we only omit flash and take others as the candidates.

Table 1 Different memory material comparison

Material	Read latency (ns)	Write latency (ns)	Volatile	Cell factor	Number of rewrites
2D DRAM	10–50	20–80	Y	6	10^{16}
3D DRAM	10–25	10–40	Y	6	10^{16}
Flash	104	10^5	N	2–5	10^4–10^5
PCM	40–70	100–300	N	4–10	10^8–10^{15}
RRAM	5–30	5–100	N	≤ 4	10^{15}
STT-RAM	10–40	10–40	N	6–40	10^6–10^{12}

Combined with these two technologies, some hybrid memory architectures have attractive properties [3, 15]. With appropriate strategies, stacking DRAM layers with NVM layers together on a logic layer can efficiently improve the memory system performance. This kind of architecture is widely used for custom applications such as graph traversal and so on. Through reasonably taking the advantage of both DRAM and NVM, such architecture can get outstanding speedup.

2.2 Related Works

There are several previous works focusing on performance test on various memory systems. Loh et al. explored more aggressive 3D DRAM organizations that make better use of the additional bandwidth that 3D stacked memory provides. The result shows that this architecture can achieve a 1.75X speedup on memory-intensive workloads on a quad-core processor [11]. Kim et al. present the results of their study of an all-PCM SSD prototype. As their result shows, PCM storage devices have promising performance for commercialize [8]. Lewis et al. survey the memristor that produced by a number of different research teams and present comparison between DRAM and RRAM. They find RRAM an interesting design alternative to DRAM technologies [10]. Kultuursay et al. explored the possibility of using STT-RAM to replace DRAM in main memory. The result indicates that an optimized, equal capacity STT-RAM can provide comparable performance to DRAM and reduce 60% energy [9].

There are also many works focusing on memory performance test for big data workloads. Dimitrov et al. based on the trends that big data technologies have rapid growth and high memory requirement; they examine how these trends intersect by characterizing the memory access patterns for big data workloads and discuss the implications on software and system design [4]. Clapp et al. present analytic equations to quantify the impact of memory bandwidth and latency on workload performance and illustrate how the values of the component in the equations to classify different workloads [2]. Park et al. explore the performance impact of large pages on in-memory big data workloads running on a large-scale NUMA system, the results

Table 2 Full system parameter

Processor	Out-of-order, Up to 8 microinstructions 192-entry ROB, 32-entry LS queue
L1 I/D cache	Private, 2-way, 32 KB/32 KB, 2 cycles
L2 cache	Shared, 8-way, 256 KB, 8 cycle

Table 3 Full system parameter

Material	Timing parameter
2D RAM	CLK: 666 MHz $tRCD = 9$, $tCAS = 10$, $tRAS = 24$, $tWR = 10$, $tCCD = 4$
3D RAM	CLK: 1333 MHz $tRCD = 9$, $tCAS = 10$, $tRAS = 24$, $tWR = 10$, $tCCD = 4$
PCM	CLK: 400 MHz $tRCD = 48$, $tCAS = 1$, $tRAS = 0$, $tWR = 60$, $tCCD = 2$
RRAM	CLK: 400 MHz $tRCD = 10$, $tCAS = 6$, $tRAS = 0$, $tWR = 4$, $tCCD = 2$
STTRAM	CLK: 400 MHz $tRCD = 14$, $tCAS = 6$, $tRAS = 0$, $tWR = 5$, $tCCD = 2$

show that large pages achieve higher performance gains because of reduction of the dataset size of the in-memory big data workloads [13].

3 Methodology

In order to have a comprehensive evaluation on the memory system, we adopt a full system simulation environment to evaluate the performance of big data workloads. We use full system simulator GEM5 [1] to simulate the CPU and cache module. In our experiment, we simulate an out-of-order CPU, attached with two-level cache hierarchy. The detail parameters are listed in Table 2.

We use NVMain simulator for the memory part [14]. The memory system is such a reconfigurable part that through fixing some timing parameters. The target architectures involve 2D DRAM, 3D DRAM, PCM, RRAM, STT-RAM. The detail timing parameters about memory configuration are shown in Table 3. These parameters are set according to the responding technique report. Note that the unit of measurement is cycles based on the memory's frequency.

We use Phoenix++ benchmarks as the big data workloads [16]. Phoenix++ is a C++ implement modular MapReduce framework that can achieve good performance, and people can write high-performance MapReduce code using this framework. It includes seven benchmarks in which we take six as the evaluation targets. These benchmarks are simple but need to process large amounts of data compared with conventional workloads. The titles and descriptions are shown in Table 4. These benchmarks are needed to be employed in an image and run using FS mode in

Table 4 Six workloads from Phoenix

Titles	Description
PCA	Computes the mean vector and covariance matrix for a randomly generated matrix
Kmeans	Groups several n-dimensional data points into a user-specified number of groups and finds the mean value of each group
Word count	Counts the frequency of occurrence of each unique word in a text document
Linear regression	Generates the summary statistics of points to give the linear approximation of all the points
String match	scrolls through a list of keys in order to determine if any of them occur in a list of encrypted words
Histogram	Generates the histogram of frequencies of pixel values in a bitmap picture

GEM5. By writing a script and the usage of checkpoint, we can avoid the effect of the operating system in the image. For every benchmark in Phoenix++, we run for 1 billion instructions to collect needful parameters.

Besides, we take 29 benchmarks from spec2K6 benchmark [5]. These conventional workloads are used as reference objects. The benchmarks were compiled using GCC/G++/GFortan 4.8.2 with O2 optimization. We use SE mode in GEM5 to run these benchmarks, and every benchmark are run for 250 million instructions to collect needful parameters. Through comparing the performance of these two different workloads, we can learn the different memory characteristic suitable for conventional workloads and big data workloads.

4 Evaluation

To test the different performance when running applications on different materials, we collect IPC, L2/L1D/L1I miss rate, and bandwidth. After careful analyses, we obtain the following conclusion.

1. L2, L1D, L2I cache miss rate only have tiny variation for different memory systems when the capacities of memory systems are same, i.e., different memory systems do not have large impact on the cache miss rate.

For SPEC benchmarks, Fig. 1 shows the L2, L1D, and L1I cache miss rate almost have no variation when running on different memory system. Setting the cache miss rate of 2DDRAM as baseline from Fig. 2, we obtain that even the largest variation, which appears when running on 3D DRAM memory system, is only 0.73%. In general, we can conclude that, for conventional applications, different memory architectures do not bring obvious improvement or degradation.

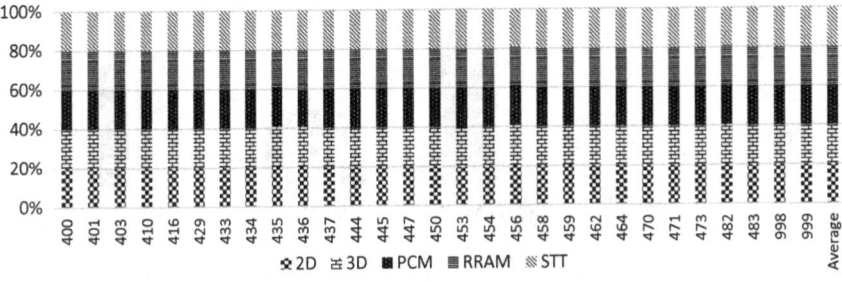

(a) L2 Miss rate Comparison for SPEC.

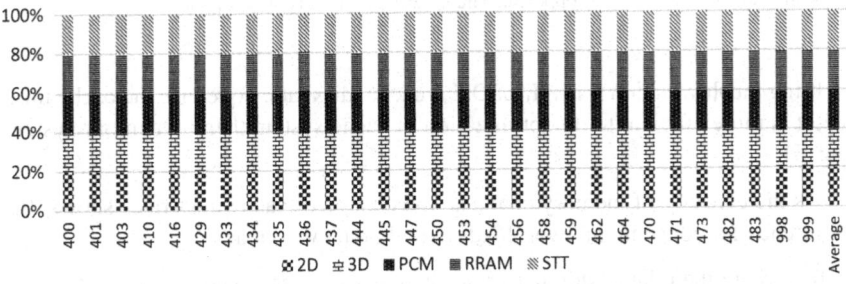

(b) L1D Miss rate Comparison for SPEC.

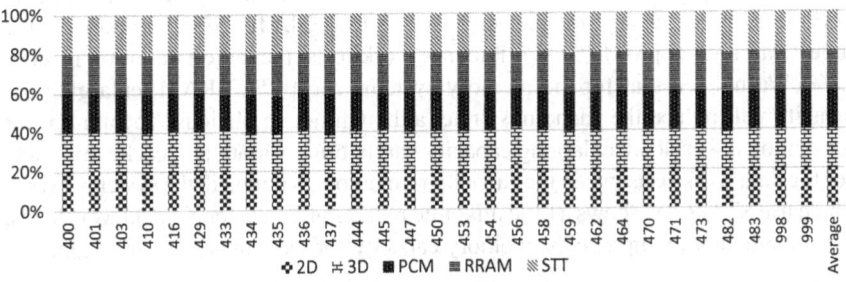

(c) L1I Miss rate Comparison for SPEC.

Fig. 1 L2, L1D, and L1I cache miss rate for SPEC. They almost have no variation when running on different memory system

For Phoenix, Fig. 3 shows that when running Phoenix benchmarks on different memory systems, the L1I, L1D, and L2 miss rate have small variation. Same as SPEC, we set the cache miss rate of 2DDRAM as baseline. From Fig. 4, we see that the largest variation reaches 3.8%. Such little variation should not be the reason for performance improvement or degradation. Thus, comparatively speaking, for big data workloads, different memory systems have relatively larger but very limited impact on cache miss rate than for conventional workloads. To sum up, for both conventional workloads and big data workloads, different memory architectures do

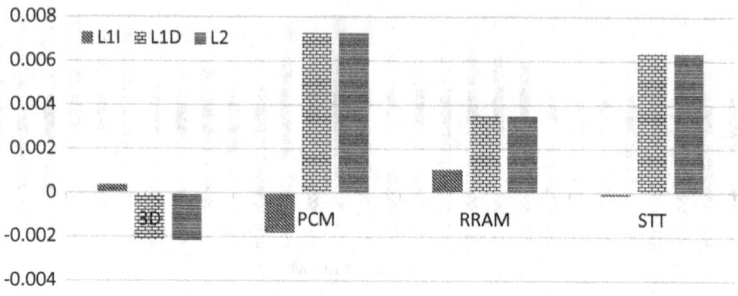

Fig. 2 Cache miss rate variation compared with 2DRAM. Positive value means larger cache miss rate while negative value means opposite. The largest variation is only 0.73%

not bring much variation for L1I, L2D, L2 cache miss rate. To reduce the cache miss rate, it is more efficient to do optimization in cache system than in memory system (Fig. 5).

2. As the capacity of memory changes, the change of cache miss rates shows difference for conventional workloads and big data workloads.

To explore the relationship between capacity of memory system and cache miss rate, we set capacity as 2 and 4 GB for comparison, and we set the value of 2 GB as baseline. Figure 6 shows the cache miss rate comparison for SPEC and Phoenix workloads. When change the capacity from 2 to 4 GB, for SPEC workloads, the largest variation is just 0.7%. For Phoenix workloads, the largest variation reaches 2.9%. We noticed that for the memory systems except STT-RAM, enlarging the capacity will reduce the cache miss rate, although not so obvious. To sum up, for conventional workload, enlarging capacity barely bring variation to cache miss rate. For big data workloads, when the capacity of memory is doubled, the systems except which use STT-RAM shows a little miss rate reduction. Thus, for big data workloads, it is not efficient to improve the memory capacity for reducing cache miss rate.

3. Given the memory system has the same capacity, the value of IPC is closely related to the total bandwidth to/from main memory.

We measured IPC and bandwidth for SPEC and Phoenix. From Fig. 5 we know that, both for SPEC and Phoenix, the IPC reaches the highest when running on 3D memory system (Consider the results of 2DDRAM as baseline, the improvement can reach 19.8% on average), and reaches the lowest when running on PCM memory system (10.8% lower than baseline). Also for bandwidth, the result of 3DDRAM is the highest (86.5% improvement) and the result of PCM is the lowest (40.6% reduction). The results of RRAM and STT-RAM show the same rules. As for Phoenix workloads, Fig. 7 illustrates the same property. Therefore, we can conclude that:

(1) IPC has strong positive correlation with bandwidth.
(2) 3D memory system is the most promising material due to that it can provide high bandwidth. In addition, we notice that when running on 3D memory system,

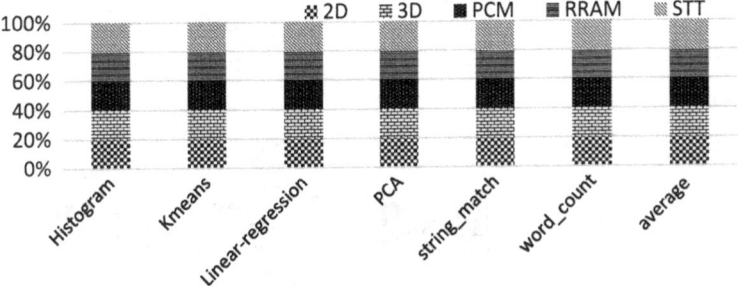

(a) L2 Missrate Comparison for Phoenix.

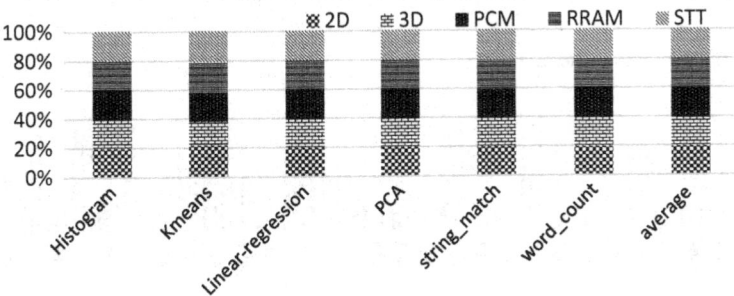

(b) L1D Missrate Comparison for Phoenix.

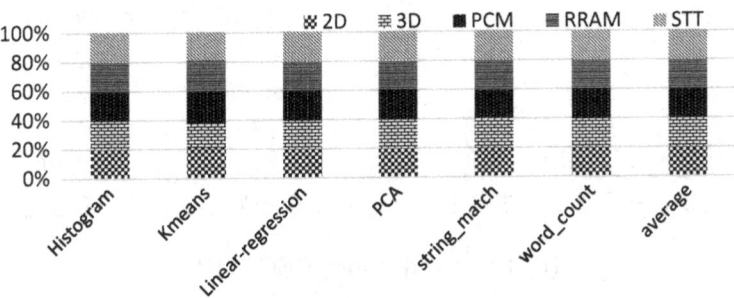

(c) L1I Missrate Comparison for Phoenix.

Fig. 3 L2, L1D, and L1I cache miss rate for Phoenix. Positive value means larger cache miss rate while negative value means opposite. The L1I, L1D and L2 miss rate have just quite small variation

the IPC improvement of Phoenix is higher than SPEC. We can infer that high-bandwidth 3D memory device like HMC and HBM can bring great performance improvement to big data workloads.

(3) Currently, individual NVM materials are not quite suitable for main memory system, both for running conventional workloads or big data workloads. Though they have very good data contention properties, not so good read/write latency may affect the overall performance. In addition, NVM materials usually do not

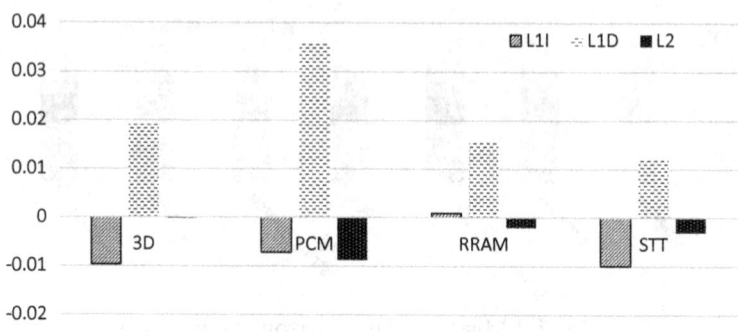

Fig. 4 Cache miss rate variation compared with 2DRAM for Phoenix. The largest variation is 3.8%

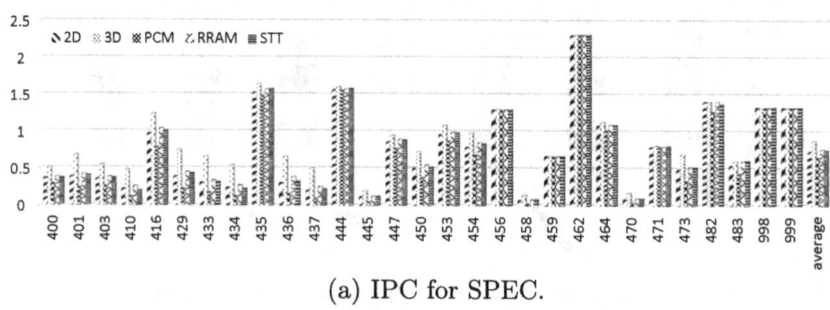

(a) IPC for SPEC.

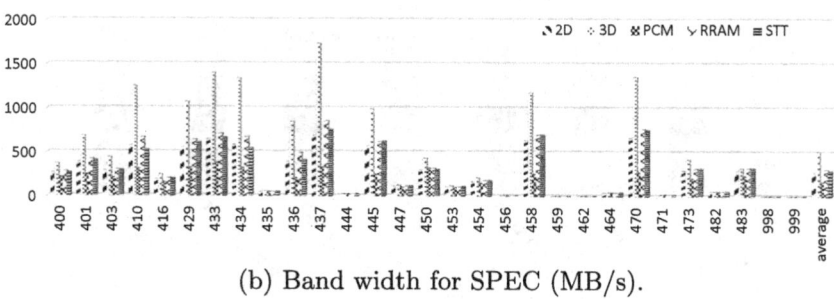

(b) Band width for SPEC (MB/s).

Fig. 5 IPC and bandwidth for SPEC. The IPC and bandwidth both reach the highest when running on 3D memory system and reach the lowest when running on PCM memory system

have advantage of bandwidth, which may be significant for IPC. It is a good way to combine NVM and DRAM as a hybrid memory system to take advantage of both DRAM and NVM.

4. For big data workloads, given the capacity of memory system is limited, the impact that bandwidth has on IPC is different with conventional workloads.

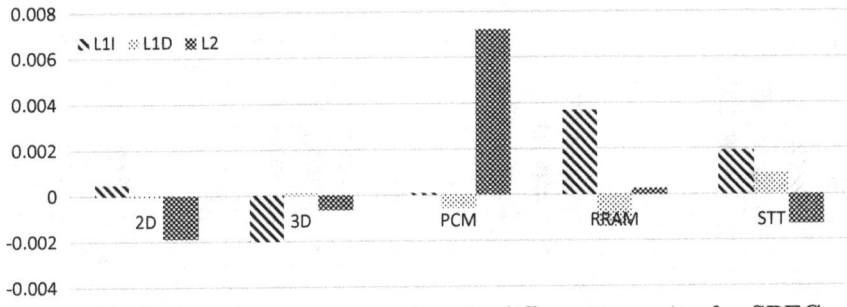

(a) Cache miss rate comparison in different capacity for SPEC.

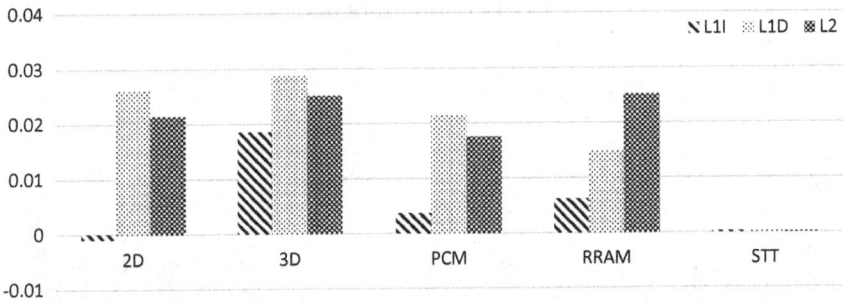

(b) Cache miss rate comparison in different capacity for Phoenix.

Fig. 6 L2, L1D, and L1I cache miss rate comparison in different capacity for SPEC and Phoenix. Positive value means smaller cache miss rate while negative

Setting 2DDRAM as baseline, the IPC and bandwidth tendency comparison for SPEC and Phoenix are shown in Fig. 8. Same as illustrated above, the IPC and bandwidth have same tendency for both SPEC and Phoenix. However, the effect degree is different for these two workloads. We noticed that, compared with baseline, when the bandwidth has positive effect, the impact on the IPC of Phoenix is more obvious than on the IPC of SPEC (i.e., for Phoenix, 53.8% bandwidth improvement brings 30.8% IPC improvement; for SPEC, 86.5% bandwidth improvement brings 19.8% IPC improvement). And when the bandwidth becomes bottleneck, the negative effect on big data workloads is less obvious than on conventional workloads (i.e., for Phoenix, 42.3% bandwidth reduction only makes 2.9% IPC improvement; for SPEC, 40.6% bandwidth improvement brings 10.9% IPC improvement). Therefore, we obtain the following conclusions:

(1) For big data workloads, the bandwidth is the leading bottleneck that affects performance. Improving the bandwidth can have outstanding improvement in workloads performance. Thus, high-bandwidth memory device such as HMC is very suitable for running big data workloads.

(2) For conventional workloads, the bandwidth also affects the performance, but less significant than big data workloads. Therefore, it is necessary to combine

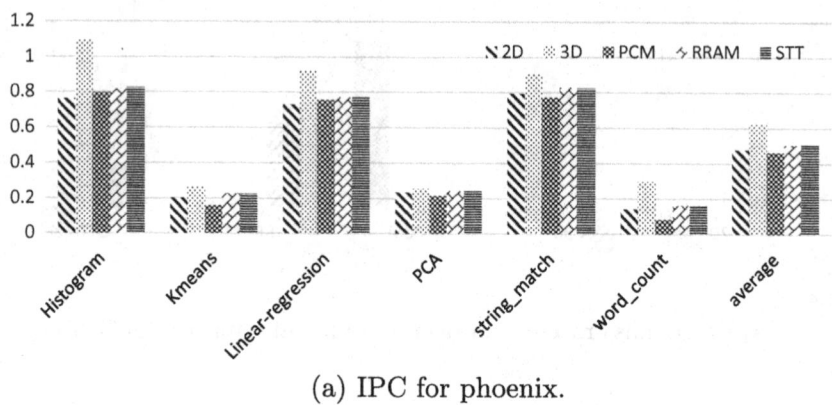

(a) IPC for phoenix.

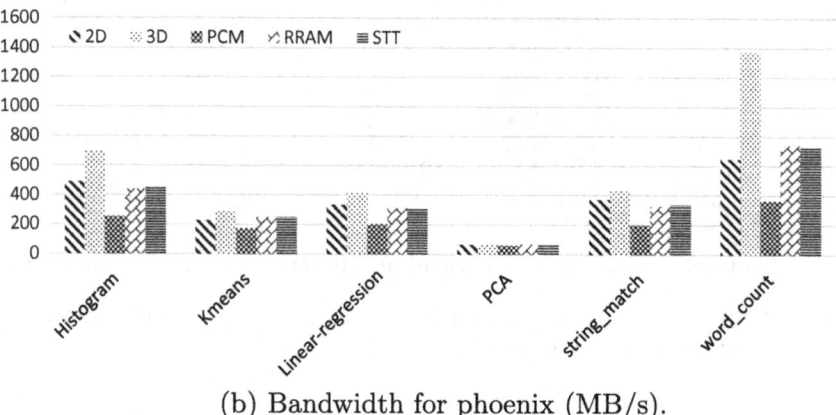

(b) Bandwidth for phoenix (MB/s).

Fig. 7 IPC and bandwidth for Phoenix. The IPC and phoenix show similar properties with SPEC

improving bandwidth and other optimization methods (such as cache replacement policy optimization, suitable prefetch strategy, and so on) to improve performance.

5. The enlargement of capacity has limited effect on improving bandwidth and IPC. For conventional workloads and big data workloads, the reason is different.

We compare the IPC and bandwidth for SPEC and Phoenix in different capacity. Given that, we set the results of 2 GB as baseline, Fig. 9 illustrated that the enlargement of capacity has just limited effect on improving IPC and Bandwidth. For Phoenix, 3DDRAM, RRAM, STT-RAM barely show difference while 2DDRAM and PCM show 2–3% improvement for IPC. This is due to big data workloads neither 2 nor 4 GB is not enough to relieve the bottleneck. For SPEC, the best IPC improvement in these memory systems is 2.6%. This is due to memory capacity is not a bottleneck for most applications.

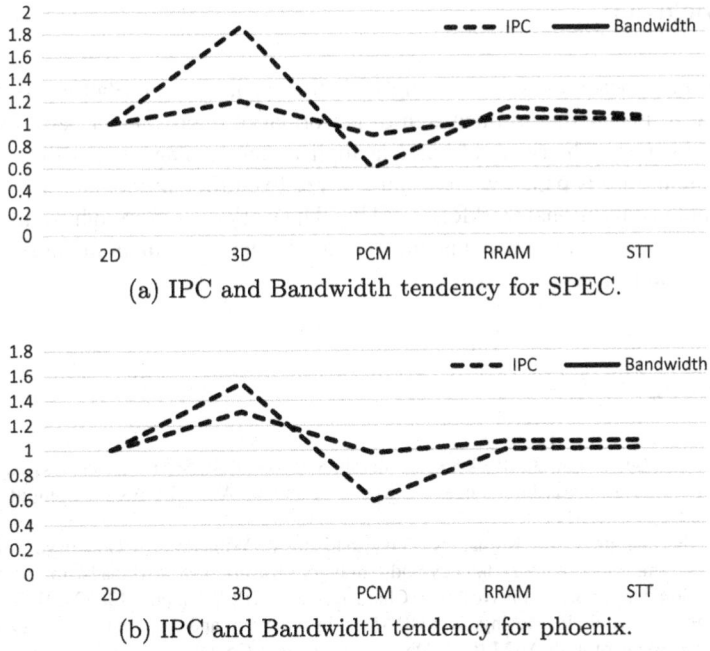

(a) IPC and Bandwidth tendency for SPEC.

(b) IPC and Bandwidth tendency for phoenix.

Fig. 8 IPC and bandwidth for phoenix. The IPC and phoenix show similar properties with SPEC. The IPC and bandwidth has same tendency for both SPEC and Phoenix. However, the effect degree is different for these two workloads

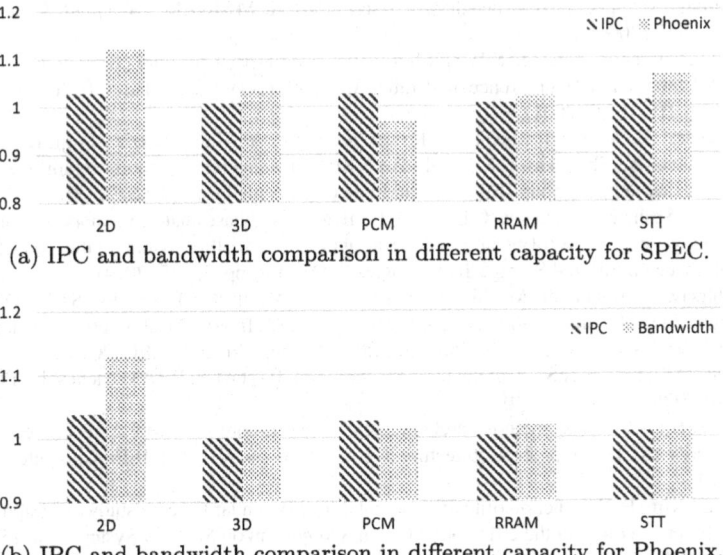

(a) IPC and bandwidth comparison in different capacity for SPEC.

(b) IPC and bandwidth comparison in different capacity for Phoenix.

Fig. 9 IPC and bandwidth in comparison with different capacity for SPEC and Phoenix. Positive value means better while negative value means opposite

5   Conclusion

Main memory becomes an increasingly significant factor contributing to overall system performance due to mismatch of performance improvement between CPU and main memory. The rising of big data applications even aggravates the memory wall problem. In this paper, we test and analyze five different memory systems by running both conventional workloads and big data workloads. Through compare and analysis, we make several conclusions to guild the design of main memory system for big data workloads.

References

1. Binkert, N., Beckmann, B., Black, G., Reinhardt, S.K., Saidi, A., Basu, A., Hestness, J., Hower, D.R., Krishna, T., Sardashti, S., et al.: The gem5 simulator. ACM SIGARCH Comput. Archit. News **39**(2), 1–7 (2011)
2. Clapp, R., Dimitrov, M., Kumar, K., Viswanathan, V., Willhalm, T.: Quantifying the performance impact of memory latency and bandwidth for big data workloads. In: 2015 IEEE International Symposium on Workload Characterization (IISWC), pp. 213–224. IEEE (2015)
3. Dhiman, G., Ayoub, R., Rosing, T.: PDRAM: a hybrid pram and dram main memory system. In: Proceedings of 46th ACM/IEEE Design Automation Conference, DAC'09, pp. 664–669. IEEE (2009)
4. Dimitrov, M., Kumar, K., Lu, P., Viswanathan, V., Willhalm, T.: Memory system characterization of big data workloads. In: 2013 IEEE International Conference on Big Data, pp. 15–22. IEEE (2013)
5. Henning, J.L.: Spec cpu2006 benchmark descriptions. ACM SIGARCH Comput. Archit. News **34**(4), 1–17 (2006)
6. Huh, J., Burger, D., Keckler, S.W.: Exploring the design space of future cmps. In: Proceedings of 2001 International Conference on Parallel Architectures and Compilation Techniques, 2001, pp. 199–210. IEEE (2001)
7. Kawahara, A., Azuma, R., Ikeda, Y., Kawai, K., Katoh, Y., Hayakawa, Y., Tsuji, K., Yoneda, S., Himeno, A., Shimakawa, K., et al.: An 8 Mb multi-layered crosspoint reram macro with 443 mb/s write throughput. IEEE J. Solid State Circuits **48**(1), 178–185 (2013)
8. Kim, H., Seshadri, S., Dickey, C.L., Chiu, L.: Evaluating phase change memory for enterprise storage systems: a study of caching and tiering approaches. In: Proceedings of the 12th USENIX Conference on File and Storage Technologies (FAST 14), pp. 33–45 (2014)
9. Kültürsay, E., Kandemir, M., Sivasubramaniam, A., Mutlu, O.: Evaluating STTRAM as an energy-efficient main memory alternative. In: 2013 IEEE International Symposium on Performance Analysis of Systems and Software (ISPASS), pp. 256–267. IEEE (2013)
10. Lewis, D.L., Lee, H.H.S.: Architectural evaluation of 3D stacked RRAM caches. In: Proceedings of 3DIC, pp. 1–4 (2009)
11. Loh, G.H.: 3d-stacked memory architectures for multi-core processors. In: Proceedings of ACM SIGARCH Computer Architecture News, vol. 36, pp. 453–464. IEEE Computer Society (2008)
12. Nai, L., Kim, H.: Instruction offloading with HMC 2.0 standard: a case study for graph traversals. In: Proceedings of the 2015 International Symposium on Memory Systems, pp. 258–261. ACM (2015)
13. Park, J., Han, M., Baek, W.: Quantifying the performance impact of large pages on in-memory big-data workloads. In: 2016 IEEE International Symposium on Workload Characterization (IISWC), pp. 1–10. IEEE (2016)

14. Poremba, M., Xie, Y.: Nvmain: An architectural-level main memory simulator for emerging non-volatile memories. In: 2012 IEEE Computer Society Annual Symposium on VLSI, pp. 392–397. IEEE (2012)
15. Ramos, L.E., Gorbatov, E., Bianchini, R.: Page placement in hybrid memory systems. In: Proceedings of the International Conference on Supercomputing, pp. 85–95. ACM (2011)
16. Talbot, J., Yoo, R.M., Kozyrakis, C.: Phoenix++: modular mapreduce for shared memory systems. In: Proceedings of the Second International Workshop on MapReduce and Its Applications, pp. 9–16. ACM (2011)

Speech Forensics Based on Sample Correlation Degree

Fang Sun, Yanli Li, Zhenghui Liu and Chuanda Qi

Abstract To address the issues of watermarking scheme by using public features, we define sample correlation degree feature of speech signal and present the embedding method based on the feature. Then, the scheme for speech forensics is proposed. We cut host speech into frames and each frame into two parts. Then convert frame number into watermark bits, which are embedded into the two parts of each frame. The integrity of each frame is testified by comparing with watermark bits extracted from the two parts. Theoretical analysis and experimental results demonstrate that the scheme is robust against desynchronization attack and effective for digital speech authentication.

Keywords Speech forensics · Digital watermark · Tamper detection · Sample correlation degree

1 Introduction

It is known that speech signal is likely to arouse the interest of attackers and caused malicious attack. For researchers, the authenticity of speech signal in hand is a question. So, the authentication of the speech signal is necessary [1, 2].

For audio, there is a large number of watermark schemes used for protecting the copyright [3, 4], and some research results on content authentication based on digital watermark technology [5–8]. In [5], authors analyzed the relationship of sample value and its Zernike moment, and obtained that the relationship between them is linear, based on a watermark embedding method is given. And then, a robust audio watermark scheme on the basis of Zernike moments (ZMs) is proposed. In [6], authors gave out a watermarking based on pseudo-Zernike moments (P-ZMs). In schemes [5, 6], authors embedded one watermark in nine segments. The embedding

F. Sun (✉) · Y. Li · Z. Liu · C. Qi
College of Computer and Information Technology,
Xinyang Normal University, Xinyang 464000, China
e-mail: ivy_xynusf@126.com

© Springer Nature Singapore Pte Ltd. 2019   173
S. K. Bhatia et al. (eds.), *Advances in Computer Communication and Computational Sciences*, Advances in Intelligent Systems and Computing 760, https://doi.org/10.1007/978-981-13-0344-9_14

capacity is very small for the schemes. In [7, 8], by using synchronization code, robust digital watermark schemes are proposed. For the schemes, synchronization codes are embedded based on openness feature of speech signal. Watermark algorithms by using openness features are vulnerable to attack; the detailed description is in [2]. Thus, for the schemes robust against desynchronization attacks by using statistical average value is insecurity.

While the integrity schemes for speech forensics are rarely [9], and existing schemes have some issues. (1) The embedding method is fragile to signal processing, and the common signal processing operations are regarded as malicious attack for the schemes [9], which limit application of speech forensics. (2) It is necessary for the watermark, generated by image, to transmit to authentication client using for speech forensics. Watermark transmission increases the bandwidth and threat of attack [10]. (3) The method based on synchronization codes embedded into energy can locate watermarked signal roughly, rather than accurately, which reduces the performance of speech forensics [11].

To address the shortcomings of present schemes, an integrity scheme is proposed in this paper. We define the sample correlation degree feature and give the corresponding watermark embedding method. In this paper, speech we cut signals into several frames and divide each frame into two parts. Frame number is converted into binary bits, which are used as watermarks and embedded into the two parts by quantifying sample correlation degree. If the attacked signal searches the integrity frame after the attacked signals and extract the watermark bits to reconstruct frame number. The difference between the number extracted before and next frame is the attacked frame. The proposed scheme improves the security and enhances the embedding capacity. Simulation results demonstrate the scheme is effective for speech forensics.

Details of the scheme will be addressed in the following sections. Section 2 introduces the definition of sample correlation degree, and the theory principle of embedding method. Section 3 illustrates the detail procedure of the scheme. The performance of the scheme is analyzed in Sect. 4. In Sect. 5, the simulation results are given, which is in used to demonstrate the effective performance of the scheme. Finally, the conclusion is concluded in Sect. 6.

2 Sample Correlation Degree

2.1 Definition

Denote $A = \{a_l, 1 \leq l \leq L\}$ as original speech, and the sample correlation degree is defined as

$$C = \sum_{l=1}^{L} |a_l \cdot a_{l+k}| \tag{1}$$

where $|a_l|$ represents the amplitude of the lth sample, k is the secret for watermark system, and C is the sample correlation degree, $C \geq 0$, $k \geq 0$. When $l + k > L$, $a_{l+k} = a_{l+k-L}$. It can be seen that sample correlation degree reflects the closeness of samples. The greater C is the closer of samples. And the smaller D is the larger difference between samples.

2.2 Performance of Feature Sample Correlation Degree

Select one speech signal randomly from library and denote it by $A = \{a_l, 1 \leq l \leq L\}$, where a_l represents the lth sample. Cut A into M segments, and represent the sample correlation degree of the jth segment as C_j, see the Eq. (1).

In this paper, $L = 10,000$, $M = 10$, and the signal selected are shown in Fig. 1, which is subjected to common signal processing and malicious attack. Figure 2 shows the signal substitution attacked, samples of the 5th and 6th segment are substituted. For the signal shown in Fig. 1, Fig. 3a, b, the variation of feature sample correlation degree after being subjected to low-pass filtering with the cutoff frequency 22 kHz and re-sampling operation with sampling frequency 16 kHz. For the signal shown in Fig. 2, Fig. 3c, the variation of feature after being subjected to substitution attack.

Test results show that the change of the feature is very tiny after being common signal operations, while it is very large for malicious attack. So, the feature defined sample correlation degree is robust against signal processing but fragile to malicious attack.

2.3 Embedding Method Based on Sample Correlation Degree

Suppose that sample correlation degree is as the feature and used to embed watermark. Denote C' as quantified feature, $\alpha = C'/C$. The correlation between C and C' is $C' = \alpha \cdot C$, which can be calculated by

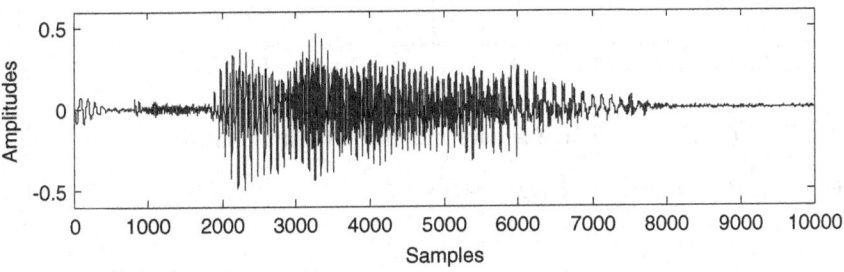

Fig. 1 Speech signal selected

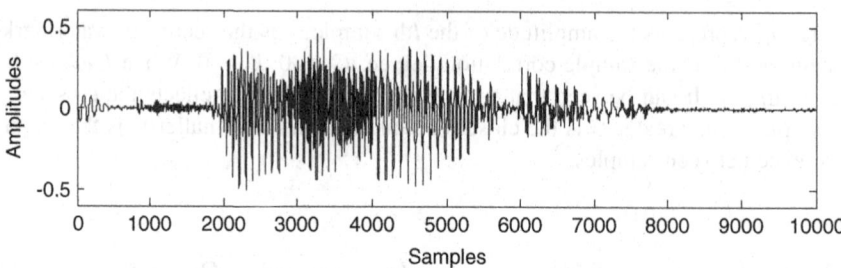

Fig. 2 Speech signal substitution attacked

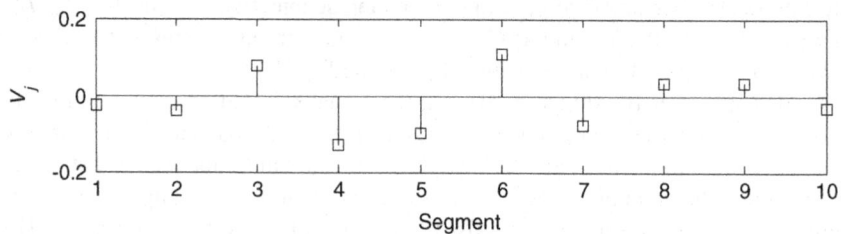

(a)The variation of feature after being subject to low-pass filtering

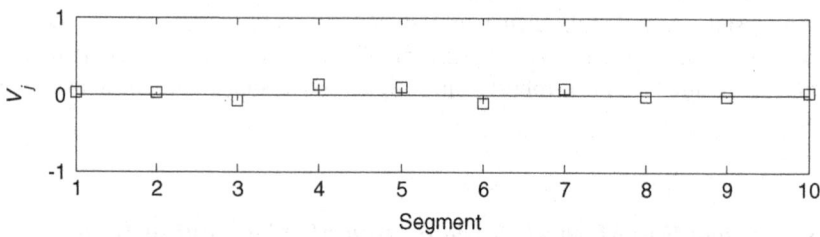

(b) The variation of feature after being subject to re-sampling operation

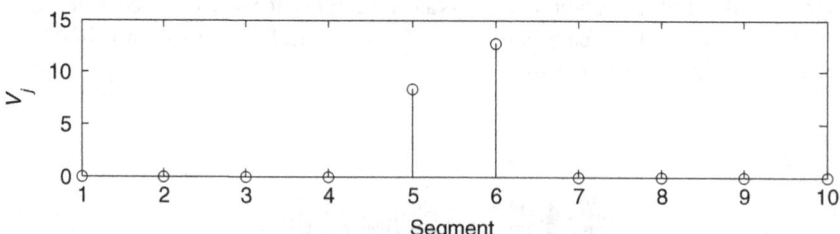

(c) The variation of feature after being substitution attacked

Fig. 3 Variation of feature after being processed

$$C' = \sum_{l=1}^{L} \left| \sqrt{\alpha} a_l \cdot \sqrt{\alpha} a_{l+k} \right| \tag{2}$$

That is, the signals corresponding to sample correlation degree C' can be obtained by magnifying the samples by using factor $\sqrt{\alpha}$, which can be regarded as watermarked signals. So, watermark can be embedded into sample correlation degree by magnifying samples value.

3 The Scheme

3.1 Watermark Embedding

(1) Cut the signal A (has L samples) into P frames, and represent ith frame as A_i, the length of A_i as L/P. For each frame A_i, it is divided into two segments, denoted by $A1_i$ and $A2_i$. The length of $A1_i$ and $A2_i$ is equal to N. Then divide $A1_i$ and $A2_i$ into T sub-segments, denoted by $A1_{i,t}$ and $A2_{i,t}$, respectively, $1 \le t \le T$.

(2) For frame A_i, by using Eq. (3), we generate watermark by frame number i

$$i = w_T \cdot 2^T + w_{T-1} 2^{T-1} + \cdots + w_1 \tag{3}$$

where $W = \{w_t, 1 \le t \le T\}$ regards the watermark in this paper and is embedded into $A1_i$ and $A2_i$, respectively.

(3) Calculate the sample correlation degree of $A1_{i,t}$ by Eq. (1), denoted by $C1_{i,t}$. Quantify $C1_{i,t}$ to embed w_t.

Step 1 Denote $S1_{i,t}$ as the second integer number of $C1_{i,t}$, in left-to-right order. Quantify $S1_{i,t}$ based on Eq. (4).

$$S1'_{i,t} = \begin{cases} 2, & w_t = 0 \\ 7, & w_t = 1 \end{cases} \tag{4}$$

Step 2 Substitute $S1_{i,t}$ by using $S1'_{i,t}$, and the result is denoted as $C1'_{i,t}$.

Step 3 Denote $\alpha = \sqrt{C1'_{i,t}/C1_{i,t}}$, and multiply α for the samples of $A1_{i,t}$ to embed w_t. For example, if $C1_{i,t} = 2.37$ and $w_t = 0$, then $S1_{i,t} = 3$, $S1'_{i,t} = 2$, and $C1'_{i,t} = 2.27$.

By this, we embed watermark W into $A1_i$ and $A2_i$, and the watermarked signal is represented by $A1'_i$, $A2'_i$. The watermark embedding process is shown in Fig. 4.

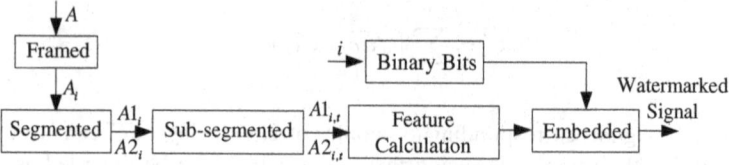

Fig. 4 Watermark embedding process

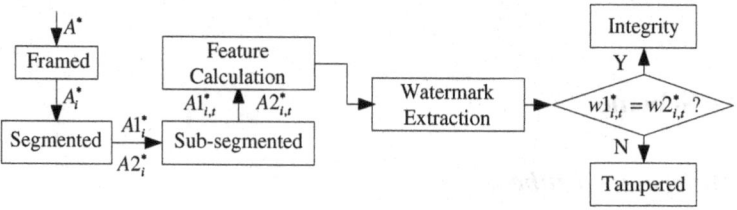

Fig. 5 Process of content authentication

3.2 Content Authentication

Denote $A^* = \{a_l^*, 1 \le l \le L\}$ as watermarked signal. Using the method shown in Section A, cut watermarked signal A^* into some frames and some segments. The ith frame is denoted by A_i^*. For the two segments, there are denoted by $A1_i^*$, $A2_i^*$. The watermark extraction and authentication method are as follows.

Step 1   Cut $A1_i^*$ in N sub-segments, represented as $A1_{i,t}^*$, $1 \le t \le N$.

Step 2   Calculate the sample correlation degree of $A1_{i,t}^*$ by using Eq. (1), and represent the result as $C1_{i,t}^*$.

Step 3   Denote $S1_{i,t}^*$ as the second integer number of $C1_{i,t}$, in left-to-right order. Then $w1_{i,t}^*$ as the watermarked bit extracted from $A1_{i,t}^*$ can be obtained by Eq. (5).

$$
w1_{i,t}^* = \begin{cases} 0, & \lfloor S1_{i,t}^*/5 \rfloor = 0 \\ 1, & \lfloor S1_{i,t}^*/5 \rfloor = 1 \end{cases}, \quad 1 \le t \le N \tag{5}
$$

Using the same method, watermark bit extracted from $A2_{i,t}^*$, can be obtained and denoted by $w2_t^*$, $1 \le t \le N$.

Step 4   Verify the authenticity of the ith frame A_i^* by comparing with the watermark extracted from $A1_{i,t}^*$, $A2_{i,t}^*$. If $w1_{i,t}^* = w2_{i,t}^*$, $1 \le t \le N$, it indicates that ith frame is intact. On the contrary, it is concluded that ith frame is attacked. The process of watermark extraction and verification is shown in Fig. 5.

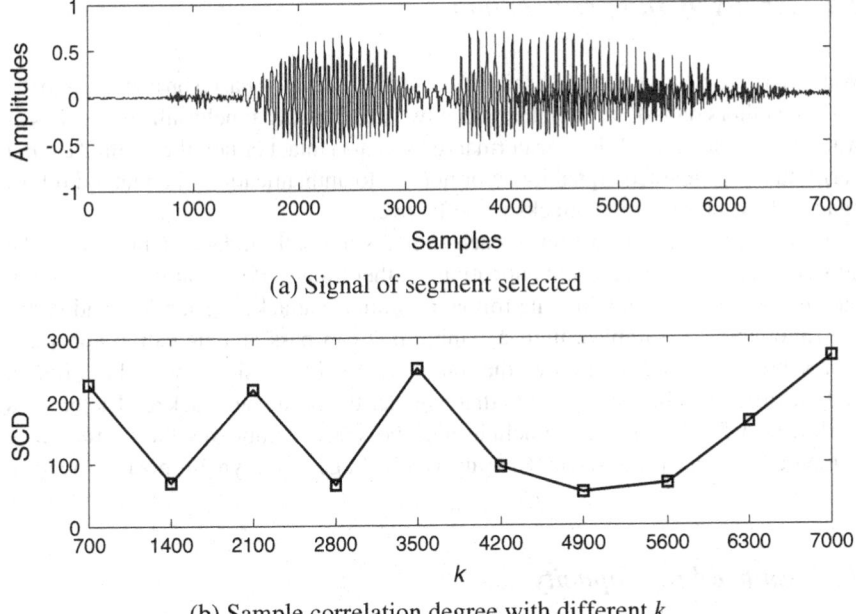

(a) Signal of segment selected

(b) Sample correlation degree with different k

Fig. 6 Signal selected and the sample correlation degree

4 Performance Analysis

4.1 Security

The watermark schemes using public features are vulnerable to attack [2]. In the paper, sample correlation degree is used as the feature for watermarking. In Eq. (1), the parameter k is the secret key. Select one segment of speech randomly, and calculate the sample correlation degree with different k. The signal selected is shown in Fig. 6a, and the values of feature for different secret keys are shown in Fig. 6b, and the feature is closely related to the key used. If attackers attack the watermarked signal without key, we can detect the attack with high probability, and the probability is calculated by Eq. (6). Therefore, based on the comparison with schemes based on openness features, the scheme improves the security.

$$P_a = 1 - \frac{1}{2^T} \qquad (6)$$

where P_a is the probability that we can detect the attacked frame, and T is the length of watermark bits.

4.2 Ability of Tamper Location

Watermarking robust to desynchronization attacks is usually based on codes. As to the papers, authors regard the signal between two neighboring synchronization codes as watermarked signal. While, if watermarked signal is intact or not, the method cannot verify the watermarked signal intact or not. As to authentication, the method robust against desynchronization attacks is ineffective.

In this paper, authentication of each frame is mutually independent, a frame be attacked does not affect the authentication of other frames. If the frames 1th to ith can pass the authentication, while the following frame is attacked. Then find and verify the following L/P samples, until a sample that can pass authentication appeared, represented as i'th frame. Extract the frame number of the ith frame and i'th frame, and the difference between them is that whether the frame be attacked. The process is shown in Fig. 7. We can conclude that the scheme improves the performance compared with that proposed in [6–8] and is robust against desynchronization attacks.

4.3 Embedding Capacity

The watermark embedding method proposed in [1] and [5, 6], authors embedded one watermark into nine segments. For the method proposed, one watermark is embedded into one segment. So, compared with the schemes [1, 5, 6], the embedding capacity in this paper increases greatly.

To sum up, we compare the performance of the proposed scheme with schemes in [5–8, 12], and the list the comparison results in Table 1, which show that the scheme improves the performance.

Fig. 7 Method robust against desynchronization attack

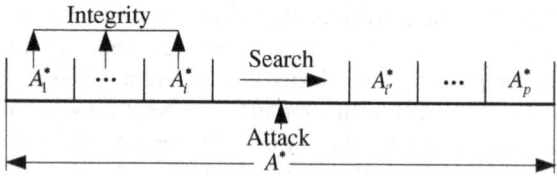

Table 1 Comparison results of the performance for some schemes

Scheme	Security	Tamper location	Embedding capacity
Refs. [5, 6]	No	No	Low
Refs. [7, 8]	No	No	High
Ref. [12]	No	No	–
Our	Yes	Yes	High

Table 2 SDG values obtained from different listeners

Listeners	SDG	Listeners	SDG
L1	−0.1206	L6	−0.1524
L2	−0.1354	L7	−0.2259
L3	−0.1623	L8	−0.1827
L4	−0.1709	L9	−0.1304
L5	−0.2181	L10	−0.1068

According to the analysis above, on the one hand, the feature used is closely related to the secret key, and it is difficult to acquire the feature for attackers without the key. So, the scheme improves the embedding security. On the other hand, the scheme improves the ability of tamper location and can locate accurately the watermarked signal relatively.

5 Experiment Results

Select 100 speech signals as the test signal, they are WAVE format and quantified sampled at signals. The parameters selected are as follows: The length of the test signal $L = 180{,}000$, $P = 30$, $N = 3000$, $T = 5$, $k = 125$.

5.1 Inaudibility Test

In this paper, the subjective difference grades (SDG) are adopted to test the inaudibility of watermarked signal. The test results are acquired from 10 listeners and listed in Table 2, where L1, L2, ..., L10 represent the ten listeners. The scores in Table 2 indicate the scheme is inaudibility.

5.2 Tamper Location Results

One watermarked signal is selected and shown in Fig. 8. Then, we perform deletion attack and give tamper location result. For other attacks, the method is similar.

Insert 10,000 samples in the signal and show attacked signal in Fig. 9a. Then we extract the frame number of attacked signal, which is shown in Fig. 9b. Figure 9b, $FR_i = 1$ indicates ith frame is veritable, and the signal between 6th and 8th frame is attacked.

According to the experimental result, it can be seen the signals attacked can be detected. So, the scheme proposed is robust to desynchronization attack and can verify the authenticity of speech signal.

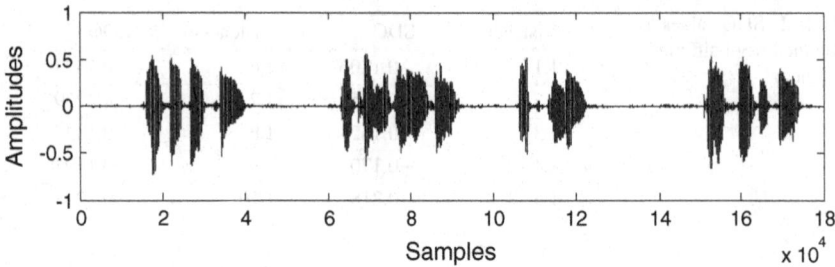

Fig. 8 Watermarked signal

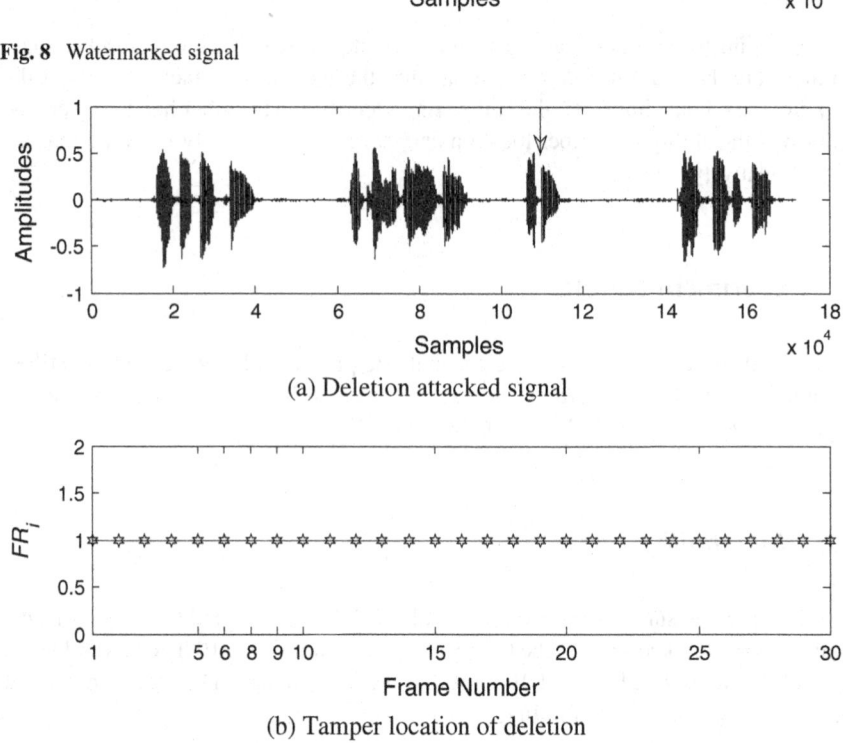

(a) Deletion attacked signal

(b) Tamper location of deletion

Fig. 9 Deletion attacked signal and frame number extracted

6 Conclusions

Considering the shortcomings of watermarking based on openness features and that robust to desynchronization attacks by using synchronization codes, an integrity scheme based on speech forensic is proposed. We define the sample correlation degree feature and give the embedding method by quantifying the feature. Frame number as watermark is embedded into the two parts of one frame. We verify the integrity of watermarked signal by comparing watermark extracted from two parts. Theoretical analysis and experiment results demonstrate that the scheme increases embedding capacity, can verify the integrity of speech signal, and increases the security.

References

1. Han, B., Gou, E.: A robust speech content authentication algorithm against desynchronization attacks. J. Commun. **9**(9), 723–728 (2014)
2. Liu, Z., Wang, H.: A novel speech content authentication algorithm based on Bessel-Fourier moments. Digit. Signal Proc. **24**(1), 197–208 (2014)
3. Wang, J., Healy, R., Timoney, J.: A robust audio watermarking scheme based on reduced singular value decomposition and distortion removal. Signal Process. **91**(8), 1693–1708 (2011)
4. Akhaee, M.A., Kalantari, N.K., Marvasti, F.: Robust audio and speech watermarking using Gaussian and Laplacian modeling. Signal Process. **90**(8), 2487–2497 (2010)
5. Xiang, S., Huang, J., Yang, R.: Robust audio watermarking based on low-order Zernike moments. In: Shi, Y.Q., Jeon, B. (eds.) IWDW 2006. LNCS, vol. 4283, pp. 226–240. Springer, Heidelberg (2006)
6. Wang, X., Ma, T., Niu, P.: A pseudo-Zernike moments based audio watermarking scheme robust against desynchronization attacks. Comput. Electr. Eng. **37**(4), 425–443 (2011)
7. Lei, B.Y., Soon, I.Y., Li, Z.: Blind and robust audio watermarking scheme based on SVD-DCT. Signal Process. **91**(8), 1973–1984 (2011)
8. Vivekananda, B.K., Sengupta, I., Das, A.: A new audio watermarking scheme based on singular value decomposition and quantization. Circuits Syst. Signal Process. **30**(5), 915–927 (2011)
9. Chen, O.T.C., Liu, C.H.: Content-dependent watermarking scheme in compressed speech with identifying manner and location of attacks. IEEE Trans. Audio Speech Lang. Process. **15**(5), 1605–1616 (2007)
10. Wang, X.Y., Shi, Q.L., Wang, S.M., Yang, H.Y.: A blind robust digital watermarking using invariant exponent moments. AEU Int. J. Electr. Commun. **70**(4), 416–426 (2016)
11. Wang, Y., Wu, S.Q., Huang, J.W.: Audio watermarking scheme robust against desynchronization based on the dyadic wavelet transform. J. Adv. Signal Process. **13**, 1–17 (2010)
12. Pun, C.M., Yuan, X.C.: Robust segments detector for de-synchronization resilient audio watermarking. IEEE Trans. Audio Speech Lang. Process. **21**(11), 2412–2424 (2013)

A Responsive Web Search System Using Word Co-occurrence Density and Ontology: Independent Study Project Search

Naruepon Panawong and Akkasit Sittisaman

Abstract This research aims to develop a responsive Web search system for the independent study project search of the department of computer and information technology, Nakhon Sawan Rajabhat University. There are four features of the proposed search system. First, the responsive Web design using bootstrap framework is used for designing the Web site. Thus, the Web sites can adapt to the user's viewing screen size; second, utilizing ontology for storing independent study project database in order to establish the data relationship and support semantic search; third, word segmentation using THSplitLib library for Thai language. Last, the proposed responsive Web search system was implemented in PHP and SPARQL for ontology search via RAP API. In addition, the retrieval decision making is based on co-occurrence density analysis. The search term can be one or more the one keyword or a sentence. The search results also display the project advisor information that is concurrent with the search term. Hence, students have more information for effective decision making. The experimental results show that the Web site can adapt to the various screen sizes and has high efficiency. The average F-measure is 95.02%, the average search precision is 94.70%, and the average recall is 95.35%.

Keywords Ontology · RAP API · Responsive · SPARQL · Words co-occurrence

1 Introduction and Related Work

Students in all three branches which are the computer science, information technology and multimedia technology, department of applied science, faculty of science and technology, Nakhon Sawan Rajabhat University must do the "bachelor final

N. Panawong (✉) · A. Sittisaman
Faculty of Science and Technology, Department of Applied Science,
Nakhon Sawan Rajabhat University, Nakhon Sawan, Thailand
e-mail: jnaruepon.p@gmail.com

A. Sittisaman
e-mail: akkasit@gmail.com

© Springer Nature Singapore Pte Ltd. 2019
S. K. Bhatia et al. (eds.), *Advances in Computer Communication
and Computational Sciences*, Advances in Intelligent Systems
and Computing 760, https://doi.org/10.1007/978-981-13-0344-9_15

project" in order to complete their bachelor degree program. The students must apply their knowledge and creativity to create a computer program. Nowadays, the amount of project reports is more than 900 and continuously increasing. The original independent study project search system support only 1 search keyword at a time. For example, if the keyword was "online" ("ออนไลน์"), the system can deliver the search results. However, if the keyword was "ออนไลน์ สินค้าคงคลัง" ("online inventory") or "ระบบสินค้าคงคลังแบบออนไลน์" ("online inventory system"), then the search system cannot find the matched projects. Furthermore, if users search projects using the project advisor's previous given name, maiden name, or family name, the search system cannot deliver the search results. Therefore, users have to search with different keywords multiple times until they have the satisfied information results. Since the word processing focus on unit of the word, therefore the word segmentation is crucial for finding the word boundary and cut off non-related words in order to obtain the keyword for the search.

In 2016, [1] have surveyed the number of users who access Web sites using smart phones or tablets and found that the number is continuously increasing. The percentage of Internet access using smart phones or tablets increased by 9% from 2015. Time usage increased from 5.7 to 6.2 h per day, and the peak usage duration is 4 p.m. to 8 a.m. It can be seen that the developing Web sites have to be able to adapt to different screen sizes for user convenience. Turan and şahin [2] have reported that, during 2009–2015, the number of smart phone users is monotonically increasing and higher than the number of computer users. In addition, more application developers have implemented responsive Web design for user convenience and easy access Web sites by various devices with different screen sizes, for instance, different screen size of smart phones and computer desktop sizes. There are many researchers that design Web application using responsive Web, e.g., [3] designed FlagShip Universities in the US Web site, [4] proposed a responsive system for agriculturists such as agricultural knowledge, suitable fertilization for plants. [5], proposed navigator within Web site using the average number of user clicks.

Therefore, the authors proposed a development of a responsive Web search system using word co-occurrence density and ontology for independent study project search. The proposed search system can adapt to various screen sizes using responsive Web. Ontology is used for constructing independent study project database. Users can use more than 1 keyword or a sentence search. The authors utilized Thai word segmentation to cut off unrelated words. Search result displaying is based on word co-occurrence density for matching the user's inquiry. Furthermore, the proposed search system shows the project advisor information search result that is concurrent with keywords. Hence, students have more information including project advisor information to take into consideration. The remainder of this paper is organized as follows. Section 2 describes methodology. The search system architecture is proposed in Sect. 3. Section 4 presents the experimental setup and result found. Finally, Sect. 5 states the conclusion of our work and future research.

2 Methodology

The proposed search system is composed of four parts, specifically responsive Web design, independent study project ontology, word segmentation, and word co-occurrence density, respectively. The details are described as the follows.

2.1 Responsive Web Design

Responsive Web Design is a recently new approach to design and develop Web sites that can adapt to various screen sizes, platform, and orientation. The advantage of the responsive Web design approach is only design once then it can be applied to any screen size. Users can neglect the screen size and access the Web site on their devices. The CSS, CSS3, and JavaScript do data prioritizing and data sorting; hence, the Web site will automatically respond to any screen size [6]. The authors implemented responsive Web in PHP and bootstrap framework. The detailed description of the design framework is as follows.

2.1.1 HTML5

<!DOCTYPE html>

2.1.2 View Port for Controlling the Display Area that Can Adapt to Different Screen Size Using Tag Meta

<meta name = "viewport" content = "width = device-width, initial-scale = 1"/>

2.1.3 Bootstrap

<link rel = "stylesheet" type = "text/css" href = "css/bootstrap.min.css">

2.1.4 Java Script of Bootstrap and JQuery Command in Tag Body

<script src = "js/jquery.js" > </script>
<script src = "js/bootstrap.min.js" > </script>

2.1.5   Check the Screen Size Using Media Queries by CSS3

CSS3 is used to determine the display area of various screen sizes and devices (platform). The controlled display area properties are "not more than", "not less than", "between" width. For example, the command suite for displaying not more than 480 pixels is as follows.

```
@media screen and (max-width:480px) {

    // CSS Code here

}
```

2.1.6   Image Displaying in Web site Using Class Img-Resposive

class img-rounded for curve-boundary pictures, class img-circle for circular picture, class img-thumbnail for framed pictures, and class center-block for center display.

```
<img class = "img-responsive img-circle center-block" src = "imagename.jpg">
```

2.2   Project Ontology

Ontology is the definition of the concepts or meaningful explanation for mutual understanding within a knowledge domain. It is used to specify properties and relationships between the concepts. It supports exchanging, searching, sharing, and reused of knowledge as well [7–9]. Ontology has been applied to diverse applications, for instances, a Thailand tourism search and recommendation system [10–12], cultivation of plants harnessing [13], ontology-based decision support system for crop selection on Android (ODESSA) [14], ontology-based annotation for semantic multimedia retrieval which improves search speed and efficiency [15], improving ontology-based text classification: an occupational health and security application [16], and location-based service via smart phone using ontology-based semantic queries which focus on indoor activities in a university context [17].

The database design and construction for the proposed independent study project search system is based on ontology. There are 945 bachelor final projects from 3 curriculums, specifically computer science, information technology and multimedia technology, department of computer and information technology, faculty of science and technology, Nakhon Sawan Rajabhat University to be stored. The design process is Protégé 5.0. The data file format is OWL, named comproject.owl as shown in Fig. 1, which has 5 classes, i.e., Advisor, Project, DBMS, ProjectType, and ProgrammingLanguage. Project is divided into 3 subclasses which are multimedia technology project (MTProject), information technology project (ITProject), and computer science project (CSProject) as shown in Fig. 2.

```
<comproj:CSProjectrdf:about="http://www.owl-
ontologies.com/Ontology1440519160.owl#CS52-2552">
    <rdf:type rdf:resource="http://www.w3.org/2002/07/owl#Thing"/>
    <comproj:hasProjectType rdf:resource="http://www.owl-
ontologies.com/Ontology1440519160.owl#WindowsApplication"/>
    <comproj:hasAuthorName xml:lang="th">นายกฤษณพงษ์ สหกิจวัฒนา
</comproj:hasAuthorName>
    <comproj:hasAdvisor rdf:resource="http://www.owl-
ontologies.com/Ontology1440519160.owl#อ.ธนากร"/>
    <comproj:hasProgrammingLanguage rdf:resource="http://www.owl-
ontologies.com/Ontology1440519160.owl#PHP"/>
    <comproj:hasNameOfProject xml:lang="th">ระบบจัดการชำระค่าทั่วภาพยนตร์ โดยใช้เทคโนโลยี RFID
</comproj:hasNameOfProject>
    <comproj:hasDBMS rdf:resource="http://www.owl-
ontologies.com/Ontology1440519160.owl#MSSQLServer"/>
    <comproj:hasGraduationYear xml:lang="th">2552</comproj:hasGraduationYear>
</comproj:CSProject>
```

Fig. 1 An example of OWL data file format for data storage

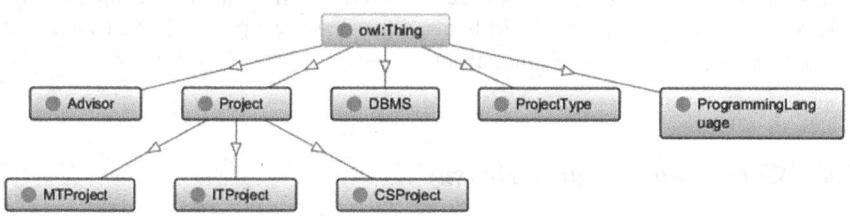

Fig. 2 Ontology-based independent study projects

```
<?php
    include(dirname(FILE) . DIRECTORY_SEPARATOR . 'THSplitLib/segment.php');
    $segment = new Segment();
    $result = $segment->get_segment_array("ระบบออนไลน์และคงคลัง");
    echo implode(' | ', $result);
?>
```

Fig. 3 An example of the Thai word segmentation using THSplitLib

2.3   Word Segmentation

Thai word in this project search system is developed by PHP. Hence, Thai word segmentation in the proposed search system is implemented by THSplitLib [18]. The example of word segmentation is shown in Fig. 3. The result from the program is "ระบบ | ออนไลน์ | และ | คงคลัง" ("system| online | inventory"). It can be seen that each word is separated by |.

Table 1 An example of the word co-occurrence density calculation

Independent Study Project Title	Word Co-occurrence Density
คำค้นหา -> ระบบ ออนไลน์ คงคลัง	
ระบบสินค้าคงคลัง ออนไลน์ บริษัท คอมพิวเตอร์ ยูเนี่ยน จำกัด	$=\dfrac{3}{3} \times 100 = 100\%$
ระบบงานพัสดุสถานีอนามัยบ้านเขาดินออนไลน์	$=\dfrac{2}{3} \times 100 = 66.67\%$
ระบบงานคาร์แคร์ คู่รักชาติคาร์แอนเซอร์วิส	$=\dfrac{1}{3} \times 100 = 33.33\%$
โปรแกรมซ่อนข้อความในไฟล์รูปภาพออนไลน์	$=\dfrac{1}{3} \times 100 = 33.33\%$

Therefore, the segmented words that are not related to the keywords are cut off specifically preposition, stop word, and adverb. The authors compile the preposition and stop word lists from [19], in order to decrease the searching database size. The result is the keywords that will be used for searching process. The example is "ระบบ ออนไลน์ คงคลัง" ("system online inventory").

2.4 Word Co-occurrence Density

The decision making for reporting the search results is based on word co-occurrence density, so the users will get the most matched results with their keywords. The word co-occurrence density is the ratio of the number of keywords appeared in the project title to the total number of keywords. If a project has the higher word co-occurrence density than the others, then the project has a higher possibility to be the search results [20–22]. The word co-occurrence density formula is shown in Eq. 1 and its implementation example is demonstrated in Table 1.

$$R = \frac{\text{Number of keywords appeared in the project topic}}{\text{Total number of keywords}} \times 100 \qquad (1)$$

3 System Architecture

The system architecture of the responsive Web search system (Fig. 4) is described as follows:

1. Users enter keywords via Web application based on responsive Web display, for example, names of the authors, advisors, graduating year. In addition, users can

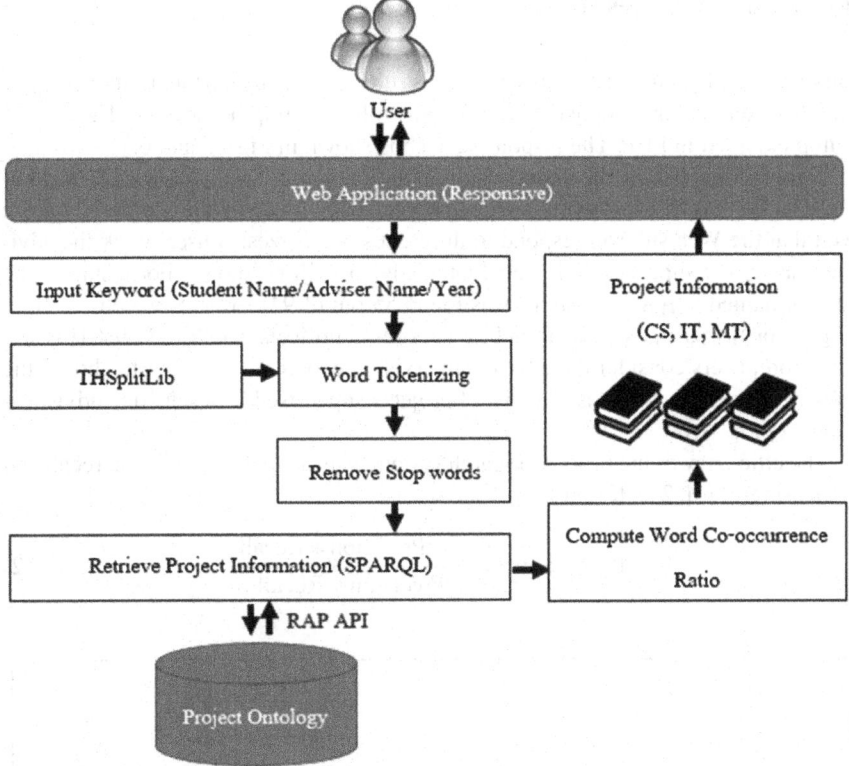

Fig. 4 System architecture

use keyword combination such as project title and graduating year, graduating year and advisor's name, or a sentence search.

2. Keywords from step 1 are segmented using THSplitLib.
3. The segmented word from step 2 is rechecked. If any words are meaningless or unrelated to users' search terms, they will be cut off, e.g., "และ (and)", "กับ (with)", and "ด้วย (too)".
4. Using words resulted from step 3 to do ontology-based search, which implemented in SPARQL. Connection between SPARQL and ontology is established by RAP API.
5. Calculating the word co-occurrence density between search results from step 4 and users' keywords.
6. Display final search results by word co-occurrence density descending order. The properties of the results are the authors, the project titles, project advisors, and graduating year.

4 Testing and Results

This section describes the experimental results of the independent study project search system using responsive Web design with bootstrap framework. The system is implemented in PHP. The responsive Web is shown in Figs. 5 and 6.

Figures 5 and 6 show the search results of the keywords "ระบบ ออนไลน์ คงคลัง" searching. The system will sort word co-occurrence density by descending order. It can be seen that the Web site can respond to different screen sizes. If users click the advisor's name, the supervised projects by the advisor will be shown. For instance, the advisor named "วัฒนาพร วัฒนชัยธรรม" was found 55 out of 945 projects which is 5.82% (Fig. 7) or "เอกวิทย์ สิทธิวะ" was found 32 out of 945 projects which is 3.39% (Fig. 8). Therefore, users consider the advisor supervision experience in order to choose the advisor who fits their needs. The list of projects supervised by a selecting advisor is shown in Fig. 9.

The efficiency of the proposed search system is measured by precision, recall and F-Measure in Eq. 2 as follows:

$$F\text{ - Measure} = \frac{2 * \text{Precision} * \text{Recall}}{\text{Precision} + \text{Recall}} \tag{2}$$

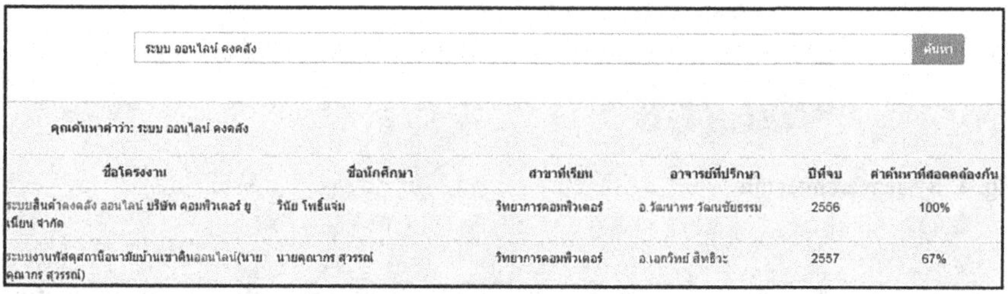

Fig. 5 Search results displayed on desktop computer

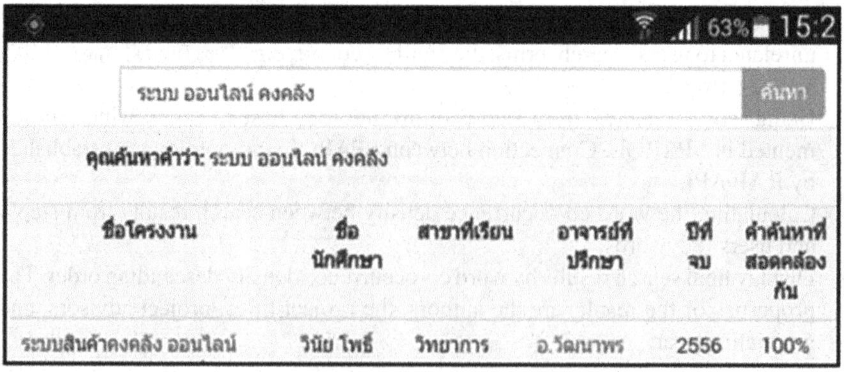

Fig. 6 Search results displayed on smart phone

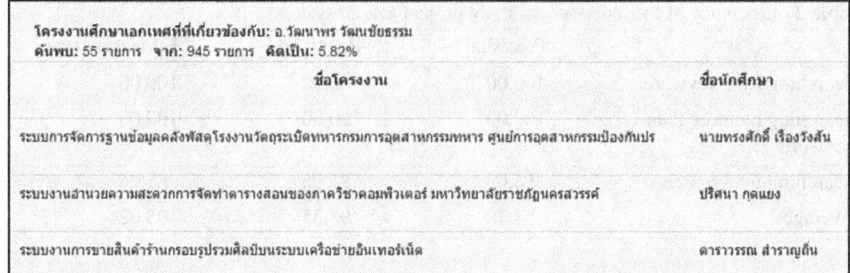

Fig. 7 List of projects supervised by วัฒนาพร วัฒนชัยธรรม (Wattanaphon Wattanachaitham)

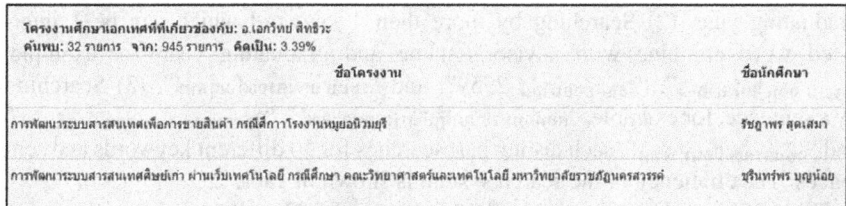

Fig. 8 List of projects supervised by เอกวิทย์ สิทธิวะ (Akkasit Sittiwa)

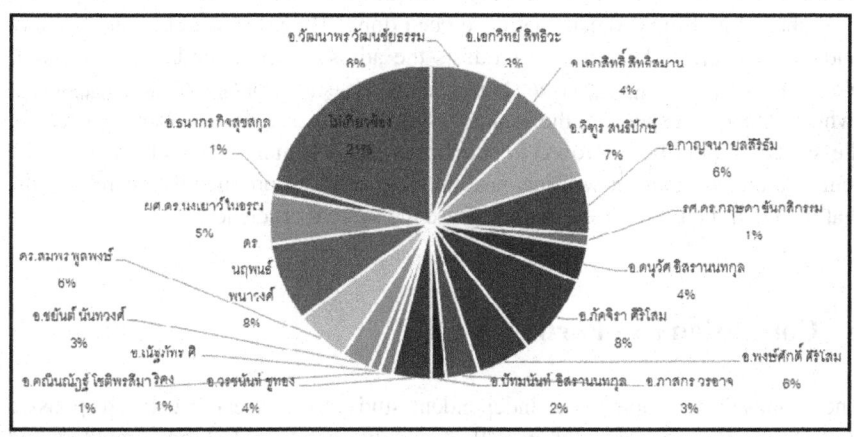

Fig. 9 Pie chart of percentage of projects found by using keywords and advisor's names

Precision   is True Positive/(True Positive + False Positive).
Recall   is True Positive/(True Positive + False Negative).
True   Positive is the right answer and displayed.
False   Positive is the wrong answer but displayed.
False   Negative is the right answer but not displayed

Table 2 Efficiency of the independent study project search system

	Precision (%)	Recall (%)	F-Measure (%)
Searching by 1 keyword	100.00	100.00	100.00
Searching by more than 1 keyword	100.00	100.00	100.00
Searching by a sentence	84.09	86.05	85.06
Average	94.70	95.35	95.02

The testing of the search system is divided into three parts: (1) searching by only 1 keywords, e.g., android, online, advisor's name, student's name, and graduating year. (2) Searching by more than 1 keyword which can be 1 interested word combine with advisor's name and graduating year, for example, "ระบบ ออนไลน์ คงคลัง", "ข้อสอบออนไลน์ 2559", and "ระบบ แอนดรอยด์ นฤพนธ์". (3) Searching by a sentence, for example, "หอพักบนระบบปฏิบัติการแอนดรอยด์", "ข้อสอบบนแอนดรอยด์มีใครทำบ้าง", and "แอนดรอยด์มีคนทำกี่คน", each testing part searches for 50 different keywords and sentences. The efficiency of the search system is shown in Table 2.

Table 2 shows that the average of F-Measure is 95.02% which is in good criteria. The average of precision is 94.70%, and the average of recall is 95.35%. Since the sentence used in searching process is in the question form, for instances, asking for a number of android projects that have been done. Then, the search system cannot find the answer. If the sentence requests the advisor recommendation, for example, "ทำโครงงานกับอาจารย์คนไหนดี (who should be my advisor?)" or "อาจารย์ท่านไหนถนัดภาษาจาวา (who is the Java expert?)", the search results may dissatisfy the users. Therefore, the system should be improved to support question sentences search by using the sentence analysis, and it will increase the possibility to produce the search results that matched the users' needs and increase the system efficiency.

5 Conclusion and Further Work

The proposed development of independent study project search using responsive Web design was implemented in PHP. The experiment on desktop computers and smart phones shows that the Web site can adapt to various screen sizes. Ontology-based database stores 945 independent study projects. Users can search using one or more than one keyword including a sentence. If the user enters keywords, the system will perform word segmentation using THSplitLib. Then, the system will cut off unrelated words and construct the SPARQL command for searching within the ontology-based database. After the system will calculate the word co-occurrence density between the search terms and the search results. The search results were displayed by the word co-occurrence density descending order. The efficiency of the project search system was measured by F-Measure. The efficiency of the proposed independent study search system is high, i.e., the average F-Measure is 95.02%, the average of the precision is 94.70%, and the average of the recall is 95.35%.

However, if users use a sentence as a search term, such as "ทำโครงงานกับอาจารย์คนไหนดี (who should be my advisor?)" or "อาจารย์ท่านไหนถนัดภาษาจาวา (who is the Java expert?)", they might obtain the dissatisfy search results. In addition, the information of the project advisor is only the project titles that are related to the search keywords. Therefore, the student might lack information to take into consideration choosing the project advisor. In future work, the analysis of the question sentence search is desired, or the development of the question and answer system is needed and applying fuzzy logic to to the effective decision making process in choosing a project advisor.

References

1. The Electronic Transactions Development Agency (Public Organization): Thailand Internet User Profile 2016. https://www.etda.or.th/content/thai-land-internet-user-profile-2016-confer ence.html (2016). Accessed 27 May 2017 (in Thai)
2. Turan, B.O., Şahin, K.: Responsive web design and comparative analysis of development frameworks. The Turkish Online J. Des. Art Commun. 7(1), 110–121 (2017)
3. Hingorani, K., McNeal, B., Bradford, J.: Mobile-friendly websites: an analysis of websites of flagship universities in the United States. Issues Inf. Syst. 17(2), 17–24 (2016)
4. Barole, K.P., Kodolikar, A.D., Marne, P.K., Joshi, S.: AgroHelp: a responsive system for agriculturists. Int. J. Adv. Eng. Innov. Technol. (IJAEIT) 2(4), 1–3 (2016)
5. Kumar, V., Jenamani, M.: Context preserving navigation redesign under Markovian assumption for responsive websites. Electron. Commer. Res. Appl. 21, 65–78 (2017)
6. Rao, T.V.N., Suhail, A.M., Rambabu, D.: Responsive web design with generic coding principle for modern devices. Int. J. Adv. Eng. Technol. Manag. Appl. Sci. 1(6), 9–15 (2014)
7. Gruber, T.R.: A translation approach to portable ontology specification. Knowl. Acquis. 5(2), 199–220 (1993)
8. Noy, N.F., McGuinness, D.L.: Ontology development: a guide to creating your first ontology. Stanford Knowledge Systems Laboratory, USA (2001)
9. Zhang, G., Jia, S., Wang, Q.: Construct ontology-based enterprise information metadata framework. J. Softw. 5(3), 312–319 (2010)
10. Namahoot, C.S., Brueckner, M., Panawong, N.: Context-aware tourism recommender system using temporal ontology and Naive Bayes. Adv. Intell. Syst. Comput. 361, 183–194 (2015)
11. Namahoot, C.S., Panawong, N., Brueckner, M.: A tourism recommendation system for Thailand using semantic web rule language and K-NN algorithm. Inf. Int. Interdiscip. J. 19(7), 3017–3023 (2016)
12. Panawong, N., Snae, C., Brueckner, M.: Ontology-driven information retrieval system for regional attractions. In: Proceedings of the 2nd International Conference in Business Management and Information Sciences, pp. 1–9, Phitsanulok, Thailand, Naresuan University (2012)
13. Panawong, N., Namahoot, C.S.: Cultivation of plants harnessing an ontology-based expert system and a wireless sensor network. J. Telecommun. Electr. Comput. Eng. (JTEC), 9(2–3), 109–113 (2017)
14. Panawong, N., Namahoot, C.S., Brueckner, M.: Ontology-based decision support system for crop selection on android (ODESSA). Inf. Int. Interdiscip. J. 19(7), 2995–3001 (2016)
15. Lakshmi, T.R., Srinivasa, R.M., Usha, K., Goudar, R.H.: Ontology-based annotation for semantic multimedia retrieval. Procedia Comput. Sci. 92, 148–154 (2016)
16. Sanchez-Pi, N., Martí, L., Garcia, A.C.B.: Improving ontology-based text classification: an occupational health and security application. J. Appl. Logic 17, 48–58 (2016)

17. Lee, K., Lee, J., Kwan, M.P.: Location-based service using ontology-based semantic queries: a study with a focus on indoor activities in a university context. Comput. Environ. Urban Syst. **62**, 41–52 (2017)
18. Phuak-im, S.: THsplitLib for PHP. https://github.com/moohooooo/thsplitlib (2017). Accessed 27 Jan 2017 (in Thai)
19. Tonglor, K.: Fundamental of Thai Language. Ruamsarn, Bangkok (2007) (in Thai)
20. Kim, S.-M., Baek, D.-H., Kim, S.-B., Rim, H.-C.: Question answering considering semantic categories and co-occurrence density. http://trec.nist.gov/pubs/trec9/papers/kuqa.pdf (2000). Accessed 27 Mar 2017
21. Chaowicharat, E., Naruedomkul, K.: Co-occurrence-based error correction approach to word segmentation. In: Proceedings of the 24th International Florida Artificial Intelligence Research Society Conference, pp. 240–244, USA (2011)
22. Shao, M., Qin, L.: Text similarity computing based on LDA topic model and word co-occurrence. In: Proceedings of the 2nd International Conference on Software Engineering, Knowledge Engineering and Information Engineering (SEKEIE 2014), pp. 199–203, Singapore (2014)

Development Design and Principle Analysis of Urban Expansion Simulation System

Ying Li, Zhichen Liu and Huimin Xu

Abstract Recently, it has become a hot spot that the forecast of urban land use change based on CA model and the city in the traditional cellular models were simulated and predicted according to some mathematical methods or artificial intelligence, so it cannot be true. Simulate the change trend of city expansion. It is rarely that the cellular space simulated the changes of city space. UML can describe that the city expansion simulation system has certain gap. Therefore, this study constructed a cellular automaton model transformation rules based on decision model of gray. We used the cellular automaton theory and decision model of gray to develop urban expansion system and combine with the unified modeling language (UML Unified, Modeling Language), use case diagram, class diagram, sequence diagram of unified expression visualization system. And UML can establish a good communication and communication bridge for the peer.

Keywords Cellular automata · Urban expansion · ArcEngine · Gray situation decision · UML

Ying Li obtained her Ph.D. degree from Tohoku University of Japan in 1997. She works as a professor and Ph.D. supervisor at Navigation College of Dalian Maritime University, the Director of the Environmental Information Institute of Dalian Maritime University.

Y. Li (✉) · Z. Liu
Navigation College, Dalian Maritime University, Dalian 116026, China
e-mail: 1838462237@qq.com

Y. Li · Z. Liu
Environmental Information Institute, Dalian Maritime University, Dalian 116026, China

Z. Liu · H. Xu
School of Urban and Environmental Sciences, Liaoning Normal University, Dalian 116029, China

© Springer Nature Singapore Pte Ltd. 2019
S. K. Bhatia et al. (eds.), *Advances in Computer Communication and Computational Sciences*, Advances in Intelligent Systems and Computing 760, https://doi.org/10.1007/978-981-13-0344-9_16

1   Introduction

Recently, the application of cellular automata to urban expansion has become a hot spot. Scholars use the "CA +" model to construct a variety of models that can simulate and predict urban growth. For example, Kheder Introduced fuzzy mathematics when using cellular automata theory and predicted the Indianapolis region with multi-temporal remote sensing data. The results show that the model improves the simulation accuracy [1]. Tang Guoan [2] using the C# programming language developed swarm intelligence geographic CA conversion rule mining tools in the integrated development environment VS2010, and he designed BCO-CA transformation rule mining algorithm.

Li Xia has artificial neural networks, ant colony algorithm into cellular automata model, the success simulation is used to predict the Dongguan area in the China [3]. Zhou Chenghu et al. used VC++ programming language, based on object-oriented based random CA-integrated GeoCA-Urban, and was used by a large number of scholars to study the actual urban growth and land value [4]. The research of these constructive cellular automata models is to introduce some mathematical models to constrain and discuss how to obtain the transformation rules of cells by artificial intelligence technology. The spatial and temporal predictive ability of CA model is analyzed by reading the predecessors' literature, but these studies are simple and simple in the development of urban cellular automata model system, and are carried out in the decentralized development environment, Some C language, VB. NET language, and some CA system must be in the GIS environment to simulate the forecast, which for the formation of a mature and unified cellular automata system caused some obstacles.

Based on the above research status, this study not only introduces the gray situation decision-cellular automata model (Gray-CA Model), but also takes into account the characteristic state of the neighborhood cells in the urban cell space, and designs the urban space expansion simulation platform. The development process of ArcEngine is integrated with C# programming language, and the whole process of visualization system design and design based on UML modeling language is carried out. The process step of the system is decomposed and fully visualized and clearly expresses the relationship between the various modules relationship. The test proves that the system can realize the expansion of urban spatial simulation.

2   UML Requirements for System Requirements

2.1   Requirements Analysis

It is difficult to describe a model of complex software, and it is difficult to describe it simply because the system has functional information and complexity that makes the system unambiguous. A system is usually described in terms of its functional deployment, static structure, dynamic behavior, concurrency, and physical deployment. These descriptions form different views of the system, each of which describes the different characteristics of the system architecture [5]. The city expansion simulation system mainly refers to the basic operation function of the layer, the loading of the album, the reading of the album pixel value, the calculation of the target comprehensive effect measure, and the generation and clarification of the urban extended image at the next time. The statistical analysis has been done on the total number of cells in the region. It can be seen that this development of the city expansion simulation system is a huge simulation system, which needs specific description like the use case diagram, class diagram, flowchart, activity diagram etc.

Urban growth simulation system for urban planning in the future, the reasonable control of future changes in development of the city, making the city's ecological benefits and economic benefits maximized. The design goal is to design the city to meet the development needs and one-button operation of the software system, and has the latter part of the maintenance and expansion of the function of the city's land concentration, land suitability, location traffic degree of these three factors to consider.

2.2   User and Use Case View

First determine the user who uses the system. As can be seen from the above, the user mainly opens the program, loads the Atlas, enters the target weight, and other key operations. Through the analysis of user and system interaction process that the basic use cases of the system, the so-called use case is a participant and the system in a dialogue in the implementation of a series of related matters [6]. The example of this simulation system is shown in Fig. 1. The ellipsoid box represents the system use case, the participant is the user in the graph, the basic operation of the system use case is divided into layers: the loading of the target atlas, the reading of the pixel value of the target set, the statistics of the number of cells, image clarification after simulation, single target effect measurement calculation, that helps city to expand the generation of layers and other functional use cases. The basic operation of the layer includes amplification, reduction, browsing, Hawkeye function, the number of cell-like statistics is divided into the number of woodland population statistics, the number of cultivated land population statistics, the number of land for construction statistics. The single-objective effect measure calculation, that is, to obtain the value of each chart after the pixel, stored in the two-bit array, the cycle of traversal, find each area around the eight cells of the land accumulation degree, land suitability level, The traffic achievement degree of the three goals of the calculation results, combined with AHP and then multi-objective comprehensive effect of the measured value. This use case diagram provides a complete overview of a city expansion simulation system. The use case model is a very critical view in the modeling process, which represents the specific functions and implementation of the system.

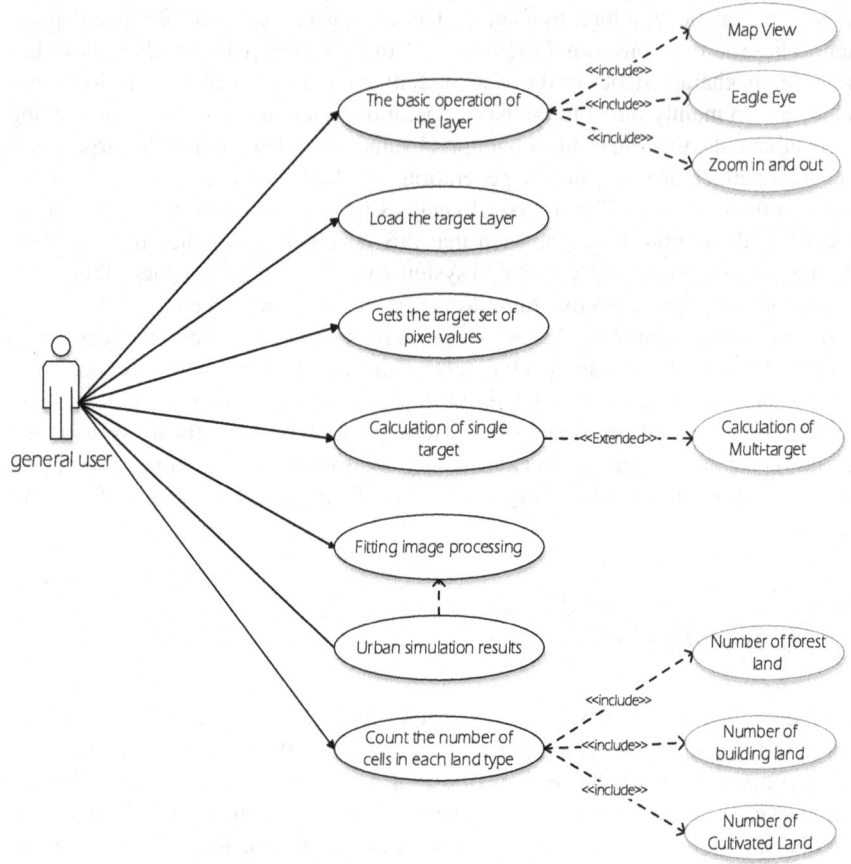

Fig. 1 Urban extension system use case diagram

3 UML Analysis of System Analysis

3.1 System Static Model Establishment

According to the requirements analysis phase, then draw the class diagram to represent the system's static model. Because the natural semantics is not very precise, we need to further analyze and modify the city to expand the simulation system class diagram. Figure 2 is the development of the system class diagram. Among them, ArcEngine package also provides a complete operation of the corresponding grid pixel components, the main use of IRaster, IRasterDataset, IRasterCursor, IRasterEdit, and other important interfaces. The assignment of the calculated raster pixel block uses the primary interface such as IPixelBlock3, IPixelBlock, and the

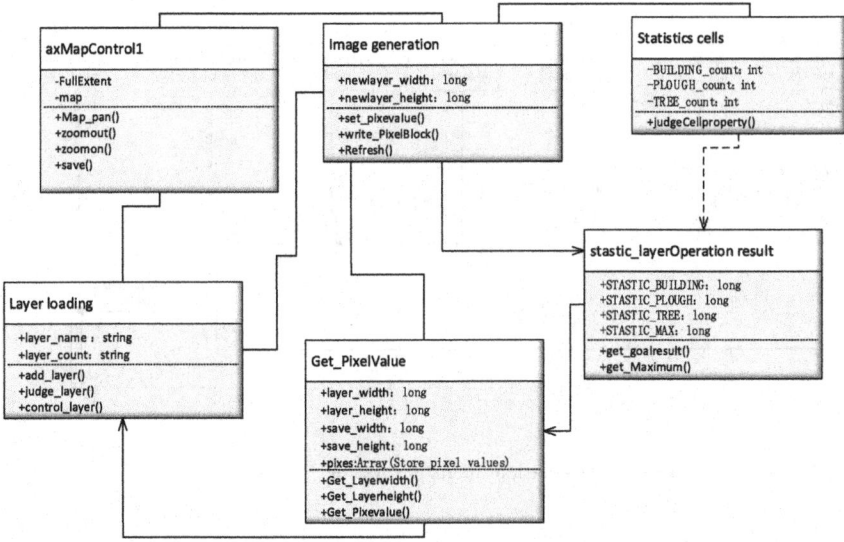

Fig. 2 Urban expansion simulation system class diagram

get_PixelDataByRef (), set_PixelData () methods to get the individual pixel values and the assigned pixel values.

3.2 System Dynamic Model Establishment

After establishing the static model, in order to express the dynamic characteristics of the system, we can establish the dynamic model of the system. The dynamic model describes the time and the change in the system, and explains when the sequence, the map, the state diagram and the activity diagram to show [7]. The selected model of the system can be developed using timing diagram and activity diagram visualization software.

(1) timing diagram model

In theory, we can develop a sequence diagram for each use case, but in practice, you can usually omit those too simple use case sequence diagram [8]. The following figure shows the timing diagram of simulating the city's extended image. The diagram mainly describes a series of operations that occur after the user enters the main interface of the system. After adding the layer information, click the button to start the program, Reading the pixel information of each layer, combining the gray situation decision, calculating the single target measure value, and providing the target weight according to the analytic hierarchy process, and then obtaining the multi-objective comprehensive effect measure value. But also can carry out the number of cells

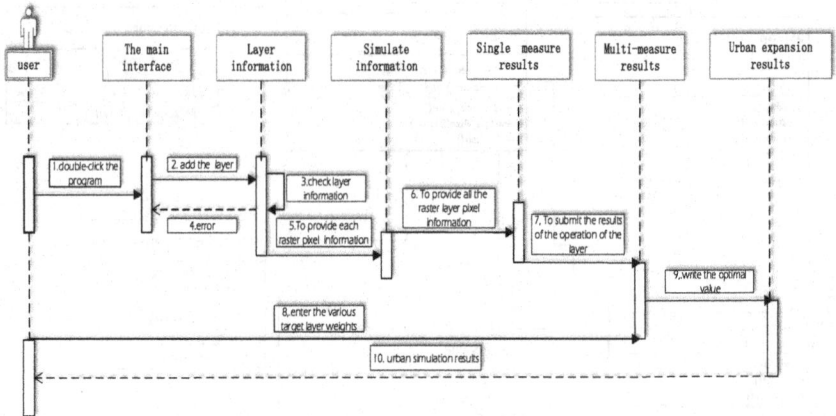

Fig. 3 Timing diagram of city expansion simulation system

around the number of statistical analysis and a series of operations. Figure 3 shows only the operation in the order of the case, the program for the exception when the processing is not too much delineation and description.

(2) Activity diagram model

The activity diagram is used to describe the processing flow or interactive flow of certain use cases of the software, similar to the specific flowchart of the system. Activity charts are often used to describe the sequence of activities, from one activity to another activity control flow, its role is to describe the use case, describe the operation of the class [9]. The following are the single-objective effect measure activity diagram, the analytic hierarchy process obtains the multi-objective effect comprehensive measure activity chart, clears the boundary value problem activity chart, and the local statistical population quantity activity chart.

(A) single target effect measure activity diagram

In order to calculate the effect of each target (land accumulation degree, land suitability level, traffic degree) corresponding to the effectiveness of the system development, a single target effect measure use case is specially designed, to facilitate the situation around the corresponding target of the quantification operation. The logic flowchart is Fig. 4.

(B) Generate the city expansion result layer activity map

This development takes into account a variety of constraints factor, the constraints factor by the single-objective calculation. Through the analytic hierarchy process to determine the corresponding weight of the target, this weight is provided to the program, and finally obtained the multi-objective comprehensive effect measure of the cell. If the comprehensive measure of the land type is the largest, then the forest land in the local area will be the same. The optimal situation is assigned. If the

Fig. 4 Activity diagram of single-objective effectiveness measures for urban expansion

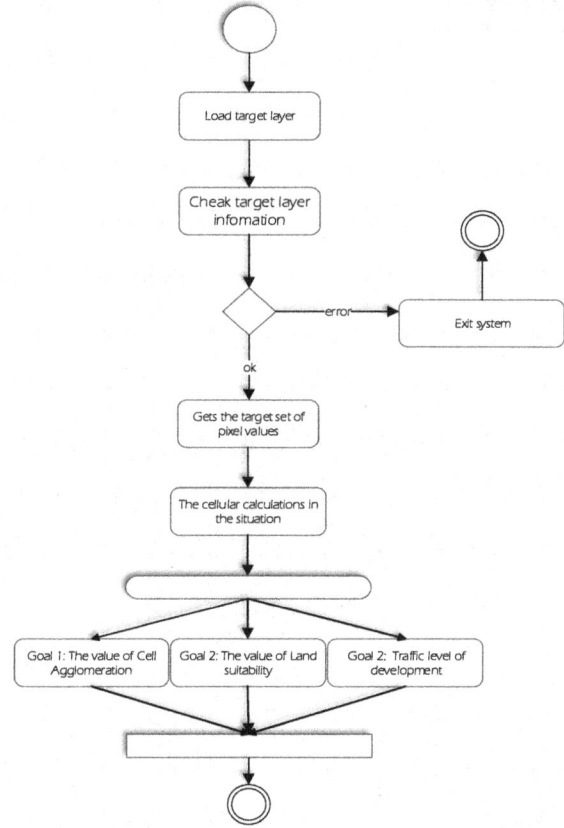

comprehensive measure of the land type is the largest, then the construction land in the local area will be assigned as the optimal situation. Finally, the attribute value will be obtained. The corresponding activity diagram is Fig. 5.

4   Principle of City Expansion Simulation System

This constraint factor is the degree of agglomeration of land per land, the degree of land suitability, and traffic level of development. The impact factor is divided into one, two, and three levels according to the level, and then the weights are 0.6, 0.2, 0.2 to participate in the multi-objective comprehensive effect measure calculation. The study divides the region into a number of small local areas, which is a cellular automata model based on the local land-type competition. It relies on the state and quantity of the surrounding cells to be the basis of local competition and obtain the final decision Score. The event set = $\{x_1, x_2, \ldots, x_n\}$, x_i is farmland, forest land.

Fig. 5 Activity diagram of
urban expansion

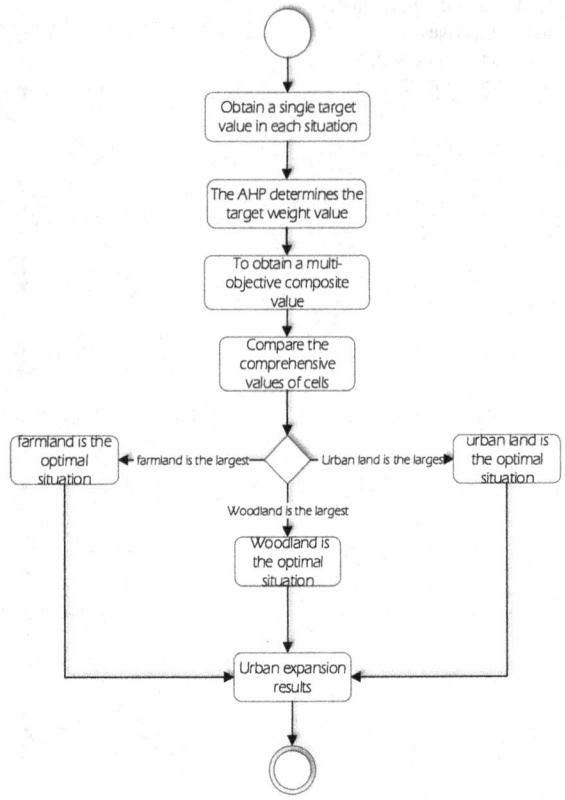

Measures set = {city land, farmland and forest land}, no waters. The situation set: $s_{(x_i,A)} = \{x_i, \text{Urban land}\}$, $\{x_i, \text{forest land}\}$, $\{x_i, \text{farmland}\}$. The next step is to study the competitive score of the cells under different goals.

The clustering degree of the cells can reflect the agglomeration efficiency of the land, and quantify the number of cells with the number of cells. As shown in Fig. 6, in the local competition scenario, Decision objective matrix (2), obtain the effect measure of each class, (3) and (4) for the target two, the target three Decision matrix. The principle of its conversion is:

$$
\begin{aligned}
\text{Sum}(x, y) = {} & \text{cell}(x, y-1) + \text{cell}(x, y+1) \\
& + \text{cell}(x-1, y-1) + \text{cell}(x-1, y) \\
& + \text{cell}(x-1, y+1) + \text{cell}(x+1, y-1) \\
& + \text{cell}(x+1, y) + \text{cell}(x+1, y+1).
\end{aligned} \tag{1}
$$

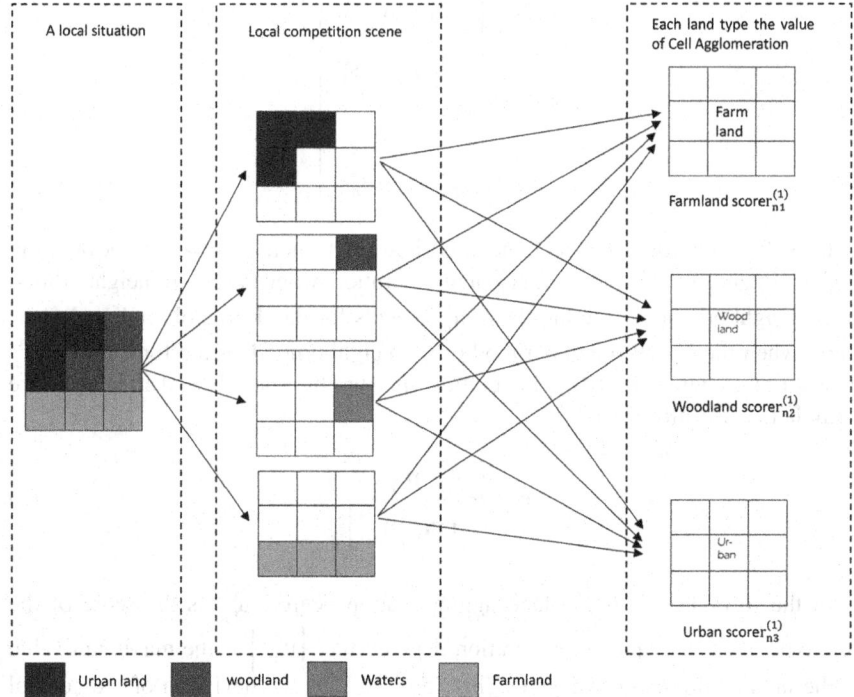

Fig. 6 Aggregate land between the fields of the target class competition scene map

$$r^{(1)} = \begin{bmatrix} u_{11}^{(1)} & u_{12}^{(1)} & u_{12}^{(1)} \\ u_{21}^{(1)} & u_{22}^{(1)} & u_{23}^{(1)} \\ \vdots & \vdots & \vdots \\ u_{n1}^{(1)} & u_{n2}^{(1)} & u_{n3}^{(1)} \end{bmatrix} \tag{2}$$

$$r^{(2)} = \begin{bmatrix} u_{11}^{(2)} & u_{12}^{(2)} & u_{12}^{(2)} \\ u_{21}^{(2)} & u_{22}^{(2)} & u_{23}^{(2)} \\ \vdots & \vdots & \vdots \\ u_{n1}^{(2)} & u_{n2}^{(2)} & u_{n3}^{(2)} \end{bmatrix} \tag{3}$$

$$r^{(3)} = \begin{bmatrix} u_{11}^{(3)} & u_{12}^{(3)} & u_{12}^{(3)} \\ u_{21}^{(3)} & u_{22}^{(3)} & u_{23}^{(3)} \\ \vdots & \vdots & \vdots \\ u_{n1}^{(3)} & u_{n2}^{(3)} & u_{n3}^{(3)} \end{bmatrix} \tag{4}$$

In the formula, when the N cell (neighborhood of the neighborhood) is the building land, $u_{n1}^{(1)}$ stands for the degree of land agglomeration; when the N cell (neighborhood of the neighborhood) is the farmland, $u_{n2}^{(1)}$ stands for the degree of land agglomeration; when the N cell (neighborhood of the neighborhood) is the forest land, $u_{n3}^{(1)}$ stands for the degree of land agglomeration; then use the upper-bound effect measure formula (5), as follows:

$$\mathbf{r}_{nj}^{(1)} = \frac{\mathbf{u}_{nj}^{(1)}}{\mathbf{max}\left\{\mathbf{u}_{nj}^{(1)}\right\}} \tag{5}$$

In the formula, $r_{nj}^{(1)}$ is the land agglomeration degree; $u_{nj}^{(1)}$ is the score of the land type under the land agglomeration degree; $\max\left\{u_{nj}^{(1)}\right\}$ is the maximum value of the land agglomeration degree. This value reflects the deviation of the general effect measure value from the maximum effect measure value. Finally, the scores of cultivated land, forest land, and building land are obtained, which is the upper-bound effect measure. Objective 2: The land suitability and target of two cells are 3 yuan. The principle of competition among different races is the same.

Multi-objective integrated cell competition is obtained by obtaining three single-objective measure value, combined with the analytic hierarchy process (AHP) to determine the relative weight of each target using the formula (6) to calculate the comprehensive measure value $r_{n1}^{(\Sigma)}, r_{n2}^{(\Sigma)}, r_{n3}^{(\Sigma)}$, and then compare the size, the maximum optimal situation as core element the cell.

$$r_{n1}^{(\Sigma)} = r_{n1}^{(1)} * P1 + r_{n1}^{(2)} * P2 + r_{n1}^{(3)} * P3$$
$$r_{n2}^{(\Sigma)} = r_{n2}^{(1)} * P1 + r_{n2}^{(2)} * P2 + r_{n2}^{(3)} * P3$$
$$r_{n3}^{(\Sigma)} = r_{n3}^{(1)} * P1 + r_{n3}^{(2)} * P2 + r_{n3}^{(3)} * P3 \tag{6}$$

$r_{n1}^{(1)}$ When the nth cell is the farmland, the land clustered score;

$r_{n2}^{(2)}$ When the nth cell is the forest land, the land suitability score;

$r_{n3}^{(3)}$ When the nth cell is urban land, the traffic degree score;

$r_{n1}^{(\Sigma)}$ The comprehensive score of cultivated land and land under multiple objectives.

P1, P2, P3 represent the degree of aggregation of land, land suitability right, traffic Accessibility of the three coefficients corresponding to the target weight.

5   Epilogue

The development and design of urban expansion simulation system, the development steps one by one mapping to the UML view, not only makes the standardization of view modeling, but also to solve the communication barriers between peer developers for the future expansion and improvement of the system to facilitate. The combination of gray situation decision and cellular automata is a very suitable method. The method is based on the different neighborhood characteristics, which makes the cell better for spatial self-organization and self-evolution.

5.1   Conclusion

The starting point of this study is to determine the state of the central cell from the neighborhood of the neighborhood (molar neighborhood). The principle is to establish the rule of cell transformation using the gray situation decision and combine the level Analysis. The cellular automata theory is used to establish the theory of urban expansion, the central cell and neighborhood cells are used to construct a number of local units, local multi-target neighborhood cell interaction is used to determine the central element. The mechanism of cellular simulation states is needed to promote the direction of cellular automata transformation to the neighborhood cell orientation.

Using UML, in the Visual Studio 2013 development platform, using C# programming language based on ArcEngine components successfully developed the city expansion simulation system. The program, after 2 h of system testing, enables simulations of future expansion of the city, generates a grid map, and ultimately counts the final area of each land type after the change. The advantages are (1) the system of natural factors, social factors, location factors for a comprehensive consideration, for different research areas, if the local planning departments have different focus on demand, the software can do some minor parameter adjustment, To meet the needs of urban planning. (2) The use of UML for the city to expand the simulation system design and development can improve the efficiency of system development and improve the reuse rate of the code for the business needs of the mobile can be dealt with. (3) UML unified modeling language, through the UML unified modeling standards, support case-driven, architecture-centric and incremental and other advantages, to achieve the city expansion simulation system from demand analysis to software design, code preparation of the transition, A large limit to reduce the difficulty of software development and price for the development of different peer design and development of a good communication and communication between the bridge, but also easy to improve the future system and expansion. (4) Using UML to make the city simulation module in the system has been decomposed in detail, the complete description of the core function of the basic simulation principle, easy to peer different development and design of the researchers to explore each other between the exchange.

5.2 Discussion and Problems

In this paper, we focus on the method of generating the next-generation cell type based on the gray-state decision-making to simulate the role of the cell in the local domain by the neighborhood state. This process is used to promote the role of exploring the transition from the cellular automata to the neighborhood, promote the role. This study needs to be discussed and solved the problem: (1) The system encountered in the simulation of protected land or by the policy intervention land, the system cannot take into account the factors, which will affect the accuracy of the simulation; (2) The system has high-performance requirements and long running time, and it needs to be further optimized for the code and algorithm to solve the redundancy. (3) The urban expansion study uses the gray situation decision to calculate each single target. But also combines the analytic hierarchy process to obtain the comprehensive effect measure, although it is very good to convert the simulation of single target into multi-objective, but there is a lot of subjective factors, and in the construction of There are still some shortcomings in the general model and the simulation results, and the next step will try to optimize the research by combining genetic algorithm and neural network.

References

1. Jokar Arsanjani, J., Helbich, M., Kainz, W.: Integration of logistic regression, Markov chain and cellular automata models to simulate urban expansion. Int. J. Appl. Earth Obs. Geoinf. **21**, 265–275 (2013)
2. Yang Jian, Y.: Research on transformation rules of geographical cellular automata based on bee swarm intelligence. Nanjing Normal University (2014)
3. Li, X., Lao, C., Liu, Y., et al.: Early warning of illegal development for protected areas by integrating cellular automata with neural networks. J. Environ. Manag. **130**, 106–116 (2013)
4. Chenghu, Z., Tao, P., Jun, X., Ting, M., Zide, F., Jianghao, W.: Urban Dynamics and GIScience. Elsevier Inc. (2013)
5. Ahmed, R.A.M., Aboutabl, A.E., Mostafa, M.S.M.: Extending unified modeling language to support aspect-oriented software development. Int. J. Adv. Comput. Sci. Appl. **8**(1), 208–215 (2017)
6. Vyas, V., Vishwakarma, R.G., Jha, C.K.: Integrate aspects with UML: aspect oriented use case model. In: 2016 Fourth International Conference on Parallel, Distributed and Grid Computing (PDGC), pp. 134–138. IEEE (2016)
7. Jäger, S., Maschotta, R., Jungebloud, T., et al.: Creation of domain-specific languages for executable system models with the eclipse modeling project. In: Proceedings of 2016 Annual IEEE Systems Conference (SysCon), pp. 1–8. IEEE (2016)
8. Akkaya, I., Derler, P., Emoto, S., et al.: Systems engineering for industrial cyber–physical systems using aspects. Proc. IEEE **104**(5), 997–1012 (2016)
9. Nayak, A., Samanta, D.: Synthesis of test scenarios using UML activity diagrams. Softw. Syst. Modeling, 101 (2011)

Meta-Model Interaction for Collaborative School Subject Management

M. Nordin A. Rahman, Abdulbasit Nuhu, Syarilla Iryani A. Saany, M. Fadzil A. Kadir and Syadiah Nor Shamsudin

Abstract Subject management is a process of developing and coordinating of teaching materials. A collaborative subject module management is used to allow participants to share and manage course modules via Internet technology. Currently, Malaysian school does not have collaborative platform to manage subject materials. Teachers usually used personalized tools to manage their subject modules. This article will discuss the modelling of cloud-based subject module management. Raquel's interaction meta-model is adopted as the central framework of the model. Based on developed model, a prototype has been developed which implemented under cloud computing platform. In the experiment, the prototype is evaluated using four cloud quality of service metrics. The results show that the model has accomplished the metrics of efficiency of 62%, effectiveness of 64%, learnability of 62%, and attractiveness of 68%. The proposed model could help to facilitate the development and management collaboration of subject module in Malaysian schools. It provides a reliable collaborative platform among teachers, students, and school management.

Keywords Subject module management · Meta-model interaction
Cloud platform

1 Introduction

Subject module management is a process of incorporation of contents in developing and coordinating teaching materials [12]. Currently, management of teaching contents is usually conducted through Web-based application such as Content Management Systems and Learning Management Systems. These applications focused on coordinating the subjects, sharing instructional materials, conducting class activities, and preparing examinations. However, most of subject module management

M. N. A. Rahman (✉) · A. Nuhu · S. I. A. Saany · M. F. A. Kadir · S. N. Shamsudin
Faculty of Informatics & Computing, University of Sultan Zainal Abidin,
Besut Campus, 22200 Besut, Malaysia
e-mail: mohdnabd@unisza.edu.my

© Springer Nature Singapore Pte Ltd. 2019
S. K. Bhatia et al. (eds.), *Advances in Computer Communication
and Computational Sciences*, Advances in Intelligent Systems
and Computing 760, https://doi.org/10.1007/978-981-13-0344-9_17

structures are conducted individually by a teacher, therefore lacking the richness of collaboration advantages including sharing activity, social advancement, and knowledge development. Collaboration concept could produce more high-quality module that can help the progressions of students and teacher capability [12].

Limited technology development fund for school contributes to lack of technological platforms used in managing subject module resources in school. This issue in particular is worrisome as it can affect the preparation of students to succeed in learning [4]. The lack of collaboration platform for teacher in subject module development and management also leads to incomprehensive of learning materials for students to support their activities [2, 8, 10, 11]. Hence, there is the need to provide a new collaborative platform that could provide subject module management effectively. The technology-based collaborative platform in developing and managing subject contents provides learning acquisition, socialization, and creation of new notions.

The limited platform for collaborative subject content development, resource sharing, bandwidth, and network facilities gives a challenge for schools to face digital management era. As reported in 2016, there are 10,154 schools in Malaysia with the overall number of students 5,120,802 and 419,820 teachers [9]. With this statistics, the government is striving to deliver best level of education services. However with the rapid information technology development, this problem can be managed by providing the teachers and school management with open data initiatives for creating advanced collaborative information platforms.

The interaction meta-model is proposed to incorporate a number of techniques used in group decision support systems for providing effective collaboration among users [5]. The model provides dynamic interaction factors for collaboration group by using contextual elements *technology, groups,* and *task.* These contextual elements could affect communication, awareness, and responsibility among participants in collaboration group. The embedding of adaptive structuration, coordination, and task–technology interaction techniques in interaction meta-model is believed that it could increase the effectiveness collaboration process in a group decision support system [5].

In today's digital era, the school management has been looking for a new approach and flexible IT platform that can transform teaching and learning process. Cloud computing technology has been explored to address the aforementioned issue. Cloud technology is a computing trend that provides elasticity, scalability, on-demand capabilities, effective cost and efficient delivery performance of services [1, 6, 7, 9]. Recently, some of cloud-based education applications are including courseware, laboratories, on-demand materials, assessment, and teaching materials publication [11].

The rest of paper is organized as follows: Sect. 2 explains the developed collaborative school subject management in the context of Malaysian education. Model deployment and its evaluation results are discussed in Sect. 3. Last section will conclude this research work.

2 Collaborative Subject Management

The proposed collaborative subject module management is modelled by using Raquel's meta-model contextual elements. It is involved group of users, task functions, and technology platform to develop the behaviors of subject module management [5]. These elements are merged as pairs to allow the identification of their relationships and variables that impose the interactions. The importance of Raquel's meta-model in this developed model context is to develop the state of activities in the area of group works, tasks activities, and technological support.

2.1 The Model

The subject management is modelled as multidimensional of *Groups–Tasks–Technology* interaction process. Interaction process involved the components of user's group, tasks functions, and technology platform. The interaction of each component is based on their intersectional variables. Three types of interactions are involved: *technology-based group interaction (T)*, *task–technology interaction (K)*, and *group–task interaction (G)*. The users that can be considered are administrators, contributors (teachers/facilitators), and the consumers (teachers, facilitator, and students). The interaction meta-model in collaboration with subject module management can be depicted as in Fig. 1. Based on the given figure, three interaction factors are created: where G ∩ T is *group–technology adaptation*, G ∩ K is *group–task coordination,* and T ∩ K is *task–technology fit*.

Groups that involved in collaborative subject module management are categorized into three and can be denoted as $U = \{u_\alpha, u_\beta, u_\delta\}$, where u_α is a group of administrator, u_β is a group of contributor, and u_δ is a group of consumer. Contributors are comprised of teacher and facilitator. Meanwhile, teacher, facilitator, and student are the consumer of subject materials. Specifically, for contributor and consumer groups

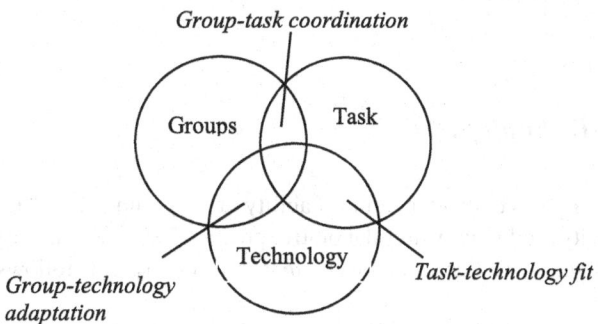

Fig. 1 Meta-model interaction in collaborative subject module management

in subject management will involve four (4) main activities and can be signed as $C = \{\sigma(\text{append}), \sigma(\text{editing}), \sigma(\text{retrieve}), \sigma(\text{archive})\}$. Administrators are included officer that manage user's grant, invitation, prioritize access activity, etc.

2.2 Group–Technology Adaptation

Group–technology adaptation interaction (GT) describes how the users use the technology which possesses on procedures and results in collaboration. In this context, users define the certain features on the advanced tool which will be employed in collaboration process. The used technology should be considered as a set of collective practices and dynamically evolve. For this constant process, users create and recreate the teaching materials or subject contents dynamically over time. Let technology used be $Y = \{y_1, y_2, \ldots, y_n\}$ and $A = (Y, \{U_i\})$ be a component of group–technology adaptation and for each component a transition $T_i = (A_i, A_i, \rightarrow_i)$; then, the interaction between group and technology can be given by:

$$GT = (A, C, T) \tag{1}$$

2.3 Group–Task Coordination

This factor is referred to the approach that the administrator uses to manage the flow of activity in collaboration with subject module management. In the proposed model, coordination activities are modelled based on a decision variable which can be signed as $V = \{v_1, v_2, \ldots, v_n\}$ where the values are from a set of domains C. Some of the examples of V are priority order, constraints, scheduling, process synchronization, and goal identification. Each coordination process indications to a creation $\mathfrak{R}$ of decision variables from decision set of χ. Therefore, the degree of coordination of each activity in the subject module collaboration process can be signed as utility function $V: \chi \rightarrow \mathfrak{R}$.

2.4 Task–Technology Fit

Task–technology fit is defined in terms of ability of the group of $Z = \{u_\beta, u_\delta\}$, $Z \subseteq U$, to deliver activity productivity in collaborative process. Let $z \in Z$, and the productivity can be formalized by resources (subjects) matrix $E = [a_{ij}]_{m \times k}$ as follows:

$$E = \begin{array}{c} \\ s_1 \\ s_2 \\ \cdots \\ s_m \end{array} \begin{array}{cccc} z_1 & z_2 & \cdots & z_k \\ \left[\begin{array}{cccc} a_{11} & a_{12} & \cdots & a_{1k} \\ a_{21} & a_{22} & \cdots & a_{2k} \\ \cdots & \cdots & \cdots & \cdots \\ a_{m1} & a_{m2} & \cdots & a_{mk} \end{array}\right] \end{array} \qquad (2)$$

where

$Z = \{z_j \mid j = 1, 2, \ldots, k\}$ are user instances;
$S = \{s_i \mid i = 1, 2, \ldots, m\}$ are subjects type in instances; and
a_{ij} $(1 \leq i \leq m, 1 \leq j \leq k)$ are subjects quantity.

From matrix E, task–technology fit (productivity) can be calculated as:

$$P(E) = \frac{\rho(E) \times \omega(E)}{\phi(E)} \qquad (3)$$

where

$\rho(E)$ is a performance metrics used (speedup);
$\omega(E)$ is quality of service metric used; and
$\phi(E)$ is the cost used for subject content development.

3 Prototype Deployment and Evaluation

The model prototype is deployed under the public cloud platform environment. A standard PHP framework and Java language are used to develop and execute the functions in the prototype.

3.1 The Architecture

The proposed application model consists of three layers that are connected and interacted with each other: *user layer, task layer,* and *technology layer*. The user layer is used by administrator, contributor, and consumer. This layer consists of dashboard and interface for users to interact with all functions. The role of administrator is to manage the contributor and consumer including inviting, granting, controlling access, and reporting. Meanwhile, task layer is referred to functionalities of the application including activity management, material development, and usage management. Technology layer represented a layer to deploy the application which consists of devices, cloud platform, and network devices. Figure 2 shows the three layers of application model architecture.

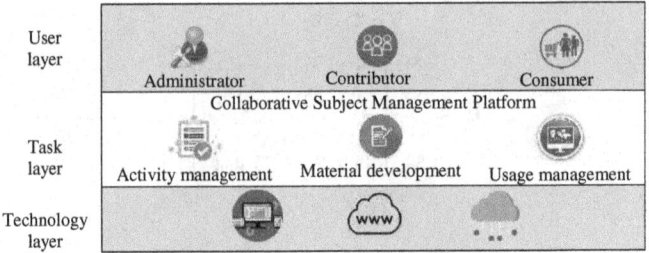

Fig. 2 Model application architecture

3.2   Prototype Evaluation

The usability perspective is used to evaluate the developed prototype. The service usability is measured using cloud-based application quality of service (QoS) metrics suggested by [1, 3]. In the process of evaluation, a questionnaire is provided and circulated to 30 respondents including teachers, students, and school management members. The evaluation process is executed in laboratory session. Four attributes of usability are used: effectiveness, efficiency, learnability, and attractiveness. The attributes are measured based on certain questions with five scales: *strongly disagree, disagree, neutral, agree, and strongly* agree. In this context, each attribute can be defined as follows:

- Effectiveness—the ability of the prototype to assist user in finalizing a range of tasks precisely and having a friendly as well as interactive interface.
- Efficiency—the ability of the prototype to assist the user to complete task in a definable time and provide a set of functions to simplify user in cooperating effectively.
- Learnability—the prototype feature that could help the user to complete task effectively with minimum supervision.
- Attractiveness—the capability of the prototype to attract the interest and engagement of the users to participate.

Four (4) specific questions are used to evaluate the effectiveness (Table 1). Here, more than 58% of respondents agreed that the prototype functions are effective for education resource sharing. The mean value of service effectiveness is 64%. The overall achievement is slightly less encouraging. This is because respondents are not fully adapted to the prototype flow of work. The effectiveness percentage is expected to be increased after the users are having continuous experience and training.

In evaluating the efficiency of the prototype, four (4) quality metric interrogations are applied. Table 2 shows respondent's feedback for efficiency evaluation. There are more than 51% of respondents agreed that the prototype features provide efficiency process in academic resources collaboration. Based on the resulting results, the capability of prototype to support user in completing task in a practical time

Table 1 Effectiveness

No.	Questions	Respondent feedback (%)				
		SD	D	N	A	SA
1	Easy to navigate and access the available functions	13	16	13	40	18
2	Meet robustness standards and usage	10	23	0	60	7
3	Provides visual display to show its intended purpose	6	13	13	50	18
4	Easy to add/edit/upload online material	1	18	18	60	3

SD strongly disagree; *D* disagree; *N* neutral; *A* agree; *SA* strongly agree

Table 2 Efficiency

No.	Questions	Respondent feedback (%)				
		SD	D	N	A	SA
1	Easy to integrate learning materials	20	10	6	64	0
2	Convenience to prepare subject material contents	16	6	16	52	10
3	Application response time is reasonable	6	6	20	48	20
4	Forum and interaction are very useful	16	20	13	38	13

SD strongly disagree; *D* disagree; *N* neutral; *A* agree; *SA* strongly agree

and making interaction effectively still needs to be improved. The mean value for efficiency quality metrics is 62%.

To evaluate the learnability of prototype, five (5) quality metric questions are used. As presented in Table 3, more than 54% of respondents agreed with the capability of the prototype features to provide the environment of learnability to users. Based on the resulting results and observation during the experiment, the user is somewhat less satisfied with the interface and it is quite difficult to move from one interface to another. The mean value for learnability features quality metrics is 62%.

In an environment of different types of users and different skill levels, application attractiveness is very important in order to maintain user trustworthiness to the application. The attractiveness feedback data is shown in Table 4. Five (5) quality metric questions are used to measure attractiveness of the developed prototype. As tabled, more than 50% of respondents agreed that the prototype has achieved the high level of attractiveness. The attractiveness quality metrics mean value is 68%. The result shows that the quality of subject content developed by contributors needs to be improved. In addition, the application interface design also needs to be enhanced

Table 3 Learnability

No.	Questions	Respondent feedback (%)				
		SD	D	N	A	SA
1	Interface is suitable for resource collaboration and subject material sharing	6	17	13	44	20
2	Practical, informative, and useful user manual	16	17	13	42	12
3	Allow the use of external resources that are relevant to application	10	10	13	60	7
4	Allow the provisioning of useful objects related to subject management	0	16	26	48	10
5	Support for knowledge acquisition in cloud environment	3	13	17	47	20

SD strongly disagree; *D* disagree; *N* neutral; *A* agree; *SA* strongly agree

Table 4 Attractiveness

No.	Questions	Respondent feedback (%)				
		SD	D	N	A	SA
1	Comfortable to share subject materials	6	10	34	40	10
2	Application architecture is suitable for non-technical user	2	2	22	40	34
3	Images, text fonts, and colors used are satisfactory	6	10	30	47	7
4	Application dashboard is user friendly	0	0	20	67	13
5	Collaboration activity can be done anywhere and anytime	0	10	20	54	16

SD strongly disagree; *D* disagree; *N* neutral; *A* agree; *SA* strongly agree

by considering the categories of users as well as the level of interactive content provisioning.

Figure 3 illustrates the spider chart of application usability based on mean value of each quality metrics measured. The results achieved may be less satisfactory. This is because the users are not familiar with the developed model. In addition, the application is the first application used by users in collaborative subject module management, especially in cloud technology environment.

Fig. 3 Application usability
quality metrics

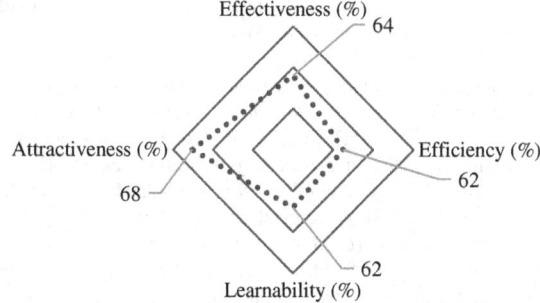

Effectiveness (%) 64

Attractiveness (%) 68

Efficiency (%) 62

Learnability (%) 62

4 Conclusions

Collaborative subject module management is concerned with group of participants in coordinating learning objects in schools. This paper discussed a cloud-based collaborative subject module management model for Malaysian schools. The model is designed based on the concept of Raquel's interaction meta-model. The process involved is basically an iterative and social approach that entailed a group who are focused on communication and coordinate the management of subjects. A working prototype has been developed using a public cloud platform. From the evaluation results, the prototype quality of services metrics considered is moderate. Hence, enhancement work is still continuing and hopefully the model will inspire school management and teachers to take advantage of cloud technology in order to increase the efficiency of teaching and learning process. In the next phase, mobile-based deployment will be the main focus for extending this research work.

Acknowledgements Special thanks to Ministry of Higher Education Malaysia and University of Sultan Zainal Abidin in providing research fund for this research project (FRGS/1/2015/ICT04/UniSZA/02/1-RR143).

References

1. Ardagna, D., Casale, G., Ciavotta, M., Pérez, J.F., Wang, W.: Quality-of-service in cloud computing: modeling techniques and their applications. J. Internet Serv. Appl. **5**(11), 1–17 (2014)
2. Barbosa, E.F., Maldonado, J.C.: IMA-CID: An integrated modeling approach for developing educational modules. J. Braz. Comput. Soc. **17**(4), 207–239 (2011)
3. Bardsiri, A.K., Hashemi, S.M.: QoS metrics for cloud computing services evaluation. Int. J. Intell. Syst. Appl. **6**(12), 27–33 (2014)
4. Bashir, G.M.M., Hoque, A.S.M.L., Majumder, S., Bepary, B., Rani, K.: An integrated online courseware: a step to globalization. In: Proceedings of the International Conference on Electrical Information and Communication Technology (2014)
5. Benbunan-Fich, R. An interaction meta-model for groupware theory and research, Proc. of the Thirtieth Hawaii International Conference on System Sciences, 2, 76–83, 1997

6. Chohan, N., Bunch, C., Krintz, C., Canumalla, N.: Cloud platform datastore support. J. Grid Comput. **11**(1), 63–81 (2013)
7. Di, S., Kondo, D., Cappello, F.: Characterizing and modeling cloud applications/jobs on a Google data center. J. Supercomput. **69**, 139–160 (2014)
8. Geyer, A., Sidman, J., Dumond, D., Rousseau, J., Monagle, R., Awiszus, T., Petty, D., Glassner, S.: Promoting STEM education through local school-industry collaboration: an example of mutual benefits. In: Proceedings of the Integrated STEM Education Conference, pp. 1–4 (2013)
9. Nordin, M.N.R., Abdullahi, N.S., Fadzil, M.A.K., Syadiah, N.S., Syarilla, I.A.S.: A gamification model for resource sharing in Malaysian schools using cloud computing platform. In: Maria, A., et al. (eds.) Intelligent Systems Design and Applications, pp. 406–416 (2017)
10. Umar, I.N., Yusoff, M.T.M.: A study on Malaysian teachers' level of ICT skills and practices, and its impact on teaching and learning. Procedia—Soc. Behav. Sci. **116**, 979–984 (2014)
11. Wang, J.: Build the college network teaching system based on cloud computing. Int. J. Digit. Content Technol. Appl. **7**(7), 1212–1218 (2013)
12. Yao, L., Xiong, X.: Design a teaching resource sharing system in colleges based on cloud computing. In: Proceedings of the International Conference on Information Technology and Applications, pp. 374–378 (2013)

A Novel Data Placement Strategy for Science Workflow Based on MCGA in Hybrid Cloud Environment

Delong Cui, Zhiping Peng, Qirui Li, Jieguang He and Feitan Huang

Abstract How to accomplish an effective data placement algorithm in hybrid cloud environment has become a crucial issue, especially that science workflow is a sophisticated compute or data-intensive application and brought new challenges by the security issues nowadays. In order to solve the security issues of data placement in hybrid cloud environment, we proposed novel data placement strategy in this paper, and the proposed strategy can be partitioned off three stages. Firstly, the initial feasible solutions are generated stage, it is generated by employing homogeneous method, and we get the initial populations of multilayer coding genetic algorithm; secondly, multi-objects optimize stage trade-off performance and cost, and we balance the resources cost and system performance by using Pareto optimal ideology; finally, generate optimal strategy stage, we estimate Pareto optimal solution and chose the optimum solution act as ultimate data placement strategy. Experiment results prove that our data placement strategy can not only guarantee data security, but also reduce data makespan time.

Keywords Data placement strategy · Hybrid cloud environment
Data placement · Multilayer coding genetic algorithm · Transfer time

D. Cui · Z. Peng (✉) · Q. Li · J. He · F. Huang
College of Computer and Electronic Information,
Guangdong University of Petrochemical Technology, Maoming, China
e-mail: pengzp@foxmail.com

D. Cui
e-mail: delongcui@gdupt.edu.cn

Q. Li
e-mail: liqirui@foxmail.com

J. He
e-mail: 1062219923@qq.com

F. Huang
e-mail: 1009299883@qq.com

© Springer Nature Singapore Pte Ltd. 2019
S. K. Bhatia et al. (eds.), *Advances in Computer Communication
and Computational Sciences*, Advances in Intelligent Systems
and Computing 760, https://doi.org/10.1007/978-981-13-0344-9_18

1   Introduction

Hybrid cloud computing environment is a cloud computing environment which uses a mix of public cloud computing services, private cloud computing services, and third-party cloud services which with orchestration between the two platforms. By allowing multi-user workloads to move between public and private clouds as computing needs and costs change, hybrid cloud computing platform gives science and businesses greater flexibility and more cost data deployment options [1]. By focusing on the science applications, Vaquero et al. design scientific workflow management system shifts away from the user tasks scheduling activities, typically considered by grid computing environments for optimizing the finished time of complex computations on predefined platform resources, to a domain-specific view of what data types, tools, or distributed virtual resources should be made available to the scientists and how can one make them easily accessible and with specific Quality of Service (QoS) requirements [2].

Due to the strong impact in big data related scientific research field, data center-oriented data placement algorithms of scientific workflow applications in hybrid cloud computing environment have gained much attention in recent years. Especially the private data security issue in hybrid cloud services. So we proposed a novel data placement strategy based on Multilayer Coding Genetic Algorithm (MCGA) in this paper. The remainder of this paper is organized as follows: We review the data placement strategy in recent years in Sect. 2; we explore the cloud system models, related definitions and problem analysis in Sect. 3; based on the proposed hybrid cloud models, a novel data placement strategy based on MCGA is proposed in Sect. 4; In Sect. 5, we present extensive performance evaluations and reach the conclusions in Sects. 5 and 6 respectively.

2   Related Works

Several studies on security data placement in clouds computing environment have been conducted in recent years [3–6]. Vaquero et al. [3] discuss more than 20 definitions of cloud computing definitions and describe the relationships between cloud computing and grid computing in details. Yu, Modi, and Subashini et al. [4–6] from side to side review the challenges of security problems in cloud computing environment. Khani et al. [7] proposed a self-organized, scalable, and distributed load balancing algorithm based on game theory in cloud computing environment. Kebert et al. [8] focus on the high mobility feature of the VMs in cloud computing, meanwhile moving of sensitive data from metadata increases the error and loss of sensitive information, so some researchers reconnoitered the man-in-the-middle attack on VMware while mitigating the VMs online.

In conclusion, lots of research works have pay attention to data placement in cloud computing environment, but a majority of works fail to pay attention to data security problem in hybrid cloud.

3 Related Definitions and Problem Analysis

3.1 Related Definitions

According to the origin of science workflow datasets, it can be divided into initial datasets and generate datasets. Initial datasets act as input of workflow tasks. Generated datasets act as not only output of workflow tasks, but also input of others workflow tasks. So, it is very intricate and complex relationship among workflow datasets. When workflow scheduling problem is described, the model of DAG is common be employed. The relationship between data and tasks is multi–multi, it means that multi-datasets are used when a task is executed, and a dataset is also used by multitasks. An example of multi-datasets workflow execute process is shown in Fig. 1 [9].

Definition 1 Workflow can be described by triple $G =< T, C, DS >$, in which $T = \{t_1, t_2, \ldots, t_n\}$ express all tasks in workflow G; C express adjacency matrix of control flow among tasks, if $C_{ij} = 1$, that means task t_j can be executed when task t_i is finished; $DS = \{d_1, d_2, \ldots, d_n\}$ express assemblage of all datasets.

Definition 2 The task of workflow can be described by double $< ID, OD >$, $i = 1, 2, \ldots, |T|$, in which ID express assemblage of task t_i input datasets, OD express assemblage of task t_i output datasets.

Definition 3 The dataset d_i can be described by $\langle ds, ptl, gt, pc \rangle$, $i = 1, 2, \ldots, |DS|$, in which ds express the size of dataset d_i; ptl express the store place of dataset d_i; gt express the task of generate dataset d_i; pc express the placement place of dataset d_i; the value of ptl and gt is given below:

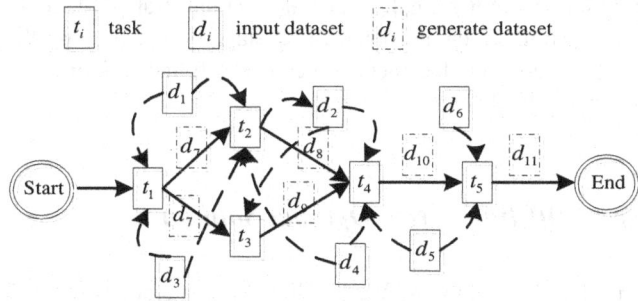

Fig. 1 An example of multi-datasets workflow execution process

$$ptl = \begin{cases} 0, & d_i \in DS_{pub} \\ pricloud(d_i), & d_i \in DS_{prc} \end{cases} \tag{1}$$

$$gt = \begin{cases} 0, & d_i \in DS_{ini} \\ Task(d_i), & d_i \in DS_{gen} \end{cases} \tag{2}$$

in which, DS_{pub} express public datasets; DS_{prc} express private datasets; DS_{ini} express initial datasets in workflow G; DS_{gen} express generate datasets in workflow G; $Task(d_i)$ express task in generate datasets d_i.

3.2 Data Placement Strategy in Hybrid Cloud Environment Problem Analysis

Data placement strategy object in hybrid cloud environment can be presented as follows: When a task can be executed in hybrid cloud environment, all input datasets must be transmitted to locate place of the task firstly, otherwise the datasets must be transmitted across the data center before the cloud task can be executed. The transfer time of datasets is commonly much larger than the scheduling time, so the object of task scheduling strategy in hybrid cloud environment schedules the task to the data center so that data can be used to minimize transfer time between various data centers.

4 Data Placement Strategy Based on MCGA

The proposed data placement strategy in this paper can be partitioned off three stages. Firstly, the initial feasible solutions are generated stage, it be generated by employing homogeneous method and we get the initial populations of multilayer coding genetic algorithm (MCGA); secondly, multi-objects optimize stage tradeoff performance and cost, we balance the resources cost and system performance by using Pareto optimal ideology; finally, generate optimal strategy stage, we estimate Pareto optimal solution which are obtained by stage two, and chose the optimum solution act as ultimate data placement strategy. The framework of data placement strategy is shown in Fig. 2.

4.1 Generate Off-line Pareto Optimal Solution

Generate initial feasible solutions. We employed homogeneous method to generate initial feasible solutions of GA. The code rule and efficient guarantee of initial solutions are described as follow:

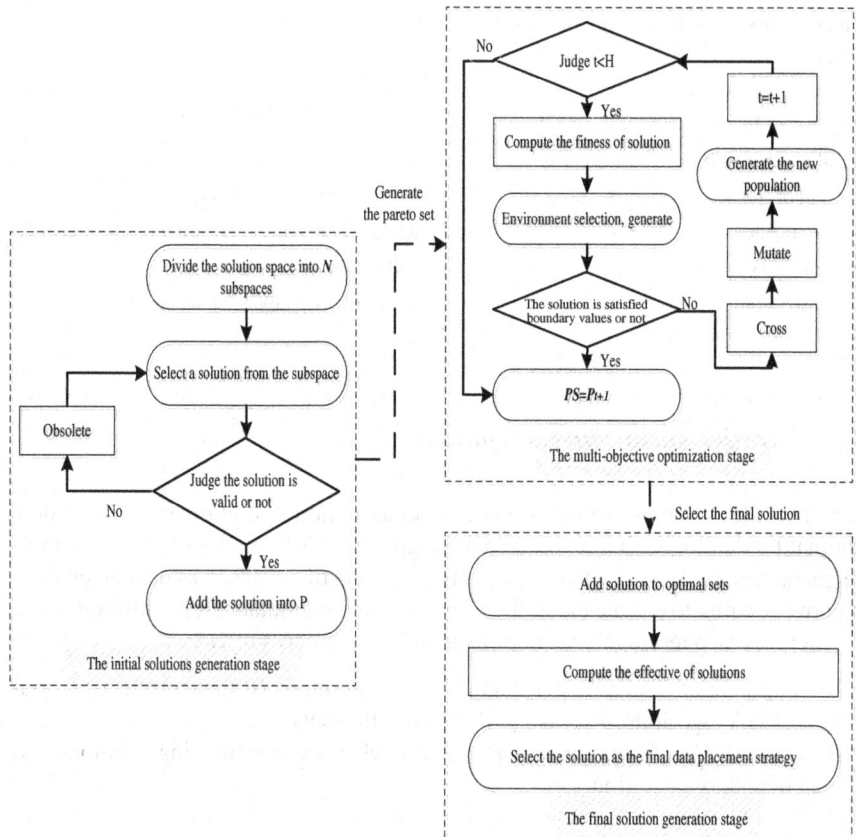

Fig. 2 Framework of proposed MCGA algorithm

- Code rule: Chromosomes are map code of data block to various data centers in hybrid cloud environment in which data placement strategy is based on multilayer coding genetic algorithm. In this paper, we supposed that one data block can only be mapped to one data center, but one data center can store multi-data block. So we employ data block-based code method, and the various data placement strategies are dissimilar map location of each data block.
- Efficient guarantee of initial solutions: Each initial feasible solutions is generated by homogeneous method ability to express data placement strategy, but the invalid solutions can be divided into three categories according to the different failure cause.

Table 1 Parameters of the experimental simulation platform

Parameter	Values
Number of data centers	4
Number of PEs requirements	5
Total number of jobs	300–1000
Length of job	10,00–20,000 million instructions
VM frequency	10,000–30,000 million instructions per second
VM memory (RAM)	2048–2048 mega byte
VM bandwidth	500–1,000 mega byte per second
Number of VM buffer	10–50

4.2 Parallel Multi-objects Optimal

Each feasible solutions of initial solutions act as an initial solution and independent optimal by employing hierarchic genetic algorithm. The validity of solutions must be guaranteed in the progress of iteration optimizing of GA (such as data blocks that is corresponding to private cloud drop out of cross and mutation operations in GA).

There are two termination criterions in iteration process of GA:

- If iterations t of GA are larger than preinstall iterations H, then termination optimizing GA algorithm, $\overline{P}_{t+1}$ is the Pareto solution sets.
- If exist $t < H$ and solution s in $\overline{P}_{t+1}$, meanwhile exist optimizing solutions less than boundary condition.

4.3 Generate Online Pareto Optimal Solution

According to the Pareto optimal solution, workflow data is distributed in hybrid cloud computing environment, but the faults of data process or operating environment are inevitable. If any fault happens, the online optimal stage is triggered. It must take certain measures and countermeasures to timely solve these problems.

5 Performance Evaluation

To evaluate the efficiency of our approach, implementations have been performed on the WorkflowSim [10] computing environment. Table 1 shows the parameters of the experimental simulation platform.

5.1 Comparison of Response Time

A comparison of makespan with different numbers of tasks for the Montage and
LIGO workflows is demonstrated in Fig. 3.

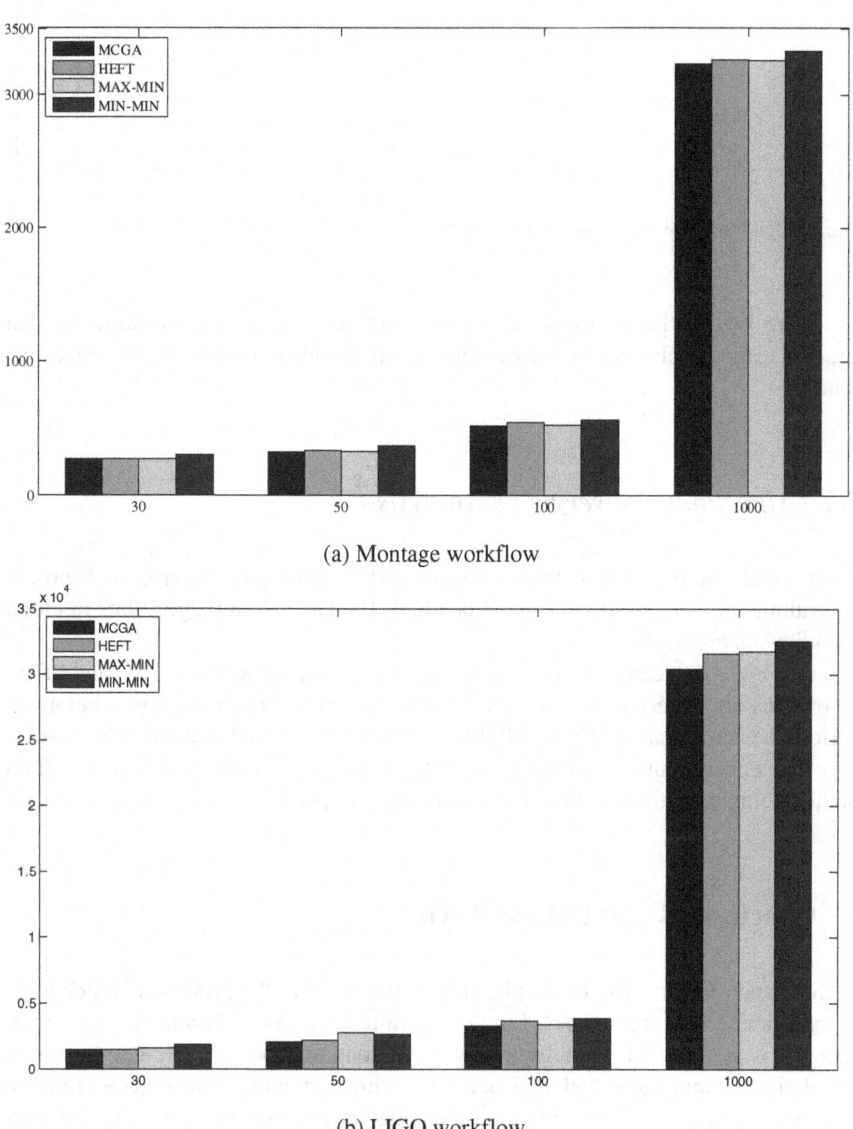

(a) Montage workflow

(b) LIGO workflow

Fig. 3 Scheduling histogram

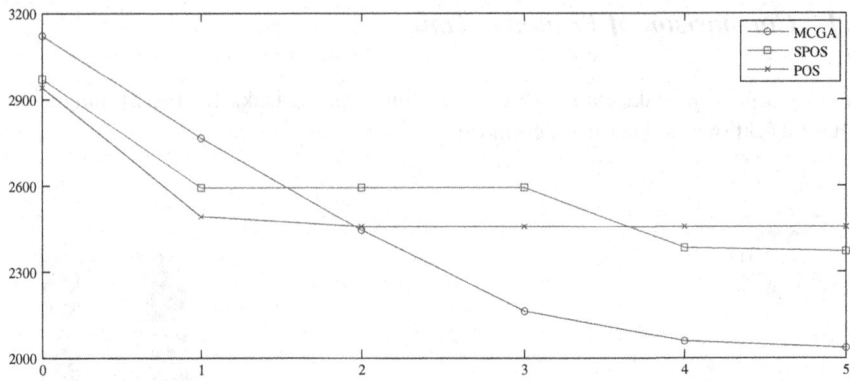

Fig. 4 Cloud workflow convergence line graph

Figure 3 shows the makespan of various workflow's execution, from which we can generalize the conclusions as our strategy can effectively reduce workflow makespan time.

5.2 Algorithm Convergence Analysis

We selected Inspiral_50.xml datasets to analyze the convergence of proposed optimal algorithm, and compared with the benchmark PSO and SPSO algorithms in cloud workflow environment.

From Fig. 4, we can obtain the in the initial stage, the benchmark PSO and SPSO algorithms are superior to proposed MCGA algorithm, but through two iterations period, the benchmark PSO and SPSO algorithms are stocked in local optimal solution. The experiment results demonstrate the superiority of the proposed algorithm in the quality of optimal solution and convergence speed.

6 Conclusions and Future Work

In this paper, we proposed data placement strategy, and the proposed strategy can be partitioned off three stages. Firstly, the initial feasible solutions are generated stage, it is generated by employing homogeneous method, and we get the initial populations of multilayer coding genetic algorithm; secondly, multi-objects optimize stage tradeoff performance and cost, and we balance the resources cost and system performance by using Pareto optimal ideology; finally, generate optimal strategy stage, we estimate Pareto optimal solution and chose the optimum solution act as ultimate data placement strategy. Experiments show that our strategy can effectively reduce data makespan time during workflow's execution.

Acknowledgements The work presented in this paper was supported by: National Natural Science Foundation of China (61672174, 61272382), Guangdong Provincial Science & Technology Program (2015B020233019, 2014A020208139), Key Project of Guangdong Province in the Research Center of Cloud Robot (Petrochemical) Engineering Technology (No. 2015B090903084), and Guangdong University of Petrochemical Technology College Students' Innovation and Entrepreneurship Training (2017pyA027).

References

1. http://searchcloudcomputing.techtarget.com/definition/hybrid-cloud
2. https://en.wikipedia.org/wiki/Scientific_workflow_system
3. Vaquero, L., Rodero-Merino, L., Caceres, J., et al.: A break in the clouds: towards a cloud definition. ACM SIGCOMM Comput. Commun. Rev. **39**(1), 50–55 (2009)
4. Yu, H., Powell, N., Stembridge, D., Yuan, X.: Cloud computing and security challenges. In: Proceedings of the 50th Annual Southeast Regional Conference, pp. 298–302. ACM (2012)
5. Modi, C., Patel, D., Borisaniya, B., Patel, A., Rajarajan, M.: A survey on security issues and solutions at different layers of cloud computing. J. Supercomput. **63**(2), 561–592 (2013)
6. Subashini, Subashini, Kavitha, Veeraruna: A survey on security issues in service delivery models of cloudcomputing. J. Netw. Comput. Appl. **34**(1), 1–11 (2011)
7. Hadi, K., Nasser, Y., Siamak, M.: A self-organized load balancing mechanism for cloud computing
8. Kebert, A., Banerjee, B., George, G., Solano, J., Solano, W.: Detecting distributed SQL injection attacks in a Eucalyptus cloud environment. In: Proceedings of the 12th International Conference on Security and Management (SAM-13), Las Vegas, NV (2013)
9. Li, X.J., Wu, Y., Liu, X., Cheng, H.M., Zhu, E.Z., Yang, Y.: Datacenter-oriented data placement strategy of workflows in hybrid cloud. J. Softw. **27**(7), 1861–1875 (2016). (in Chinese)
10. Chen, W., Deelman, E.: WorkflowSim: a toolkit for simulating scientific workflows in distributed environments. In: Proceedings of the IEEE International Conference on E-Science, Bangalore, India, pp. 1–8. IEEE (2012)

Adaptive Resource Allocation in Interoperable Cloud Services

S. Anithakumari and K. Chandrasekaran

Abstract Interoperable cloud computing is the one in which the services or resources of one cloud can be accessed by another cloud. The implementation of interoperable cloud architecture is a challenging one because various characteristics of the cloud computing environment need to be considered for its achievement. The aim of this work is to implement interoperable cloud computing with the awareness of service-level agreements and to provide adequate resources when shortage of resources occurs at one cloud while providing the agreed services to the user. To achieve this, we proposed a methodology of interoperability-based flexible resource management. Initially, the SLA templates of private and public cloud are mapped using the Soft TF-IDF metric with case-based reasoning (CBR) approach. Then, based on the mapped SLAs, different clusters of cloud providers are formed with the help of K-means clustering technique. And finally, if one of the cloud in a cluster faces the problem of resource shortage, the flexible resource allocation is provided through the adaptive dimensional search algorithm.

Keywords Cloud computing · Service-level agreement (SLA) · Resource allocation · Interoperability · SLA mapping

1 Introduction

Cloud computing technology is based on different computing techniques, coupled with virtualization which defines the operating system (OS) images, middleware, and applications procreated and pro-allocated to physical machines or slices of a server stack [3]. One of the challenges need to be faced by cloud computing is related to the data interoperability and portability. An interoperable cloud infrastructure

S. Anithakumari (✉) · K. Chandrasekaran
NITK Surathkal, Surathkal, Karnataka, India
e-mail: lekshmi03@gmail.com

K. Chandrasekaran
e-mail: kchnitk@gmail.com

© Springer Nature Singapore Pte Ltd. 2019
S. K. Bhatia et al. (eds.), *Advances in Computer Communication and Computational Sciences*, Advances in Intelligent Systems and Computing 760, https://doi.org/10.1007/978-981-13-0344-9_19

management standard focuses on interoperability among cloud service providers and their customers or developers [5]. Different researchers have proposed a number of solutions for the interoperability problem in distributed systems. The aim of those solutions is finding a universal standard that enabling all distributed services to communicate with one another [9]. The highest financial profit in cloud computing can be achieved by allocating the cloud resources in a powerful manner [10].

Here, we have established a mechanism for flexible resource allocation between the cloud service providers based on SLA mapping and clustering techniques to propose an interoperable cloud computing environment. This interoperability is made possible with the use of multiple techniques such as adaptive dimensional search algorithm (ADS), clustering, and SLA mapping. The remaining part of the paper is organized as follows. Section 2 describes the related research work conducted in this field, and Sect. 3 explains the resource allocation for interoperable clouds. Section 4 describes the experimental evaluation, and finally, Sect. 5 gives the conclusion of the paper.

2 Related Work

The very recent works related to the resource allocation problem in cloud computing environment are listed below:

Huang et al. [15] had proposed an adaptive resource management system in cloud computing environment. They employed support vector regression (SVR) to determine the number of resource utilization by prediction with respect to the service-level agreement (SLA) of each process. Based on this, the resources were redistributed according to the current status of all of the virtual machines (VM) installed in the physical machine. In order to find the reallocation of resources, they used genetic algorithm, and with this, they optimally allocated the resources in the cloud. At the end, they had reached an agreement between physical machines resource utilization which was monitored by the physical machine monitor as well as SLA between virtual machine operators and cloud service providers.

Addis et al. [16] had proposed a scalable distributed hierarchical framework based on a mixed-integer nonlinear optimization of resource management acting at multiple timescales in a very large cloud platform. The run time management framework of their proposed method assumed that the PaaS provider supported multiple transactional services execution with a set of heterogeneous servers. They mainly focused on the CPU and RAM as representative resources among many other physical resources for the resource allocation problem. The main objective of their resource allocation problem is to maximize the profit such that the difference between revenues from SLA contracts and costs associated with servers switching and VM migrations.

Shen and Liu [17] had proposed a resource sharing platform for the collaborative cloud computing (CCC) called harmony which integrated the resource management and reputation management in a balanced manner. Their work incorporated three innovations; they are integrated multi-faceted resource and reputation management, multi-QoS-oriented resource selection, and price-assisted resource/reputation control. Shen and Liu [17] had proposed dynamic resource management system in heterogeneous mobile cloud computing (HMC) environment with the awareness of the quality of service (QoS). They designed the network by integrating the mobile access part and the cloud computing part with the utilization of inherent heterogeneity. They also proposed a novel cross-network radio and cloud resource management scheme for HMC networks by aiming at the maximization of the tenant revenue while satisfying the QoS requirements.

Lu et al. [18] had proposed a fairness evaluation framework for the resource allocation scheme in cloud computing. In this paper, they proposed Dynamic Evaluation Framework for Fairness (DEFF) to evaluate the fairness on allocating the resources using a resource allocation algorithm. They had proposed two submodels to describe the resource demand, with dynamic characteristics and the computing node number in cloud environment, named as dynamic demand model (DDM) and dynamic node model (DNM). They had employed two typical resource allocation algorithms such as utility-based algorithm—fairness in order to show the effectiveness of their proposed fairness evaluation framework.

3 Resource Allocation for Interoperable Clouds

The proposed methodology contains three phases: First, an automatic SLA matching is performed to form an interoperable cloud environment by considering the SLA templates of different cloud service providers using case-based reasoning (CBR) for the classification and regression process. The matching of the SLA templates is performed with the use of a string similarity distance called Soft Term Frequency-Inverse Document Frequency (Soft TF-IDF). After that clustering of different cloud service providers is done based on the mapped SLAs in the second phase with the help of K-means clustering technique. Finally, to optimally allocate the resources between different cloud computing providers in the same cluster, we design a flexible resource management system with the use of the population-based metaheuristic approach like adaptive dimensional search (ADS) in the last stage. The resource allocation is performed based on hypervisor standardization and provides an enhanced adaptive resource management system in cloud computing with different cloud providers. The proposed methodology is depicted in the form of a block diagram as shown in Fig. 1. In this proposed methodology, automatic SLA mapping is performed initially based on the SLA templates of different cloud providers with the help of the string similarity metric called TF-IDF distance metric. According to the results of this SLA mapping, cloud service providers are grouped into multiple clusters using K-means clustering algorithm. In a single cluster, when the shortage of resources occurs with a single cloud service provider, the resources of another cloud service provider are allocated optimally through the adaptive dimensional search (ADS) algorithm when the request for the resources arrived at the InterGrid Gateway present in each cloud.

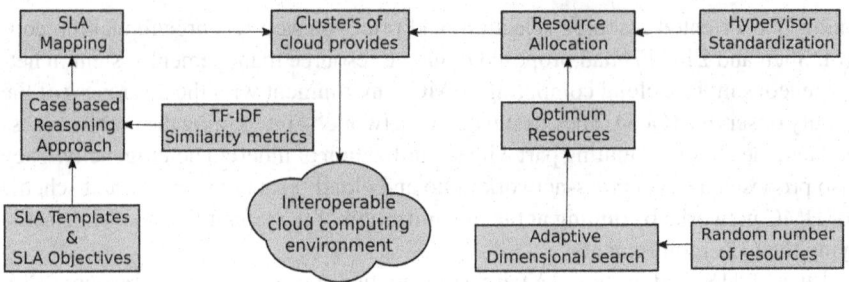

Fig. 1 Schematic diagram of the proposed methodology

3.1 SLA Mapping of Multiple Cloud Service Providers

Normally, the cloud service providers provide the services to the cloud users based on the contract duly signed between them called service-level agreements (SLAs) which include the type of service provided by the service provider. While accessing the services from the intended service provider, a shortage of resources may occur and this can be tackled with the help of an interoperable cloud computing environment. The process of SLA mapping is done here by using case-based reasoning (CBR) approach. An SLA template contains three parts; they are as follows: (i) service-level objectives (SLO), (ii) SLA metric, and (iii) SLA parameters. The SLO defines the type of the service requested from the user, and the quantity can be measured by the specification from SLA metric. The SLA parameters represent the name or unit used to denote the SLA metric. The SLA templates of private cloud will differ from that of public clouds in terms of either the SLA metric defined for the measurement of SLOs or the SLO itself.

3.2 Clustering of Cloud Service Providers Based on Mapped SLAs

After mapping the SLA templates, based on the knowledge available from the CBR approach, clustering of the cloud service providers is performed with the use of K-means clustering algorithm. Through this clustering method, finally, the clusters of cloud providers are formed that are having the most similar SLA templates.

K-means clustering algorithm is the generally used clustering technique to form different clusters with elements of similar nature, and we used this algorithm here to form the clusters of different service providers as this is the algorithm to produce the clusters with minimum overhead. The steps of the algorithm are given below.

Algorithm 1 K-means clustering algorithm

Input: Existing cloud service providers
Output: Clusters of cloud service providers
step1: Select K number of initial centroids
step2: Calculate the squared distance from each centroid to all other points
step3: Assign the point to the centroid which is closest to them to form clusters
step4: Update the centroids in each cluster if the centroids formed are similar to previous ones
step5: Stop

3.3 Resource Allocation Using IGG and Adaptive Dimensional Search

The resource allocation process we are considered here is useful at the time when one of the cloud providers in the cluster faces the problem of shortage of resources. This can be tackled with the help of InterGrid Gateway (IGG) present at each of the cloud which employs the optimum resource allocation mechanism where the resource provider will provide the resources more efficiently. In our paper, the resource allocation between different cloud providers within the same cluster is done through the use of optimal resource allocation mechanism. This resource allocation mechanism is used here to provide the fair amount of resources to the intended cloud provider when it faces the problem of resource scarcity, and it is accomplished by means of the resource request sent by them. The optimal resource allocation mechanism used here is equipped with the new metaheuristic approach called adaptive dimensional search (ADS) algorithm. Before discussing this algorithm for resource provisioning, first, we analyze about the operation of IGG in the clouds.

3.3.1 Resource Allocation Through Inter Grid Gateway (IGG)

In the above Fig. 2, how the resources are optimally allocated when the request for resources arrived at the IGG is clearly mentioned. At first, while providing a service to the intended user, shortage of resources for providing the service may occur. On that particular moment, the service provider 1 in the private cloud will send the request to the public cloud present in that same cluster through the IGG and is forwarded to the service provider 2 in the public cloud. The request for resources may be of the form in Eq. 1.

$$Req = \{R_T, R_n, T_u\} \tag{1}$$

where

R_T Type of resources requested (e.g., bandwidth, memory)
R_n Total number of resources requested
T_u Resource utilization time.

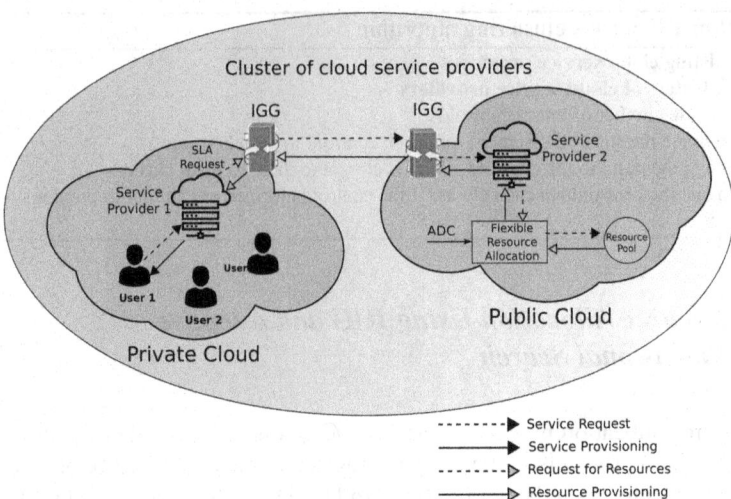

Fig. 2 Flexible Resource allocation through IGG and ADS algorithm

Based on the request as given in Eq. 1, the service provider 2 will employ the resource allocation mechanism to optimally allocate the resources in the form of virtual machines (VMs) to the service provider 1 from the resource pool. Note that, the resources will be allocated only the requested resources are available otherwise, the service provider 1 has to be waited for a certain period of time until the resources used in another location becomes free. To achieve the optimal allocation of resources to the requesting provider, the adaptive dimensional search algorithm is used here and is explained in the following section.

3.3.2 Adaptive Dimensional Search Algorithm

Adaptive dimensional search (ADS) algorithm [21] is a new metaheuristic approach, but it differs from nature-inspired metaheuristic approaches in the sense that it does not employ any metaphor for its implementation. The candidate solutions for the current population in this algorithm can be generated from the best solutions obtained for previous population and is generally referred as $1 + \mu$ type selection. In [21], the algorithm is used for the discrete truss optimization and we modified this algorithm in our approach for the flexible resource allocation between the clouds based on the resources requested from one cloud to another. To provide the optimal resource allocation with the fairness utilization of resources and this can be achieved by minimizing the Skewness and this is taken as the objective function of the resource allocation algorithm.

Skewness

Skewness [22] is the parameter which is used to enumerate the inequality in the utilization of resources. Consider that the number of resources as m, and the utilization of jth resource as u_j, then the skewness of the nth service provider can be calculated as,

$$skewness(n) = \sqrt{\sum_{j=1}^{m} \left(\frac{u_j}{\bar{u}} - 1\right)^2} \qquad (2)$$

where

$\bar{u}$ Average resource utilization for nth service provider.

4 Experimental Evaluation

The proposed methodology is implemented using CloudSim tool with Intel i5 Personal Computer with 2.99 GHz CPU, 8 GB RAM. Here, we are considering six private clouds and three public clouds with three SLA templates for each public cloud and two SLA templates for each private cloud for the mapping phase. The number of virtual machines considered is 500 to provide the resources to the requested cloud provider. Initially, we performed the mapping of SLA templates of public and private clouds through the CBR approach. Then, to enable the interoperability, different clusters of cloud service providers are formed based on the SLA mappings with the help of K-means clustering technique in which a single cluster contains a number of cloud providers with similar SLA templates. The performance achieved by our proposed method is depicted in the following paragraphs.

The utility and cost values for the proposed SLA mapping technology with N (N = 10) number of SLA mappings are given in the Figs. 3 and 4.

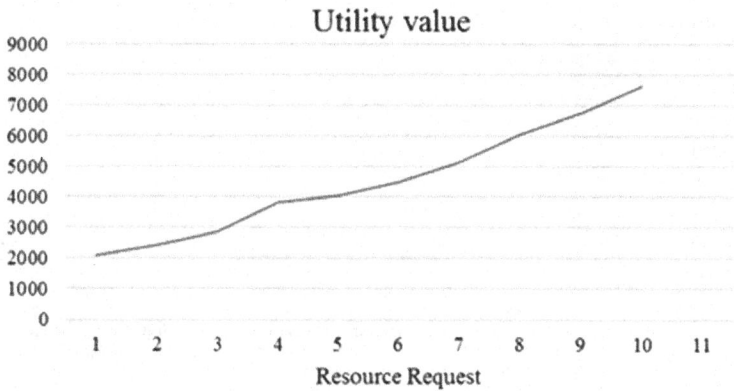

Fig. 3 Utility Values for N SLA mappings (N = 10)

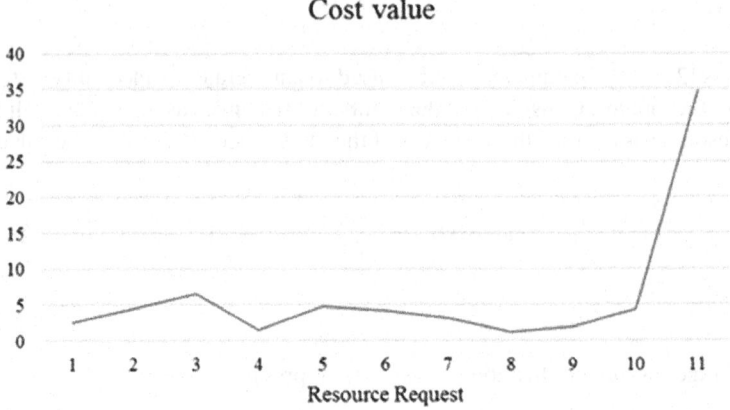

Fig. 4 Cost Values for N SLA mappings (N = 10)

The performance of our proposed resource allocation is compared with other existing works for SLA mapping [23] and resource allocation [24] in cloud computing. In [23], the authors used different methods to measure the overall utility and overall cost occurs in the SLA mapping process. Similarly, while allocating the resources based on user requests, different strategies are followed and compared. Our proposed methodology is also compared with the results obtained in [23, 24], and the evaluation of performance is done, and this is given below. The efficiency of SLA mapping by our proposed methodology with other existing works namely maximum method, threshold method and significant-change method [23] in terms of the parameters such as overall net utility and overall cost are compared. The results of comparison are given in the Figs. 5 and 6. The average weighted response time (AWRT) and slow-

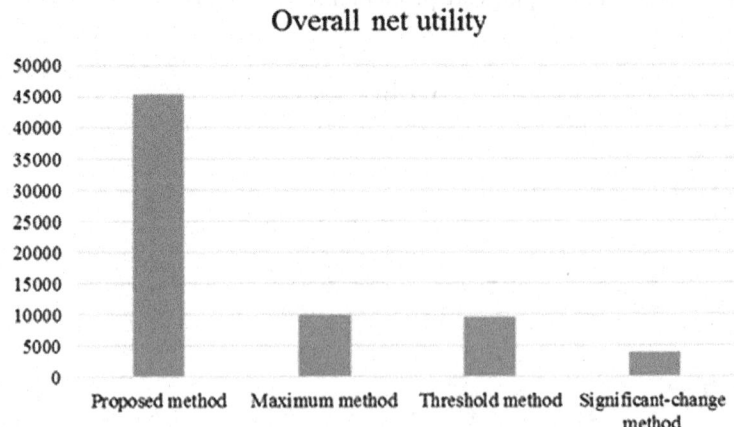

Fig. 5 Overall net utility of proposed methodology with existing works

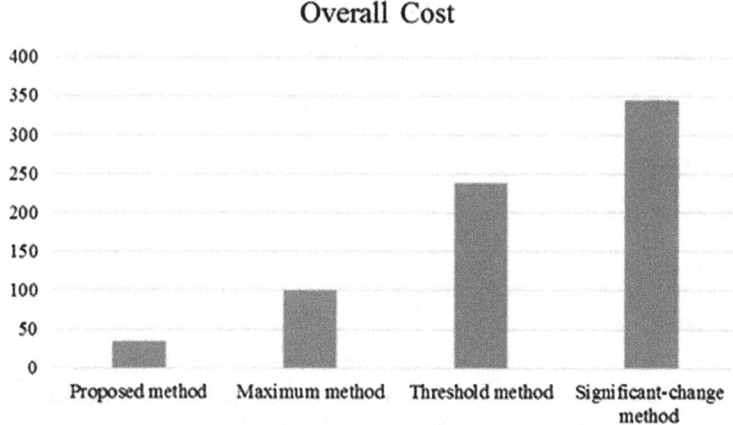

Fig. 6 Overall cost of proposed methodology with existing works

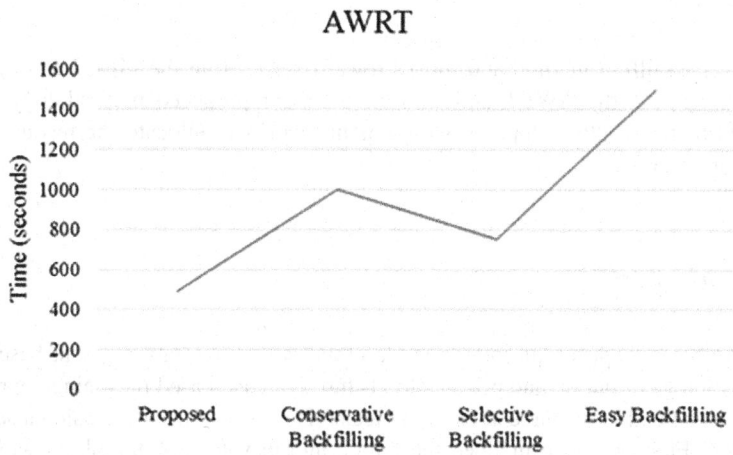

Fig. 7 Performance comparison of proposed method (AWRT)

down of the proposed methodology while responding to the request for resources by the resource allocation mechanism are compared with other existing techniques such as conservative backfilling, selective backfilling, and easy backfilling which are the techniques used for the provision of cloud resources in [24], and the comparison is given in the form of graphs in Figs. 7 and 8, respectively.

Figure 7 represents that the AWRT of our proposed methodology yields minimum result than the selective backfilling technique. Similarly, the slowdown incurred during our proposed resource allocation scheme is also minimum compared to the

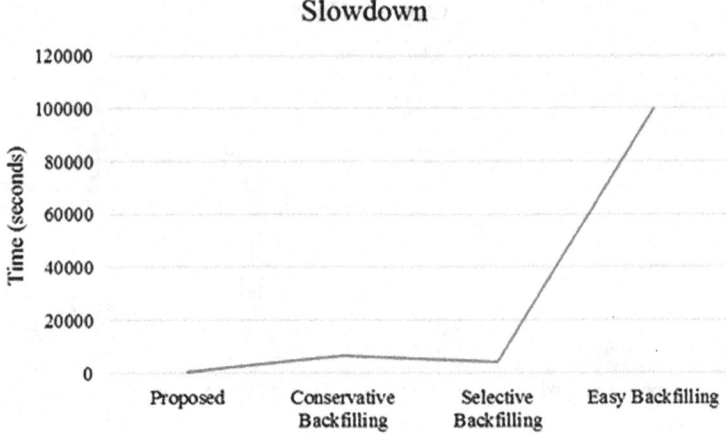

Fig. 8 Performance comparison of proposed method (Slowdown)

selective backfilling method and moves toward zero. From the Figs. 7 and 8, it is proved that both the AWRT and slowdown of our proposed methodology have achieved better results compared with existing works to allocate the resources in cloud computing.

5 Conclusion

Here, we have proposed an interoperable cloud computing environment based on SLA mappings of public and private clouds through case-based reasoning approach to achieve a flexible resource allocation mechanism by adaptive dimensional search algorithm. First, we have mapped the public and private SLA templates with the help of a string similarity metric called Soft TF-IDF, an effective metric used to find the similarity between set of strings. And this knowledge is used in the automatic mapping of SLA templates by CBR approach. After creating the mapping, the clouds that are having similar SLA templates are grouped into same cluster with the use of K-means clustering technique. Finally, the flexible resource allocation mechanism is optimally allocated the resources to the cloud provider with scarcity of resources through ADS algorithm. The experimental results shown that better output is obtained by our proposed methodology and the uneven utilization of resources by certain user also gets minimized.

References

1. Katsaros, G., Kousiouris, G., Gogouvitis, S.V., Kyriazis, D., Menychtas, A., Varvarigou, T.: A self-adaptive hierarchical monitoring mechanism for clouds. J. Syst. Softw. **85**(5), 1029–1041 (2012)
2. Dikaiakos, M.D., Katsaros, D., Mehra, P., Pallis, G., Vakali, A.: Cloud computing: distributed internet computing for it and scientific research. IEEE Internet Comput. **13**(5), 10–13 (2009)
3. Buyya, R., Yeo, C.S., Venugopal, S., Broberg, J., Brandic, I.: Cloud computing and emerging it platforms: vision, hype, and reality for delivering computing as the 5th utility. Future Gener. Comput. Syst. **25**(6), 599–616 (2009)
4. Marston, S., Li, Z., Bandyopadhyay, S., Zhang, J., Ghalsasi, A.: Cloud computing the business perspective. Decis. Support Syst. **51**(1), 176–189 (2011)
5. Kaufman, L.M.: Data security in the world of cloud computing. IEEE Secur. Priv. **7**(4), 61–64 (2009)
6. Petcu, D., Macariu, G., Panica, S., Crăciun, C.: Portable cloud applicationsfrom theory to practice. Future Gener. Comput. Syst. **29**(6), 1417–1430 (2013)
7. Ranjan, R.: The cloud interoperability challenge. IEEE Cloud Comput. **1**(2), 20–24 (2014)
8. Hofmann, P., Woods, D.: Cloud computing: the limits of public clouds for business applications. IEEE Internet Comput. **14**(6), 90–93 (2010)
9. Blair, G., Grace, P.: Emergent middleware: tackling the interoperability problem. IEEE Internet Comput. **1**, 78–82 (2012)
10. Rochwerger, B., Breitgand, D., Levy, E., Galis, A., Nagin, K., Llorente, I.M., Montero, R., Wolfsthal, Y., Elmroth, E., Caceres, J., et al.: The reservoir model and architecture for open federated cloud computing. IBM J. Res. Dev. **53**(4), 535–545 (2009)
11. Beloglazov, A., Abawajy, J., Buyya, R.: Energy-aware resource allocation heuristics for efficient management of data centers for cloud computing. Future Gener. Comput. Syst. **28**(5), 755–768 (2012)
12. Silaghi, G.C., Şerban, L.D., Litan, C.M.: A time-constrained sla negotiation strategy in competitive computational grids. Future Gener. Comput. Syst. **28**(8), 1303–1315 (2012)
13. Goudarzi, H., Pedram, M.: Multi-dimensional SLA-based resource allocation for multi-tier cloud computing systems. In: 2011 IEEE International Conference on Cloud Computing (CLOUD), pp. 324–331. IEEE (2011)
14. Abu Sharkh, M., Jammal, M., Shami, A., Ouda, A.: Resource allocation in a network-based cloud computing environment: design challenges. IEEE Commun. Mag. **51**(11), 46–52 (2013)
15. Huang, C.-J., Guan, C.-T., Chen, H.-M., Wang, Y.-W., Chang, S.-C., Li, C.-Y., Weng, C.-H.: An adaptive resource management scheme in cloud computing. Eng. Appl. Artif. Intell. **26**(1), 382–389 (2013)
16. Addis, B., Ardagna, D., Panicucci, B., Squillante, M.S., Zhang, L.: A hierarchical approach for the resource management of very large cloud platforms. IEEE Trans. Dependable Secure Comput. **10**(5), 253–272 (2013)
17. Shen, H., Liu, G.: An efficient and trustworthy resource sharing platform for collaborative cloud computing. IEEE Trans. Parallel Distrib. Syst. **25**(4), 862–875 (2014)
18. Lu, D., Ma, J., Xi, N.: A universal fairness evaluation framework for resource allocation in cloud computing. China Commun. **12**(5), 113–122 (2015)
19. Unger, T., Leymann, F., Mauchart, S., Scheibler, T.: Aggregation of service level agreements in the context of business processes. In: 12th International IEEE Enterprise Distributed Object Computing Conference, EDOC'08, pp. 43–52. IEEE (2008)
20. Cohen, W.W., Ravikumar, P.D., Fienberg, S.E., et al.: A comparison of string distance metrics for name-matching tasks. IIWeb **2003**, 73–78 (2003)
21. Hasançebi, O., Azad, S.K.: Adaptive dimensional search: a new metaheuristic algorithm for discrete truss sizing optimization. Comput. Struct. **154**, 1–16 (2015)
22. Xiao, Z., Song, W., Chen, Q.: Dynamic resource allocation using virtual machines for cloud computing environment. IEEE Trans. Parallel Distrib. Syst. **24**(6), 1107–1117 (2013)

23. Breskovic, I., Maurer, M., Emeakaroha, V.C., Brandic, I., Dustdar, S.: Cost-efficient utilization of public SLA templates in autonomic cloud markets. In: 2011 Fourth IEEE International Conference on Utility and Cloud Computing (UCC), pp. 229–236. IEEE (2011)
24. Javadi, B., Thulasiraman, P., Buyya, R.: Cloud resource provisioning to extend the capacity of local resources in the presence of failures. In: 2012 IEEE 14th International Conference on High Performance Computing and Communication and 2012 IEEE 9th International Conference on Embedded Software and Systems (HPCC-ICESS), pp. 311–319. IEEE (2012)

Expert System for Diagnosing Disease Risk from Urine Tests

Nisanart Tachpetpaiboon, Kunyanuth Kularbphettong and Satien Janpla

Abstract The research presents the expert system for diagnosing the risk from urine tests by collecting diagnostic information from documents, books, experts, and the results of urine tests. The sample was 9,961 to create knowledge under decision tree technique. The result of the study was shown that the systems could approximately diagnose the risk of 12 types of disease, including diabetes, infected, gallstone, tumor, bladder inflammation, urological disorders, kidney disease, SLE, jaundice, G6PD deficiency, infectious disease, and diabetes insipidus. Also, the model gained from the creation of determinability tree in diagnosing the diseases created 96 IF-THEN Rules that used the strategy of forward chaining inferences in diagnosing the risk.

Keywords Expert system · Diagnose · Urine test · Decision tree

1   Introduction

Nowadays, healthcare trends are getting more and more fashionable according to health information, including a disease prevention. Many departments whether government agencies and private states give priority to employees' health by arranging welfares and annual health check. The annual health check, especially urinalysis in medical technique, believes that it can be very effective to analyze and treat diseases because the kidney excretes the wastes from the body. Therefore, the urinalysis can specify how well the kidney's function and other organ systems are. There are many

N. Tachpetpaiboon · K. Kularbphettong (✉) · S. Janpla
Computer Science Program, Suan Sunandha Rajabhat University,
Bangkok 10300, Thailand
e-mail: kunyanuth.ku@ssru.ac.th

N. Tachpetpaiboon
e-mail: Nisanart.te@ssru.ac.th

S. Janpla
e-mail: satien.ja@ssru.ac.th

© Springer Nature Singapore Pte Ltd. 2019
S. K. Bhatia et al. (eds.), *Advances in Computer Communication and Computational Sciences*, Advances in Intelligent Systems and Computing 760, https://doi.org/10.1007/978-981-13-0344-9_20

kinds of chemicals excreted in urine. If the urine is diagnosed daily, it can diagnose parasites and diseases. However, the urinalysis must be processed by experts. Unfortunately, the depletion of medical person and the limit of time and place cause those who would like to have the health check unable to have a medical treatment.

With the rapid growth of information technology, expert systems have advantages for decision-making in diagnosis and treatment in medicine and it can provide information based on the stored medical knowledge to a possible diagnose and to the adapted treatment of the diseases. There are much research of intelligent healthcare systems to facilitate medical knowledge base was represented as production rules [1, 2]. To reduce the diagnosis time and improve the diagnosis accuracy, this research is applied to the idea as expert system to analyze the annual health check of urinalysis for finding the relative risks of diseases by taking information for blood test results document to be used as a guideline of manners protection or taking advice from medical personnel. MYCIN [3] is the earliest medical expert system to provide information of infectious disease by using a set of rules. Majali and et al. [4] proposed a diagnosis and prognosis system to predict the possibility of breast cancer by using FP (frequent pattern mining) and decision tree algorithms.

Most of the research in data classification focuses on the modeling of classification without implementing the expert system. This research is applied in expert system approach to analyze the annual health check of urinalysis for finding the relative risks of diseases by using blood test information. To set up the expert system, Bayesian classification and decision tree algorithm were evaluated to create model for this project and compared performance between two models. After the expert system was implemented for diagnosing the risk from urine tests.

2 The Related Works and Methodologies

There are many various methodologies to benefit from using data mining technique. Data mining is the process of analyzed data and information from different perspectives, including machine learning, information science, and statistic, to discover and present knowledge in an easily comprehensible form [5]. B. Adhi Tama et al. used medical records of patients to check diabetes [6]. Also, many proposed expert systems are applied to diagnose diseases based on medical data and data mining techniques [7–10]. Decision tree was used to extract the valuable data in health care for the diagnosis from urine analysis values [11].

Bayesian classification is the creation of a data discriminator by applying statistical values that can indicate the probability of data. This method enables rapid and accurate classification of data. Naïve Bayes is one of the important classification approaches, and it is simple probabilistic theory.

A decision tree is one of the most well-known classification approaches that are commonly used to examine data and induce the tree in order to make predictions [12]. A decision tree is a classifier tool that uses a treelike graph or model of decisions to generate set of rules and their possible consequences, including chance event outcomes, resource costs, and utility [13, 14]. Decision support systems [15] used single decision tree (SDT), boosted decision tree (BDT), and decision tree forest (DTF) for the detection of breast cancer. Also, decision trees are generally used in decision analysis to help identify a strategy most likely to reach a goal, but are also a popular tool in machine learning [16]. The C4.5 is a decision tree algorithm used for classification from a set of training data, and C4.5 uses gain ratio as an attribute selection measure to build a decision tree.

Rapid application development (RAD) is the significant process of consideration of desire, suitability, and possibility in the real situation usage by understanding problems, arranging resolution, specifying form of advice, and gathering knowledge and information to develop the system.

3 Experimental Setup

In our research, we collected the patient's urine tests and the number of data of patient's urine tests was 9,961. The data is composed of personal records and the results of urine tests as shown in Table 1.

From the study, it was found that the diagnosis of relative risks of disease can be analyzed by urinalysis by using the form of urinalysis from CLINIC OF TEACH-ERS' COUNCIL. There are two sections of the urinalysis form: the urinalysis and microscope examination. These two sections are considered for the results for diagnosis of relative risks of diseases as following lists: color, turbidity, pH, specific

Table 1 Data for the development system

Feature name	Value
Age	<30, 30–40, 40–50, >50
Sex	Male, female
Weight	$\leq$50, 51–99, $\geq$100
Height	<150, 151–170, >170
BMI	Thin, normal, obesity
Systolic blood pressure	Normal, high-normal, hypertension
Diastolic blood pressure	Normal, high-normal, hypertension
Pulse	$\leq$60, 61–79, $\geq$80
Urine color	Colorless, cloudy, milky, red, orange, green-blue, brown, brown-black
Spot PCR	<0.5, <0.2, 0.5–2.0/0.2–2.0, >2
Dipstick	Negative, trace, 1+, 2+, 3+, 4+
24-HUP	<4 mg/m^2/h, 4–40 mg/m^2/h, >40 mg/m^2/h
Creatinine	Normal, high-normal, high
GPT	Normal, high-normal, high

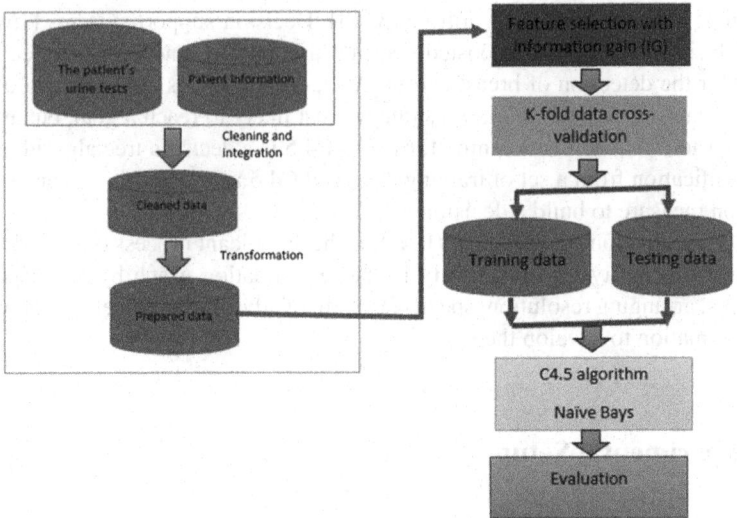

Fig. 1 Data mining process of this research

gravity, albumin, sugar/glucose, acetone, blood, white blood cells, red blood cells, epithelial cells, mucous threads, microorganism, and crystals/casts. The result could be summarized that the scales of acetone, sugar/glucose, and albumin/Alb/protein can be a sign of relative risk of diabetes by abnormal scales. Besides, other abnormal scale results reflect diseases related to the kidney and abnormal urinary tract, for example bacteria, causing the urinary tract inflection; blood, causing the relative risk of stone chopper tools; tumor, urinary bladder inflammation. And the abnormal scales of epithelium cells, mucous thread, and RBC can diagnose the problem of urinary tract.

There are many types of data in this research, and data must be preprocessed and transformed to be appropriated format in order to apply data mining technique to discover important rules. The classification algorithms were used to discover valuable patterns when the preparation data process was finished. The data was divided into two parts: learning set and testing set, and Fig. 1 displayed the classification process of this project.

To evaluate the accuracy, the effectiveness of the experiment was measured based on the statistical data obtained from the analysis and testing, such as accuracy, precision, and F-measure.

$$Recall = TP/(TP + FN) \tag{1}$$

$$Precision = TP/(TP + FP) \tag{2}$$

$$F\text{-}measure = (2 * TP)/(2 * TP + FP + FN) \tag{3}$$

Table 2 Confusion matrix

	Positive prediction	Negative prediction
Actual positive class	True positive (TP)	False negative (FN)
Actual negative class	False positive (FP)	True negative (TN)

Confusion matrix displays a summary of the assessment of the ability to classify data from a test as present in Table 2.

To evaluate the expert system for diagnosis the relative risks of diseases that can diagnose the risks of diseases by considering the accuracy of overall model. Satisfaction evaluation is inspired by the evaluation called acceptance test to evaluate for the quality of the system that can separate into four sections as follows: functional requirement test; functional test; performance test; and usability test. The criterion of the quantitative score is the evaluation of rating scale and the criterion of data reading from the evaluation by considering from the average scale as the following: excellent quality, good quality, moderate quality, fair quality, and less quality needed to be improved. The result of the system's process, the users can answer the questions from the system, and get the diagnosis result of the relative risks of diseases.

4 Experimental Results

The result gained from the experiment by using Bayesian classification and decision tree algorithm is shown in Table 3. The data from this project was collected from 9,961 patients, and accuracy, precision, and F-measure were used to evaluate the classification models.

The result was shown that the C4.5 was highest the classification performance and it created 96 Rules used the strategy of forward chaining inferences in diagnosing the risk and two physicians compare the decision tree with the original result. The decision tree result was easily readable as presented in Fig. 2.

Also, after created model, the prototype system was implemented, Fig. 3 presented the example of user interface of this system, and users can use this diagnosis to prevent the disease that could change one's behavior in everyday life for the better health or to be a guideline to get advice from medical personnel.

To assess the satisfaction of users with this prototype system, the questionnaire was the significant approach to survey from 20 users in related field. The standard

Table 3 Classification performance

Recall		Precision		F-measure	
C4.5	Naïve Bayes	C4.5	Naïve Bayes	C4.5	Naïve Bayes
83.2	74.9	80.5	81.7	81	77.6

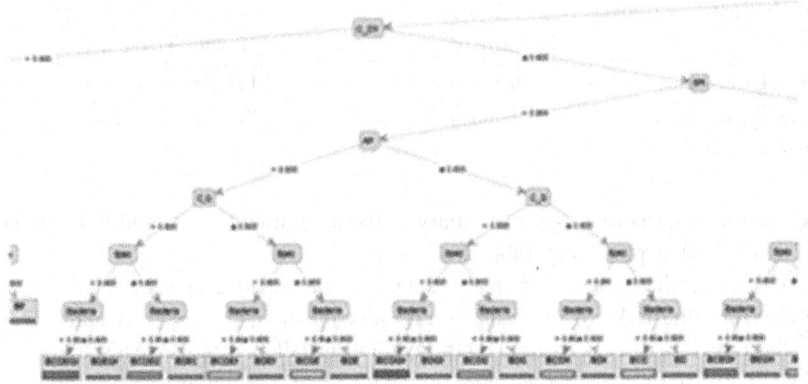

Fig. 2 Decision rules

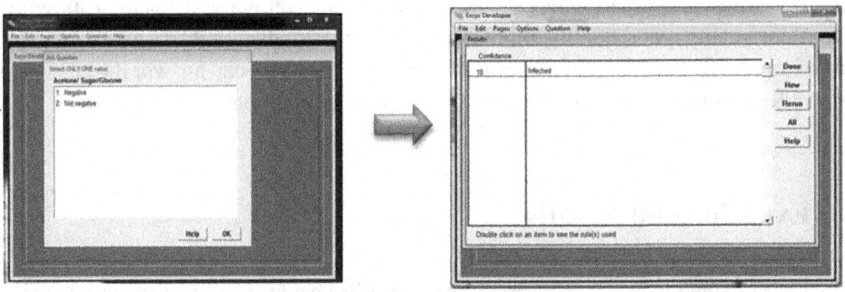

Fig. 3 Display of diagnosis result of the relative risks of diseases

Table 4 Result of satisfaction evaluation of the system

Lists of evaluation result	Users		Rate
	$\bar{x}$	SD	
Section: Work as users' desire	4.15	0.68	Good
Section: System usage	4.05	0.69	Good
Section: System quality	4.15	0.71	Good
Section: Design and simplicity	4.25	0.72	Good
Total average	**4.15**	**0.70**	**Good**

deviation equals 0.70. So, it could be summarized that the system has been developed. The users are well satisfied, and the system is available to be used, as shown in Table 4.

5 Conclusion and Future Works

In conclusion, this paper presented the expert system for diagnosing the risk from urine tests by collecting diagnostic information from medical information of patients. The result of the study was shown that decision tree was selected to create model because of the highest precision and the prototype system could approximately diagnose the risk of 12 types of diseases. The expert system for diagnosing the relative risks of diseases by urinalysis can be applied in diagnosis for the risks of other diseases by holding the criterion in considering the risks of clinic's areas in order to improve and support people's health or even classify patients before getting diagnosis by doctors.

Acknowledgements The authors would like to be grateful to the financial subsidy provided by Suan Sunandha Rajabhat University, and we would like to acknowledge friends and family.

References

1. Kitporntheranunt, M., Wiriyasuttiwong, W.: Development of a medical expert system for the diagnosis of ectopic pregnancy. J. Med. Assoc. Thai. **93**(Suppl 2), S43–S49 (2010)
2. Kumar, D.S., Sathyadevi, G., Sivanesh, S.: Decision support system for medical diagnosis using data mining. Int. J. Comput. Sci. Issues **8**(Issue 3, No. 1), 147–153 (2011)
3. Buchanan, B.G., Shortliffe, E.H.: Rule Based Expert Systems: The Mycin Experiments of the Stanford Heuristic Programming Project (The Addison-Wesley Series in Artificial Intelligence). Addison-Wesley Longman Publishing Co., Inc., Reading, MA, USA (1984)
4. Majali, J., Niranjan, R., Phatak, V., Tadakhe, O.: Int. J. Comput. Sci. Inf. Technol. (IJCSIT) **5**, 6487–6490 (2014)
5. Fayyad, U.M., Pitatesky-Shapiro, G., Smyth, P., Uthurasamy, R.: Advances in Knowledge Discovery and Data Mining. AAAI/MIT Press (1996)
6. Tama, B.A., Rodiyatul, F.S., Hermansyah: An early detection method of Type-2 diabetes mellitus in public hospital. In: Proceeding of the International Conference on Informatics, Cybernetic, and Computer Applications. Bangalore, vol. 9, no. 2, pp. 287–294 (2010)
7. Chen, T.C., Hsu, T.C.: A GAs based approach for mining breast cancer pattern. Expert Syst. Appl. **30**, 674–681 (2006)
8. Tomar, P.P., Singh, R., Saxena, P.K.: A medical multimedia based clinical decision support system for operational chronic lung diseases diagnosis and training. Int. J. Comput. Appl. **49**(8), 1–12 (2012)
9. Joshi, J., Doshi, R., Patel, J.: Diagnosis and prognosis breast cancer using classification rules. Int. J. Eng. Res. General Sci. **2**, 315–323 (2014)
10. Kularbphettong, K., Waraporn, P., Tongsiri, C.: Analysis of student motivation behavior on e-learning based on association rule mining. World Acad. Sci. Eng. Technol. (2012)
11. Topaloğlu, M., Malkoç, G.: Decision tree application for renal calculi diagnosis. Int. J. Appl. Math. Electron. Comput. Special Issue, 404–407 (2016)
12. Edelstein, H.: Introduction to Data Mining and Knowledge Discovery, 3rd edn. Two Crows Corporation, Potomac, MD, USA (1999)
13. https://en.wikipedia.org/wiki/Decision_tree
14. Azar, A.T., El-Metwally, S.M.: Decision tree classifiers for automated medical diagnosis. Neural Comput. Appl. (2013)

15. Azar, A.T., El-Metwally, S.M.: Decision tree classifiers for automated medical diagnosis. Neural Comput. Appl. **23**(7–8), 2387–2403 (2012)
16. Chen, C., He, B., Zeng, Z.: A method for mineral prospectivity mapping integrating C4.5 decision tree, weights-of-evidence and m-branch smoothing techniques: a case study in the eastern Kunlun Mountains, China. Earth Sci. Inform. (2014)

Incremental Constrained Random Walk Clustering

Ping He, Tianyu Jing, Xiaohua Xu, Huihui Lin, Zheng Liao and Baichuan Fan

Abstract In many real-world application scenarios, data usually incrementally update over time. In such cases, traditional constrained clustering algorithms become unsuitable for dealing with incremental data because of high computational cost. In this paper, we propose a novel incremental constrained random walk clustering algorithm (ICC), which not only efficiently deal with the incremental data but also utilize the incremental constraints. To reduce the time complexity, it updates the influence range of each selected data point and utilizes the intermediate structure of the previous time step. Extensive experiment results on datasets demonstrate that our algorithm is both effective and efficient.

Keywords Incremental data · Random walk · Constrained clustering

1 Introduction

As an effective data analysis technique in data mining, clustering has been widely studied in recent years. Each cluster generated by clustering algorithms can be regarded as the set of data objects which share high similarity with each other but have large difference from other sets [1]. In the current real-world application scenarios, traditional clustering methods cannot appropriately deal with online incremental data [2]. Therefore, researchers are paying increasing attention on it in recent years.

Unlike static data, the incremental data are unpredicted. If we apply the traditional clustering methods to incremental data, the clustering results may not be what we expect. Hence, incremental data have been a hot area to researchers. Besides, in real-world problem, we can get a small amount of priori knowledge such as pairwise constraints of incremental data [3, 4]. Now that the priori knowledge has been

P. He · T. Jing · X. Xu (✉) · H. Lin · Z. Liao · B. Fan
Department of Computer Science, Yangzhou University, Yangzhou, China
e-mail: arterx@gmail.com

P. He
e-mail: angeletx@gmail.com

© Springer Nature Singapore Pte Ltd. 2019
S. K. Bhatia et al. (eds.), *Advances in Computer Communication and Computational Sciences*, Advances in Intelligent Systems and Computing 760, https://doi.org/10.1007/978-981-13-0344-9_21

provided, it is unnecessary for us to insist adopting unsupervised learning. How to make full use of the prior knowledge becomes a popular problem. For that purpose, it is necessary to develop a constrained clustering algorithm which is expected to deal with incremental data efficiently and accurately [5, 6].

Inspired by the work of SCRAWL [7], we propose a novel incremental constrained random walk clustering algorithm (ICC), which is used to deal with incremental data with a small number of incremental pairwise constraints. SCRAWL is composed of two steps. In the lower-level random walk, we first compute the influence of the selected constrained data point on the other data points. In the higher-level random walk, each pairwise constraint is propagated to the estimated influence range with adaptive strength. At last, we integrate all the constraint influences into a cluster-indicating matrix to obtain the clustering results. Regarding the increasing number of incremental data, we update the estimated influence range of each selected data point incrementally and utilize the information of previous time step. We validate the performance of our algorithm on the processed real-world datasets and demonstrate both of its effectiveness and efficiency.

The rest of this paper is organized as follows: Sect. 2 briefly introduces the semi-supervised clustering via random walk algorithm. In Sect. 3, the procedure of ICC is described in detail. The experimental results on the UCI datasets and real-world datasets are discussed in Sect. 4. In the end, Sect. 5 concludes the whole paper.

2   Semi-supervised Clustering via Random Walk

He et al. [7] proposed an algorithm named semi-supervised clustering via random walk (SCRAWL) which can propagate the influence of pairwise constraints to other unconstrained edges locally and smoothly. It also can take a full account of strength of influence to other unconstrained pints. Must-link constraint and cannot-link constraint are used in SCRAWL as constrained information.

Suppose that there is a dataset $\mathcal{X} = \{x_1, x_2, \ldots, x_n\}$ and a set of pairwise constraints $\mathcal{C} = \mathcal{C}_= \cup \mathcal{C}_{\neq}$, where $\mathcal{C}_=$ is the set of must-link constraints and $\mathcal{C}_{\neq}$ is the set of cannot-link constraints, respectively. x_i and x_j belong to the same cluster, $c_=(x_i, x_j) \in \mathcal{C}_=$. Similarly, $c_{\neq}(x_i, x_j) \in \mathcal{C}_{\neq}$ denotes the different clusters. Constrained clustering refers to segmenting $\mathcal{X}$ into multiple clusters and meet the pairwise constraints condition as much as possible.

In the next step, it maps $\mathcal{X}$ to weighted undirected graph $\mathcal{G} = (\mathcal{V}, \mathcal{E}, A)$, where $\mathcal{V} = \{\upsilon_1, \upsilon_2, \ldots, \upsilon_n\}$ is vertex set and υ_i corresponds to data point x_i. $\mathcal{E}$ is the set of edges, and a is the similarity function defined on the $\mathcal{E}$. Similarity matrix A is composed of elements a. On basis of the given constraints set $\mathcal{G}$, add the constrained information into similarity matrix. The new similarity matrix is obtained, $\tilde{A} = (\tilde{a}(i, j))_{n \times n}$. If $\exists c_=(\upsilon_i, \upsilon_j)$, $\tilde{a} \approx a(i, j)^q$. If $\exists c_{\neq}(\upsilon_i, \upsilon_j)$, $\tilde{a} \approx a(i, j)^{1/q}$. Otherwise, $\tilde{a} \approx a(i, j)$. And the parameter $q \in (0, 1]$ measures the similarity updating.

Suppose that each vertex represents the different states on a time-homogenous Markov chain in lower random walk level. Let a small amount of the vertices in $\mathcal{C}$

be absorbing state and the rest of them be transient. After random walk beginning, the particle on each vertex transfers from v_i to v_j with possibility p_{ij}. If it reaches any absorbing state, it will stay there. And if not, it will continue to walk until all the particles reach absorbing state. The criterion we obeyed is $V_{C_{\neq}} \prec V_{C_{=}} \prec V_u$, where $V_{C_{=}}$ is the set of must-link constrained vertices, $V_{C_{\neq}}$ is the set of cannot-link constrained vertices, and V_u is set of the unconstrained vertices.

Suppose V_s is vertex set of absorbing state and V_r is vertex set of transience state. Under the condition of $V_s \cup V_s = V$, we rewrite the transition possibility matrix $P = \tilde{D}^{-1}\tilde{A}$ as the following, where $\tilde{A}$ is similarity matrix after updating according to constraints and $\tilde{D} = \mathrm{diag}\left(\tilde{A}1_n\right)$:

$$P = \begin{bmatrix} P_{ss} & P_{sr} \\ P_{rs} & P_{rr} \end{bmatrix} \tag{1}$$

where P_{ss} and P_{rr} are transition possibility sub-matrices of V_s and V_r, respectively. P_{sr} and P_{rs} are mutual transition possibility sub-matrices of V_s and V_r, respectively.

The possibility matrix absorbed by s different absorbing states $Y = \left(y_{ij}\right)_{n \times n}$ when each particle reaches steady state can be written as:

$$Y = \begin{bmatrix} Y_s \\ Y_r \end{bmatrix} = \begin{bmatrix} y_s^0 \\ (I - P_{rr})^{-1} P_{rs} Y_s^0 \end{bmatrix} \tag{2}$$

where Y_s is the possibility sub-matrices of absorbed particle that starts from V_s and Y_r is the possibility sub-matrices of absorbed particle that starts from V_r.

The information of influence in lower level is encapsulated in an intermediate structure called "component." Random walk in higher-level adaptive propagates with constraints in component and integrates the influence on each vertex into a cluster-indicating matrix.

Let T denotes the set of components and $|T|$ denotes the number of components. The jth component T_j represents that it is composed of the jth vertex of absorbing state and the fragments of vertices influenced by it. It can be formulated as follows:

$$T_j = \{y_{ij}v_i | y_{ij} > 0\} \tag{3}$$

where y_{ij} is the element of Y. It indicates the membership grade of v_i to T_j. Thus, it can define the similarity matrix between any two components as $A_c = Y^T \tilde{A} Y$.

The similarity matrices of components are used to increase the influence of constraints and then obtain the clusters of components. SCRAWL integrates membership grade of vertex to component and membership grade of components to clusters to acquire the cluster-indicating matrix.

3   Incremental Constrained Random Walk Clustering

Compared with other constrained clustering algorithms, SCRAWL propagates pair-wise constraints to unconstrained edges locally and proportionally without any metric parameters. However, this static incremental constrained clustering algorithm is not suitable for incremental data, because not only the number of data increases with time, but also the constraint updates accordingly. To solve this problem, we proposed incremental constrained random walk clustering algorithm (ICC).

3.1   Notations

Given a dataset $\mathcal{X} = \{\mathcal{X}^1, \mathcal{X}^2, \dots, \mathcal{X}^t\}$, where $\mathcal{X}^t$ is the dataset that arrives during the time interval $(t - 1, t]$. Following the arrival of $\mathcal{X}^t$, there may be accompanying related pairwise constraint set $\mathcal{C}^t = \mathcal{C}^t_= \cup \mathcal{C}^t_{\neq}$, where $\mathcal{C}^t_=$ indicates the must-link constraint that arrives at time step t and $\mathcal{C}^t_{\neq}$ indicates the cannot-link constraint that arrives at time step t.

We build an undirected weighted graph $\mathcal{G}^t = (\mathcal{V}^t, \mathcal{E}^t, A^t)$ for the dataset and constraint set at time step t, where $\mathcal{V}^t$ is the vertices set, whose elements correspond to the data points that appear before and at time step t, i.e., $\forall_{v_i} \in \mathcal{V}^t$, the vertex v_i corresponds to $x_i \in \mathcal{X}^1 \cup \dots \cup \mathcal{X}^t$, $\mathcal{E}^t$, is the edge set among $\mathcal{V}^t$ at the time step t, and A^t is the similarity matrix among $\mathcal{V}^t$ at time step t.

On the basis of the following two assumptions, we proposed our ICC algorithm: (1) There is no conflict among different pairwise constraints, and (2) data distribution does not change or drift over time. It is composed of two stages: lower-level random walk at time step t and higher-level random walk at time step t, which are described as follows.

3.2   Lower-Level Random Walk

Assume that every vertex in $\mathcal{G}^t$ reveals different states in a homogeneous Markov chain. $\mathcal{V}^t_s$ is the vertex of absorbing states constrained by C^t in Markov chain. $\mathcal{V}^t_r$ represents the vertex of transient states. Every vertex is distributed with a particle after the random walk starts. Each particle walks randomly from an arbitrary vertex v^t_i to v^t_j with the probability p^t_{ij}. If it reaches any absorbing state, it stays; otherwise, it continues walking until it reaches an absorbing state. Based on the assumption that the data distribution keeps unchanged, we regard the absorbing states as the vertex defined in the initial time:

$$\mathcal{V}^t = \begin{bmatrix} \mathcal{V}^t_s \\ \mathcal{V}^t_r \end{bmatrix} = \begin{bmatrix} \mathcal{V}^{t-1} \\ \mathcal{V}^t_r \end{bmatrix} \tag{4}$$

where $\mathcal{V}^t$ denotes the data from initial time to time t and $\mathcal{V}_s^t$ denotes the vertices set of absorbing state at time t. $\mathcal{V}_r^t$ is the set of transience state at time t.

So, the transition possibility matrix at time t is $P^t = \left(\tilde{D}^t\right)^{-1}\tilde{A}^t$. It can be written as follows:

$$P = \begin{bmatrix} P_{ss}^t & P_{sr}^t \\ P_{rs}^t & P_{rr}^t \end{bmatrix} \tag{5}$$

where P_{ss}^t and P_{rr}^t are transition possibility sub-matrices of $\mathcal{V}_s^t$ and $\mathcal{V}_r^t$, respectively. P_{sr}^t and P_{rs}^t are mutual transition possibility sub-matrices of $\mathcal{V}_s^t$ and $\mathcal{V}_r^t$, respectively. The elements of updating similarity matrix $\tilde{A}^t = (\tilde{a}^t(i, j))n \times n$ are:~

$$\tilde{a}^t = \begin{cases} a^t(i, j)^q & \text{if } \exists\, c_{=}^t\left(v_i^t, v_j^t\right) \\ a^t(i, j)^{1/q} & \text{if } \exists\, c_{\neq}^t\left(v_i^t, v_j^t\right) \\ a^t(i, j) & \text{otherwise} \end{cases} \tag{6}$$

Due to the settings of $\mathcal{V}_s^t$ and $\mathcal{V}_r^t$, we can learn $P_{ss}^t = I$ and $P_{sr}^t = 0$. On the bases above, we can obtain the state distribution as follows:

$$Y^t = \begin{bmatrix} Y_s^t \\ Y_r^t \end{bmatrix} = \begin{bmatrix} Y^{t-1} \\ \left(I - P_{rr}^t\right)^{-1} P_{rs}^t\, Y^{t-1} \end{bmatrix} \tag{7}$$

3.3 Higher-Level Random Walk

Let the set of components at time t be T^t, and the jth component T_j^t can be defined as:

$$T_j^t = \left\{ y_{ij}^t v_i^t \,|\, y_{ij}^t > 0 \right\} \tag{8}$$

The similarity matrix between components is:

$$A_c^t = \left(Y^t\right)^T \tilde{A}^t Y^t = \tilde{A}_c^{t-1} + 2\left(Y_r^t\right)^T \tilde{A}_{rs}^t\, Y_s^t + \left(Y_r^t\right)^T \tilde{A}_{rr}^t\, Y_r^t \tag{9}$$

$D_c^t = \text{diag}\left(A_c^t 1_{|T^t|}\right)$ is the metric matrix of A_c^t, and we normalize similarity matrix between components as follows.

$$\tilde{A}_c^t = \left(D_c^t\right)^{-\frac{1}{2}} A_c^t \left(D_c^t\right)^{-\frac{1}{2}} \tag{10}$$

Similar to SCRAWL, we take the same updating strategy q/q^{-1} to add the constraint $\mathcal{C}$ into $\tilde{A}_c^t$:

$$
\tilde{a}_c^t(\alpha, \beta) = \begin{cases} \tilde{a}_c^t(\alpha, \beta)^{q_{\alpha\beta}} & \text{if } \exists\, c_{\underline{\underline{}}}^{\leq t}\left(v_i^\alpha, v_j^\beta\right) \\[2mm] \tilde{a}_c^t(\alpha, \beta)^{1/q_{\alpha\beta}} & \text{if } \exists\, c_{\neq}^{\leq t}\left(v_i^\alpha, v_j^\beta\right) \\[2mm] \tilde{a}_c^t(\alpha, \beta) & \text{otherwise} \end{cases} \tag{11}
$$

Under the assumption of none conflictions to pairwise constraints, we can define $\mathcal{C}_{\underline{\underline{}}}^{\leq t}$ as $\mathcal{C}_{\underline{\underline{}}}^{\leq t} = \mathcal{C}_{\underline{\underline{}}}^1 \cup \mathcal{C}_{\underline{\underline{}}}^2 \cup \ldots \cup \mathcal{C}_{\underline{\underline{}}}^t$ and $\mathcal{C}_{\underline{\underline{}}}^{\leq t} = \mathcal{C}_{\neq}^1 \cup \mathcal{C}_{\neq}^2 \cup \ldots \cup \mathcal{C}_{\neq}^t$. But if we consider the confliction, we can integrate the time information by referring the parameter thoughts. Apply exponential decay onto updating parameters:

$$
q_{\alpha\beta} = 1 - \frac{1}{e^{\lambda(t-t_c)} + \delta} \tag{12}
$$

in which λ sets 1 and δ takes a small value to prevent $q_{\alpha\beta} = 0$. t is the current moment, and t_c is the moment of the propagated constraints. After adaptive propagating pairwise constraints among components, we can compute and gain the transition possibility matrix as $P_c^t = \left(D_c^t\right)^{-1}\tilde{A}_c^t$, in which $D_c^t = \text{diag}\left(\tilde{A}_c^t 1\right)$.

By virtue of the theory proposed by Shi and Malik [8], the minimum cut of $\tilde{A}_c^t$ is $P_c^t u_i^t = \lambda_i u_i^t$. We can obtain the cluster-indicating matrix composed of d eigenvectors of corresponding maximum eigenvalue:

$$
U^t = \left[u_1^t, u_2^t, \ldots, u_d^t\right] \tag{13}
$$

The cluster-indicating matrix of vertices G^t is:

$$
G^t = Y^t U^t \tag{14}
$$

A common technique [9] is adopted to map the row vectors of cluster-indicating matrix of vertices G^t onto a hypersphere:

$$
\tilde{G}^t = \left(D_G^t\right)^{-\frac{1}{2}} G^t \tag{15}
$$

where $D_G^t = \text{diag}\left(\text{diag}\left(G^t\left(G^t\right)^T\right)\right)$. At last, we adopt k-means methods to obtain the clustering results.

4 Experiments

In this section, we validate ICC by using UCI datasets and real-world datasets. Firstly, we consider the probability of applying improving off-line algorithm SCRAWL to cluster incremental data. So, we only compare ICC with SCRAWL to evaluate the clustering results. Next, we measure the influence of selected constrained sets to algorithm performance. At last, we will analyze the time complexity.

Table 1 lists the basic attributes of UCI datasets and real-world datasets, in which Iris is UCI dataset and 20 Newsgroups is a text dataset.

In the preprocessing, delete the missing data and select data which are disposing least in 20 Newsgroups dataset.

Because of the lack of incremental data, we may use dataset in Table 1 to simulate incremental data with the dynamic variation of time. For Iris dataset, we add the extra Gaussian noise which satisfies $N(0, 0.5)$ to approach the incremental data. With regard to the real-world dataset, we are randomly and equally divided it into six segments to simulate the incremental data.

Figures 1 and 2 show the comparison of the clustering performance of ICC and SCRAWL. T1 to T6 on the x axis represent different time points. We can easily learn that the ICC performs better than static SCRAWL. With the time changes, the clustering performance tends to be stable.

Table 1 Description of datasets

Dataset	Samples	Features	Classes
Iris	150	4	3
20 Newsgroups	2774	24253	3

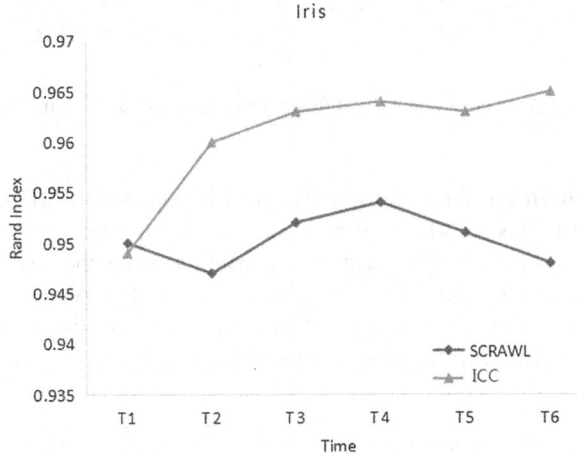

Fig. 1 Comparison of clustering performance on Iris dataset

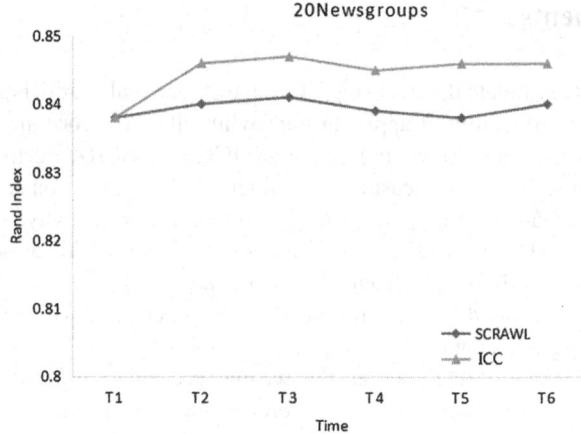

Fig. 2 Comparison of clustering performance on 20 Newsgroup dataset

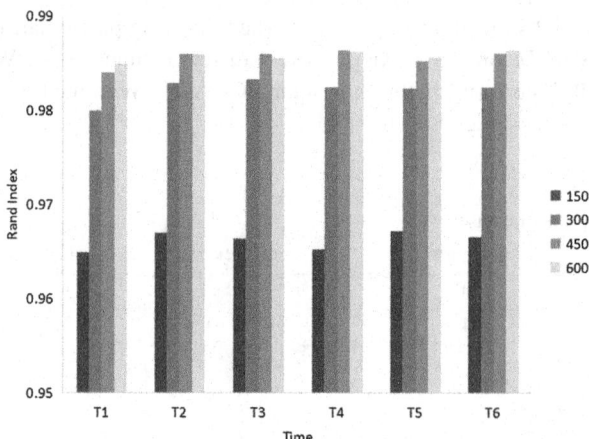

Fig. 3 Clustering performances of ICC for different number of pairwise constraints on Iris dataset

In Fig. 3, there are clustering performances on Iris dataset with different quantities of pairwise constraints at different times. The notations in Fig. 2 are the number of pairwise constraints. We can find that more constraints make the results better. But until increasing to 450 and 600, the performances of clustering are very closer. Therefore, with the increasing quantity of pairwise constraints the clustering results get better in a certain number and then come to be stable.

5 Conclusion

In this paper, we proposed a novel incremental constrained random walk clustering algorithm (ICC). It not only accepts incremental data that increase over time, but also can utilize incremental pairwise constraints. It is composed of two levels of random walk, where the lower level updates the intermediate structure at each time step, while the higher level adjusts the clustering result. The extensive evaluations on UCI and real-world datasets demonstrated the effectiveness and efficiencies of ICC.

Acknowledgements This research was supported in part by the Chinese National Natural Science Foundation under Grant Nos. 61402395, 61472343, 61379066, and 61502412, Natural Science Foundation of Jiangsu Province under contracts BK20140492, BK20151314, and BK20150459, Jiangsu overseas research and training program for university prominent young and middle-aged teachers and presidents, Jiangsu government scholarship funding, and practice innovation plan for college graduates of Jiangsu Province under contracts SJLX16_0591.

References

1. Likas, A., Vlassis, N., Verbeek, J.J.: The global k-means clustering algorithm. Pattern Recogn. **36**(2), 451–461 (2003)
2. Halkidi, M., Spiliopoulou, M., Pavlou, A.: A semi-supervised incremental clustering algorithm for streaming data. In: Proceedings of Pacific-Asia Conference on Advances in Knowledge Discovery and Data Mining, vol. 7301, pp. 578–590 (2012)
3. Yu, Y., Wang, Q., Wang, X., Wang, H., He, J.: Online clustering for trajectory data stream of moving objects. Comput. Sci. Inf. Syst. **10**(3), 1293–1317 (2013)
4. Young, S., Arel, I., Karnowski, T.P., et al.: A fast and stable incremental clustering algorithm. In: International Conference on Information Technology. New Generations, pp. 204–209. IEEE (2012)
5. Song, Y., Yang, Y., Dou, W., Zhang, C.: Graph-based semi-supervised learning on evolutionary data (2015)
6. Kulis, B., Basu, S., Dhillon, I., Mooney, R.: Semi-supervised graph clustering: a Kernel approach. In: International Conference on Machine Learning, vol. 74, pp. 457–464 (2005)
7. He, P., Xu, X., Hu, K., Chen, L.: Semi-supervised clustering via multi-level random walk. Pattern Recogn. **47**(2), 820–832 (2014)
8. Shi, J., Malik, J.: Normalized cuts and image segmentation. IEEE Trans. Pattern Anal. Mach. Intell. **22**(8), 888–905 (2000)
9. Ng, A.Y., Jordan, M.I., Weiss, Y.: On spectral clustering: analysis and an algorithm. In: International Conference on Neural Information Processing Systems: Natural and Synthetic, vol. 14, pp. 849–856 (2001)

Evolutionary Tree Spectral Clustering

Xiaohua Xu, Zheng Liao, Ping He, Baichuan Fan and Tianyu Jing

Abstract Existing evolutionary clustering algorithms are mostly based on temporal smoothness. Yet it is still a difficult problem to find out the rule from the evolutionary data. In this paper, we try to solve this problem by using an evolutionary tree to describe the evolution process of data. Based on evolutionary tree, the structural smoothness is proposed. By taking account of both the existing temporal smoothness and the structural smoothness, we propose an evolutionary tree clustering framework and an evolutionary tree spectral clustering algorithm. Moreover, we analyze the cost function and the solution of evolutionary tree spectral clustering algorithm. Promising experiments on both artificial and real-world datasets demonstrate the superior performance of our proposed method.

Keywords Evolution · Spectral clustering · Tree

1 Introduction

Evolutionary clustering has attracted increasing amount of attention from machine learning and data mining researchers. Different from traditional static clustering, the features of objects in evolutionary clustering change with time. We can easily find different application scenarios for evolutionary clustering, such as video analysis [1], financial data analysis [2], scientific data processing [3].

By far, various methods for solving evolutionary clustering problems have been proposed by many researchers. For instance, Wang et al. [4] present an evolutionary clustering algorithm that combines matrix factorization-based clustering and low-rank matrix approximation. Wang et al. [5] present an evolutionary clustering algorithm based on Gaussian mixture model. Yu [6] propose an implicit semi-Markov

X. Xu · Z. Liao · P. He (✉) · B. Fan · T. Jing
Department of Computer Science, Yangzhou University, Yangzhou, China
e-mail: angeletx@gmail.com

X. Xu
e-mail: arterx@gmail.com

© Springer Nature Singapore Pte Ltd. 2019
S. K. Bhatia et al. (eds.), *Advances in Computer Communication and Computational Sciences*, Advances in Intelligent Systems and Computing 760, https://doi.org/10.1007/978-981-13-0344-9_22

model algorithm. At each time step, the clustering model can only take finite states, which is suitable for the data of obvious periodicity.

Temporal smoothness is frequently used in the above method, which means compared with the most recent clustering results, the current clustering results cannot deviate too drastically. There are generally two ways to implement temporal smoothness [7]. (1) Consider both the current data and the most recent data when clustering and (2) Set up a sliding window, and consider the historical data from a certain period of time to the current time. However, the present temporal smoothness assumption only considers the smoothness of the temporal linear structure, but ignores the evolutionary structure inside data. In the real world, evolutionary data with nonlinear structures are often present. For instance, (1) in the long-term development of academic research topics, the old topics may become popular again because of some special events, and (2) the atavism phenomenon exists in the family pedigree. In such cases, the current evolutionary clustering algorithms with temporal smoothness would not capture the structure of the data. If we want to capture the structure of the data, we should use data-driven approaches to obtain the internal structure from data itself, and this can improve our understanding of real-world system. In this paper, an evolutionary tree extracted from the evolutionary data is used to describe the evolutionary structure of data.

There is a well-known concept of evolutionary tree in biology. Each node on the tree represents a taxonomic unit (genus, population, individual, etc.), and the edges between nodes represent the degree of change in the evolutionary process. In biology, evolutionary tree is used to describe the evolution of organisms. So, this paper also uses tree structure to describe the evolution of data. Each node on the tree represents the set of data at a given time, and the edges between nodes represent the evolutionary distance.

In this paper, we first propose the concept of the evolutionary tree under the concept of evolutionary data. Then based on evolutionary tree, we propose a new evolutionary tree clustering framework, named ETC, which takes account of both temporal smoothness and structural smoothness. Based on ETC, an evolutionary tree spectral clustering is proposed, named ETSC. Next, the cost functions for the ETSC are proposed, and we solved it. Finally, we execute experiments on both artificial datasets and real-life datasets in comparison with the state-of-the-art evolutionary spectral clustering algorithm.

The remainder of this paper is arranged as follows: Sect. 2 presents some definitions and the ETC framework, and Sect. 3 proposes ETSC algorithm under the framework of ETC. The performance of ETSC is evaluated in Sect. 4. In the end, Sect. 5 is conclusion.

2 Evolutionary Tree Clustering (ETC) Framework

First, some definitions that will be used are given:

Definition 1 *Evolutionary Data.* Using X_t to denote dataset at time t, an evolutionary data with T time points can be represented by $X = \{X_1, X_2, \ldots, X_T\}$.

Definition 2 *Evolutionary Tree.* An evolutionary tree is a tree structure that describes how data evolves. An evolutionary tree with T nodes consists of two parts $Tree = \{V, p(\cdot)\}$, V is a set of nodes made up of evolutionary data $V = \{X_1, X_2, \ldots, X_T\}$,, and $p(\cdot)$ is a mapping function that finds the parent node of X_t. The data of parent node is represented by $X_p = p(X_t)$.

Definition 3 *Structural Smoothness.* Structural smoothness considers that the current data to the data of the parent node in the evolutionary tree is a smooth transition, which means compared with the clustering results of parent node in the evolutionary tree, the current clustering results cannot deviate too drastically.

Definition 4 *Branch of the Evolutionary Tree.* All the nodes and edges passing from the root to any leaf constitute a branch of the evolutionary tree.

On the basis of these definitions, we propose a new framework that considers both structural smoothness and temporal smoothness. The framework considers three criteria: (i) clustering quality of the current time; (ii) compared with the most recent clustering results, the current clustering results cannot deviate too drastically; (iii) compared with the clustering results of parent node in the evolutionary tree, the current clustering results cannot deviate too drastically.

Then, a cost function is defined to quantify the results of evolutionary clustering. The cost function consists of three parts. The first part, the CS (Cost Snapshot), is used to measure the quality of clustering result F_t under the data X_t at current time t. The second part, the CE (Cost tEmporal), is used to measure the similarity between the clustering result F_t of the current time and the clustering result F_{t-1} of the previous time. The third part, the CR (Cost tRee), is used to measure the similarity between the clustering results of the current time results F_t and the parent results F_p in the evolutionary tree. Then, we propose the cost function which is composed by three parts mentioned above:

$$cost = \alpha \cdot CS + (1 - \alpha) \cdot [\beta \cdot CE + (1 - \beta) \cdot CR] \tag{1}$$

where $0 \leq \alpha \leq 1, 0 \leq \beta \leq 1$ are smoothing parameters which values are customized by the user. When $\alpha = 1$, the results only consider the current time clustering quality. When $0 < \alpha < 1$, the quality of clustering and the influence of two smoothness terms are considered. When $\beta = 1$, only consider temporal smoothness. When $0 < \beta < 1$, both structural smoothness and temporal smoothness are taken into account. Simultaneous regulation of α, β can control three criteria in order to achieve the optimal equilibrium and find the optimal clustering results.

3 Evolutionary Tree Spectral Clustering (ETSC)

3.1 Spectral Clustering

Two common spectral clustering methods are maximizing the average association and minimizing normalized cut [8]. We use A that denote the similarity matrix, and $\mathcal{V}_a$ and $\mathcal{V}_b$ that denote subsets of the node set $\mathcal{V}$. Then, the association between $\mathcal{V}_a$ and $\mathcal{V}_b$ can be defined as

$$asso(\mathcal{V}_a, \mathcal{V}_b) = \sum_{i \in \mathcal{V}_a, j \in \mathcal{V}_b} A(i, j) \tag{2}$$

Let $\mathcal{V} \backslash \mathcal{V}_l$ denote the complement of $\mathcal{V}_l$, then k-way normalized cut can be written as

$$NC = \sum_{l=1}^{k} \frac{asso(\mathcal{V}_l, \mathcal{V} \backslash \mathcal{V}_l)}{asso(\mathcal{V}_l, \mathcal{V})} \tag{3}$$

and k-way average association can be written as

$$AA = \sum_{l=1}^{k} \frac{asso(\mathcal{V}_l, \mathcal{V}_l)}{|\mathcal{V}_l|} \tag{4}$$

but both of maximizing the average association and minimizing the normalized cut are NP-hard [9]. Therefore, the original problem in the spectral clustering algorithm is generally relaxed. And relaxed optimization problem is solved by (1) the eigenvectors Y of some variations of similarity matrix A are calculated, (2) use Y to construct the eigenvector space, and all data points are projected to this subspace, and (3) k-means or other classical clustering algorithms are used to cluster the eigenvectors in the eigenvectors space to obtain the clustering result. In this paper, we focus on how to calculate the eigenvectors Y of some variations of similarity matrix A of step (1). And for the remainder, we follow the step (2) and step (3) to obtain the clustering results.

3.2 Average Association

For a given partition F_t at time t, the cost function of AA in our framework is:

$$\begin{aligned} Cost_{AA} &= \alpha \cdot CS_{AA} + (1 - \alpha) \cdot [\beta \cdot CE_{AA} + (1 - \beta) \cdot CR_{AA}] \\ &= \alpha \cdot AA_t|_{F_t} + (1 - \alpha) \cdot \beta \cdot AA_{t-1}|_{F_t} + (1 - \alpha) \cdot (1 - \beta) \cdot AA_p|_{F_t} \end{aligned} \tag{5}$$

where $AA_t|_{F_t}$ measures the clustering quality of the current time. $AA_{t-1}|_{F_t}$ promotes temporal smoothness by penalizing the current clustering result that does not fit the historic data X_{t-1} well. $AA_p|_{F_t}$ promotes structural smoothness by penalizing the current clustering result that does not fit the data of the parent node in the evolutionary tree X_p well. This cost function corresponds to three parts in our frame, respectively.

And then, we start solving the problem of maximizing $Cost_{AA}$. The average association in Eq. 4 is equivalent to the following form

$$Cost_{AA} = Tr(\tilde{F}^T A \tilde{F}) \tag{6}$$

where $\tilde{F}$ denote column normalized F, and we get $\tilde{F}$ by dividing each column of F by the size of nodes per column. Then, we write the entire cost Eq. 5 as

$$\begin{aligned} Cost_{AA} =\ & \alpha \cdot [Tr(\tilde{F}_t^T A_t \tilde{F}_t)] \\ & + (1 - \alpha) \cdot \beta \cdot [Tr(\tilde{F}_t^T A_{t-1} \tilde{F}_t)] \\ & + (1 - \alpha) \cdot (1 - \beta) \cdot [Tr(\tilde{F}_t^T A_p \tilde{F}_t)] \\ =\ & Tr[\tilde{F}_t^T [\alpha \cdot A_t + (1 - \alpha) \cdot \beta \cdot A_{t-1} + (1 - \alpha) \cdot (1 - \beta) \cdot A_p] \tilde{F}_t] \end{aligned} \tag{7}$$

And as shown in Eq. 7 that maximizing $Cost_{AA}$ is equivalent to maximizing the trace $Tr[(\tilde{F}_t^T [\alpha \cdot A_t + (1 - \alpha) \cdot \beta \cdot A_{t-1} + (1 - \alpha) \cdot (1 - \beta) \cdot A_p] \tilde{F}_t)]$. But maximization of average association is NP-hard, so we relax $\tilde{F}_t$ to $Y_t \in \mathbb{R}^{n \times k}$ with $Y_t^T Y_t = I_k$. The solution is the matrix Y_t whose columns are the $top - k$ eigenvectors of matrix $\alpha \cdot A_t + (1-\alpha) \cdot \beta \cdot A_{t-1} + (1-\alpha) \cdot (1-\beta) \cdot A_p$ [10]. Next, we can use Y_t to construct the eigenvector space and all data points are projected to this subspace. Then, k-means or other classical clustering algorithms are used to cluster the eigenvectors in the eigenvectors space to obtain the clustering result.

3.3 Normalized Cut

Analogously, the cost function of NC in our framework is defined as

$$\begin{aligned} Cost_{NC} &= \alpha \cdot CS_{NC} + (1 - \alpha) \cdot [\beta \cdot CE_{NC} + (1 - \beta) \cdot CR_{NC}] \\ &= \alpha \cdot NC_t|_{F_t} + (1 - \alpha) \cdot \beta \cdot NC_{t-1}|_{F_t} + (1 - \alpha) \cdot (1 - \beta) \cdot NC_p|_{F_t} \end{aligned} \tag{8}$$

Similar to the meaning of the cost function of AA, these three items correspond to the three items in our framework. The normalized cut defined in Eq. 3 is equivalent to the following form [11]

$$Cost_{NC} = k - Tr[Y^T (D^{-\frac{1}{2}} A D^{-\frac{1}{2}}) Y] \tag{9}$$

where D is a diagonal matrix with $D(i, i) = \sum_{j=1}^{n} A(i, j)$. $Y \in \mathbb{R}^{n \times k}$ and satisfies two conditions: (1) $D^{-1/2}Y$ is equal to partition F and (2) $Y^T Y = I_k$. This is an NP-hard problem, and the relaxed optimization problem is obtained by removing condition (1)

$$
\begin{aligned}
Cost_{NC} \approx {} & \alpha \cdot k - \alpha \cdot Tr[\hat{F}_t(D_t^{-\frac{1}{2}} A_t D_t^{-\frac{1}{2}})\hat{F}_t] \\
& + (1 - \alpha) \cdot \beta \cdot k - (1 - \alpha) \cdot \beta \cdot Tr[\hat{F}_t(D_{t-1}^{-\frac{1}{2}} A_{t-1} D_{t-1}^{-\frac{1}{2}})\hat{F}_t] \\
& + (1 - \alpha) \cdot (1 - \beta) \cdot k - (1 - \alpha) \cdot (1 - \beta) \cdot Tr[\hat{F}_t(D_p^{-\frac{1}{2}} A_p D_p^{-\frac{1}{2}})\hat{F}_t] \\
= {} & \gamma \cdot k - Tr[\hat{F}_t(\alpha \cdot D_t^{-\frac{1}{2}} A_t D_t^{-\frac{1}{2}} + (1 - \alpha) \cdot \beta \cdot D_{t-1}^{-\frac{1}{2}} A_{t-1} D_{t-1}^{-\frac{1}{2}} \\
& + (1 - \alpha) \cdot (1 - \beta) \cdot D_p^{-\frac{1}{2}} A_p D_p^{-\frac{1}{2}})\hat{F}_t \quad\quad\quad\quad (10)
\end{aligned}
$$

where $\hat{F}_t \in \mathbb{R}^{n \times k}$, $\hat{F}_t^T \hat{F}_t = I_k$. And first term $\gamma \cdot k$ is a constant, so this is also a trace maximization problem. Again the solution of this relaxed optimization problem is the matrix $\hat{F}_t$ whose columns are the $top - k$ eigenvectors of matrix $\alpha \cdot D_t^{-\frac{1}{2}} A_t D_t^{-\frac{1}{2}} + (1 - \alpha) \cdot \beta \cdot D_{t-1}^{-\frac{1}{2}} A_{t-1} D_{t-1}^{-\frac{1}{2}} + (1 - \alpha) \cdot (1 - \beta) \cdot D_p^{-\frac{1}{2}} A_p D_p^{-\frac{1}{2}}$.

Next, we follow the step (2) and step (3) mentioned in Sect. 3.1 to obtain the final clusters.

4 Experiments

4.1 Experiments on Artificial Dataset

First, we use two-dimensional Gaussian distribution to generate a synthetic dataset. This dataset consists of nine evolutionary stages, each with 1000 two-dimensional data points. There are two types of data evolution: shifting and splitting. The stages of evolution are as follows: 1–2 stages shifting, 2–4 stages splitting, and 4–9 stages shifting. But it is worth noting that 7–9 stages are evolved from stage 4. So we get two branches: one is from stage $1 \rightarrow 6$, and other one is $1 \rightarrow 4 \rightarrow 7 \rightarrow 9$. We conduct experiments on this dataset and try to find out the evolutionary structure.

Figure 1a shows the evolutionary tree obtained from our framework. We can see that in the stage 4 branching phenomena occur in the evolutionary tree, and the evolutionary tree captures the change points of the structure exactly. This structure information will be lost if only temporal smoothness is taken into account. And the loss of this information will be directly reflected in the clustering results.

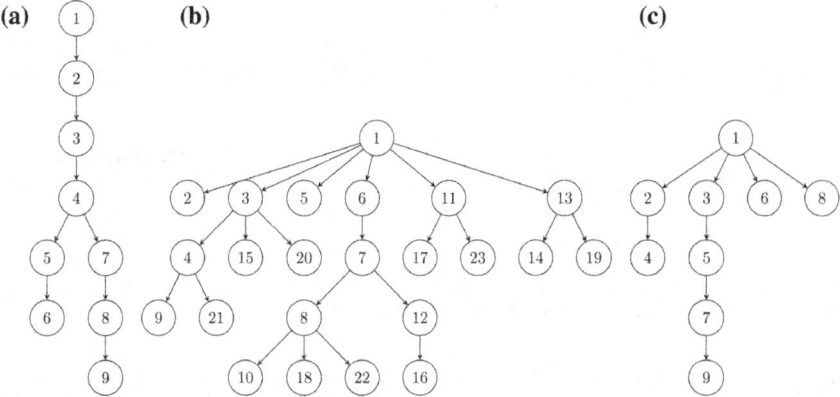

Fig. 1 a Evolutionary tree of synthetic dataset, **b** evolutionary tree of Hospital dataset, and **c** evolutionary tree of High-school dataset

4.2 Experiments on Real Dataset

In this section, we validate our ETSC on two real-world datasets, namely Hospital ward dynamic contact network and High-school contact and friendship networks. The basic information of datasets is shown in Table 1. And we call them Hospital dataset and High-school dataset. Hospital dataset contains 46 medical staffs and 29 patients [12]. This dataset describes the contact network between patients and medical staff from 1:00 p.m. on December 6, 2010 to 2:00 p.m. on December 10, 2010. High-school dataset included 329 students [13]. This dataset describes the contact network between students in the December 2013. In data preprocessing, we refer to the methods proposed by Gauvin et al. [14] to accept a parameter that divides dataset into evolutionary data, and the parameter in our experiments is equal to 4 h.

In this experiment, our evolutionary tree spectral clustering (ETSC) is compared with evolutionary spectral clustering (ESC) under PCQ frame [15]. We use FScore, NMI, and Purity as the evaluation index and Kullback–Leibler divergence as evolutionary distance. About arguments, we set α to 0.9 in ESC, which is same as the setting in PCQ frame. And in our framework, we set α to 0.6 and β to 0.5. We validate the AA on Hospital dataset and validate the NC on High-school dataset.

Figure 2 shows experiments of AA on Hospital dataset, and Fig. 3 shows experiments of NC on High-school dataset. They indicate that ETSC performs better than

Table 1 Basic information of datasets

Dataset	Record number	Sample size	Cluster number
Hospital	32424	75	4
High-school	188508	329	9

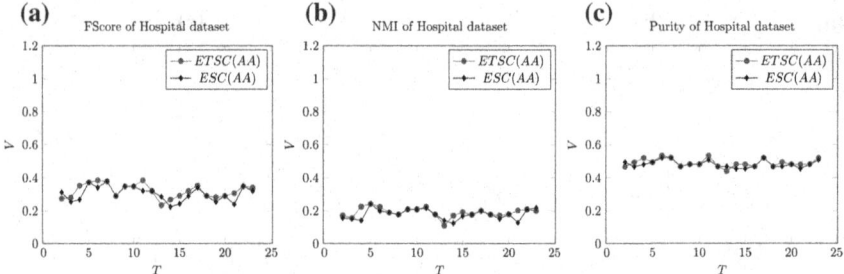

Fig. 2 FScore, NMI, and Purity of AA on Hospital dataset

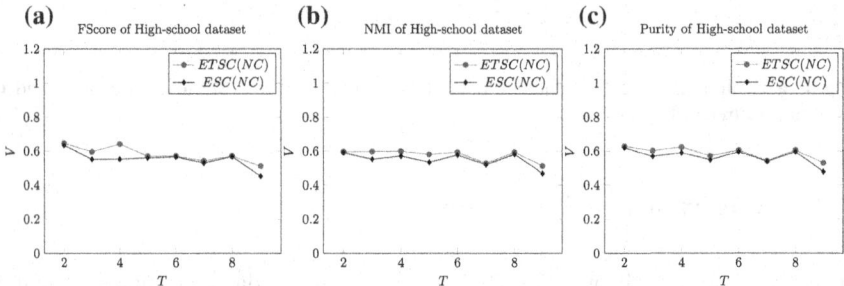

Fig. 3 FScore, NMI, and Purity of NC on High-school dataset

ESC at most of time. Then, the evolutionary trees of Hospital dataset and High-school dataset are shown in Fig. 1b and Fig. 1c, respectively. These evolutionary trees suggest that ETSC leads to better clustering result most of the time, when it acquired nonlinear structural during data evolution.

5 Conclusion

In this paper, we propose a novel evolutionary tree clustering framework, which effectively utilizes the structural information of data and uncovers the branch phenomena in the evolution of data. On the basis of evolutionary tree clustering framework, we develop a new evolutionary tree spectral clustering algorithm that considers structural smoothness and temporal smoothness at the same time. Experimental results on both synthetic and real-world datasets demonstrate that our proposed algorithm outperforms the existing evolutionary spectral clustering algorithm.

Acknowledgements This research was supported in part by the Chinese National Natural Science Foundation under Grant nos. 61402395, 61472343, and 61502412, Natural Science Foundation of Jiangsu Province under contracts BK20140492, BK20151314, and BK20150459, Jiangsu overseas research and training program for university prominent young and middle-aged teachers and presidents, Jiangsu government scholarship funding, and practice innovation plan for college graduates of Jiangsu Province under contracts SJLX16_0591.

References

1. Gaina, R.D., Liu, J., Lucas, S.M., et al.: Analysis of vanilla rolling horizon evolution parameters in general video game playing. In: European Conference on the Applications of Evolutionary Computation, pp. 418–434. Springer, Cham (2017)
2. Gidea, M., Katz, Y.: Topological data analysis of financial time series: landscapes of crashes. Papers (2017)
3. Liu, C., Wu, C., Jiang, L.: Evolutionary clustering framework based on distance matrix for arbitrary-shaped data sets. IET Signal Proc. **10**(5), 478–485 (2016)
4. Wang, L., Rege, M., Dong, M., et al.: Low-rank kernel matrix factorization for large-scale evolutionary clustering. IEEE Trans. Knowl. Data Eng. **24**(6), 1036–1050 (2012)
5. Wang, Y., Liu, S.X., Feng, J., et al.: Mining naturally smooth evolution of clusters from dynamic data. In: Proceedings of SDM, pp. 125–134 (2007)
6. Yu, S.Z.: Hidden semi-Markov models. Artif. Intell. **174**(2), 215–243 (2010)
7. Corne, D., Handl, J., Knowles, J.: Evolutionary clustering. In: Twelfth ACM SIGKDD International Conference on Knowledge Discovery and Data Mining, pp. 554–560, Philadelphia, PA, USA, Aug 2006
8. Ng, A.Y., Jordan, M.I., Weiss, Y.: On spectral clustering: analysis and an algorithm. In: International Conference on Neural Information Processing Systems: Natural and Synthetic, pp. 849–856. MIT Press (2001)
9. Shi, J., Malik, J.: Normalized cuts and image segmentation. IEEE Trans. Pattern Anal. Mach. Intell. **22**(8) (2000)
10. Golub, G., Loan, C.V.: Matrix Computations, 3rd edn. Johns Hopkins University Press (1996)
11. Bach, F.R., Jordan, M.I.: Learning spectral clustering, with application to speech separation. J. Mach. Learn. Res., 7 (2006)
12. Vanhems, P., et al.: Estimating potential infection transmission routes in hospital wards using wearable proximity sensors. PLoS ONE **8**(9), e73970 (2013)
13. Mastrandrea, R., Fournet, J., Barrat, A.: Contact patterns in a high school: a comparison between data collected using wearable sensors, contact diaries and friendship surveys. PLoS ONE **10**(9), e0136497 (2015)
14. Gauvin, L., Panisson, A., Cattuto, C.: Detecting the community structure and activity patterns of temporal networks: a non-negative tensor factorization approach. PLoS ONE **9**(1), e86028 (2014)
15. Chi, Y., Song, X., Hino, K., et al.: Evolutionary spectral clustering by incorporating temporal smoothness. US 7831538 B2 (2010)

The Study on Intelligent Decision and Personalized Recommendation of Job Information Service

Chen Li and Cheng Yang

Abstract With the rapid development of the Internet, the vast majority of college students have been applying for jobs online mainly. For the applicants, while the upside of the Internet is to provide considerable and various job information, it may take them a great deal of time and energy to pick the fitting post from as well. Currently, job-hunting Web sites demonstrate all sorts of information through integration and classification, requiring users to manually retrieve to find the best for them. Combining the key technology of personalized recommendation service and online job-seeking of graduates, the paper designs and implements the employment information recommendation service for college students. The concrete procedures include adopting recommendation algorithm and similarity calculating method that cater to job information based on data mining, intelligent decisions, analysis of users' preference and information integration and recommendation and testing in terms of certain situations or data features. It achieves the desired recommendation result.

Keywords Job information · Intelligent decisions · Personalized recommendation · Similarity

1 Introduction

Plentiful information is needed for students, employers, governments, and universities to choose and make decisions. The problem is how to explore, analyze, and process information with efficiency and accuracy specific to their demands. It is of high significance to provide basis for the decision-making of governments and schools on talent cultivation and give students access to integration and recommen-

C. Li (✉) · C. Yang
Communication University of China, Beijing, China
e-mail: lichengood@cuc.edu.cn

C. Yang
e-mail: cafeeyang@163.com

© Springer Nature Singapore Pte Ltd. 2019
S. K. Bhatia et al. (eds.), *Advances in Computer Communication and Computational Sciences*, Advances in Intelligent Systems and Computing 760, https://doi.org/10.1007/978-981-13-0344-9_23

dation service to meet their demands. In terms of the employment service, it can offer intelligent and personalized technical support.

The paper researches on statistical analysis of the college students' employment, improving the monitoring and forecasting service and showing all data with regard to graduation employment to policymakers and universities. It also focuses on the analysis and prediction on the trend of job market demands, beneficial to adjusting the talent development for universities and getting feedback on talent training quality of the colleges for governments to make better decisions. The paper, at the same time, introduces statistical intelligence technology, intelligent decision-making [1, 2], and information integration and recommendation through analyzing the personalization and trend of job-hunting, mining the data and users' preference. Thus, helpful information from database can be obtained to serve as a guideline for graduates. Section 2 displays intelligent decision models and personalized recommendation method; Sect. 3 evaluates our approach, and Sect. 4 concludes this work and present perspectives.

2   Intelligent Decisions and Personalized Recommendation

What graduates majored in is vitally important for graduates to seek job, especially comparing the similarity of two university students. In order to increase the probability of successful matches, more scientific and optimized methods need to be combined and adopted. For example, "accounting" and "sociology" both have two same Chinese characters, bearing high resemblance of 66.7% literally via cosine similarity, but they are quite different. Cosine similarity indicates the similarity of "accounting", and "financial management" is zero, but students of these two majors often need the same jobs actually.

Besides optimization on similarity, sometimes, major requirements that employers' posts are not clear enough. For instance, if an IT company advertises for engineers' post that marks "computer science and other majors is needed", the student who majors in "software engineering" and is qualified for the job may find it difficult in finding it using conventional tools for information retrieval. Here, "software engineering" may belong to "other majors" without being highlighted by employers for the sake of brevity. Students often need to browse more information to judge if the major "software engineering" matches. The paper studies the intelligent decisions and personalized recommendation that can spare a thought for users.

2.1 The Process of Intelligent Decision Algorithm

Designing the Function of Algorithm

Suppose the frequent item set one and k − 1 and candidate item set k can be figured out. By comparing the proportion of some major's occurrence in the item set and minimum support, whether the major in or out of "k" will be decided [3].

- Read each item set with regard to the certain major from the database and add them using "list" mode in some particular order.
- As all frequent and candidate item sets are in "Treeset" [4], all items are in order and without repetition [5].
- Setting minimum support and confidence [6].

Reading Data

(a) Suppose each unit "Key" and the "major" data "Value" and store all these into a "map" typed "IdentityHashMap<String, String>" [7].

(b) The next step is to further process the data under each unit and single out all "major" data by setting each major data "Key" and its default value 1. Here, the first iteration begins, with all "major" data stored into another "map" typed "Map<Stranger, Integer>" and named "allzhuanye". After the second iteration finishes, another sub-map (typed "Map <String, Map<String, Integer≫" and named "company") will come into existence. In the course of the second, each particular major is "value" and its original unit is "Key". Then, check if the random map and its "Key" have been established or not in the map "company". If yes, then add it in place specific to its key; if no, then establish and add.

(c) The third step is to integrate each Key and its "major" map involved into a new data, then gathering to be a data set. The concrete implementation starts with getting a sub-map given the Key in the map "company". Then, the function "Key-set()" can help sort out each key-major set for each sub-map. In terms of each key-major set, all elements will be spliced with separators (/) between any two. Finally, all elements will be included in the general list typed "List<String>" and the list reshaped to the final data set composed of character strings.

Analyzing the Apriori Algorithms

(a) The first is to list all majors in the candidate item set one; then, compare each data in the final data set mentioned above with each element in the candidate item set. If the element can be found in the data set, plus one. When the matching finishes, the percent of the element's occurrence needs to be calculated to decide whether the element is put into the frequent item set one or not. The element that has its percentage below the minimum support shall be kept out.

(b) If there are two elements or more in the frequent item set one, it is possible to calculate the frequent item set two and candidate item set two. The candidate item set two is implemented by combining every two elements of the frequent item set one. What needs to be mentioned is to put the combinations in ascending order in case of the unnecessary double-counting. The procedures leading to the frequent item set two are incredibly similar to those of frequent item set one, as mentioned above.

(c) Continue all the way through "$k - 1$", then to candidate item set and frequent item set "k", using the similar method until the empty set emerges.

(d) Bring all the frequent item sets together, from the second to the "k", into a maximum set.

(e) Going through each element of the maximum and exploring the Apriori algorithm among majors in each element. The confident coefficient can also be calculated.

(f) Getting out the candidate item set, frequent item set, and the Apriori algorithm.

The paper mainly studies data about the career path and major distribution of the graduates to help discover the association rules among different majors through Apriori algorithm [8], providing the solid ground for the further similarity calculation of recommendation algorithms.

2.2   Personalized Recommendation

In the course of study, the data used in the paper mainly involves job introduction and CV. They are all in text form, and the advantage of text form is its easier data extraction. The content-based recommendation algorithms [9] and the project-based collaborative filtering algorithms [10] here will be combined to design a better service catering to graduates. The detailed way includes feature extraction, combination, and supplement.

Major Similarity

What matters is to exclude the illogical association rules. The difference between confident coefficients normally exists, but it can be negligible in the case of job information. Job-seekers who have similar majors often compete for the same post [11]. Basically, employers may pay more attention to their personal experience or competency rather than their majors under such circumstance but major may act as a yardstick when HR selects CV. Hence, the confident coefficient difference can be dismissed and only the bigger value is chosen.

$$\text{sim}(m_1, m_2) = \begin{cases} c(m_1 \to m_2) \\ c(m_2 \to m_1) \\ max(c(m_1 \to m_2), c(m_2 \to m_1)) \\ 0 \\ 1 \end{cases} \qquad (1)$$

Suppose there are two majors, major "m1" and "m2" and the parameter "similarity" between m1 and m2, which is "sim(m1, m2)". The sim(m1, m2) also indicates confident coefficient of the association rule. When major "m1" and major "m2" are the same, the similarity value will be 1.

The Vector Model of Post and CV Information

The main idea of vector space model, put forward by Saton in 1970s [12, 13], is committed to reduce the processing of the text to vector space operation and put the similarity that was semantically presented into the special manner.

Although CV and post information will be stored in the database of the graduates' employment information service in text form [14], the text form of post information first needs to be conversed to spacial vector mode [15]. Then, similarity is calculated through the cosine distance based on vector special mode. The post information involves a range of aspects [16], such as job title, working place, type of work, and CV information includes gender, educational background, schools, and political affiliation. The former and the latter are represented as "$\vec{J} = (j_1, j_2, j_3)$" and "$\vec{S} = (s_1, s_2, s_3, s_4)$".

The formula of similarity is as follows

$$\text{simBasic}(J_1, J_2) = cos\left(\theta_j\right) \tag{2}$$

$$f(s_{m1}, s_{m2}) = \begin{cases} 0, \ s_{m1} \neq s_{m2} \\ 1, \ s_{m1} = s_{m2} \end{cases} \tag{3}$$

If the value is 1, it indicates two students have the same information; if 0, it means students' information is totally different. Here "m" can be 1, 2, 3, 4, and each indicates one part of the four.

Similarity Between the Posts

Before calculating the similarity between any two employment notes, it is necessary to take into account two things, which is the job field and education required. Hence not only in similarity calculation, but also in algorithms, the comparison of the two items comes first. If there is any difference on either one for two notes, then the similarity between two notes will be set 0. The similarity formula is listed below.

Here,

$$\text{sim}(J_1, J_2) \begin{cases} 0, & J_1 \text{ and } J_2 \text{ belong to} \\ & different \ fields \\ 0, & J_1 \text{ and } J_2 \text{ academic} \\ & requirements \ are \ different \\ simBasic(J_1, J_2), & J_1 \text{ and } J_2 \text{ belong} \\ & to \ same \ fields, \\ & J_1 \text{ and } J_2 \text{ academic} \\ & requirements \ are \ same \end{cases} \tag{4}$$

indicates the vector similarity of post information.

3 Test Results

The database for test covers 85,616 data of career paths for graduates from the year 2011 to 2013, given by Counseling and Career Guidance Center for Information of China's Higher Learning Students. Each data includes students' name, school, major, and career.

More details are as follows. name, gender, political affiliation, school, major, student number, educational degree, specialty field, training orientation, hometown, name of company, public or private, industry, working region, post information.

From the above, some data, as they are superfluous or meaningless, needs to be removed beforehand. For example, data removal leads to 80,826 data available. For the relevant data, more parameter adjustment tests help determine the minimum support is 2 and minimum confident coefficient must above 20% line. The following table shows the related similarity was kept at a reasonable level under the association rules.

The item which shows higher confident coefficient has its major A and B more relevant and vice versa. All these researches gear up for similarity service in the recommendation algorithm, and all data includes 262,819 posts, 898,651 CVs, and 1,687,986 operation records (Table 1).

To evaluate the performance of the recommendation service better, three parameters "Precision", "Recall", and "F-measure" are introduced.

Precision

The percentage of items "hit" by recommendation service against all items recommended.

Table 1 Confident coefficient of majors

Item	Major A	Major B	Coefficient
1	Chemical engineering	Polymeric chemistry and physics	1
2	Oncology	Clinical medicine	0.67
3	Logistic management	Accounting	0.67
4	Mechatronic engineering	Measurement and control technology and instrumentation	0.63
5	Engineering management	Civil engineering	0.6
6	Acupuncture and massage	Clinical discipline of Chinese and Western integrated medicine	0.6
7	Signal and information processing	Communication and information system	0.8
8	Stomatology	Clinical medicine	0.5
9	Applied chemistry	Mechatronic engineering	0.5
10	Midwifery	Nursing	1
……	……	……	……

Table 2 Test result of recommendation service

Number recommended	Precision (%)	Recall (%)	F-measure (%)
Value = 1	50.05	38.88	43.76
Value = 2	45.77	35.03	39.69
Value = 3	46.55	33.33	38.85
Value = 4	37.60	33.18	35.25
Value = 5	33.25	28.65	30.78
Value = 6	28.76	26.68	27.68
Value = 7	26.89	25.99	26.43
Value = 8	25.97	26.01	25.99

$$P_u = \frac{|hits_u|}{|recset_u|} \tag{5}$$

The "hit" ones refer to items which are recommended rightly. All items recommended refer to all in the service.

Recall

The percentage of items "hit" by recommendation service against the maximum theoretical hit

$$R_u = \frac{|hits_u|}{|testset_u|} \tag{6}$$

The former refers to items which are recommended rightly, while the latter means items which can be recommended rightly in theory.

F-measure

F-measure requires all factors (Precision and Recall) taken into account, and the way is to work out the weighted average of the Precision and Recall. It attempts to strike a balance between Precision and Recall by being closer to the one with a lower value.

$$F1 = \frac{2 * P * R}{P + R} \tag{7}$$

The test recommends eight posts (Value = 1, 2, ..., 8), and the figures are shown below.

The table shows as the number recommended increases, the recommendation will play a less important role. In fact, three to six posts are the best to meet the needs of college students. In addition, three to six posts keep its high Precision and Recall (Table 2).

4 Conclusion

This paper adopts the Apriori algorithm to mine data from the career and major information of the college students to study the association rules among majors in the recommendation service. It also optimizes the statistical analysis technique, facilitating the further analysis and prediction, which is beneficial to government, universities, and graduates. It is also helpful to enhance the monitoring and warning mechanism on job market.

With the help of information integration and recommendation technique based on users' preference, the service retrieves the characteristic data from the huge database to make it more personalized and intelligent and efficient. Knowing the characteristic of the data also contributes to the improvement and optimization of the existing algorithm, which means put forward a mixed recommendation algorithm. The algorithms help establish the content model featured by textual data of posts and calculate the similarity among students' information, posts, and their majors, respectively. In the course of test, the accuracy of the Precision, Recall, and F-measure result are all improved markedly and graduates know better how the post fits their majors.

Besides, the prototype system for job personalized recommendation also rolled out the online sharing function. It can be operated well through PC browser and smart terminals in the case of massive concurrent visits. The smart terminal, framed by C/S, encapsulates HTML5, JS, CSS by virtue of Phonegap and supports HTML5, enabling graduates to benefit from quality recommendation and browsing history.

Acknowledgements The work is supported by grants from National Science and Technology Supporting Program of China (2014BAH10F00) and University Research Program of Communication University of China (3132015XNG1522). We thank the reviewers and editor for their helpful comments.

References

1. Azuaje, F., Witten, I.H., Frank, E.: Data mining: practical machine learning tools and techniques. Biomed. Eng. Online **5**(1), 1–2 (2006)
2. Lee, W., Stolfo, S.J.: Data mining approaches for intrusion detection. Heat Mass Transf. **48**(2), 291–300 (2012)
3. Liu, D.R., Shih, Y.Y.: Integrating AHP and data mining for product recommendation based on customer lifetime value. Inf. Manag. **42**(3), 387–400 (2005)
4. Chen, C.: Using data mining technology to provide a recommendation service in the digital library. Electr. Libr. **25**(25), 711–724 (2013)
5. Tsai, C.: Using adaptive resonance theory and data-mining techniques for materials recommendation based on the e-library environment. Electron. Libr. **26**(3), 287–302 (2008)
6. Wang, J., Jia, B., Zhang, W., et al.: Study on the data mining web service recommendation engine. In: International Conference on E-Business & E-Government, pp. 1081–1084 (2012)
7. He, B.: Personalized web information recommendation based on data mining. Adv. Mater. Res. **225–226**, 546–549 (2011)
8. Smyth, B., Wilson, D., O'Sullivan, D.: Data mining support for case-based collaborative recommendation. Lect. Notes Comput. Sci. **2464**, 111–118 (2002)

9. Walter, F.E., Battiston, S., Schweitzer, F.: A model of a trust-based recommendation system on a social network. Auton. Agent Multi-Agent Syst. **16**(1), 57–74 (2008)
10. Kim, J.K., Kim, H.K., Oh, H.Y., et al.: A group recommendation system for online communities. Int. J. Inf. Manag. **30**(3), 212–219 (2010)
11. Tuan, C.C., Hung, C.F., Tseng, K.W.: A relational compound collaborative filtering recommendation system. In: International Conference on Broadband and Wireless Computing, Communication and Applications, pp. 411–415. IEEE Computer Society (2011)
12. Chen, H.C., Chen, A.L.P.: A music recommendation system based on music and user grouping. J. Intell. Inf. Syst. **24**(2), 113–132 (2005)
13. Smirnov, A., Kashevnik, A., Ponomarev, A., et al.: Recommendation system for tourist attraction information service. In: Conference of Open Innovations Association, pp. 148–155 (2013)
14. He, J., Du, J., Zhang, Y., et al.: A recommendation system for a web portal. In: International Conference on Progress in Informatics and Computing, pp. 12–13. IEEE (2014)
15. Yang, D., Zhang, D., Yu, Z., et al.: A sentiment-enhanced personalized location recommendation system. In: ACM Conference on Hypertext and Social Media, pp. 119–128 (2013)
16. Sakamoto, T., Kitamura, Y., Tatsumi, S.: A competitive information recommendation system and its rational recommendation method. IEICE Trans. Inf. Syst. **38**(9), 74–84 (2007)

Part II
Intelligent Image Processing

Enhancing Source Camera Identification Using Weighted Nuclear Norm Minimization De-Noising Filter

Mayank Tiwari and Bhupendra Gupta

Abstract Photo-response non-uniformity noise (PRNU) is widely accepted as fingerprint (FP) of digital camera. However, extraction of PRNU from given images is still a challenging task. In the previous literature, number of de-noising filters has been used for PRNU extraction. However, it is observed that PRNU extracted by existing de-noising filters contains high-frequency (edges and texture) details of the image. This increases false rejection rate in source camera identification (SCI) process. In this work, we have used weighted nuclear norm minimization (WNNM)-based de-noising filter for PRNU extraction. The PRNU extracted by WNNM-based de-noising filter contains least amount of scene details. Experimental results demonstrate the proposed method outperforms, or at least performs comparably to, the state-of-the-art methods.

Keywords Digital image forensics · Photo-response non-uniformity noise
Source camera identification

1 Introduction

The advancement in technologies has replaced traditional film imaging device from digital camera. Nowadays, every digital multimedia device contains high-quality digital cameras for capturing images. The digital images are also used as trustworthy evidence in the courtroom. But with the help of image sharing and image editing, software's images can easily be shared or altered. Due to this, determination of image origin and integrity has become an important task. The digital image forensic

M. Tiwari · B. Gupta (✉)
Department of Mathematics, Indian Institute of Information Technology,
Design & Manufacturing, Jabalpur 482005, Madhya Pradesh, India
e-mail: gupta.bhupendra@gmail.com
URL: http://www.bhupendragupta.com

M. Tiwari
e-mail: mayanktiwariggits@gmail.com

© Springer Nature Singapore Pte Ltd. 2019
S. K. Bhatia et al. (eds.), *Advances in Computer Communication
and Computational Sciences*, Advances in Intelligent Systems
and Computing 760, https://doi.org/10.1007/978-981-13-0344-9_24

techniques are used to perform various forensic tasks such as SCI and image integrity verification. The SCI tries to find out the imaging device that captured the image, while tempering information of the image is determined by integrity verification.

2 Related Work

In the given literature, a number of techniques have been developed to perform SCI and integrity verification tasks. These techniques try to discover the inherent normal for digital camera left in the image. There are numerous elemental traces left in the image by the image preparing segments (software or hardware) of the image acquisition pipeline. These follows are for the most part color interpolation artifacts (CFA), these artifacts are discussed by [1, 2]. Saturated pixels-based SCI task is discussed by [3, 4]. JPEG compression artifacts are discussed by [5, 6]. Lens distortion-based SCI work is discussed by [7, 8]. The previously mentioned procedures are not steady in various natural conditions, and now and again they are likewise not exceptional for singular cameras of same models. Because of this, sensor pattern noise (SPN)-based strategies pull in many considerations. The SPN which for the most part comprises PRNU noise is steady in various environmental conditions and unique for singular cameras of same models.

In typical PRNU-based SCI, a PRNU extraction technique (de-noising filter) is connected to extricate the noise (η) part as it is the difference between the given image and its de-noised version. This noise residue (η_R) contains noteworthy qualities of the PRNU and yet in the meantime can without much of a stretch be influenced by the image contents (high-frequency details). PRNU-based SCI work was initiated by [9]. Later by the same research team, PRNU-based SCI and integrity verification was initiated. In the newer work, [10] also introduced PRNU enhancement techniques. These enhancement techniques enhance PRNU by reducing non-unique artifacts shared among different camera models. Another PRNU enhancement technique was developed by [11]. This idea was based on the fact that stronger signal component present in the PRNU is basically scene details and hence it should be less trustworthy. Another method to deal with the color interpolation artifacts in the PRNU-based processing was developed by [12]. PRNU enhancement using filtering distortion removal was done by [13].

Apart from all these mentioned methods, it was observed that PRNU can be well extracted from given image by choosing suitable de-noising filters. Hence, a number of de-noising filters have been used for PRNU extraction. Few of the widely used filters for PRNU extraction are Mihcak's filter [14], Perona–Malik diffusion filter (PMD) [15], context adaptive-guided image filter (CAGI) [16], total variation-based filter [17], and context adaptive interpolator (CAI) [18]. The advantages and disadvantages of these de-noising filters have been systematically discussed by [19].

2.1 Our Contribution

In the proposed work, we are using weighted nuclear norm minimization (WNNM)-based de-noising filter for PRNU extraction. On the basis of performance of WNNM filter suggested by [20], we observe that WNNM de-noising filter does not distort high-frequency details of the image; if high-frequency details are preserved in the restored image, then these details will not be present in the extracted PRNU. This means PRNU extracted by WNNM filter will contain least amount of scene details and therefore the extracted PRNU will be more suitable for SCI work.

The rest of the paper is organized as: Sect. 2 contains a brief description of related work. Section 3 covers some basic concepts required for proper understanding of our work. Section 4 covers an introduction to WNNM de-noising filter. A detailed description of proposed method is shown in Sect. 5. Sections 6 and 7 cover experimental setup and experimental results, respectively. At last, Sect. 8 concludes our work.

3 Preliminaries

This section covers three basic concepts that are required for proper understanding of the proposed work.

3.1 Camera Model

In the proposed work, we are using pixel output model as suggested by [21]. Based on this model, the final image generated by a digital camera is given by:

$$I = I_0 + K \cdot I_0 + N_t, \tag{1}$$

where N_t signifies the combination of independent random η components. The model is received in the vast majority of the current PRNU-based methods. Also, numerous methods demonstrate $K \cdot I_0 + N_t$ joined as white Gaussian η. Few authors recognize K by the PRNU variable and $K \cdot I_0$ by the PRNU signal. In any case, K is the genuine unique FP of a camera, and every one of the strategies certainly or expressly looks to gauge this amount or a scaled variant of it—which we essentially allude to be the camera PRNU.

3.2   The Standard Procedure of PRNU Estimation

The camera PRNU can be estimated by utilizing L of its images: $I_l \in R^{M \times N}$, $l = 1, \ldots, L$. In like manner, rehearsed L images of low-frequency regions (sky areas or flat surface) are utilized for PRNU extraction. The de-noised version of each image is subtracted from the original image to get its η_R:

$$\eta_R = I - F(I), \tag{2}$$

where $F(\cdot)$ is de-noising operation. The η_R contains the PRNU η and different sorts of arbitrary η that are not appropriate for scientific applications. The random η shown in the PRNU is suppressed by averaging all noise residuals. Now, the camera FP is calculated as:

$$PRNU = \sum_{i=1}^{L} \frac{\eta_{R_i}}{L} \tag{3}$$

3.3   Performance Metric

Correlation between η_R and camera PRNU ζ_{PRNU} is calculated to check whether an image I is taken from camera ζ or not.

$$corr(\eta_R, \zeta_{PRNU}) = \frac{(\eta_R - \overline{\eta_R}) \cdot (\zeta_{PRNU} - \overline{\zeta_{PRNU}})}{||\eta_R - \overline{\eta_R}|| \cdot ||\zeta_{PRNU} - \overline{\zeta_{PRNU}}||} \tag{4}$$

where $\overline{\eta_R} = mean(\eta_R)$ and $\overline{\zeta_{PRNU}} = mean(\zeta_{PRNU})$. Now experimentally, we are able to determine whether a given image I is taken from given camera ζ or not. If the correlation $corr(\eta_R, \zeta_{PRNU})$ satisfies a threshold T, then we can conclude that the image I is taken from given camera ζ.

4   Weighted Nuclear Norm Minimization De-Noising Filter

This method is a significant extension of the nuclear norm minimization problem [20]. Here, authors [20] have studied different types of weights for the weighted nuclear norm minimization (WNNM). The WNNM method is based on three facts which are:

1. When the weights are in a non-ascending order, WNNM is as yet convex and it utilizes the expository ideal arrangement.
2. When the weights are in a subjective order, WNNM utilizes an iterative calculation to comprehend it.

3. When the weights are in a non-descending order, WNNM demonstrates that the iterative calculation can bring about a systematic settled point arrangement, which can be proficiently figured.

The WNNM method preserves better image's local structures and generates less visual artifacts. Algorithm 1 describes working of WNNM method.

Algorithm 1 image de-noising by WNNM

Require: noisy image φ

initialize $\widehat{\psi}^{(0)} = \varphi$, $\varphi^{(0)} = \varphi$

$k = 1$;

while $k \leq K$ **do**

    iterative regularization $\varphi^{(k)} = \widehat{\psi}^{(k-1)} + d(\varphi - \widehat{\varphi}^{(k-1)})$

    for each patch φ_j in $\varphi^{(k)}$ **do**

        search for similar patch group φ_j

        find weight vector w

        singular value decomposition $[U, \Sigma, V] = SVD(\varphi_j)$

        estimate: $\widehat{\chi}_j = U S_w(\Sigma) V^T$

    end for

    aggregate χ_j to form the clean image $\widehat{\psi}^{(k)}$

    k = k + 1;

end while

return clean image $\widehat{\psi}^{(K)}$

---

5 The Proposed Method

The Algorithm 2 describes complete working of proposed method. For each camera model C^i, it first calculates PRNU (C^i_{PRNU}) of that camera by using flat-field images C^i_{FF}. After that for each natural image I (which are provided in C^i_{NI}), the method performs SCI task by using Eqs. (2) and (4).

Algorithm 2 source camera identification using WNNM

Require: C^i_{FF} contains total len^i_{FF} number of flat-field images for camera model C^i.

Require: C^i_{NI} contains total len^i_{NI} number of natural-images for camera model C^i.

Require: T a predefined threshold.

$CNT^i = 0$.

de-noise all len^i_{FF} images of C^i_{FF} using Algorithm 1 and calculate PRNU for C^i_{PRNU} by using Eq. (2)-(3).

for each image I of C^i_{NI} **do**

    calculate noise residual n of each I using Eq. (2).

    find correlation value $CORR(n, C^i_{PRNU})$

    if $CORR(n, C^i_{PRNU}) \geq$ T **then**

        $CNT^i = CNT^i + 1$.

    end if

end for

true positive rate $TPR = \frac{CNT^i}{len^i_{NI}}$.

return TPR.

Table 1 List of cameras from Dresden image database used in our experiment

Camera model	# Devices	Camera model	# Devices
Canon IXUS55	1	Pentax OptioW60	1
Casio EXZ150	5	Praktica DCZ5.9	5
FujiFilm FinePixJ50	3	Ricoh GX100	5
Nikon CoolPixS710	5	Rollei RCP7325XS	3
Nikon D200	2	Samsung L74wide	3
Nikon D70	1	Samsung NV15	3
Olympus Mju1050SW	5	Sony DSCH50	2
Panasonic DMCFZ50	3	Sony DSCT77	4
Pentax OptioA40	4	Sony DSCW170	2

6 Experimental Setup

For a fair evaluation of the proposed work, we have used the 'Dresden image database' [22]. Totally, fifty-seven cameras listed in Table 1 are used by us in the proposed work. For each camera, we have cropped two sub-images of size 128×128, and 64×64 from center of the original image without affecting its content. Now, the total number of images is $9083 \times 2 = 18166$. The flat-field (FF) images of the database are used to generate camera FP. Then, the natural images (NI) are used to perform the SCI task.

In the proposed work, we have worked on the green color channel of the image since it contains the greater part of the PRNU data and smallest amount of interpolation η [12]. The proposed technique can likewise be utilized for three color channels of the image. This should be possible be utilizing technique recommended by [12]. At long last, it merits saying that similar parameter settings, as proposed in original papers, have been utilized as a part of our trials unless generally expressed.

7 Experimental Results

In this section, a number of experiments have been performed to judge performance of the proposed method. Here, performance of WNNM's filter is judged with eight other PRNU extraction methods. The first two methods are Model 3 and Model 5 from the work of [11]; these models were shown to be delivered the best results by the authors [11]. Other six models are filtering distortion removal [13], the Mihcak's filter [9], context adaptive interpolator [18], context adaptive-guided filter [16], Perona–Malik diffusion [15], and total variation filter [17]. Here for each PRNU extraction method $\sigma = 5$ is used.

Results of Table 2 show that the WNNM filter is able to work better than all eight PRNU extraction methods. This is due to the reason that PRNU extracted by WNNM

Table 2 True positive rate (in %) for false positive rate 0.01.

Methods	64×64	128×128
Model 3 [11]	12.33	26.97
PMD [15]	41.55	48.62
TVSTEP [17]	44.74	46.06
Model 5 [11]	45.98	53.32
Mihcak [9]	48.57	53.89
CAI [18]	48.91	45.73
FDR [13]	49.80	54.31
CAGI [16]	52.07	54.94
WNNM	52.30	59.61

filter contains least amount of scene details. The only one limitation of WNNM filter is its computation time. The WNNM filter takes significantly more time as compared to other de-noising filters. This is due to the iterative nature of WNNM filter.

8 Conclusion

In this paper, we have used the WNNM de-noising filter for PRNU-based source camera identification. PRNU extracted using WNNM de-noising filter contains least amount of high-frequency details of the image, and hence it improves efficiency of the SCI. Based on experimental results, the WNNM filter is able to work better than all 8 PRNU extraction methods. In future performance of WNNM, de-noising filter can be improved by using different PRNU enhancement methods such as removing and sharing components given by [10], color decoupled PRNU given by [12], phase-only method given by [23], and spectrum equalization method given by [24].

Acknowledgements Authors thank [19, 20] for many helpful suggestions and sharing of their *MATLAB* code with us.

References

1. Bayram, S., Sencar, H., Memon, N., Avcibas, I.: Source camera identification based on CFA interpolation. In: Proceedings of the IEEE International Conference on Image Processing vol. 3, pp. III-69–III-72 (2005)
2. Swaminathan, A., Wu, M., Liu, K.J.R.: Nonintrusive component forensics of visual sensors using output images. IEEE Trans. Inf. Forensics Secur. 2(1), 91–106 (2007)
3. Geradts, Z.J., Bijhold, J., Kieft, M., Kurosawa, K., Kuroki, K., Saitoh, N.: Methods for identification of images acquired with digital cameras. Proc. SPIE **4232**, 505–512 (2001)

4. Kurosawa, K., Kuroki, K., Saitoh, N.: CCD fingerprint method identification of a video camera from videotaped images. Proc. IEEE Int. Conf. Image Process. **3**, 537–540 (1999)
5. Alles, E.J., Geradts, Z.J., Veenman, C.J.: Source camera identification for heavily JPEG compressed low resolution still images. J. Forensic Sci. **54**(3), 628–638 (2009)
6. Sorrell, M.J.: Digital camera source identification through JPEG quantization. In: Multimedia Forensics and Security. Hershey, NY, USA: Information Science Reference, pp. 291–313 (2008)
7. Choi, K.S., Lam, E.Y., Wong, K.K.: Source camera identification using footprints from lens aberration. Proc. SPIE **6069**, 60–69 (2006)
8. Van, L.T., Emmanuel, S., Kankanhalli, M.S.: Identifying source cell phone using chromatic aberration. In: Proceedings of the IEEE International Conference on Multimedia Expo, pp. 883–886 (2007)
9. Lukas, J., Fridrich, J., Goljan, M.: Digital camera identification from sensor pattern noise. IEEE Trans. Inf. Forensics Secur. **1**(2), 205–214 (2006)
10. Chen, M., Fridrich, J., Goljan, M., Lukas, J.: Determining image origin and integrity using sensor noise. IEEE Trans. Inf. Forensics Secur. **3**(1), 74–90 (2008)
11. Li, C.-T.: Source camera identification using enhanced sensor pattern noise. IEEE Trans. Inf. Forensics Secur. **5**(2), 280–287 (2010)
12. Li, C.-T., Li, Y.: Color-decoupled photo response non-uniformity for digital image forensics. IEEE Trans. Circuits Syst. Video Technol. **22**(2), 260–271 (2012)
13. Lin, X., Li, C.-T.: Enhancing sensor pattern noise via filtering distortion removal. IEEE Signal Process. Lett. **23**(3), 381–385 (2016)
14. Mihak, M., Kozintsev, I., Ramchandran, K.: Spatially adaptive statistical modeling of wavelet image coefficients and its application to denoising. In: Proceedings of the IEEE International Conference on Acoustics, Speech, Signal Processing, vol. 6, pp. 3253–3256 (1999)
15. Houten, W.V., Geradts, Z.: Using anisotropic diffusion for efficient extraction of sensor noise in camera identification. J. Forensic Sci. **57**(2), 521–527 (2012)
16. Zeng, H., Kang, X.: Fast source camera identification using content adaptive guided image filter. Forensic Sci. **61**(2), 520–526 (2016)
17. Gisolf, F., Malgoezar, A., Baar, T., Geradts, Z.: Improving source camera identification using a simplified total variation based noise removal algorithm. Digit. Invest. **10**(3), 207–214 (2013)
18. Kang, X., Chen, J., Lin, K., Peng, A.: A context-adaptive SPN predictor for trustworthy source camera identification. EURASIP J. Image Video Process. **2014**(1), 19–30 (2014)
19. Al-Ani, M., Khelifi, F.: On the SPN estimation in image forensics: a systematic empirical evaluation. IEEE Trans. Inf. Forensics Secur. **12**(5), 1067–1081 (2017)
20. Gu, S., Zhang, L., Zuo, W., Feng, X.: Weighted nuclear norm minimization with application to image denoising. In: 2014 IEEE Conference on Computer Vision and Pattern Recognition CVPR, Columbus, OH, pp. 2862–2869 (2014). https://doi.org/10.1109/CVPR.2014.366
21. Healey, G.E., Kondepudy, R.: Radiometric CCD camera calibration and noise estimation. IEEE Trans. Pattern Anal. Mach. Intell. **16**(3), 267–276 (1994)
22. Gloe, T., Bhme, R.: The dresden image database for benchmarking digital image forensics. J. Digit. Forensic Pract. **3**(2–4), 150–159 (2010)
23. Kang, X., Li, Y., Qu, Z., Huang, J.: Enhancing source camera identification performance with a camera reference phase sensor pattern noise. IEEE Trans. Inf. Forensics Secur. **7**(2), 393–402 (2012)
24. Lin, X., Li, C.-T.: Preprocessing reference sensor pattern noise via spectrum equalization. IEEE Trans. Inf. Forensics Secur. **11**(1), 126–140 (2016)

An Efficient Fingerprint Matching Using Continuous Minutiae Template Learning

Kamlesh Tiwari, Vandana Dixit Kaushik and Phalguni Gupta

Abstract Biometric samples acquired by any sensor are highly dependent on the behavior of the user with that scanner. No two fingerprint samples acquired at different instant of time are to be exactly same. This could be due to the sensor properties, translation, and rotation along with smudging effects. However, some of these effects can be minimized by using multiple images of the same trait and thereby improving the template or feature vector by assigning higher weights to the prominent features. This paper deals with the problem of improving the quality of the template of fingerprint with the help of multiple fingerprint samples acquired from the same finger of the subject at different instant of time. It has used the method of continuous minutiae template learning. The proposed method has been tested on FVC-2004 DB2 A database and has shown significant increase in the accuracy of the fingerprint-based biometric system if one adopts the proposed template learning method.

Keywords Biometrics · Fingerprint recognition · Minutiae template
Matching · ROC curve

1 Introduction

Fingerprint is an impression that gets developed on a surface *that comes in contact* to the lower skin of a human. The impression is a representation of furrow on finger skin. This impression appears on nearly all surfaces that come in direct contact with

K. Tiwari
Department of CSIS, Birla Institute of Technology and Science,
Pilani, Pilani 333031, Rajasthan, India

V. D. Kaushik (✉)
Department of CSE, Harcourt Butler Technical University, Kanpur 208002, India
e-mail: vandanadixitk@yahoo.com

P. Gupta
National Institute of Technical Teachers' Training & Research,
Kolkata, Kolkata 700106, India

© Springer Nature Singapore Pte Ltd. 2019
S. K. Bhatia et al. (eds.), *Advances in Computer Communication
and Computational Sciences*, Advances in Intelligent Systems
and Computing 760, https://doi.org/10.1007/978-981-13-0344-9_25

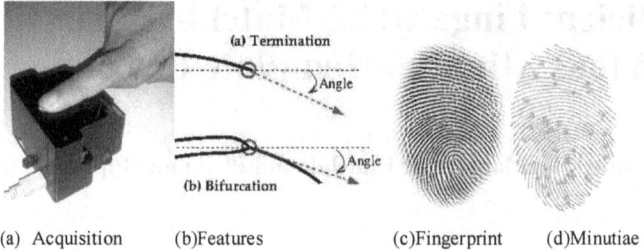

(a) Acquisition   (b)Features   (c)Fingerprint   (d)Minutiae

Fig. 1 Fingerprint acquisition and feature

the finger. Some of them are directly visible such as the one on a water glass or cup, whereas some are not such as the one on wooden surfaces. Fingerprint needs to be collected from the surfaces for further useful application. Sophisticated forensic techniques, such as spreading specialized powder and fixing, have been used to acquire fingerprint from quite a long time. A fingerprint, which is covertly obtained in this way from an already touched surface, is called as a latent fingerprint. Alternatively, a specialized digital fingerprint scanner can also be used if the user cooperates and can place his finger on its surface. These scanners directly acquire the digital fingerprint image. A typical digital fingerprint scanner is shown in Fig. 1a, and the acquired fingerprint image is shown in Fig. 1c.

Print pattern of a fingerprint contains a lot of black lines called ridges. At the end point, a ridge can either terminate or join with other ridges. Both of this kind of points are of special interest and are called minutiae points. A direction is also associated with the minutiae points as shown in Fig. 1b. Location along with the direction of the minutiae point contributes toward the individuality of a fingerprint. Figure 1d shows the location of each minutiae point determined by an automatic minutiae extractor.

Fingerprint patterns have spawn curiosity in the human mind for quite a long time. They are found to be distinct for two different fingers; even within monozygotic twins who bear similar faces but these patterns do not bear similarity. Usefulness of fingerprint has been increased due to the development of automatic techniques that enabled to answer queries like "whether the two different fingerprints are of the same person or not?" Although its accuracy is not 100% till date, but it is found to be convincing enough to be accepted in a broad perspective. Fingerprint has become one of the endorsed forensic evidences to establish person's identity in the court of law.

Initially reported literature on fingerprint includes a study conducted by William Herschel in 1858 while he was working as an officer in the Bengal region of India. He has identified that the fingerprints are unique and can be used for personal identification [1]. First paper on fingerprint was published in 1880 in the scientific journal Nature by Henry Fauld. Galton initially discovered the classification of fingerprints in 1892 [2] as arches, loops, and whorls. This classification is extended further by Henry [3] to arch, tent arch, left loop, right loop, and whorl. First research paper on automatic recognition of fingerprint images is presented by Trauring [4] in 1963 in

Nature. However, even before that people were aware of the patterns of their fingers. Evidence from the ancient artifacts like clay models [5] which contain identifiable fingerprint patterns of Bayzilian period of fourth or fifth century A.D. have shown that the people were curious about fingerprint patterns.

Fingerprint-based recognition can broadly be categorized into texture-based [6] and minutiae-based [7] methods, and there exist several such methods in the literature. The efficiency of such systems is not yet saturated, despite fingerprints being one of the most popular and widely accepted biometric traits. Especially for large-scale deployment where general population is targeted to use such a system, it has often shown poor performance in terms of efficiency and accuracy. There is a need to improve the performance of the fingerprint-based system under this environment.

This paper deals with the problem of improving the efficiency of a fingerprint-based system by utilizing multiple samples from multiple fingers, ensuring quality, stable feature extraction, efficient matching, and suitable fusion. It proposes a minutiae confidence-based fingerprint template learning approach *that would help in* compensating the aging effects of the fingerprint template. The paper is organized as follows. Next section provides details about a general biometric system. In Sect. 3, an efficient matching algorithm has been proposed. Experimental results have been analyzed in Sect. 4. Last section presents the concluding remarks.

2 Biometrics

In today's world, interest is toward systems that are capable of automatic authentication. Traditionally, these systems are based on either token or knowledge. Token provides security of level 1 which is dependent on something in possession, such as an access card, passport, driving license, ATM card, key, dongle. The user is asked to present the token at the time of authentication. One major drawback of this type of systems is that the security of the token itself is fishy. They can easily be misplaced, lost, or stolen. A bit higher level of security, called level 2, is offered by techniques using knowledge. Authentication depends upon something known only to the user like personal identification number (PIN) or password. The problem here is again the leakage of the secret. The user for the sake of his reference may like to store the piece of information in some secret place, but the attacker could trace it. Or the user can share the secret information with others. It may also be possible that the user forgets the secret. The adversary can guess easy secrets. Another major problem arises when the user uses the same secret at two different authentication places. There also exists risk of collusion that refers to the possibility of reconstructing fingerprint template even having with partial information. However, there is a possibility of using both level 1 and level 2 security means simultaneously, but even then they do not overcome their inherent individual shortcomings.

Security measure of level 3 introduces more complexity to the authentication system by using personal characteristics of the user. It utilizes the physiological or behavioral characteristics, called biometrics, of the individual. The solutions offered to identity management are reliable by utilizing fully automated or semi-automated schemes to recognize an individual.

The physiological characteristics such as face, fingerprints, iris, ear, palmprint, hand geometry, and behavioral characteristics such as keystrokes, signature, voice, gait are the exclusive properties of a person. These characteristics are referred as traits. It has been found that these properties are not only binding but also are capable of uniquely identifying a person. By using a characteristic feature extraction and its suitable matching technique, one can construct an automatic human recognition system that uses biometric. Some of the advantages of biometric-based systems over any token or knowledge-based authentication systems are:

- **Uniqueness**: Biometric characteristics are believed to be unique. They are exclusive to an individual and are found to be distinctive enough to identify a person uniquely.
- **Convenient**: Use of biometric is convenient as the user does not need to carry any authentication token or have to memorize any secret information. Physiological or behavioral characteristics of the user are always available to him which cannot be misplaced, lost, or forgotten.
- **Hard to forge**: Biometric characteristics are hard to forge. An adversary can use a spoofing technique to perform an attack on it, but the simultaneous use of more than one biometric trait massively reduces the chance of forgery.
- **Requires Physical Presence**: A biometric system captures live biometric sample at the time of authentication.

Therefore, the physical presence of the user is essential. It inculcates a sense of belongingness with the advantage of non-repudiation; that is, the user cannot deny his participation in authentication at some later point in time. Some of the physiological and behavioral traits are shown in Fig. 1. Each biometric trait has its advantages and disadvantages.

The properties of a biometric trait that determine its suitability for the applications are discussed in [8]. They are:

- Universality: Every individual must possess the characteristic/attribute for authentication. The attribute must be one that is universal and seldom lost to accident or disease.
- Permanence: Over an extended period of time, the biometric characteristics should be constant. These should not have significant differences in age or disease.
- Measurability: The characteristics should be suitable to capture and quantitatively measurable.
- Acceptability: The capturing of biometric characteristics should be acceptable to a large percentage of the population.
- Uniqueness: Features in the trait should be distinct from individuals.
- Performance: The recognition accuracy, speed, and robustness should be within the limitation of the application.
- Circumvention: The attribute should be difficult to mask or manipulate. The process should ensure high reliability and low reproducibility.

Biometrics characteristics require various types of imaging technology to capture the evidences. Applications of these broad ranges of biometrics traits vary considerably. We would briefly discuss some of the most commonly used physiological traits such as face, fingerprint, hand geometry, palmprint, iris, ear, vein pattern, DNA and behavioral traits such as signature, voice, keystroke, and gait.

2.1 Architecture of Biometric System

Pattern recognition techniques [9] can be used to design a biometric system. It consists of four major stages that are as follows:

- Data Acquisition: The process of acquiring raw biometrics data using a suitable reader or scanner is known as data acquisition. The data acquiring process is initiated by a sensing device such as optical fingerprint scanner, video camera, or iris scanner. The quality of the captured image plays a very important role to provide true features from the trait that are used in matching.
- Preprocessing: In this stage, the region of interest is extracted from the raw biometric trait acquired using techniques like filtering, morphological, and segmenting operations. After preprocessing, the extracted region of interest is enhanced for the correction of non-uniform illumination and to improve the features.
- Feature extraction: Distinctive information known as features is extracted from the preprocessed trait biometric samples at this stage. These features possess low intra-class and high inter-class difference and form a feature vector or biometrics template.
- Matching and Decision: To make the decision on recognition, features that are obtained from the test sample are compared against those available in the database at this stage.

Based on the application, a biometric system operates on either enrollment mode, verification mode, or identification mode. During enrollment, the feature vector extracted from the biometric trait is stored or registered in a storage medium for future comparison against a query sample during verification or identification. This mode registers each subject for recognition. In verification, user is validated by comparing his query biometric characteristics against the ones stored in the database. More clearly, when the user claims his identity and presents the biometric trait for verification, the system compares captured biometric features with features of the claimed user stored in the database to verify the user. In identification mode, a user presents his biometric sample at the acquisition device and the system identifies him without any extra information about his identity. More clearly, when the user presents the biometric characteristics, the system compares biometric features with features of all individuals registered with the system.

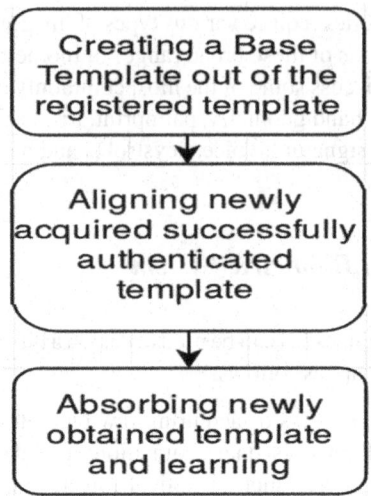

Fig. 2 Block diagram of the proposed system

3 Proposed Approach

This section describes the proposed fingerprint learning approach which consists of three main steps: (1) creating a base template out of the registered template, (2) aligning newly acquired successfully authenticated template, (3) absorbing newly obtained template and learning. Block diagram is shown in Fig. 2. MINDTCT of NIST [10] has been used to extract minutiae points from the fingerprint images.

3.1 Base Template Creation

The minutiae points of a fingerprint stored in the database for a particular use are first extracted. Initially, a black image of dimensions bigger than the fingerprint is created to accommodate possibility of distorted fingerprints. For each minutiae point, a circle is drawn with 1 pixel radius on the dark image keeping the coordinate of minutiae as a center of the circle. These circles are used to pinpoint the location of minutiae points.

3.2 Authenticated Template Alignment

Fingerprint presented at the time of verification is generally not same as the one provided at the time of registration. This happens because of the creativity in

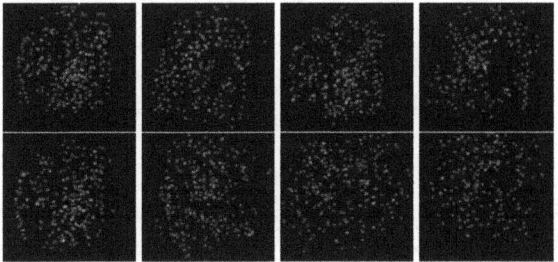

Fig. 3 Templates

human interaction. Major changes in the newly acquired fingerprint may be due to displacement and rotation which can be corrected by determining the angle by which the image might have been rotated by the most common angle of difference between the minutiae points of both fingerprints and by rotating the newly acquired fingerprint by that angle. The mean displacement between the minutiae points of the base image and the rotated fingerprint is used to find the updated image and to get the base template.

3.3 Absorption of New Template and Learning

This adjusted fingerprint is used to create a template of this image and is added this template to update the base template. Thus, a template containing more true minutiae points with higher weights can be obtained from multiple fingerprint samples of the same finger. A sample of such template is shown in Fig. 3.

4 Experimental Results

The proposed system has been tested on FVC-2004 DB2 A. It contains fingerprints from 100 individuals, and each individual has provided eight fingerprints. There is no artificial deformation added to the fingerprints. All fingerprints are of size approximately 21 kb and with dimensions 328×364 (width $\times$ height). Some of the fingerprints of the database are shown in Fig. 4. Randomly chosen five fingerprints of each subject are used to make the base template. Remaining three images are used for testing. Matching among the minutiae points obtained from fingerprints is done on the newly obtained based template, instead of NIST Bozorth3 matching strategy. Matching scores are converted to dissimilarity scores, and all statistics have been calculated using the dissimilarity scores only.

Results for CRR, EER, and DI are also presented in Table 1. The proposed algorithm has a correct recognition rate (CRR) of 98.66%. That is, a clear improvement

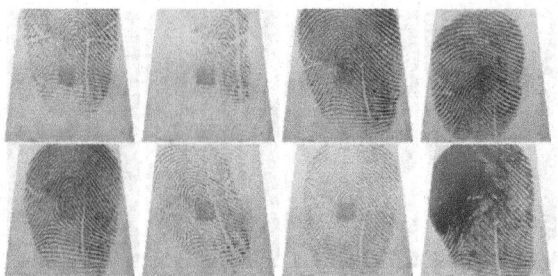

Fig. 4 Sample images from FVC-2004

Table 1 Comparison with NIST

	NIST matching	Proposed method
CRR (%)	95.62	98.66
EER (%)	9.01	2.18
DI	1.24	2.32

over the NIST Bozorth3 matching that has CRR of 95.62% for $n2$ matching using one verses all strategy. Equal error rate (EER) of the proposed system has been found to be 2.18% as compared to the general setting of NIST with EER of 9.01% on the specified database. Receiver operating characteristic (ROC) curves of both the systems are shown in Fig. 5. Histogram of genuine and imposter matching scores of both the system is shown in Fig. 6. It can be observed that the score distribution is well separated for the proposed system. For every fingerprint, if we only consider best genuine and imposter scores and plot them, then we get a plot as shown in Fig. 7. It can be seen that the separation between the two values is more for the proposed system. Discrimination index of each of the ROC curve for proposed and the NIST one is found to be 2.32% and 1.24%, respectively. It shows that the proposed method has more score discrimination power as compared to NIST one.

5 Conclusions

This paper has proposed a method of creating base template of fingerprint from multiple fingerprints of the same finger. The base template provides better chances of getting more true minutiae points with higher weights. The matching algorithm which has used this base template for matching has shown better accuracy of matching. It has been tested on FVC-2004 DB2 A database and has shown higher correct recognition rate, lower equal error rate, and higher score discrimination power of the proposed system.

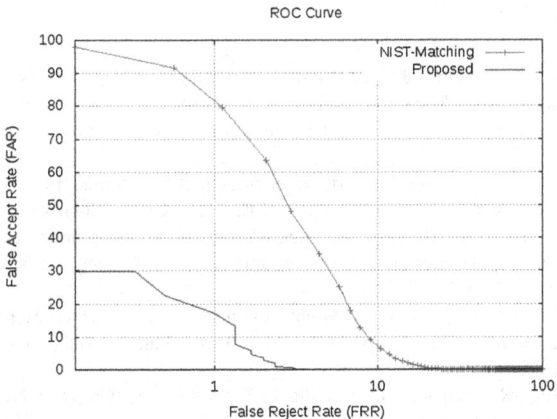

Fig. 5 ROC curves

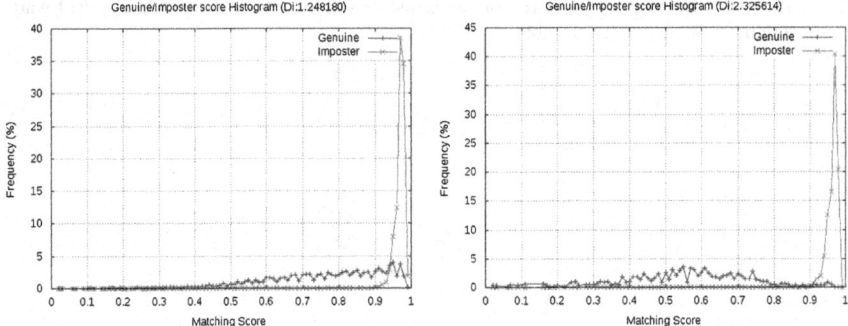

Fig. 6 Histogram of genuine and imposter scores of NIST-based and proposed system

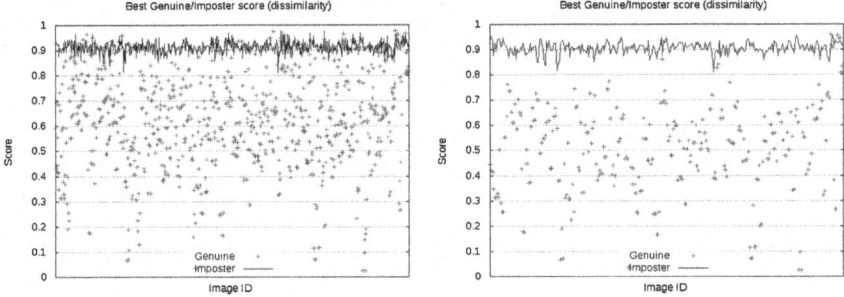

Fig. 7 Best dissimilarity score for genuine and impostor matching for NIST and proposed system

References

1. Herschel, W.J.: The Origin of Finger-Printing. H. Milford, Oxford University Press (1916)
2. Galton, F.: Finger Prints. Macmillan and Company (1892)
3. Henry, E.R.: Classification and Uses of Finger Prints. George Routledge and Sons, London (1900)
4. Trauring, M.: Automatic comparison of finger-ridge patterns. Nature **197**, 938–940 (1963)
5. Cummins, H.: Ancient finger prints in clay. J. Crim. Law Crim. **32**(4), 468–481 (1941)
6. Marana, A.N., Jain, A.K.: Ridge-based fingerprint matching using hough transform. In: 18th Brazilian Symposium on Computer Graphics and Image Processing, 2005. SIBGRAPI 2005, pp. 112–119. IEEE (2005)
7. Cappelli, R., Ferrara, M., Maltoni, D.: Minutia cylinder-code: A new representation and matching technique for fingerprint recognition. IEEE Trans. Pattern Anal. Mach. Intell. **32**(12), 2128–2141 (2010)
8. Jain, A.K., Bolle, R., Pankanti, S.: Biometrics: Personal Identification in Networked Society. Kluwer Academic (1999)
9. Pankanti, S., Bolle, R.M., Jain, A.K.: Biometrics: The future of identification. Computer **33**(2), 1–13 (2000)
10. NIST biometric image software (nbis). https://www.nist.gov/services-resources/software/nistbiometric-image-software-nbis

A Software-Supported Approach for Improving Visibility of Backlight Images Using Image Threshold-Based Adaptive Gamma Correction

Mayank Tiwari, Subir Singh Lamba and Bhupendra Gupta

Abstract The backlight images have complex structure that makes them different from other types of images. In such images, light source is at the background of the objects; as a result, the object (foreground) becomes darker. We observed that the conventional contrast enhancement methods do not work well for the enhancement of backlight images as the enhancement takes place at equal level for background as well as foreground. To solve this problem, we have developed the simplest method. The proposed method is best suited for low-contrast images having uniform background. The proposed method initially applies image threshold operation to extract the background of the image, then it applies contrast enhancement method on the image, and at last the extracted background is merged in the processed image. We have also developed a tool that implements our method. Experimental results show that the proposed method is able to enhance visibility of low-contrast backlight images without distorting its color details.

Keywords Visibility improvement · Low-contrast backlight images · Otsu threshold · Adaptive gamma correction

1 Introduction

Unfavorable environmental conditions at the time of image capturing degrade quality of the image. This is due to many reasons such as poor lightening conditions,

M. Tiwari · S. S. Lamba (⊠) · B. Gupta
Department of Mathematics, Indian Institute of Information Technology,
Design & Manufacturing Jabalpur, Jabalpur 482005, Madhya Pradesh, India
e-mail: subirs@gmail.com

M. Tiwari
e-mail: mayanktiwariggits@gmail.com

B. Gupta
e-mail: gupta.bhupendra@gmail.com
URL: http://www.bhupendragupta.com

© Springer Nature Singapore Pte Ltd. 2019
S. K. Bhatia et al. (eds.), *Advances in Computer Communication and Computational Sciences*, Advances in Intelligent Systems and Computing 760, https://doi.org/10.1007/978-981-13-0344-9_26

relative motion between object and camera, bad weather conditions. One such unfavorable environment is backlight effect. This is due to the reason that at the time of capturing of the image the light source is at the back of the objects; as a result, the object (foreground) becomes darker, whereas rest of the region of the image becomes brighter. The low-contrast backlight images cannot be directly applied for further post-processing operations such as computer vision and pattern recognition. Hence, it becomes necessary to apply some form of pre-processing like image enhancement for low-contrast backlight images before applying post-processing operations on them.

Histogram equalization (HE)-based methods are widely used to enhance contrast of all types of images. HE-based methods are simple to implement, and these methods are computationally inexpensive than other methods. It was observed that the conventional HE method is not able to enhance contrast of the image when image contains regions that are significantly darker or brighter than other parts of the image. To solve this problem, adaptive histogram equalization (AHE) method was developed by Pizer [14]. The AHE method enhances contrast of the given image, but in many situations, this method enhances noise too. It is observed by Zuiderveld [21] that by controlling contrast enhancement function of AHE method, the noise enhancement can be reduced. By using this idea, [21] proposed an improved version of AHE method known as contrast-limited AHE. Next to solve the 'mean-shift' problem of the conventional HE method, a number of variation of conventional HE method have been proposed. These variation are discussed by BBHE [19], DSIHE [18], MMBEBHE [1], RMSHE [2], RSIHE [16], RSWHE [11], HSQHE [17], and LQSWDHE [6]. All mentioned methods are developed for contrast enhancement with sufficient brightness preservation.

1.1 Our Contribution

It is very well known that the image enhancement methods are problem-oriented [5]. In other words, the best methods designed to enhance images of '$type - X$' cannot work well for images of '$type - Y$'. This is the main reason that existing methods cannot be used to enhance low-contrast backlight images. This is due to the reason that structure of low-contrast backlight images is different from other types of natural images. Hence in this work, we have developed the simplest method to enhance visibility of low-contrast backlight images. Our work is motivated by the fact that in low-contrast backlight images light source is at the background of the objects; as a result, the object (foreground) becomes darker. Hence, we need a method that enhances foreground of the image separately. For doing this, we have performed background detection and subtraction using the Otsu method suggested in Otsu [13]. To the best of our knowledge, the proposed method is the simplest method to enhance low-contrast backlight images. Few other methods such as Chiu et al. [3], Gayathri et al. [4], and Shi et al. [15] are developed for image contrast enhancement. Although these methods are using either the Otsu method suggested in Otsu [13] or gamma correction method in their processing, the proposed method is quite different as we use foreground and background detection and then we use foreground enhancement concept. Hence, working methodology of these methods are totally different from the proposed method. For practical use of our method, we have also developed a tool which is based on our method. Experimental results show that the proposed method improves visibility of low-contrast backlight image without distorting its other details.

The rest of this paper is organized as follows: Sect. 2 describes some basic concepts that are required for proper understanding of the proposed work, Sect. 3 describes the proposed method completely. Section 4 describes experimental setup used in our work. In order to evaluate the effectiveness of the proposed method, experimental results are shown in Sect. 5. Finally, Sect. 6 concludes our work.

2 Preliminaries

This section covers few basic concepts that are required for proper understanding of the proposed work.

2.1 HSV Color Model

The hue–saturation–value color model is more suitable for human visual system [8]. This color model can decouple chromatic and achromatic information of the input image in order to maintain color distribution. In HSV color model, hue indicates spectrum color, saturation indicates amount of white color diluted in hue, and value indicates the luminance intensity value of that color. The color image can be enhanced by preserving H and S while enhancing only V. Hence, the AGC method was applied to the V component for color contrast enhancement.

2.2 Adaptive Gamma Correction

In the proposed method, we are enhancing overall visibility of given low-contrast backlight image using 'adaptive gamma correction' method. This is an efficient method for visibility improvement of all types of images. In this method, weight for gamma correction method is determined by cumulative distribution function of normalized histogram of input image. For a better understanding of this method, one may refer the work suggested by Gupta and Tiwari [7].

Algorithm 1 AGC

```
Require: IN_IMG;
Require: ALPHA;
    ROW = SIZE(IN_IMG, 1); COL = SIZE(IN_IMG, 2);
    M_HIST = ZEROS(256, 1); U_HIST = (ROW * COL) / 256;
    while I ≤ 256 do
        M_HIST(I) = ALPHA * IMG_HIST(I) + (1 - ALPHA) * U_HIST;
    end while
    M_HIST = M_HIST / (ROW * COL);
    SUM = 0; M_CDF = ZEROS(256,1);
    while I ≤ 256 do
        SUM = SUM + M_HIST(I); M_CDF(I) = SUM;
    end while
    I ← 1; J ← 1;
    while I ≤ ROW do
        while J ≤ COL do
            L = UINT8(IN_IMG(I, J)); GAMMA = 1 - M_CDF(L+1);
            P_IMG(I, J) = (255 * ((L/255)^GAMMA));
        end while
    end while
    P_IMG IS OUTPUT OF THE 'AGC' METHOD;
```

2.3 Otsu Method

The Otsu method suggested in Otsu [13] is used to perform automatic clustering-based image threshold operation. This algorithm assumes bi-model histogram of pixel that makes it suitable for low-contrast backlight images. In our work, we are considering low-contrast backlight images having uniform background (minimum texture region). Hence, the Otsu method suggest in Otsu [13] is the best choice to perform image threshold operations.

3 The Proposed Method

The proposed method utilizes total five-step processing. These steps are defined as follows:

1. Initially, the input image I_{RGB} is converted to equivalent image I_{HSV}.
2. In second step, the Otsu method suggested in [13] is applied in the V part of I_{HSV} image. This step is applied to detect background (B) of the image.
3. In next step, 'adaptive gamma correction' method is applied to enhance overall visibility of the V part of I_{HSV} image.
4. In fourth step, the enhanced V part of I_{HSV} image is merged with background (B) of the image. As a result, we get enhanced image I^e_{HSV} in HSV color domain.

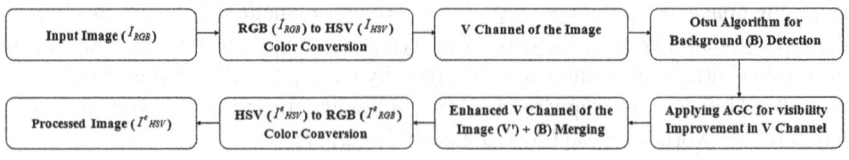

Fig. 1 Flowchart of the proposed method

5. At last, the I^e_{HSV} is converted to equivalent image I^e_{RGB}.

Figure 1 shows flowchart of the proposed method.

4 Experimental Setup

In this work, we have considered low-contrast backlight images having uniform (minimum texture) background. Because even for small amount of texture information present in the background of the given low-contrast backlight image, the Otsu method suggested in Otsu [13] may detect the background as part of the foreground in its threshold operation. Hence in the processed image, some portion of the background will be enhanced by the proposed method, whereas rest of the part of background will remain unchanged. This will reduce structural similarity in between input and processed images. Figure 2 shows set of images used in our work.

All the experiments are performed using our proposed tool (Sect. 5.1). That means, each image is processed and saved using the proposed tool. For numerical evaluation, $MATLAB - 2016(b)$ is used. For 'adaptive gamma correction' [7] method, same parameters settings as suggested by the original authors have been used in our work. For the Otsu method suggested in Otsu [13], we have developed our code in java programming language with same parameters settings as suggested in the original work.

Fig. 2 Set of images used in our work

5 Experimental Results

This section covers experimental results that are performed on low-contrast back-light images for evaluation of effectiveness of the proposed method. Figure 3 shows intermediate results generated by the proposed method at different stages of its processing. In Fig. 3c, black-and-white regions indicate foreground and background of the given low-contrast backlight image, respectively.

Figure 4 shows results produced by the proposed method on all test images used by us in our work. It is clear that the proposed method is able to improve visibility of all types of images without distorting their original features.

Figure 5 shows a comparison of results of the proposed and other image contrast enhancement methods. A careful examination of Fig. 5 reveals that the proposed method is able to improve visibility of given low-contrast backlight image without distorting its other visual details. Also as the proposed method does not apply any form of enhancement on the background of the input low-contrast backlight image, it is able to maintain color information of input low-contrast backlight image in the processed high-contrast backlight image more accurately than other methods. It is clear that enhancement results of other methods are distorting color information of the image. On the other hand, if we check results of the proposed method, then we can clearly observe that the proposed neither distorts color information nor introduces any amount of noise at any part of the image. This makes the proposed method a superior choice over state of the art.

Fig. 3 Intermediate results generated by the proposed method at different stages of its processing. Here, **a** input image (I_{RGB}), **b** V channel of I_{HSV}, **c** threshold image, **d** enhanced V channel of image, (e) enhanced V channel after background merging, **f** processed image in RGB form (I_{RGB}^e)

Fig. 4 Visibility improvement results of the proposed method for few other images used by us in our work. Readers are encouraged to zoom in on the electronic document, as fine details may not be visible in a printed copy

Fig. 5 Comparison of results of the proposed and other image contrast enhancement methods. Here, a is input low-contrast backlight image, and b-i are results of HE, BBHE [19], DSIHE [18], MMBEBHE [1], RMSHE [2], RSIHE [16], RSWHE [11], and proposed method, respectively

For evaluating color difference in between input and processed images, we have used ΔE_{00}^S measure [20]. This measure is a spatial extension of the $CIEDE2000$ [12] formula. The ΔE_{00}^S measure uses an opponent color space and an approximation of the contrast sensitivity function to mimic contrast masking in human vision. A small value of ΔE_{00}^S shows better color preservation. Table 1 shows average results of ΔE_{00}^S for all images used in our work. A careful examination of Table 1 reveals that, on the average basis, the proposed method has least color reproduction error ΔE_{00}^S. This clearly shows that the proposed method is able to maintain color information of input image in the processed image more accurately than other methods.

5.1 The Proposed Tool

Figure 6 shows a snapshot of the proposed tool. This tool is developed in java programming language. The tool is fully functional and easy to use. It works on the image

Table 1 Average value of ΔE_{00}^S for all 12 images used in our work

Methods	Average ΔE_{00}^S
HE	19.06
BBHE	11.53
DSIHE	11.02
MMBEBHE	11.53
RSWHE	2.32
RMSHE	6.48
RSIHE	6.71
PROPOSED	2.27

Fig. 6 A snapshot of the proposed tool

formats supported by the java programming language [10]. In the tool, first image (left) is the input image selected by the user. After selecting the required parameters, the tool generates two more images: threshold image (middle) and processed image (right). The user can save these images at any desired place.

6 Conclusion

In this paper, we present a method for visibility improvement of low-contrast backlight images. The proposed method works in HSV color domain. Initially, it applies the Otsu method suggested in Otsu [13] to detect foreground and background regions of the image. After that, it applies the 'adaptive gamma correction' method to improve visibility of the image. At last, it merges the detected background in the enhanced image to generate the processed image. Experimental visibility improvement results demonstrate that our proposed method performed well compared with other state-of-the-art methods. In future, the proposed method can be improved in many ways such as

1. In the proposed work, we have considered images that have uniform background. If this method is applied for image having texture regions in background, then after applying the Otsu method suggested in Otsu [13], some form of morphological

operations can be performed in the threshold image to reduce the dark spots from the background.

2. Instead of 'adaptive gamma correction,' any other visibility improvement method can be applied.

Permission to Use Personal Images: Few images used in this work are our personal images, and we give permission to others to use these images in their work. The permission is given for use in research purpose only. Use of these images for any other purpose is totally illegal.

Acknowledgements Authors thank [9] for providing free access of $MATLAB$ code of ΔE_{00}^S matrix [20].

References

1. Chen, S., Ramli, A.R.: Minimum mean brightness error bi-histogram equalization in contrast enhancement. IEEE Trans. Consum. Electron. **49**(4), 1310–1319 (2003)
2. Chen, S., Ramli, R.: Contrast enhancement using recursive mean-separate histogram equalization for scalable brightness preservation. IEEE Trans. Consum. Electron. **49**(4), 1301–1309 (2003)
3. Chiu Y.S., Cheng F.C., Huang S.C.: Efficient contrast enhancement using adaptive gamma correction and cumulative intensity distribution. In: Proceedings of IEEE Conference on Systems Man and Cybernetics, pp. 2946–2950 (2011)
4. Gayathri, S., Mohanapriya, N., Kalaavathi, B.: Efficient contrast enhancement using gamma correction with multilevel thresholding and probability based entropy. Int. J. Res. Eng. Technol. **3**, 7 (2014)
5. Gonzalez, R.C., Woods R.E.,: Digital Image Processing, 3rd edn, Addison-Wesley, Reading, MA, pp. 85–103 (2008)
6. Gupta B., Agarwal T.K.: Linearly quantile separated weighted dynamic histogram equalization for contrast enhancement. Comput. Electr. Eng. (2017)
7. Gupta, B., Tiwari, M.: Minimum mean brightness error contrast enhancement of color images using adaptive gamma correction with color preserving framework. Int. J. Light Electron Opt. **127**, 1671–1676 (2015)
8. Huang, S.C., Cheng, F.C., Chiu, Y.S.: Efficient contrast enhancement using adaptive gamma correction with weighting distribution. IEEE Trans. Image Process. **22**(3), 1032–1041 (2013)
9. Jaramillo, B.O., Kumcu, A., Philips, W.: Evaluating Color Difference Measures in Images, Accepted QoMEX (2016)
10. $Java^{TM}$ Tutorials. http://docs.oracle.com/javase/tutorial/index.html. Accessed 8 July 2017
11. Kim, M., Chung, G.C.: Recursively separated and weighted histogram equalization for brightness preservation and contrast enhancement. IEEE Trans. Consum. Electron. **54**(3), 1389–1397 (2008)
12. Luo, M., Cui, G., Rigg, B.: The development of the CIE 2000 colour-difference formula: CIEDE2000. Color Res. Appl. **26**, 340–350 (2001)
13. Otsu, N.: A threshold selection method from gray-level histograms. IEEE Trans. Syst. Man. Cybern. **91**, 62–66 (1979). https://doi.org/10.1109/TSMC.1979.4310076
14. Pizer, S.M.: Adaptive histogram equalization and its variations. Comput. Vis. Graph. Image Process. **39**, 355–368 (1987)
15. Shi, C., Wang, Y., Xiao, B., Wang, C.: Otsu guided adaptive binarization of captcha image using gamma correction. In: 2016 23rd International Conference on Pattern Recognition ICPR (2016)

16. Sim, K.S., Tso, C.P., Tan, Y.: Recursive sub-image histogram equalization applied to gray scale images. Pattern Recogn. Lett. **28**, 1209–1221 (2007)
17. Tiwari, M., Gupta, B., Shrivastava, M.: High-speed quantile-based histogram equalisation for brightness preservation and contrast enhancement. Image Process. IET **91**, 80–89 (2015)
18. Wang, Y., Chen, Q., Zhang, B.: Image enhancement based on equal area dualistic sub-image histogram equalization method. IEEE Trans. Consum. Electron. **45**(1), 68–75 (1999)
19. Yeong, T.K.: Contrast enhancement using brightness preserving bi-histogram equalization. IEEE Trans. Consum. Electron. **43**(1), 1–8 (1997)
20. Zhang, X., Wandell, B.: A spatial extension of CIELAB for digital color-image reproduction. J. Soc. Inf. Disp. **5**, 61–63 (1997)
21. Zuiderveld, K.: Contrast limited adaptive histogram equalization. In: Heckbert, P.S. (ed.) Chapter VIII, Graphics Gems, I.V, pp. 474–485, Academic Press, Cambridge, MA (1994)

Sidelobe Minimization of Uniform Linear Array by Position- and Amplitude-Only Control Using WDO Technique

Santosh Kumar Mahto, Arvind Choubey, Rashmi Sinha and Prakash Ranjan

Abstract This paper describes efficient optimization wind-driven optimization (WDO) technique for pattern synthesis of uniform linear array (ULA) having maximum sidelobe level (SLL) suppression, constrained on DRR, beam width and null control by controlling the amplitude-only and position-only of array element. The single, multiple (double and triple), and broad nulls are placed in the direction of maximum interference while receiving signal from the desired direction. The WDO is a new nature-inspired evolutionary algorithm derived from to the point movement of the air parcel in the earth's atmosphere. It uses a new learning strategy to update the velocity and position of air packets based on their present pressure values to accelerate the convergence. The six examples are considered, and the results are compared with those obtained by others such as PSO, comprehensive learning PSO, differential evolution (DE), bacterial foraging optimization (BFO), plant growth simulation algorithm (PGSA), and BEES algorithm. The simulation study demonstrates that improves performance of WDO algorithm than twelve reported algorithms particularly in terms of maximum SLL suppression, beam width control, null control, and convergence rate.

Keywords Antenna array · PSO · WDO · Null control · Minimum SLL
Evolutionary programming · Interference · Dynamic range ratio (DRR)

S. K. Mahto (✉) · A. Choubey · R. Sinha · P. Ranjan
Department of Electronics and Communication Engineering,
National Institute of Technology Jamshedpur, Jamshedpur, India
e-mail: ec51236@nitjsr.ac.in

A. Choubey
e-mail: achoubey.ece@nitjsr.ac.in

R. Sinha
e-mail: rsinha.ece@nitjsr.ac.in

P. Ranjan
e-mail: 2014rsec001@nitjsr.ac.in

© Springer Nature Singapore Pte Ltd. 2019
S. K. Bhatia et al. (eds.), *Advances in Computer Communication
and Computational Sciences*, Advances in Intelligent Systems
and Computing 760, https://doi.org/10.1007/978-981-13-0344-9_27

1 Introduction

Presently, researchers are concerned to design an antenna with larger coverage, maximum directivity having minimum SLL and placing desired nulls in direction of maximum interference. But the single antenna does not meet this requirement. Hence, an array of antennas is used. Nulling methods in array pattern synthesis are used to suppress signal not of interest while receiving the signal of interest. It has been extensively studied in the last decade and a great area of interest for researchers, today [1–24]. The null steering can be achieved by various techniques such as controlling the position-only [1–16], the excitation amplitude-only [17–21], the phase-only [1, 22, 24], and complex weight [23] of array antenna. In modern wireless environment, pattern nulling methods become very important to nullify unwanted interfering signals for higher quality and better efficiency. These nulling methods have their own advantages and inherent demerits, and none of them entirely fulfill desired requirement.

A variety of evolutionary computational techniques have emerged such as genetic algorithm (GA) [3], neural network (NN), PSO [7], DE [8], biogeography-based optimization (BBO) [9], cuckoo optimization algorithm (COA) [10], cuckoo search algorithm (CSA) [11], variant of PSO [12, 13], comprehensive learning particle swarm optimization (CLPSO) [16], bacterial foraging optimization (BFO) [19], BEES algorithm [18], plant growth simulation algorithm [20], ant colony optimization (ACO) [22, 23], Tabu search (TS) [24], and many others effectively implemented for complex problem of array pattern synthesis. The WDO algorithm was proposed and implemented to uniformly excite linear array synthesis with maximum SLL suppression by controlling the array element position-only by Warner et al. in [6] and Santosh et al. in [14].

In this paper, WDO technique is used for nulling pattern synthesis having minimum SLL and beam width by optimizing the excitation amplitude-only and position-only of array element. The nulls (single, multiple, and broad null) are placed in the initial array pattern to minimize the undesired interference effect. Chebyshev linear array having a broad null is considered and the excitation amplitude is optimized using WDO algorithm. The simulated results are compared with those obtained by other evolutionary algorithms such as BFO, PGSA, and BEES algorithms. Further, five design examples of linear array pattern synthesis having 10, 28, 32, 64, and 128 elements are considered and the position of array element is optimized. The results are compared with those obtained by other evolutionary methods such as PSO, CLPSO, DE, BBO, CS, IWPSO, COA, ILPSO, and CPSO algorithms.

2 Wind-Driven Optimization Algorithm

The WDO is a new evolutionary algorithm inspired from the movement of wind in the earth's atmosphere where it blows to equalize horizontal pressure similar to

Fig. 1 Flowchart of WDO

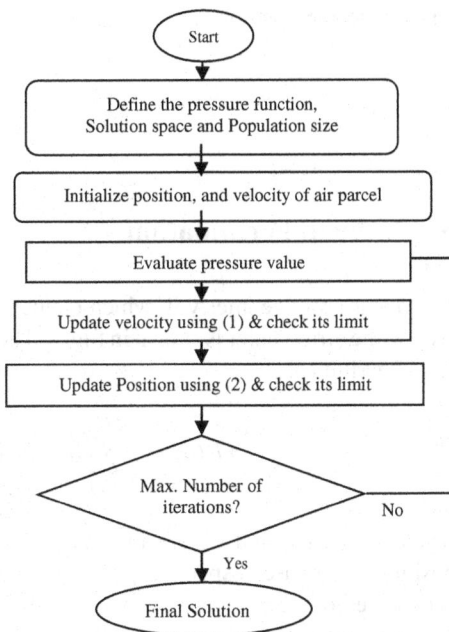

particle-based algorithm [6]. The vital force that initiates the motion of air parcels is pressure gradient force. The detail description of WDO is given in [6]. The four major forces such as pressure gradient, frictional, gravitational, and Coriolis force are considered to describe the motion of air parcel. The velocity and position update equation of air parcel is given as,

$$u_{new} = (1 - \alpha) . u_{cur} - g.x_{cur} + RT . \left| \frac{1}{i} - 1 \right| . \left(x_{opt} - x_{cur} \right) + \frac{c . u_{cur}^{otherdim}}{i} \quad (1)$$

where R and T are the gas constant and the absolute temperature. α and u are friction coefficient and velocity of the air parcel, respectively. The gravitational constant is g, u_{cur} is the current velocity vector, i represents the rank of particle in the population based on their pressure value, and x_{opt} is optimum position of air parcel, respectively.

The velocity of the air packet is updated, and then the new position is given as

$$x_{new} = x_{cur} + u_{new} \quad (2)$$

The flowchart of WDO algorithm is given below (Fig. 1).

Fig. 2 Linear array antenna
(2 M elements)

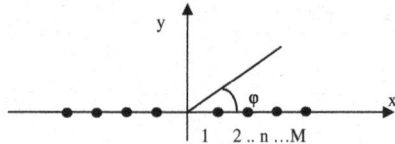

3 Problem Formulation

The linear array geometry in which elements are symmetrically placed and excited around the array center is shown in Fig. 2. The array factor of a 2 M-identical elements can be written as,

$$AF(\theta) = 2 \sum_{n=1}^{M} a_n \cdot \cos\left(\frac{2\pi}{\lambda} d_n \cos\theta + \emptyset_n\right) \tag{3}$$

where λ, a_n, d_n, and $\emptyset_n$ are the nth element, wavelength, excitations amplitude, position, and phase, respectively. d_n, the distance between two consecutive elements, is considered $\lambda/2$.

For a uniformly excited array ($\emptyset_n = 0$), Eq. (3) becomes

$$AF(\theta) = 2 \sum_{n=1}^{M} a_n \cdot \cos\left(\frac{2\pi}{\lambda} d_n \sin\theta\right) \tag{4}$$

To control the maximum SLL suppression and nulls at desired location, pressure function [7, 9, 15, 16] is given as,

$$Pressure I = w1. \sum_{i=1}^{M} \frac{1}{\Delta\theta_i} \int_{\theta_{li}}^{\theta_{ui}} \left|AF^{\bar{x}}\right|^2 + w2. \sum_{i=1}^{N} \left|AF^{\bar{x}}(\theta_k)\right|^2 \tag{5}$$

where $[\theta_{li}, \theta_{ui}]$ is region for maximum SLL suppression, $\Delta\theta_i = \theta_{ui} - \theta_{li}$. The other constants M, N, and θ_k are the number of maximum suppressed SLL, the number of nulls, and kth null direction, respectively, and $\bar{x}$ is the position vector. On the right-hand side of (5), the first and second terms are employed to the minimized SLL-only and null control-only, respectively.

In order to increase the directivity of the array, the beam width is controlled to a desired value. The pressure function to control the beam width [7, 9, 16] is given as,

$$Pressure II = w3 \cdot max\{0, |BW_c - BW_d| - 1\} \tag{6}$$

where BW_c and BW_d are calculated and desired beam width, respectively.

The interference effect can be minimized by controlling the NDL [16] and is given as,

$$Pressure III = w4 \cdot \sum_{k=1}^{K} max\{0, AF_{dB}^{\bar{x}}(\theta_k) - C_{dB}\} \tag{7}$$

where K, C_{dB}, and θ_k are the no. of essential null locations, NDL desired (dB), and kth null place, respectively.

Finally, a single-objective (pressure) function is formulated by (5)–(7) using an exact penalty method [25],

$$Pressure = Pressure I + Pressure II + Pressure III \tag{8}$$

where $w1$, $w2$, $w3$ and $w4$ are weighting factor used in the pressure function.

In (8), the first and second terms are used to minimize the SLL and placing nulls at desired direction, respectively. However, the third term is used to implement constraint on beam width and the last term is used to control NDL to a desired value. It must be pointed out that same objective function was used in [16]; however, in this paper, the weighting factor is introduced in (8) and WDO is given task to optimize the linear array antenna having maximum SLL suppression and null control, while keeping the main beam to a desired value within $\pm 1°$ by position-only and amplitude-only control. The weighting factor is used in the pressure function and should be selected by experience such that pressure (objective) function is capable of guiding potential solution to obtain array pattern with specified constraints. By properly adjusting the value of weighting factors, it can be possible to obtain very reasonable approximations and trade-offs between achievable and the desired pattern. By proper selection of these factors, the desired pattern can be obtained. The main objective is minimizing pressure (cost) function in array pattern synthesis by optimizing the parameter of array element such as excitation amplitude-only and position-only.

4 Simulation Results

To illustrate the effectiveness of WDO for synthesizing linear array, six examples are considered and results are compared with others reported in the literature.

4.1 Amplitude-Only Control

One design example of 20 identical elements' Chebyshev array is considered and a broad null is placed in the direction of maximum interference to minimize the interference effect (Fig. 3).

The parameters of WDO algorithm are same as reported in [14]. The position of air packet in solution domain represents the parameters of array antenna such as

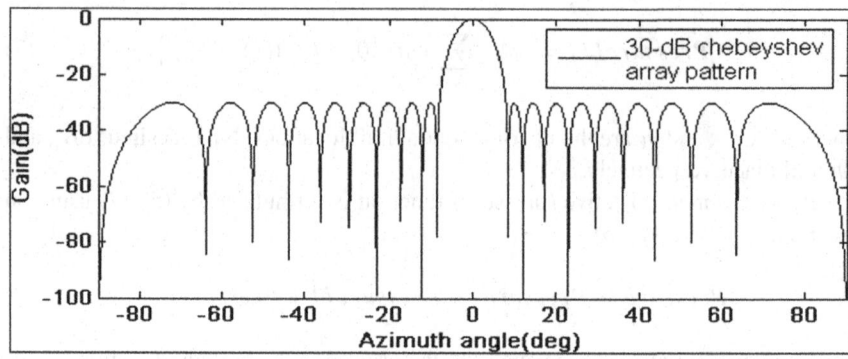

Fig. 3 Normalized initial pattern of Chebyshev array (N = 20)

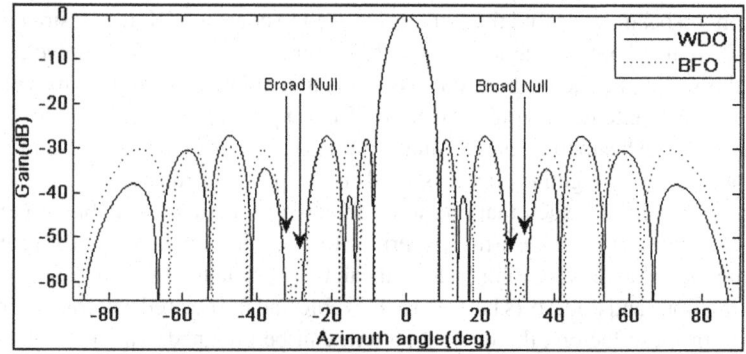

Fig. 4 Normalized patterns of ULA null at 30° with $\Delta\theta = \pm5°$ obtained using WDO and BFO

excitation amplitude and the position. A broad null is placed at 30° with $\Delta\theta = \pm5°$ in Chebyshev pattern as shown in Fig. 4.

The results are compared with the BFO, PGSA, and Bees algorithms. The corresponding optimized excitation amplitude is mentioned in Table 1. It achieves minimum SLL of −25.4 dB while preserving beam width 9.01° and NDL are greater than −64 dB over concern region as compared to −55 dB in [11, 19] mentioned in Table 2. This illustrates the efficacy of WDO algorithm for placing broad null in array pattern synthesis to minimize the interference effect.

4.2 Position-Only Control

In this application, five design examples of uniformly excited linear array are considered and the position of array element is optimized. The second example demonstrates that synthesis of 10-element array with desired beam width equal to 23° and

Table 1 Normalized amplitude of the (N = 20) elements Chebyshev linear array obtained by WDO algorithm

Element		Initial Chebyshev		Obtained by the WDO	
±1	±6	1.00000	0.62034	1.000000	0.543097
±2	±7	0.97010	0.50461	1.043963	0.550084
±3	±8	0.91234	0.39104	0.998943	0.422996
±4	±9	0.83102	0.28558	0.916700	0.224038
±5	±10	0.73147	0.32561	0.791017	0.21790

Table 2 Optimum results obtained for Chebyshev array (N = 20)

Algorithm	BEES [18]	BFO [19]	PGSA [20]	WDO
MSLL (−dB)	–	30	27	25.4
NDL (−dB)	More than 55	More than 55	More than 55	More than 64
FNBW	–	–	–	9.01°
DRR	4.16	4.18	3.92	4.79

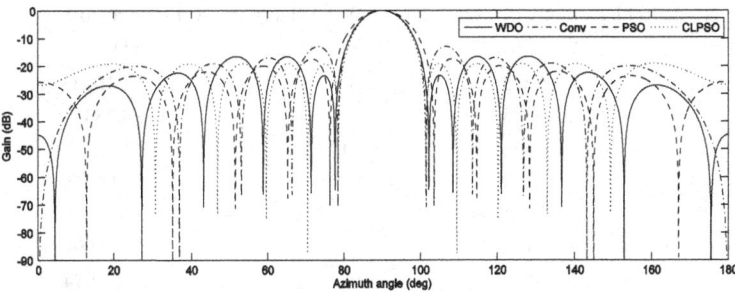

Fig. 5 Normalized patterns of (N = 10) elements array with maximum SLL suppression in the spatial region [0°, 82°] and [98°, 180°]

Table 3 Normalized geometry of array obtained (N = 10) using WDO algorithm (with respect to λ/2)

Conv. (±)	0.50	1.50	2.50	3.50	4.50
WDO (±)	0.4009	1.4012	2.3515	3.7104	4.9990

two spatial regions [0°, 82°] and [98°, 180°] are of interest for maximum SLL suppression by optimizing the elements position-only. The optimized radiation pattern is shown in Fig. 5, and the corresponding geometry is mentioned in Table 3. The simulation results are compared and summarized in Table 4. The simulation study demonstrates that this algorithm achieves maximum SLL suppression of −26.4 dB compared to −17.44 dB by PSO [7] and −19.07 dB by CLPSO [16]. In this case, WDO method provides a maximum SLL suppression of 50.12% compared to the conventional method.

Table 4 Optimum result for the (N = 10) array antenna

Algorithm	Conventional	PSO [7]	CLPSO	WDO
MSLL (−dB)	12.96	17.44	19.07	26.20
Beam width	–	–	23°	23°

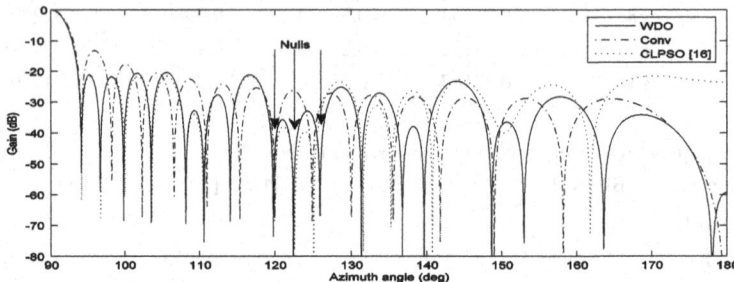

Fig. 6 Normalized pattern of (N = 28) ULA with SLL suppression in the region [0°, 180°]

Table 5 Normalized geometry of (N = 28) using WDO algorithm (with respect to λ/2)

Element	WDO	Element	WDO
1	±0.5379	8	±7.1666
2	±1.3325	9	±8.3832
3	±2.2860	10	±11.3332
4	±3.2317	11	±12.9406
5	±4.2734	12	±14.5096
6	±5.4117	13	±14.5096
7	±6.3331	14	±15.9974

The third example demonstrates the synthesis of 28-element array antenna with six prescribed nulls at 55°, 57.5°, 60°, 120°, 122.5°, and 125°, desired beam width equal to 8.35° and maximum suppression of SLL in [0°, 180°]. The radiation pattern is shown in Fig. 6, and the corresponding array geometry is mentioned in Table 5. The value of performance parameter such that MSLL and NDL obtained by WDO algorithm is compared with other existing evolutionary algorithms such PSO, CLPSO, DE, BBO, CS, COA, ILPSO, and CPSO and summarized in Table 6.

The comparison results demonstrate that WDO algorithm achieves a minimum SLL of −21.4 dB and the NDL greater than −67 dB, respectively, which is much better than other algorithms reported in Table 6; however, minimum SLL is almost equal to CLPSO and COA. The simulation study shows that this reduced maximum SLL by 36.92% as compared to the conventional method.

Fourth example demonstrate 32-element array is synthesized with two prescribed nulls at 81° and 99°, desired beam width equal to 7.1° and region of concern for SLL suppression are [0°, 87°] and [93°, 180°]. The radiation pattern obtained by WDO algorithm is shown in Fig. 7, and the corresponding optimized geometry is

Table 6 Optimum results found by different algorithms (N = 28) ULA

Algorithm	MSLL (−dB)	NDL (−dB)		
		120°	122.5°	125°
PSO [7]	13.22	52.73	51.65	61.46
BBO [9]	17.13	63.66	64.68	61.36
CS [11]	15.30	62.27	61.09	63.98
CLPSO	21.60	60.45	60.00	60.64
WDO	**21.4**	**73.50**	**94.74**	**67.00**

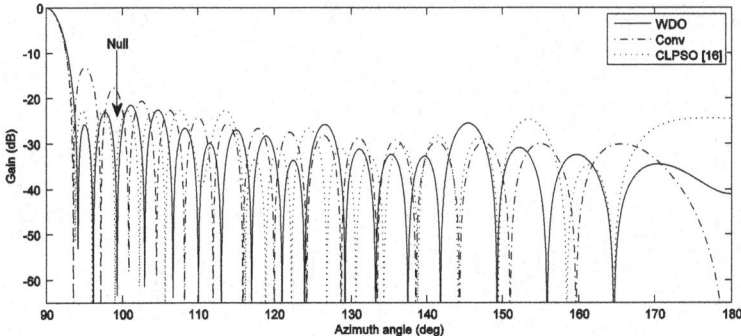

Fig. 7 Normalized patterns of the 32-element linear array obtained using three different methods

Table 7 Normalized geometry (N = 32) ULA obtained using WDO algorithm

Element	WDO	Element	WDO
1	± 0.3469	9	± 7.1802
2	± 1.1941	10	± 8.1284
3	± 2.0094	11	± 9.2803
4	± 2.8240	12	± 10.5561
5	± 3.6072	13	± 12.1189
6	± 4.4692	14	± 13.7438
7	± 5.3503	15	± 15.3508
8	± 6.3063	16	± 16.8678

mentioned in Table 7. The simulation results is compared with others such as PSO [7], DE [8], BBO [9], COA [10], CSA [11], CPSO [12], ILPSO [13], and CLPSO [16] and summarized in Table 8. The comparison results show that WDO algorithm achieves minimum SLL of −25.8 dB and better null depth level at 99° is −63.5 dB as reported in Table 8; however, the NDL obtained by WDO algorithm is slightly less than [10] and [16]. The simulation study demonstrates that WDO algorithm reduced the maximum SLL by 48.84% compared to the conventional method.

To show the efficacy and versatility of WDO for linear array synthesis, the array length is very large. Two examples of 64- and 128-element array antenna are considered in fifth and sixth cases.

Table 8 Optimum results obtained by different algorithm for (N = 32) ULA

Algorithm	MSLL (−dB)	NDL at 99° (−dB)	Beam width
PSO [7]	18.80	62.12	–
DE [8]	22.81	60.03	–
BBO [9]	16.93	61.73	–
COA [10]	23.81	79.85	–
CS [11]	22.81	60.03	–
CPSO [12]	23.17	63.16	–
ILPSO [13]	23.81	73.30	–
CLPSO [16]	22.73	60.00	7.10
WDO	**25.80**	63.51	7.10

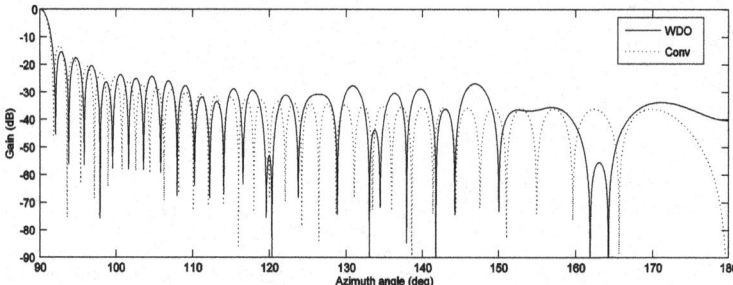

Fig. 8 Normalized patterns of (N = 64) ULA obtained using WDO algorithm and conventional (Conv) method

In the fifth design example, 64-element array is synthesized using WDO technique with two prescribed nulls at 98° and 100°, desired beam width equal to 3.1°, and two spatial regions of interest for maximum SLL suppression are [0°, 88.5°] and [91.5°, 180°]. The simulation results show that this achieves minimum SLL of −15.46 dB and better null depth level at 98° and 100° are 75.7 dB and 57.7 dB, respectively. In this case, the near- and far-field array pattern is almost better than the conventional method as shown in Fig. 8.

In the last design example, 128-element array is synthesized using WDO technique with two prescribed nulls at 70° and 110°, desired beam width equal to 1.6°, and two spatial regions of concern for SLL suppression are [0°, 89.2°] and [90.8°, 180°]. The simulation results show that WDO algorithm achieves minimum SLL of −14.2 dB and null depth level at 110° is 58.3 dB. In this case, the near- and far-field array pattern is almost same as the conventional method as shown in Fig. 9.

The simulation results demonstrate that the performance parameter such as null control and minimum SLL is improved for all design examples. Such an improvement is obtained by increasing the weighting factors w1 and w2 in (8). The weighting factors used gives antenna designer greater flexibility and control over the actual pattern. The convergence graph of pressure function versus number of iterations for

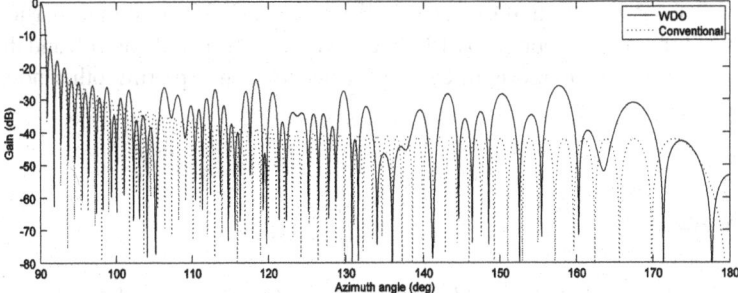

Fig. 9 Normalized patterns of (N = 128) ULA obtained using WDO algorithm and conventional method

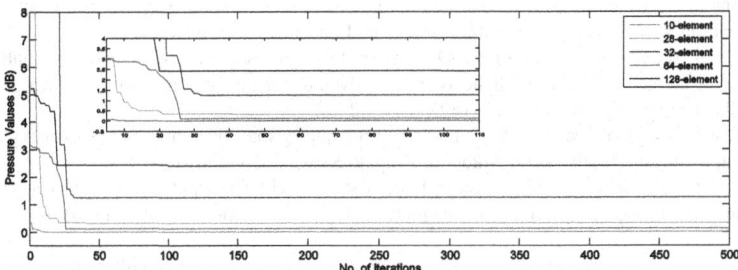

Fig. 10 Convergence characteristics for (N = 10, 28, 32, 64, and 128) element array antenna using WDO algorithm

(N = 10, 28, 32, 64, and 128) element array is shown in Fig. 10 and is summarized in Table 10. The learning characteristics depict that the pressure value rapidly converges to a final value in about 60 iterations compared to 300 iterations in PSO [7] and about 100 iterations in CLPSO [16] for (N = 10, 28, and 32) linear array synthesis. In case of 64- and 128-element array, the learning characteristic depicts that the pressure value rapidly converges in about 30 and 105 iterations, respectively. Thus, the comparative simulation studies show better performance of WDO algorithm compared to others such as PSO, CLPSO, DE, BBO, CSA, COA, ILPSO, and CPSO in terms of minimum SLL, beam width control, null control, and the rate of convergence.

5 Conclusion

This paper illustrates efficacy of the WDO algorithm for placing single, multiple, and broad nulls at desired direction in the array having maximum suppressed SLL, and null control while imposing minimum beam width constraints. The position-only and excitation amplitude of array element are optimized using WDO algorithm. The simulation study demonstrates that WDO algorithm outperforms the BFO, PGSA,

Bees, PSO, CLPSO, DE, BBO, CS, COA, ILPSO, and CPSO algorithms particularly in terms of maximum suppressed SLL, beam width control, null control, and the rate of convergence. WDO algorithm can be further used in exploring other nonlinear array geometry.

References

1. Steyskal, H., Shore, R.A., Haupt, R.L.: Methods for null control and their effects on the radiation pattern. IEEE Trans. Antennas Propag. **34**, 404–409 (1986)
2. Er, M.H.: Linear antenna array pattern synthesis with prescribed broad nulls. IEEE Trans. Antennas Propag. **38**, 1496–1498 (1990)
3. Tennant, A., Dawoud, M.M., Anderson, A.P.: Array pattern nulling by element position perturbations using a genetic algorithm. Electron. Lett. **30**, 174–176 (1994)
4. Akdagli, A., Guney, K., Karaboga, D.: Pattern nulling of linear antenna arrays by controlling only the element positions with the use of improved touring ant colony optimization algorithm. J. Electromagn. Waves Appl. **16**, 1423–1441 (2002)
5. Lin, A.C., Qing, Y., Feng, Q.Y.: Synthesis of unequally spaced antenna arrays by using differential evolution. IEEE Trans. Antennas Propag. **58**(8), 2553–2561 (2010)
6. Bayraktar, Z., Komurcu, M., Bossard, J.A., Werner, D.H.: The wind driven optimization technique and its application in electromagnetics. IEEE Trans. Antennas Propag. **61**(3), 771–779 (2013)
7. Khodier, M.M., Christodoulou, C.G.: Linear array geometry synthesis with minimum sidelobe level and null control using particle swarm optimization. IEEE Trans. Antennas Propag. **53**, 2674–2679 (2005)
8. Yang, S.W., Gan, Y.B., Qing, A.Y.: Antenna-array pattern nulling using a differential evolution algorithm. Int. J. RF Microwave Comput. Aided Eng. **14**, 57–63 (2004)
9. Singh, U., Kumar, H.T., Kamal, S.: Linear array synthesis using biogeography based optimization. Prog. Electromagn. Res. M **11**, 25–36 (2010)
10. Singh, U., Rattan, M.: Design of linear and circular antenna arrays using cuckoo optimization algorithm. Prog. Electromagn. Res. C **46**, 1–11 (2014)
11. Khodier, M.: Optimization of antenna arrays using the cuckoo search algorithm. IET Microwaves Antennas Propg. Res **7**, 458–464 (2013)
12. Wang, W.B., Feng, Q., Liu, D.: Application of chaotic particle swarm optimization algorithm to pattern synthesis of antenna arrays. Prog. Electromagn. Res. **115**, 173–189 (2011)
13. Liu, D., Feng, Q., Wang, W.B., Yu, X.: Synthesis of unequally spaced antenna arrays by using inheritance learning particle swarm optimization. Prog. Electromagn. Res. **118**, 205–221 (2011)
14. Mahto, S. K., Choubey, A., Suman, S.: Linear Array Antenna with reduced side lobe level using WDO Algorithm. IEEE conference, SPACES, KL university, India, 191–195 (2015)
15. Bhargav, A., Gupta, N.: Multiobjective Genetic Optimization of Non-uniform Linear Array with Low Sidelobes and Beamwidth. IEEE Antennas Wirel. Propag. Letts. Res. **12**, 1547–1549 (2013)
16. Goudos, S.K., Moysiadou, V., Samaras, T., Siakavara, K., Sahalos, J.N.: Application of a comprehensive learning particle swarm optimizer to unequally spaced linear array synthesis with sidelobe level suppression and null control. IEEE Trans. Antennas Propag. Res. **9**, 125–129 (2010)
17. Liao, W.P., Chu, F.L.: Array pattern synthesis with null steering using genetic algorithms by controlling only the current amplitudes. Int. J. Electron. **86**, 445–457 (1999)
18. Guney, K., Onay, M.: Amplitude-only pattern nulling of linear antenna arrays with the use of bees algorithm. Prog. Electromagn. Res. **70**, 21–36 (2007)
19. Guney, K., Basbug, S.: Interference suppression of linear antenna array by amplitude-only control using a bacterial foraging algorithm. Prog. in Electromagn. Res. **79**, 475–497 (2008)

20. Guney, K., Durmus, A., Basbug, S.: A plant growth simulation algorithm for pattern nulling of linear antenna arrays by amplitude control. Prog. Electromagn. Res. B **17**, 69–84 (2009)

21. Babayigit, B., Akdagli, A., Guney. K.: A clonal selection algorithm for null synthesizing of linear antenna arrays by amplitude control. J. Electromagn. Waves Appl. Res. **20**, 1007–1020 (2006)

22. Karaboga, N., Guney, K., Akdagli, A.: Null steering of linear antenna arrays by using modified touring ant colony optimization algorithm. Int. J. RF and Microwave Comput. Aided Eng. **12**, 375–383 (2002)

23. Karaboga, D., Guney, K., Akdagli, A.: Antenna array pattern nulling by controlling both the amplitude and the phase using modified touring ant colony optimization algorithm. Int. J. Electron. **91**, 241–251 (2004)

24. Pillo., G.D.: Exact penalty methods. In: Spedicato, E. (ed.), Algorithms for Continuous Optimization: The State of the Art, Norwell, vol. 434. Kluwer, MA (1994)

25. Mahto, S K., Choubey, A., Suman, S., Sinha. R.: Synthesizing nulling pattern of linear array by amplitude control using a wind driven optimizer for faster convergence. n: IEEE International Conference, SAI London, UK, 10–11 Nov 2015

Multi-atlas Segmentation: Label Propagation and Fusion Based Approach

Shruti Karkra and Janak Kumar B. Patel

Abstract To understand the profundity of the subject in importance, revision of previous work has always played a vigorous role in developing interest and curiosity. For morphological assessment and measurement of quantitative parameters of biomedical structures, segmentation is done. Number of segmentation techniques has been widely used in the field of image processing since four decades. However, the problems related to segmentation still remain candid providing no optimum solution. Segmentation process is always considered as difficult due to a variation in medical images, image resolution, pixel intensity, signal variability, noise, and other artifacts. From the previous study, multi-atlas segmentation (MAS) techniques have proven to be a flexible and robust approach for medical images. Multi-atlas segmentation works in two steps: first propagation of the manually labeled images to the target image and then combining the transfer images to get the best segmentation result. Label propagation and label fusion using multiple atlases have made multi-atlas segmentation approach as forefront of segmentation research. This survey paper provides a snapshot of the current progress in the field of segmentation, registration, and label propagation.

Keywords Multi-atlas segmentation · Label propagation · Label fusion
Registration · Atlases

1 Introduction

Segmentation is the process that divides the image into various different regions of interest. Segmentation separates the pixels of an image into groups that intensely correlate with the entities in an image, and a label is assigned to every pixel in a way that

S. Karkra (✉) · J. K. B. Patel
Amity University Haryana, Gurgaon, India
e-mail: Skarkra@ggn.amity.edu

J. K. B. Patel
e-mail: jkbpatel@ggn.amity.edu

© Springer Nature Singapore Pte Ltd. 2019
S. K. Bhatia et al. (eds.), *Advances in Computer Communication and Computational Sciences*, Advances in Intelligent Systems and Computing 760, https://doi.org/10.1007/978-981-13-0344-9_28

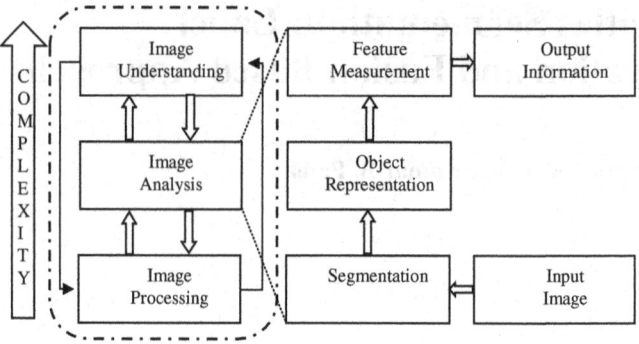

Fig. 1 Image engineering [19]

all pixel of same label will have some certain chromatic characteristics [1]. The purpose of segmentation is to improve the process of visualization to handle the detection process more effectively and efficiently [2]. There are number of segmentation techniques. Selection of appropriate type of technique for a special image is very difficult task as not all techniques are suitable for all type of images. Specially in case of medical image segmentation which is considered as an active research area, it is very challenging to find suitable technique due to poor contrast noise and diffusing boundaries [3]. Automatic interpretation and analysis of medical images make the clinician free from the manual and intensive aspects of their work and also helps in increasing the consistency and accuracy of outcome [4]. Segmentation of medical images covers all the factors that stimulus the analysis of a disease. It helps to analyze, diagnose, quantify, monitor, and plan the navigation of a disease [5]. In general, image segmentation algorithms categorize as low-level and high-level segmentation. Low-level segmentation includes similarity-based, discontinuity-based, and pixel-based classification methods which do not include priori information. These algorithms are based upon the intensity, texture sharpness, contour, and other related significant features [6]. Examples for low-level segmentation are thresholding, point-based, line, edge, and region-based approaches. Various segmentation problems especially medical images require the use of a priori knowledge to increase the robustness and accuracy called high-level segmentation techniques. For example, clustering-based techniques, atlas-based segmentation, Bayesian approach, statistical approaches, deformable models, machine learning-based classifications, and many more [7]. High-level segmentation methods specially atlas-based methods can do the segmentation when there is no well-defined relation between the region and the pixels intensities like the objects of same textured or structured to be segmented and when the prior information is available in the form of spatial relationship between the objects [8] (Fig. 1).

1.1 Various Segmentation Schemes

There are various techniques developed for image segmentation perform well and analogous to the methods used in practice. Accuracy and analysis of any segmentation method purely depends upon the type of image, image intensity, texture, image content, and other relevant factors. Hence, a single segmentation method cannot be applied to all type of images and also all methods do not perform well for one kind of image. There are number of segmentation techniques which are categorized into three generations (Fig. 2).

Techniques are placed into different categories according to the development and progress in the field of medical image segmentation. First generation contains the techniques and methods that need little prior information for processing the image and hence involves low-level techniques. Second generation includes the methods based on optimization and uncertainty models. Third generation techniques are highly reliant on information known prior and recommend experts defined models and rules for classification of an image. The comparison between some of the techniques has been mentioned below which can be extended further for the selection of best suitable method the given type of image and situation (Table 1).

In writing this survey paper, we will focus on the techniques that do not depend only on the gray-level information but also embrace the available a priori information [4]. Atlas-based segmentation is the technique that makes use of a priori information in form of a set of manually labeled images [9]. In this paper, we will focus on multi-atlas segmentation which certainly is one of the most prominent methods for biomedical images especially for heart images. This paper has further organized in five more sections that include multi-atlas segmentation (MAS) technique and the related literature review, atlas generation and selection, image registration, label propagation and label fusion, and last section is about the general discussion and future scope.

Fig. 2 Generations of medical segmentation approaches [7]

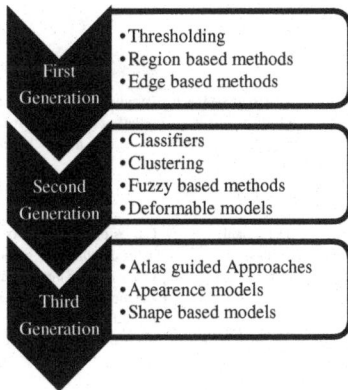

Table 1 Comparison between various segmentation schemes [1, 7, 11, 12, 29–31]

S. No.	Methods	Process	Advantages	Limitations	Applications	Memory Usage
1	Thresholding	Based on the histogram peaks of the image to find particular threshold values	Does not require prior information of the image. Computationally easy, simple, inexpensive and can be used in real-time applications	Highly dependent on peaks. Does not work for the broad and flat valleys. Highly noise sensitive. Selection of threshold is crucial	Structures have divided intensity allotment	Fastest
2	Region growing	Based on partitioning the image into homogeneous regions	Assure the piecewise continuity. Less sensitive to noise. Proper selection of seed. Gives accurate result	Expensive method in terms of time and memory. Deciding stopping criteria and selection of seed is very difficult task	Work well for the structures with high contrast boundaries	Fast
3	Fuzzy C-mean method	Based on division into homogeneous clusters	Better than K means. unsupervised and converge very well	Sensitive to noise. Computationally expensive. Determination of fuzzy membership is not very easy	For MR images	Medium
4	Clustering	A set of class continuum points of membership grades with no sharp boundary	Easy to implement. More useful for real problems. Eliminates noisy spots. Reduces false blobs. More homogeneous regions are obtained	Determining membership. Function is not easy. Computationally expensive. Difficult to predict 'k' with fixed number of clusters. Sensitive to noise	Mainly work well for MR images	Medium
5	Classifiers	Executed on the basis of spectrally defined features, e.g., density, texture	Works well for images having good contrast among regions. Result reliable for second order differential operator	Computationally complex. Slow approaches. Size of operator and computational complexity are proportional to each other. Generally, boundaries determined are discontinuous	Work well for MR and CT images	Slow
6	Bayesian approach	Permits the use of prior knowledge concerning the situation under study	Suitable situation for an extensive variety of models	Difficult to figure out a priori and makes use of prior dependent posterior distributions	These are mainly applicable to the verification problems	Fast
7	Deformable methods	Curves or surfaces defined within an image domain that can move under the influence of internal forces	Assure the piecewise continuity. Noise insensitive. Provide sub-pixel accuracy	Requires the tuning of parameters and thus can affect speed of the system	Work best with statistical regional information of the image	Medium

(continued)

Table 1 (continued)

S. No.	Methods	Process	Advantages	Limitations	Applications	Memory Usage
8	Atlas-guided approaches (Multi-atlas Segmentation)	Number of atlas are potentially used for segmenting the novel image	Assure an optimum solution for two class segmentation	Composition is difficult	They are mainly applied to MR images	Fast
9	Edge-based methods	Based on discontinuity detection	Easy to implement. Effective computational factor. Good for images having better contrast between objects	Not good for wrongly detected and more number of edges	For medical	Fast
10	Watershed method	Based on topological interpretation	Results are more stable. Detect continuous boundaries	Complex calculation	Good for traffic monitoring	Moderate
11	PDE-based method	Based on the working of differential equations	Fastest method. Best for real-time applications	More computational complexity	Cell biology	Fast
12	ANN-based methods	Based on the simulation of the learning process for decision making	No need to write complex programs	More wastage of time in training	Medical images	Slow

2 Multi-atlas-Based Segmentation

Segmentation partitions the image region or volume into non-coinciding, connected and strongly correlated regions, being consistent with respect to some image characteristics. Image segmentation is an essential tool in medical image processing. Atlas[1]-based segmentation performs classification and tagging of desired voxels[2] at target image. Final label at target image is the result of joining all voxels from all training atlases. Segmentation of medical images using multi-atlas approach helps in automated interpretation of images that helps the radiologist in identification of important disease characteristics and early detection of disease progression. For assigning labels to the voxels relationship between image intensities and segmentation labels is considered. Multiple atlases are registered to a target image, and the delineated atlas segmentations are fused together using label fusion [10]. The key assumption of multi-atlas. segmentation is that multiple atlases incorporate much more anatomical variability than a single atlas. Therefore, target image can be labeled

---

[1] Atlases refers to model for a collection of images with parameters that learned from training data set. These are the certain type of manually delineated images, which are labeled by experts. Atlas is known with various names as training images, reference images, template images.

[2] Voxels is a unit of graphic information. It defines a point in three-dimensional space with its x, y, and z coordinates. These coordinates are position, color, and density.

more accurately by transforming the label information from the atlas images that have the most similar characteristics [11].

2.1 Literature Review

Compiling information about the subject in interest at one place work as aid and encouragement for beginners in that field. It also gives general idea about the knowledge, and understanding needs required for certain problem solving and caution that need to be taken care before we apply the results to test images. In this section, we intend to explore the detailed references of past research activity in MAS. Table below has composed the different techniques and their brief description (Table 2).

From the Table 2, it is cleared that authors have done work either on atlas selection techniques or on various registration strategies. As increase in registration speed might not work well for medical images but accuracy and good selection of atlas are crucial. The choice of technique truly depends upon the type of problem under consideration. The basic building block for MAS is elaborated in preceding sections (Fig. 3).

Success of MAS technique depends on the selection of relevant atlases and maintaining the spatial correspondence with the novel image coordinates. To avoid the registration error number of voxel in the local neighborhood which is strongly correlated with the target images are considered. To make this technique more accurate and robust, the voxels are replaced by patches. Patches are weighted according to the similarity matrices. Label from the wrapped atlas patches is fused to give the label estimate in target images.

Fig. 3 General building block for MAS [27]

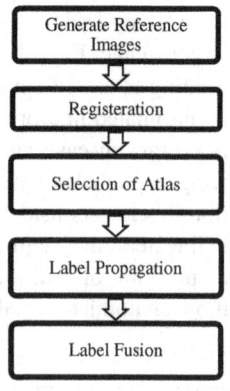

Table 2 Techniques and methods proposed by various authors

S. No.	Author	Technique	Description
1	Wyawahare et al. [5]	Use adaptive multi-atlas segmentation and adaptive multi-local atlas segmentation	Selection of suitable atlases for a novel image and registration process automatically stops when no further improvement is expected
2	Bai et al. [10]	Probabilistic patch-based label fusion model	Use multiple number of patches to weight the label fusion process. correlation matrices between target voxel and atlas voxel decide the weight of the label fusion process
3	Zhuang et al. [16]	Local affine registration methods	Use registration algorithms for automatic segmentation of CMR images
4	Subunch et al. [23]	Based on non-parametric probabilistic model	Use probabilistic framework to stimulate label fusion methods as segmentation tool
5	Wang and Yushkevich [24]	Use weight-based label fusion approaches	Atlases are weighed according to the similarity measures between the atlas and target images. Most similar atlas mapped with more weight
6	Zhuang et al. [28]	Use patch-based label fusion method	Registration accuracy improved by using registration refinement process
7	Rohlfing et al. [32]	Based on atlas selection methods	Atlas selected on the basis of alignment to individual atlas, to average atlas, most similar atlas and to all atlas in the data set
8	Alvén et al. [33]	Use Uber atlases technique	Find and improve the relation between the feature points to speed up the registration process

3 Atlas Generation

Generation of atlases, the manually labeled reference images are the scrupulous, careful, and costly effort of an expert. Prior knowledge on the problem statement is required to select the atlases. This prior knowledge is incorporated in atlas construction (shape, orientation, atlas size and atlas fitting procedure energy function, and optimization procedure) [12]. MAS use multiple number of atlases to average out error and improve the accuracy. A very common approach to build atlas is to choose the reference image from the given samples and transform other images onto that target. This method is used when atlas has been generated from number of images. The resulting atlas using this approach may biased toward the chosen reference image [4]. Other approach that can help in atlas generation is to choose the sample which closest to mean. This method reduces the biasing. Tung et al. [13] use neointima segmentation method which neglect low-quality atlases and increase the accuracy of segmentation, but it is not preferred as this method is application specific that may not work for all type of images. Zhuang et al. [14] use population-level processing (co-registration of atlases) to increase SNR ratio to yield a more accurate registration and segmentation of the novel image. Jia et al. [15] used principal component analysis (PCA) a statistical model to synthesis deformations. This method is not reliable much as synthetic deformations might not always be anatomically probable. Bai et al. [10] used weighted label fusion process (atlas weighted by its similarity with target image). The subset from more similar atlases can be chosen as reference atlases.

4 Registration

The images need to be aligned geometrically for better observation. This procedure of aligning the corresponding points in one image to another is called image registration. The image which are mapped are taken from same scene but at different angles and times and from different sensors. Registration is the difficult task to perform the spatial correspondence between images. The registration process is done to precise the spatial misalignment between the images. The problem of registration becomes complicated in the presence of inter-image noise, missing features, and outliers in the images [16]. To perform MAS, the approach is to transform the single or average atlas over the target image. Using more advanced techniques of registration, we can improve the accuracy and computation time of MAS algorithm. Using single atlas for registering over the target image always gives better result than if number of atlases that need to be register are more as multiple atlases increases the computation time but at the same time using multiple atlases increase the accuracy [17]. Image registration algorithm has three basic building blocks, the deformation model, objective function, and optimizer. The deformation model is the spatial transformations such as rigid transform, affine transform, or any non-parametric model [4]. The objective function depends upon the similarity and alignment measures. For maximum correspondence, the parameters like cross-correlation, mutual information can be considered. The optimization method is an iterative training process. Methods like conjugate gradient, gradient descent can be used. Stavros et al. [18] integrate registration and label fusion process and this helps to increase the accuracy. The registration is time crucial step in MAS. The accuracy depends upon that how accurately the registration process has been done. Computational complexity must be reduced by choosing appropriate atlas and by discarding the low-quality atlases [14] (Fig. 4).

Fig. 4 Registration process: alignment of two images. Energy function defines the mapping between the images and transformation function and ensures the maximum correspondence [28]

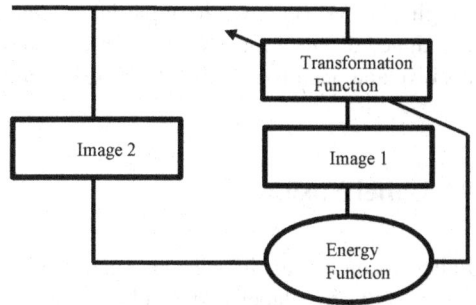

5 Atlas Selection

Atlas selection is the last step before the final segmentation. There are number of ways which help in selection of atlases. These can be selected by 1. Choosing an atlas from the data set. 2. An average shape is selected. 3. The most similar to the target is chosen. 4. More than one number of images used as atlases [16]. Various other approaches are based on mean value, median value, and least distances can also be used for selection of atlases [19]. It is always not necessary to use all the available atlases because the computational efficiency can be improved by discarding the irrelevant atlases or by reducing the number of atlases, for example, by selecting half of the atlases the speed may be increased by double and memory needs can be reduced by half. For atlases which are non-linearly registered to the novel image. The atlas selection is purely based on similarity of images and for non-registered images the selection. Tung et al. [13] use neointima segmentation method which neglect low-quality atlases and increase the accuracy of segmentation. Asman et al. [20] introduce new approach based on principle component analysis. This technique performs well while selection of atlases. Nouranian et al. [21] proposed the concept of clustering to increase the accuracy of atlas selection procedure. He used K-means clustering to analyze and identify the clusters similar cases.

6 Label Propagation

After the selection of relevant atlases, a spatial rapport established between the atlases and the testing images. Now the atlases label propagates to the novel image coordinates and label from the atlases transfer to each novel image voxel. There are various models and learning algorithms on propagation of label from atlases to target image. Since early MAS methods (Heckemann et al. 2009) [22] use nearest neighbor interpolation method to transform label from chosen atlases over each voxel in novel image. Sabuncu et al. [23] further refined the nearest neighbor interpolation and named it as linear interpolation where each atlas extent over multiple labels, with associated weights that reflect the ratio of partial volumes. Xu et al. [9] proposed method which use signed distance map where each label has an associated signed distance to take positive and negative value within the structure and outside the structure.

7   Label Fusion

Label fusion means different independent labels of same object are combined into a single discrete label and results is much more accurate, than that the individual input. In MAS technique, the labels are assigned to target image with help of delineated labels of atlases if the target image and the atlases share similar characteristics in terms of shape appearance, texture. Hence, all the atlases are registered over the target image before label fusion process. To reduce segmentation error produced by registration label fusion is applied to combine all candidate segmentations. Effective label propagation will give more accurate label fusion results [24]. There are three type of label fusion techniques STAPLE, voting, shape-based averaging (SBA), and a hybrid technique called SVS. These techniques are based on expectation-maximization (EM) algorithm, summing of each pixel/voxel and label calculation of Euclidean distance for input label, respectively. Heckemann et al. [22] used single atlas and observed the match between the registered atlas and novel image intensities using the following methods as mutual information, Sum of squared differences, normalized cross-correlation. Artaechevarria et al. [25] used forest method for label fusion in which weights are linear to the mutual information between the atlas image and novel image intensities. Guorong wu et al. [26] used patch-based label fusion method. In this method, the label at the target image is determined by using patch wise similarity between the atlases and the target image. Bai et al. [10] considered patch rather than a voxel for labeling the atlases. An anatomical context considered to improve the quality of the patch matches.

8   Conclusion and Future Scope

Quantitative analysis is always better than visual image interpretation in terms of good analysis, understanding, and diagnosis which is further beneficial to decide upon the appropriate treatment. In this paper, we have shared the expert views of various authors for three fundamental steps in MAS algorithms which are registration, label propagation, and label fusion. These steps further explore ideas from computer vision, statistical model, machine learning, and various other fields. As the quality of atlases and the number of relevant atlases are the two prominent factors to measure segmentation accuracy, so manual delineation of atlases should be done very carefully by the experts, which is both expensive and time consuming. Increasing the speed of registration process is another factor that helps in reducing the computational time. A single ideal algorithm that can give optimum results in terms of complexity, accuracy, and computation time is a very complex to develop because of change in each problem statement the unique objective and constraint to problem statement changes. For example, in some biomedical applications, high accuracy is more critical rather than computational cost and time and on the other hand for some applications computational time is more critical factor than the accuracy of the algorithm. In conclusion, we can say for good segmentation work can be done either to find the good techniques to increase registration accuracy and speed or to find the techniques to select better atlases. However, the multi-atlas segmentation technique explained above seems to be more suitable approach for medical image segmentation because this technique does not depend upon single modality image data and image intensities, rather it depends upon spatial location of voxels and can be applied even when there is no well-defined relation between the region and the pixels intensities (e.g., objects of same textured or structured such as heart tissue can be segmented). In spite of so many advantages, MAS also has some shortcomings. As processing and working on all the training data set led to computationally expensive and time-consuming affair. But there is a huge scope that MAS can be implemented successfully on medical images. We suggest that further work can be done in the direction to develop the algorithms which are more robust against change in imaging hardware, image intensity, resolution. and other imaging parameters.

References

1. Arti, T., Ranjan, P., Ujjlayan, A.: A Performance Study of Image Segmentation Techniques. 3978-1-4673-7231-2/15/$31.00 ©2015 IEEE
2. Elnakib, A., Gimel'farb, G., Suri, J., El-Baz, A.: Medical Image Segmentation: A Brief Survey. Springer Science Business Media, LLC (2011). https://doi.org/10.1007/978-1-4419-8204-9_1
3. Maksoud, E.A.A., Elmogy, M., Al-Awadi, A.M.: MRI Brain tumor segmentation system based on hybrid clustering techniques in advance machine learning technologies and applications. In: Series Communications in Computer and Information Science, vol. 488, pp. 401–412. ©springer international publishing Switzerland 2014
4. Kalini, H.: Atlas based Segmentation: A survey IF 0.73. https://doi.org/10.2174/1573405611 01150423103441 (email: hrvoje.kalinic@fer.hr)
5. Wyawahare, M.V., Patil, P.M., Abhyankar, H.K.: Image registration techniques: an overview. Int. J. Signal Process. Image Process. Pattern Recognit. 2(3) (2009)
6. Pham, D.L., Chenyang, X., Prince, J.L.: Current methods in medical image segmentation. Annu. Rev. Biomed. Eng. 2, 315–337 (2000)
7. Masood, S., Sharif, M., Masood, A., Yasmin, M., Raza, M.: A survey on medical image segmentation. Curr. Med. Imaging Rev. 11, 3–14 (2015)
8. Iglesias, J.E., Sabuncu, M.R.: Multi-Atlas Segmentation of Biomedical Images: A Survey. © Elsevier 12 June 2015
9. Xu, Z., Li, B., Panda, S., Asman, A.J., Merkle, K.L., Shanahan,P.L., Abramson, R.G., Landman, B.A.: Shape-constrained multi-atlas segmentation of spleen in CT. In: SPIE Medical Imaging, pp. 903446–903446. International Society for Optics and Photonics (2014)
10. Bai, W., Shi, W., O'Regan, D.P., Tong, T., Wang, H., Jamil-Copley, S., Peters, N.S., Rueckert, D.: A Probabilistic patch-based label fusion model for multi-atlas segmentation with registration refinement: application to cardiac mr images. IEEE Trans. Med. Imaging 32(7) (2013)
11. Shruti, K., Janak B.P.: Atlas based segmentation techniques a review. Cur. Med. Imaging Rev. 3(5) (2015), ISSN: (2321-1717)

12. Zuva, T., Oludayo, O.O., Ojo, S.O., Ngwira, S.M.: Image segmentation, available techniques, developments and open issues. Can. J. Image Process. Comput. Vis. **2**(3), 20–29 (2011)
13. Tung, K.P., Bei, W.J., Shi, W.Z., Wang, H.Y., Tong, T., De Silva, R., Edwards, E., Rueckert, D.: Multi-atlas based neointima segmentation in intravascular coronary OCT. In: 2013 10th International Symposium on Biomedical Imaging (ISBI), pp. 1280–1283. IEEE
14. Zhuang, X., Leung, K., Rhode, K., Razavi, R., Hawkes, D., Ourselin, S.: Whole heart segmentation of cardiac MRI using multiple path propagation strategy. In: Medical Image Computing and Computer-Assisted Intervention, MICCAI 2010, pp. 435–443. Springer
15. Jia, H., Yap, P.T., Shen, D.: Iterative multi-atlas-based multi-image segmentation with tree-based registration. Neuroimage **59**, 422–430
16. Zhuang, X., Rhode, K., Arridge, S., Razavi, R., Hill, D., Hawkes, D., Ourselin, S., Metaxas, D.: An Atlas-Based Segmentation Propagation Framework Using Locally Affine Registration—Application to Automatic Whole Heart Segmentation, pp. 425–433. Springer, Berlin, Heidelberg (2008)
17. Langerak, T.R., Berendsen, F.F., Van der Heide, U.A., Kotte, A.N.Pluim, J.P.: Multi-atlas-based segmentation with preregistration atlas selection. Med. Phys. **40**(9) (2013). https://doi.org/10.1118/1.4816654
18. Stavros, A., Aristeidis, S., Nikos, P.: Discrete multi atlas segmentation using agreement constraints. In: British Vision Conference (2014)
19. Chijindu, V.C., Inyiama, H.C., Uzedhe, G.: Medical image segmentation methodologies—a classified overview. Afr. J. Comput. ICT 2012 **5**(5), (2012). All Rights Reserved IEEE. ISSN 2006-1781
20. Asman, A.J., Bryan, F.W., Smith, S.A., Reich, D.S., Landman, B.A.: Group wise multi-atlas segmentation of the spinal cord's internal structure. Med. Image Anal. **18**, 460–471 (2014)
21. Nouranian, S., Mahdavi, S., Spadinger, I., Morris, W., Salcudean, S., Abolmaesumi, P.: A multi-atlas-based segmentation framework for prostate brachytherapy. IEEE Trans. Med. Imaging **34**(4), 950–961 (2015)
22. Heckemann, R.A., Hajnal, J.V., Aljabar, P., Rueckert, D., Hammers, A.: Automatic anatomical brain MRI segmentation combining label propagation and decision fusion. Neuroimage **33**, 115–126 (2006)
23. Sabuncu, M.R., Yeo, B.T.T., Van Leemput, K., Fischl, B., Golland, P.: A generative model for image segmentation based on label fusion. IEEE Trans. Med. Imag. **29**(10), 1714–1729 (2010)
24. Wang, H., Yushkevich, P.A.: Multi-atlas segmentation with joint label fusion and corrective learning. In: An Open Source Implementation, 22 Nov 2013. 7:27. https://doi.org/10.3389/fninf.2013.00027. Collection 2013
25. Artaechevarria, X., Munoz-Barrutia, A., Ortiz-de Solorzano, C.: Combination strategies in multi-atlas image segmentation: application to brain MR data. IEEE Trans. Med. Imaging **28**, 1266–1277 (2009)
26. Wu, G., Kim, M., Sanroma, G., Wang, Q., Munsell, B.C., Shen, D.: Hierarchical label fusion with multi scale feature representation and label specific patch partition. In: MICCAI 2014, Part 1, LNCS8673, pp 299–306. Springer International Publishing, Switzerland (2014)
27. Van Rikxoort, E.M., Isgum, I., Arzhaeva, Y., Staring, M., Klein, S., Viergever, M.A., Pluim,.P.W., van Ginneken, B.: Adaptive local multi-atlas segmentation: application to the heart and the caudate nucleus. Med. Image Anal. **14**(1), 39–49 (2010)
28. Zhuang, X., Rhode, K.S., Razavi, R.S., Hawkes, D.J., Ourselin, S.: Registration-based propagation framework for automatic whole heart segmentation of cardiac MRI. IEEE Trans. Med. Imaging **29**(9) (2010)
29. Mustafa, I.D., Hassan. M.A.: A Comparison between different segmentation techniques used in medical imaging. Am. J. Biomed. Eng. **6**(2), 59–69. ISSN: 2163-1050, © 2016 Scientific & Academic Publishing
30. Hu, G.: Survey of Recent Volumetric Medical Image Segmentation Techniques. Memorial Sloan, Kettering Cancer Center, New York. www.intechopen.com
31. Tokas, N., Karkra, S., Pandey, M.K.: Comparison of Digital image segmentation techniques-a research review. IJCSMC **5**(5), 215–220 (2016)

32. Rohlfing, T., Brandt, R., Menzel, R., Maurer, C.R.: Evaluation of atlas selection strategies for atlas-based image segmentation with application to confocal microscopy images of bee brains. NeuroImage **21**(4), 1428–1442 (2004)
33. Alvén, J., Norlén, A., Enqvist, O., Kahl, F.: Überatlas: Robust Speed-Up of Feature-Based Registration and Multi-Atlas Segmentation, vol. 9127 © Springer International Publishing Switzerland 2015

Color Image Retrieval Using Statistically Compacted Features of DFT Transformed Color Images

Sushila Aghav-Palwe and Dhirendra Mishra

Abstract Feature extraction of images are crucial in image retrieval systems. Many approaches are stated and proved by researchers for image feature extraction and processing. Research is being done from low-level feature extraction toward high-level feature extraction. This paper discusses the feature extraction from the DFT transformed color images in multiple color planes. DFT image transform provides effective way to differentiate the image textures. For dimensionality reduction statistical parameters such as kurtosis, standard deviation, and variance are used for feature vector generation. Euclidian distance is used in the proposed approach. Four different types of feature vectors are created and tested for each image class. The images are retrieved based on the image pixel values of DFT phase information and DFT magnitude information of different color spaces like RGB, YIQ, HSV, and YCbCr similar to that of image class. Image retrieval performance of the proposed approach is compared for database of 1000 images of ten different categories. Precision of image retrieval is above 60% for all classes and more than 80% for some of the image classes.

Keywords Image retrieval · DFT · Feature vector · DFT phase · DFT magnitude

1 Introduction

Content-based image retrieval is most researched area in the field of image processing. Many attempts for feature extraction have been proposed and evaluated for the accurate retrieval of the similar images [1–5]. High-level features and low-level features are exaggeratedly accessed to create image signatures which precisely represent

S. Aghav-Palwe (✉) · D. Mishra
Department of Computer Engineering, SVKM's NMIMS Mukesh Patel School of Technology
Management and Engineering, Mumbai, India
e-mail: sushila.aghav@mitcoe.edu.in

D. Mishra
e-mail: Dhirendra.mishra@gmail.com

© Springer Nature Singapore Pte Ltd. 2019   337
S. K. Bhatia et al. (eds.), *Advances in Computer Communication
and Computational Sciences*, Advances in Intelligent Systems
and Computing 760, https://doi.org/10.1007/978-981-13-0344-9_29

the image in short and optimized way. High-level feature extraction mainly focuses on object detections and faces the limitation of huge object corpus size. Low-level feature extraction is still in the researchers' loop to find the optimal way for image representation which is to be used for accurate image retrieval. Many approaches like color-based feature extraction, edge and shape-based feature extraction, texture-based feature extraction have been proposed and evaluated by researchers for their usage in the image retrieval [3, 4, 6–9]. While considering the feature extraction image, the main focus was on identifying the way that represents the image in minimal and precise way. That is the challenge of feature vector optimization for image retrieval [5, 7–18]. In typical CBIR system, the visual contents of the image are extracted and represented in terms of feature vector. The feature vectors of the images in the database form a feature database. To retrieve images, users provide the retrieval system with example images. The system then converts these images into its feature vectors. Similarity between the feature vectors of the query example and those of the images in the database are then calculated, and retrieval is performed with the aid of an indexing scheme. The indexing scheme provides an efficient way to search for the image database. Some of image feature extraction techniques are color-based, edge- and texture-based, and shape-based [1]. Color histogram describes the distribution of colors within image. As a pixel-wise characteristic, the histogram is invariant to rotation, translation, and scaling of the image and is easy and inexpensive to calculate [2]. Similarly, shape and edge feature are used for high-level feature extraction, but due to ambiguity in shapes and edge components, the image retrieval is still need to be researched. As discussed in [19–21], image retrieval with feature vector of combination of object detection, color feature, and classification approach is effective for the image retrieval, but it may face the problem of diverse database.

Major limitations of current approaches are complex computations, impact on performance for diverse database, large feature vectors, and scalability. To overcome the above limitations, there is a need of preparing the compact feature vector, with key elements of image, which are uniform in similar image and will perform well for diverse database. This paper discusses the experimented approach for image retrieval using feature vector containing DFT transformed image information. We prepared the feature vector that uses the DFT transformed information effectively. The main aim behind using the DFT is that the pixel-to-pixel computation can be reduced using image transforms. Image transforms possess very useful property like energy compaction in higher coefficients [22]. This property is useful to reduce the feature vector size of the image. DFT image transform provides effective way to differentiate the image textures.

Magnitude of 2D DFT is an array which components determine the intensities of image; the corresponding phase is an array of angles that carry much of the information about where the discernable objects are located in image [22]. These properties of DFT are used in proposed approach of feature vector creation. In our CBIR approach, the phase angle and magnitudes are effectively used to improve the retrieval performance. Efforts are made to test the performance of retrieval against various minimal feature vectors. Thousand color images of various classes are tested for image retrieval, and performance is compared. Main contribution of this paper is

1. To use DFT phase and magnitude for feature vector creation.
2. To compare performance for image retrieval of DFT phase and magnitude and compare their performances for image retrieval.
3. To use different statistical parameters for feature vector reduction.

2 Feature Vector Creation Using DFT Phase and Magnitude

Transform domain feature extraction provides the way to study the internals of image contents like image intensities of spectrum, edges, and objects. In this section, we explore the usage of DFT for feature extraction DFT and way to extract the internals of image in terms of phase and magnitude (transform spectrum). In the literature [22], the phase property is used to detect the edges and line. As shown in Fig. 1, observation is done by reobtaining original image using inverse DFT with phase and inverse DFT with magnitude only [22]. When both magnitude and phase collectively used, we are able to achieve to original image. But when we tried to analyze both the DFT image parts, for comparing the two images for finding similarity, the usage of any one part of DFT image is sufficient for generating image signatures, which are more compact and optimized.

Each image firstly preprocessed for decided ROI. Then, the image is transformed to DFT image using DFT. As we work on color image, each color plane of transformed image is treated separately for feature vector creation. Magnitude and phase parts of DFT image are calculated separately.

Following methodology described in Fig. 2, each image is represented by feature vector. Feature vector for n classes is created as follows. Feature vector with phase

Fphase (Color Plane1) = {SC (CP1_Phase)}
Fphase (Color Plane2) = {SC (CP2_Phase)}
Fphase (Color Plane3) = {SC (CP3_Phase)}
FV_Phase = {Fphase (CP1), Fphase (CP2), Fphase (CP3)}

Fig. 1 Inverse DFT with phase and magnitude

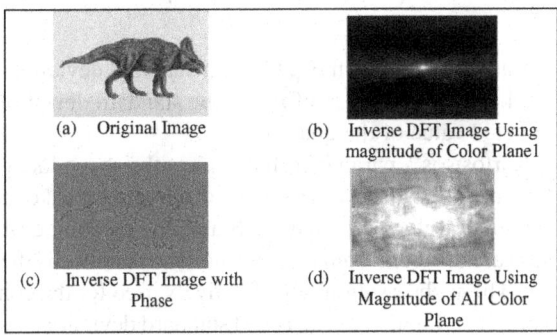

(a) Original Image

(b) Inverse DFT Image Using magnitude of Color Plane1

(c) Inverse DFT Image with Phase

(d) Inverse DFT Image Using Magnitude of All Color Plane

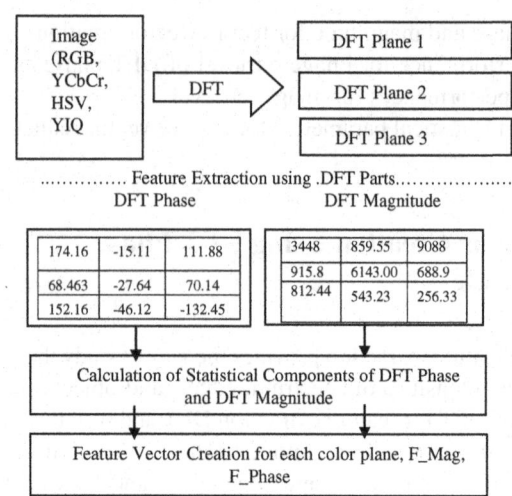

Fig. 2 Feature vector preparation

Feature vector with Magnitude
Fmag (Color Plane1) = {SC (CP1_mag)}
Fmag (Color Plane2) = {SC (CP2_mag)}
Fmag (Color Plane3) = {SC (CP3_mag)}
FV_magnitude = {Fmag (CP1), Fmag (CP2), Fmag (CP3)}
Here, CP = color plane of different color spaces.
SC = statistical components.

We used statistical components: kurtosis, variance, and standard deviation [23] of the phase and magnitude parts to extract image representation in optimized way

$$k = \frac{E(x - \mu)^4}{\sigma^4} \tag{1}$$

$$s = \sqrt{\frac{1}{n}\sum_{i=1}^{n}(x_i - {}^-x)^2} \tag{2}$$

Kurtosis is given in Eq. (1), and standard deviation is given in Eq. (2)
where μ is the mean of x, σ is the standard deviation of x, and E(t) represents the expected value of the quantity.

Kurtosis is a parameter that depicts the shape like peakedness—broad or narrow of a value distribution. Here, we observe the peakedness or narrowness phase and magnitude values distribution. Standard deviation evaluates group of sample values those are different from mean value. The variance is the square of the standard deviation. These observations are effectively used for the comparison of image signatures. Using Eqs. 1 and 2, kurtosis and standard deviation of phase and magnitude for each

Table 1 Sample images of different classes

Wild people		Elephant	
Beach		Flower	
Monuments		Horse	
Bus		Mountain	
Dinosaur		Food items	

color bin are calculated and collected in feature vector. Separate feature vector for each of three color bins is created. Combined feature vector of all colors is also created, and image retrieval and performance are tested.

3 Image Database Creation

Thousand images from WANG database [24] are tested for the discussed approach. We evaluated the retrieval performance for four different color spaces: RGB, HSV, YCbCr, and YIQ. For performance measurement purpose, 10 classes of images with 100 images for each class are considered. Ten classes of images, each of 100, are used for testing as shown in Table 1.

As the main focus was to compare the usability of magnitude and phase with respect to color planes, for image representation, eight feature vectors, four for DFT phase and four for DFT magnitude for each image, are stored in database for comparison purpose. Both of DFT parts contribute for image representation. We effectively used these parts of the DFT for image representation. And performance of retrieval compared using them individually and collectively. Precision measure is used for testing performance. Different color spaces like RGB, YCbCr, HSV, YIQ are evaluated for retrieval performance measurement. Following are feature vectors evaluated for each color space in performance measurement.

FV_PhaseN:: feature vector containing phase of color plane N.

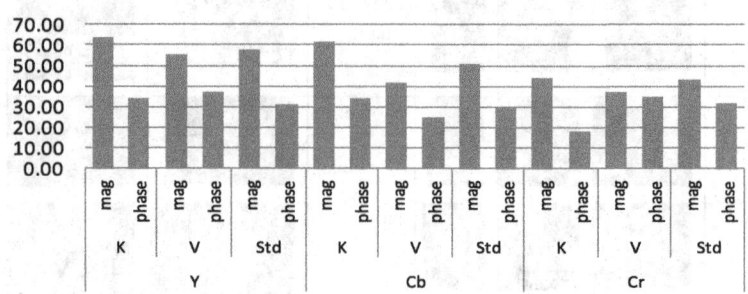

Fig. 3 Precision graph for phase and magnitude comparison of YCbCr

FV_MagN:: feature vector containing magnitude part of each color plane N.
FV_magnitude:: feature vector containing magnitude of all color planes.
FV_phase:: feature vector containing statistical measures of phase of all color planes.

4 Performance Measurement

Euclidian measure is popular measure for image similarity measurement [6, 21].
Using Euclidian measure performance of image retrieval is measured for different
feature vectors. Following performance measures are used to test the performance of
the image retrieval, Using Eq. 3, precision is evaluated for each color plane against
phase and magnitude.

$$\text{Precision} = \text{no of Images correctly retrieved}/\text{Total no of retrieved images} \quad (3)$$

Figures 3, 4, 5, and 6 are different precision graphs for each color space, for
magnitude and phase w.r.t kurtosis, variance, and standard deviation (Fig. 7).

Figure 3 shows the precision graph of image retrieval for YCbCr, with
F_Phase, F_Magnitude feature vectors for all color planes (Y, Cb, Cr) individually. F_Magnitude feature vector showing retrieval performance up to 65% for some
classes of images with first color planes Y. Feature vector with kurtosis performs better as compared to other statistical measure. For some image classes like wild people,
Horse, both magnitude and phase perform equally for image retrieval. Particularly
for Cr color plane, the phase and magnitude both give equal results for some image
classes like Dinosaur, Food. Figures 4, 5 and 6 are the precision graph for RGB, HSV,
and YIQ. Here, we can observe that first plane of each color space provides better
results than other color planes. Feature vector with kurtosis measure performs above
60% of precision for image retrieval using YIQ, HSV, and YCbCr, but for RGB the
standard deviation performs better than other statistical measures. It provides 65%
precision R color plane.

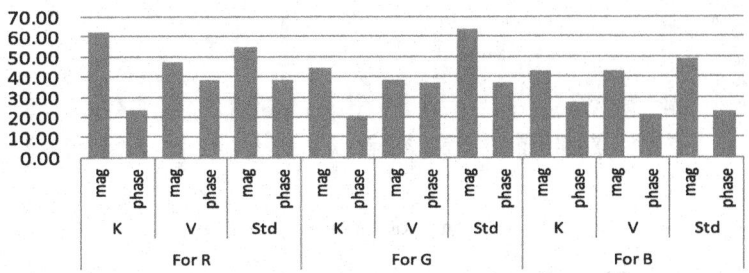

Fig. 4 Precision graph for phase and magnitude comparison of RGB

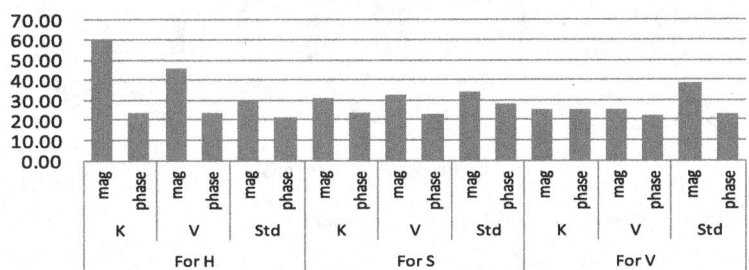

Fig. 5 Precision graph for phase and magnitude comparison of HSV

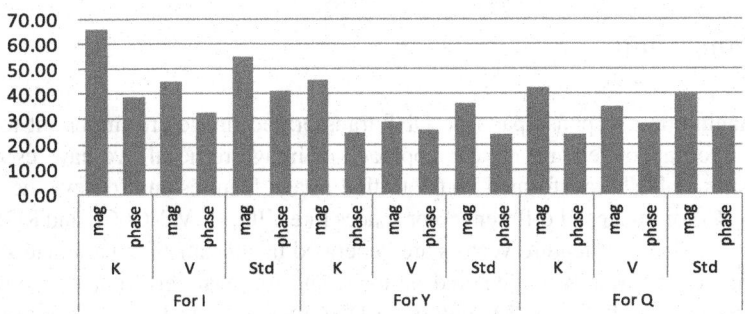

Fig. 6 Precision graph for phase and magnitude comparison of YIQ

Precision graph comparing combined feature vector using phase and magnitude is shown in Fig. 7. All classes of images are evaluated separately for RGB color space. In above graph, the magnitude of DFT images is observed performing better

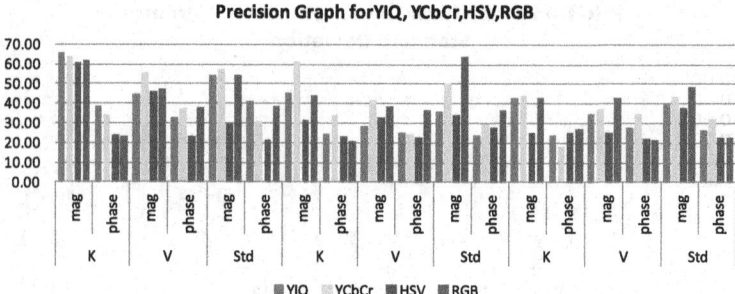

Fig. 7 Precision graph for phase and magnitude comparison

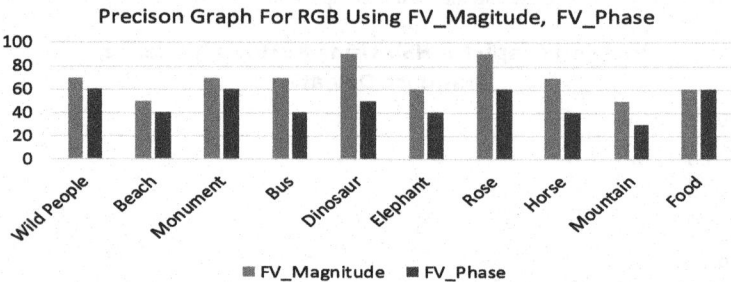

Fig. 8 Precision graph for phase and magnitude comparison

than the phase part of the images. As we can see for some image classes like Food, Wild people, both parts of DFT images are performing similar (Fig. 8).

5 Conclusion

The magnitude and phase parts on DFT transformed image are important to find image internals or features. In our approach of image retrieval, we have used the both parts of DFT transformed image to find image features and retrieve the color images. Color images of different color spaces like YIQ, HSV, YCbCr, and RGB are tested. The compact feature vectors are generated using statistical parameters such as standard deviation, variance, and kurtosis, and retrieval performance are tested using precision calculation. Magnitude of DFT image showed with better retrieval performance for image classes: Wild people, Rose, Dinosaurs, etc. For classes like Horse, Monument, Food, the DFT phase performed equally that of DFT magnitude for image retrieval. Further to reduce the feature vector size, we have successfully used the statistical measure. This has been observed that the color planes R, Y, H perform well as compared to other color plane of color spaces. When compared the performance for all feature vectors combined, by selecting random images of any class, magnitude feature vector gives average 75% precision. F_Phase performance observed as lower than other magnitude feature vector. It is on average 60%. With F_Mag1, F_Phase1 the first color planes R, Y, H perform better than other color plane. Some of the class images, like Monuments, Wild people, Food, image retrieval performance is less than 45%. When overall image performance is observed for all classes, the Dinosaur, Bus, and Flowers classes are observed performing more than 60%. And in classes like Beach, Mountains, retrieval performance is as low as 40%. By observing the above graphs, the conclusion can be drawn that the DFT magnitude part is more useful to compare image signatures.

References

1. Kekre, H.B., Mishra, D., Kariwala, A.: Survey of CBIR techniques and semantics. Int. J. Eng. Sci. Technol.
2. Kekre, H.B., Sonawane, K.: Retrieval of images using DCT and DCT wavelet over image blocks. (IJACSA) Int. J. Adv. Comput. Sci. Appl. **2**(10) 2011
3. Kekre, H.B., Thepade, S.D., Sanas, S.P., Iyer, S.: Shape content based image retrieval using LBG vector quantization. (IJCSIS) Int. J. Comput. Sci. Inf. Secur. **9**(12) (2011)
4. Kekre, H.B., Mishra, D.: CBIR using upper six FFT sectors of color images for feature vector generation. Int. J. Eng. Technol. **2**(2) (2010)
5. Kekre, H.B., Mishra, D.: Sectorization of walsh and walsh wavelet in CBIR. Int. J. Comput. Sci. Eng. **3**(6) (2011)
6. Shih, J.L., Chen, L.H.: Colour image retrieval based on primitives of colour moments. IEE Proc. **149**(6), 370–374 (2002)
7. Memon, I., Chen, L., Majid, A.: Travel recommendation using geo-tagged photos in social media for tourist. Wirel. Pers. Commun. **80**(4) (2015)
8. Di Sciascio, E., Celentano, A.: Similarity Evaluation in Image Retrieval Using Simple Features (2010)
9. Datta, R., Joshi, D., Li, J., Wang, J.Z.: Image retrieval: idea, influences and trends of the new age. ACM Comput. Surv. **40**(2), Article 5 (2008)
10. Kekre, H.B., Mishra, D.: Performance comparison of density distribution and sector mean in Walsh transform sectors as feature vectors for image retrieval. Int. J. Image Process. (IJIP) **4**(3) (2010). ISSN 1985-2304
11. Kato, T.: Database architecture for content based image retrieval in image storage and retrieval systems. Proc. SPIE **2185**, 112–123 (1992)
12. Kekre, H.B., Mishra, D.: Content based image retrieval using weighted hamming distance image hash value. In: The Proceedings of International Conference on Contours of Computing Technology, pp. 305–309 (Thinkquest 2010)
13. Afifi, A.J., Ashour, W.M.: Image retrieval based on content using color feature. ISRN Comput. Graph. (2012)
14. Porat, M., Zeevi, Y.Y.: The generalized Gabor scheme of image representation in biological and machine vision. IEEE Trans. Pattern Anal. Mach. Intell. **10**(4), 452–468 (1988)
15. Jhanwar, N., Chaudhuri, S., Seetharaman, G., Zavidovique, B.: Content based image retrieval using motif cooccurrence matrix. Image Vis. Comput. **22**(14), 1211–1220 (2004)
16. Rao, M.B., Rao, B.P., Govardhan, A.: CTDCIRS: content based image retrieval system based on dominant color and texture features. Int. J. Comput. Appl. **18**(6), 40–46 (2011)
17. Boyle, R., Sonka, M., Hlavac, V.: Image Processing, Analysis, and Machine Vision, 2nd edn. University Press, Cambridge (2001)
18. Goshtasby, A.A.: Similarity and dissimilarity measures. In: Image Registration. Advances in Computer Vision and Pattern Recognition. Springer, London (2012)

19. Memon, M.H., Li, J.P., Memon, I., et al.: Efficient object identification and multiple regions of interest using CBIR based on relative locations and matching regions. In: 2015 12th International Computer Conference on Wavelet Active Media Technology and Information Processing (ICCWAMTIP). IEEE (2015)
20. Arain, Q.A., Memon, H., Memon, I.: Intelligent travel information platform based on location base services to predict user travel behavior from user-generated GPS traces. Int. J. Comput. Appl. (2017)
21. Memon, M.H., Li, J.P., Memon, I.: GEO matching regions: multiple regions of interests using content based image retrieval based on relative locations. Multimed. Tools Appl. 76(14) (2017)
22. Kekre, H.B., Sarode, T.K., Thepade, S.D., Sanas, S.: Assorted color spaces to improve the image retrieval using VQ codebooks generated using LBG and KEVR. IJCA (2011)
23. Seletchi, E.D., Duliu, O.G.: Image processing and data analysis in computed tomography, Rom. J. Phys. 52(5–7), 667–675 (2007)
24. Wang, J.Z.: Wang Database (2010)

Discrete Cosine Transform-Based Feature Selection for Marathi Numeral Recognition System

Madhav Vaidya, Yashwant Joshi, Milind Bhalerao and Ganesh Pakle

Abstract Optical character recognition system is hotcake for the researchers since last four decades. Recognition of handwritten Devanagari characters and digits is comparatively a tough task as compared to recognition other scripts like English or Latin. In this manuscript, a novel feature extraction and selection method is proposed for the recognition of isolated handwritten Marathi numbers based on one-dimensional *Discrete Cosine Transform* (1-D DCT) algorithm for reducing the dimensionality of feature space. The scanned document is preprocessed and segmented to create isolated numerals. Features for each numeral can be calculated after normalizing the numeral image to 32×32 size. Based on these reduced features, the numerals are classified into appropriate groups. Database of 6000 numerals size is used for the proposed work. Neural network is used for classification of numerals based on the extracted and selected features. Experimental results show accuracy observed for the method is 90.30%.

Keywords Feature extraction · Marathi · Numeral · DCT · DWT · ANN

1 Introduction

Optical character recognition (OCR) systems primarily convert the scanned document into editable text, audio, or speech format which can be stored in digital format. OCR converts scanned text/character images into editable text format. In this paper, focus is given on Marathi numeral recognition. Marathi language uses Devanagari script for writing. It is mainly spoken by most of the peoples from Maharashtra state. As Marathi is using decimal numbering system, there are ten distinct numerals from 0 to 9. The structure of Marathi numerals is different from English number system. Thus, the structural features obtained are different as compared to English, but there

M. Vaidya (✉) · Y. Joshi · M. Bhalerao · G. Pakle
Shri Guru Gobind Singhji Institute of Engineering and Technology,
Vishnupuri, Nanded, Maharashtra, India
e-mail: mvvaidya@gmail.com

© Springer Nature Singapore Pte Ltd. 2019   347
S. K. Bhatia et al. (eds.), *Advances in Computer Communication and Computational Sciences*, Advances in Intelligent Systems and Computing 760, https://doi.org/10.1007/978-981-13-0344-9_30

are some numerals which seem similar in handwritten format though it is having some differentiation in printed form. Basically, recognition systems are developed for either online or offline character or numeral recognition. Most of the researchers worked on offline OCR systems. In this paper, we have considered offline numeral database for experimentation.

DCT is used for feature space reduction. First summation vectors of normalized numeral image are calculated in horizontal, vertical, and diagonal direction (projection profile vectors). These vectors are concatenated. Also, histogram of oriented gradients (HOG) is calculated for the same numeral. DCT is applied on these two vectors: PFV and HOG. First, 25% coefficients are selected to form the vector and to form a new selected feature containing less number of dimensions.

The remaining paper is organized as follows. Section 2 describes the related work where DCT is used in research for different purposes. Section 2 describes the detailed method for recognition of Marathi numerals which consist of image acquisition, preprocessing, segmentation, and size normalization and the proposed approach for feature selection using 1-D DCT. Experimental results along with comparisons with existing methods are discussed in Sect. 3. In the last Sect. 4, we conclude the work.

1.1 About Marathi Language

Marathi language is very much popular in the state of Maharashtra (India). Nearly, 71 million people speak the Marathi language. Maharashtri was one of the Prakrit languages. Marathi is thought to be originated from Maharashtri language. Many inscriptions were found on the stones and the copper plates in the eleventh century.

The *Modi* alphabets were used to write in Marathi till twentieth century. Thereafter, Devanagari script is being used to write the Marathi language. Marathi language is having 13 vowels and 36 consonants as shown in Fig. 1. Marathi language is written using the mixture of these basic characters composed of vowels and consonants.

Character recognition in European countries has been popular area of research since 1870s. There are many scripts and languages in India but not much work has been done for the recognition of Marathi characters.

In ancient India, it was the fourth century B.C., Panini (a well-known Sanskrit linguist, grammarian, and a revered scholar) was giving his best to the Sanskrit language, in the same era Aadi-Prakrut -the language of common man was also taking a new form as Magadhi which is also referred as Magadhi of Lord Buddha era.

Main OCR research application is aiming to reduce human efforts and produce error-free system running in post office work [1], also to reduce the work in different areas like automation systems for reading numerals [2]. As India is having different scripts in different regions, recognition methods studied so far are discussed by Pal [3, 4].

Niranjan Das used topological features for classification [5]; likewise, statistical features are also used for feature extraction [6, 7]. Combination of horizontal and

अ आ इ ई उ ऊ ऋ ए ऐ ओ औ अं अः
(a) Vowels

क ख ग घ ङ
च छ ज झ ञ
ट ठ ड ढ ण
त थ द ध न
प फ ब भ म
य र ल व श
ष स ह ळ क्ष
ज्ञ

(b) Consonants

१ २ ३ ४ ५ ६ ७ ८ ९ **0**

(c) Numerals

Fig. 1 Basic Marathi characters and numbers

vertical summation vector is used to form feature vector. Work is also carried out for numeral recognition in many scripts like Bangla [8], Oriya [9], and Kannada [10]. Among all the emerging standards for image and video compression, JPEG (Joint Photography Expert Group), for compression of still images [11], MPEG (Motion Picture Expert Group), for compression of motion video are widely accepted. All above-mentioned standards employ a basic technique known as the *Discrete Cosine Transform* (DCT) [12]. The mathematical formulation by Ahmed is used in most of the literature.

In most of the comparison with the template methods, the normalization of the position and the size of the feature are carried out for comparing the features for matching purpose [13, 14], and the performance of the normalization is very important in order to get remedy from the situation for sampling in advance. Fuzzy logic also used for preprocessing and segmentation in some research [15]. Most of the clustering and classification algorithms used for recognition are elaborated in [16]. Segmentation-based approach was proposed by Shaw et al. [17]. Template matching is also very straightforward approach for recognition which is used for recognition whole word or string of digits using holistic approach [18].

On the basis of the above observations, we were motivated by the need to develop progressive matching approach on the basis of the DCT for the recognition of handwritten Marathi digits.

In our approach feature, vector is obtained by using projection profiles and histogram of oriented gradients for normalized numeral binary image of size 32×32. 1-D DCT is applied on the obtained feature vector. As per as DCT is concerned, according to energy compaction property, signal has a tendency that most of the energy portion lies in the low-frequency components. Thus, low-frequency components of all samples generated after application of DCT are further explored for recognition purpose. Thus, we can recognize an unknown numeral by finding the prototype numeral whose DCT coefficients are best fitted for that unknown numeral. The accuracy of the method depends upon the number of coefficients selected after application of DCT. Experimental results show that our approach performs well in this application domain.

2 Proposed Method

In this paper, DCT-based features considered for Devanagari numeral recognition flowchart for system are shown in Fig. 2. First, the document image scanners connected to the machine is used to acquire the document images Fig. 3.

Image enhancement techniques are used to get better quality image for enhanced features. Segmentation algorithm, bounding box method can be applied, so that interested part or block from the whole image can be separated to get required numeral. Segmented parts are then normalized to get consistency in the feature dataset in generalized format.

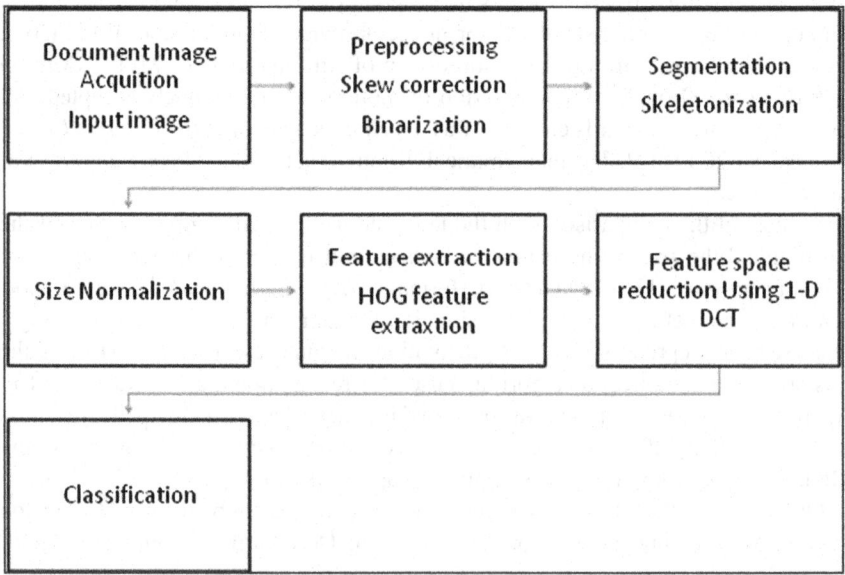

Fig. 2 System architecture for numeral recognition

Fig. 3 Sample document image

2.1 Preprocessing

Preprocessing step in recognition includes noise removal, skew correction, slant removal, and binarization. The scanned document mat contains some noise related to background which is removed using median filter. After noise removal, skew of the document is detected. Hough transform is easy and efficient for detecting and correcting the skew from the document and gives better results.

In Marathi numeral recognition, noise is mainly a very small pixel made by some dirt and writing style of individual. After receiving source document image, it has to be preprocessed. In some cases, noise reduction may remove some punctuation symbols.

Image is firstly converted into grayscale image and then to the binary format by using some threshold algorithm. The OSTU [19] method is employed to convert the grayscale image into binary image as shown in Fig. 4. The advantage of using this method is that it chooses threshold automatically to reduce the intraclass difference between black and white pixels.

The edge and the bounding box of the image are computed. There are various ways of obtaining edge of image such as Robert, Sobel, Laplace, and Canny. We have used Canny edge detection method in our system. This is because it can obtain background information more effectively.

2.2 Segmentation

Many algorithms to partition the required object from the given image are available. Internal segmentation and external segmentation are used by many researchers for

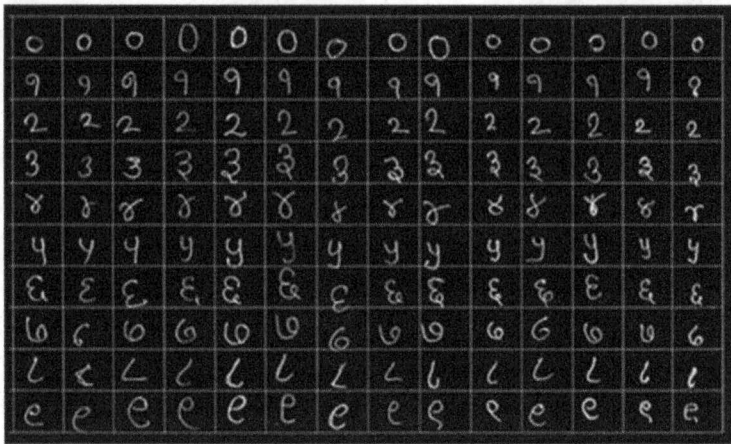

Fig. 4 Binarization of document image

creating individual sets of characters or numerals. Generally, a skewed document segmentation is difficult using projection profile method. To get better segmented characters, skew correction algorithm is applied on the document. Hough transform is globally accepted algorithm for skew correction which gives comparable results is used for removing the skew from the documents. Thus, corrected document image is processed further for segmentation.

Horizontal and vertical projection profiles are used to segment the individual numerals from the document. First, horizontal projections are calculated to separate the horizontal lines from the document image. Based on the projection, histogram minimum value is calculated to separate the lines. After getting the separated horizontal series of numbers, vertical projection profile is applied to separate the digits. Vertical projections are also scanned to find the zero or minimum value of histogram to separate the individual numerals to create the database as shown in Fig. 5.

In the process of database storage, the individual sample points indicating numerals are cropped and inserted into database. For printed number, it is a simple to recognize the numbers which requires no additional modifications in the algorithm as the size and orientation of the number is the same. But the documents having handwritten numerals, written by different persons with different writing style it becomes quite difficult to apply the same OCR system. In addition, the person with the personal writing style to make radical changes in the normal partition is very cumbersome.

These segmented numbers may vary in either shape or size. As the segmented image size changes, feature related to that number changes, which will lead to an unexpected result. To solve the problem, segmented characters are scaled to proper dimensions to get standard results in training as well as testing. Thus, any size numeric identification is possible with minimal error. Normal size for the selected basic sample has overall greater impact on system performance. Size is bigger, the

Fig. 5 Segmented
individual numerals

validation error is low, but time complexity has adverse effects and will be more. To get optimum performance, size of segmented numeral is normalized to 32×32 pixel.

2.3 Proposed Feature Extraction Selection

The binary numeral image with normal size of 32×32 is used for generating summation vectors in horizontal, vertical, and diagonal directions.

$$H(j) = \sum_{i=1}^{n} I(i, j) \forall j = 1, 2, 3 \ldots n \tag{1}$$

$$V(i) = \sum_{j=1}^{n} I(i, j) \forall i = 1, 2, 3 \ldots n \tag{2}$$

$$D(i) = \sum_{j=1}^{2n-1} I(i, j) \forall i = 1, 2, 3 \ldots j \tag{3}$$

$$D'(j) = \sum_{i=1}^{2n-1} I(j, i) \forall j = 1, 2, 3 \ldots i \tag{4}$$

Pixels in horizontal vector, vertical vector, and diagonally scanned vectors are calculated. These vectors obtained $H(j)$, $V(i)$, $D(i)$, and $D'(j)$ are used for further processing.

Histogram of oriented gradients (HOG) vector is calculated for the normalized numeral image. Cell of 3×3 is considered and moved throughout the image. The histogram of oriented gradients (HOG) is a method to represent the shape and local portions of objects in the image which can be described by using oriented gradients of intensity or by the edge directions.

Two features mainly strength and angle of gradient are calculated. The direction of gradient is calculated as given in Eq. 6. These directions are quantized into 32 directions, and each direction is accumulated with strength of gradient.

$$\text{Strength of gradient}: g(u, v) = \sqrt{(\Delta x)^2 + (\Delta v)^2} \tag{5}$$

Angle of gradient is

$$\theta(u, v) = \tan^{-1} \frac{\Delta y}{\Delta x} \tag{6}$$

where

$$\Delta x = f(x + 1, y + 1) - f(x, y)$$
$$\Delta y = f(x + 1, y) - f(x, y + 1) \tag{7}$$

$f(x, y)$ is intensity value at point (x, y).

After calculation of these two features, a separate feature called as histogram of oriented gradient (HOG) is obtained. HOG is a feature descriptor which counts total number of occurrences of gradient; i.e., depending upon the direction of pixel intensity, it counts intensity value. A feature vector of 1×392 is obtained. The above features calculated using summation vector technique are combined with HOG features to form a final feature vector of dimension 582.

Thus, all features calculated using above two methods are concatenated into one vector. After getting this vector, 1-D DCT is applied to get transformed coefficients for the same vector. Only first 25% of the coefficients from the transformed vector are selected. Based on these features, numerals are classified by comparing the respective features and calculating minimum distance. In this paper, only forward 1-D DCT is used to extract the features vector using. In most of the research work, inverse DCT is used for further processing, but in the presented work, only forward DCT is sufficient to create the feature vector.

The equation for 1-D DCT [22] is defined by the following equation

$$F(u) = \left(\frac{2}{N}\right)^{\frac{1}{2}} \sum_{I=0}^{N-1} A(i) \cdot \cos\left[\frac{\pi \cdot u}{2 \cdot N}(2i + 1)\right] f(i) \tag{8}$$

where $A(i) = \begin{cases} \frac{1}{\sqrt{2}} & for\ \xi = 0 \\ 1 & otherwise \end{cases}$

2.4 Classification

Classification deals with the sorting or separating the patterns having similar characteristics. In OCR, different types of structural and statistical features are used to classify the numerals or characters. In literature, many algorithms are used for classification like decision trees, Bayesian network. Based on some distance criteria, also the patterns can be classified.

The multilayer networks to solve the problem of classification by using hidden layers, including neurons which are not directly connected to the output layer. The Perceptron composed of one or several layers of artificial neurons, the entries, are directly fed to the outputs via a series of weight. In this way, it can be considered to be the most simple type of the feedforward network. The multilayer networks are used to overcome many of the limitation of the single layer network. The capacity of networks multi-layers of the nonlinearity used with the units. Each neuron in the network receives the data from other neurons in the network or receives inputs from the outside. The outputs of the input or hidden layer neurons are connected to other neurons. Each input is fed to the neurons using a weight. The weighted sum of the inputs is calculated by the neuron which is passed through a transfer function with nonlinear property to produce the actual output for that neuron.

We utilized a feedforward neural network system that maps sets of features to recognize the samples in MATLAB. This system is utilized with three layers including input layer, output layer, and hidden layers for two unique sorts of element set. Output of one layer neuron is fed forward to the next layer. Every hidden layer has corresponding weight $w0$ to be adjusted. In ANN, every node in the system is neuron with a nonlinear actuation function. In this procedure, supervised learning is utilized. Input layer is used to feed the raw data without any processing. Every hidden node determines the weighted sum of corresponding inputs and manipulates the output of the node by using threshold function which can be either a step function or a sigmoid function. Output node is computed based on type of problem given to neural network as shown in Fig. 6.

Each neuron in the proposed network is formed separately for fifty classes out of hundreds. Linear decision surface is defined by the weight vector of each neuron. The rule of learning plays a very important role for the behavior of linear. Thus, for nonlinearly separable samples, the generalized Delta rule minimizes a root mean squared error and converges to unique solution. The generalized Delta rule using sigmoid function guarantee about finding a proper hyperplane, if hyperplane exists. There are no local minima in the MSE criterion, by initializing the weights to zero and choosing the learning rate carefully. In the case of generalized Delta rule, this hyperplane is positioned in such a way as to maximize the distance for the samples

Fig. 6 Neural network for classification

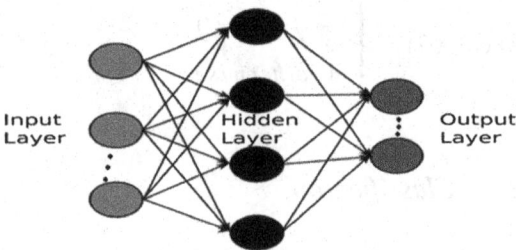

of the class. The free parameter of the training procedure, rate of learning, is fixed permanently before training. At the time of the formation of neurons on the problem of the recognition of characters, author has not observed in the effect of the overspecialization of the classifier for the training set.

In the classification of numerals, the feature vector generated is fed to the neural network. The neural network consists of ten nodes in the output layer representing ten classes of ten numerals. In training phase, the weights are adjusted to get required class output for the numeral. Thus, trained neural network is used for testing purpose. In the experimentation, half of the numerals are used for training and remaining half numeral database is used for testing or recognition purpose.

3 Results

Training set of the first 50% normalized samples in a dataset is used for classification using artificial neural network. In the training, weights are adjusted to get specified output. Thus using this data and trained ANN, the numeral features from testing data are fed to the ANN. Classification rate of individual Marathi numeral from zero to nine is calculated the using ANN for all samples. After experimentation, it is found that minimum accuracy is found for numeral 3 and the maximum accuracy is investigated for numeral 4.

Thus, the average classification rate of 90.30% is observed on the test set. As per as time complexity is concerned, 25% features are selected in final stage for classification and recognition. The number of comparisons required for finding nearest match depends upon number of features. If the selected number of features is less, then the number of comparisons is also less. For large dataset having 30,000 samples, it will have great impact on time complexity. This gain in the time complexity is due to DCT application.

Comparison with existing approaches

Lot of work is done previously by researchers in this field for classification of numerals and characters in Devanagari script along with non-Indian scripts. In this paper, the author compared some of the methods where discrete cosine transform is used for feature extraction or feature selection. Table 1 shows the comparison of accuracy

Table 1 Comparison with existing methods

Sr. No.	Proposed by	Methodology	Image size	Classifier	Accuracy %
1	Birajdar and Subhedar [21]	JPEG DCT	60 × 60	Euclidean distance	55.00
2	Lwin [24]	DWT	32 × 32	ANN Back propagation	71.64
3	Sethy and Patra [25]	DCT	20 × 20	ANN, MSE	80.40
4	Mishra et al. [23]	DCT	256 × 256	BPNN	84.60
5	Proposed method	1-D DCT	32 × 32	ANN	90.30

of proposed approach with existing transform-based approaches. JPEG algorithm for image compression is used by [20]. All steps for JPEG are used including quantization which requires large computational time for extracting the feature vector. It is obvious that larger the resolution of image greater will be the details. By using this property, [21] proposed better accuracy for same JPEG-based approach. After application of 2-D DCT on input image, zonal DCT coefficients are selected for classification [22]. The size of input image [23] used is also very large as compared to our proposed method. DWT is also used for feature extraction but the accuracy found using this transform is not comparable. Sethy [25] used very small size of input image as 20 × 20 but the accuracy obtained is for the method is very less .

4 Conclusion

In the proposed approach, few components having more information are selected for feature vector generation which reduces the number of samples in the vector in turn the number of comparisons required is also reduced. The overall complexity of the system is the sum of complexity required for recognition as well as the time required to calculate the features. Average classification rate for the proposed approach is found to be 90.30% on given samples in the test dataset.

In future, other feature extraction techniques can be applied for achieving better accuracy. Also, more number of features lead to better result, so two or more approaches can be combined for further exploration. The validation process also plays very important role as better segmentation methods which can be employed to separate the numerals without any overlap or without any noise. The paper discusses the system which is developed for dataset of Marathi handwritten numerals only. In future, it can be extended to Devanagari characters or conjugate character recognition. In this approach, database is created and then offline recognition is done. Thus, the development can be further extended to the online recognition system.

Acknowledgements The authors acknowledge their thanks to Umapada Pal, Indian Institute of Statistics, Kolkata, and Vikas Dongre for providing the databases support for the experimentation.

References

1. Roy, K., Vaidya, S., Pal, U., Chaudhuri, B.B., Belaid, A.: A system for indian postal automation. In: Proceedings of the 8th International Conference Document Analysis and Recognition, Seoul, Korea, pp. 1060–1064 (2005)
2. Wen, Y., Lu, Y., Shi, P.F.: Handwritten Bangla numeral recognition system and its application to postal automation. Pattern Recogn. **40**(1), 99–107 (2007)
3. Pal, U., Chaudhari, B.: Indian script character recognition: a survey. Pattern Recogn. **37**, 1887–1899 (2004)
4. Jayadevan R., Kolhe, S.R., Patil, P.M, Pal, U.: Offline recognition of devanagari script: a survey. IEEE Trans. Syst. Man, and Cybern. Part C (Appl. Rev.) **41**(6), 782–796 (2011)
5. Das, N., Reddy, J.M., Sarkar, R., Basu, S., Kundu, M., Nasipuri, M., Basu, D.K.: A statistical–topological feature combination for recognition of handwritten numerals. J. Appl. Soft Comput. **12**, 2486–2495 (2012)
6. Vaidya, M., Joshi, Y.V.: Marathi Numeral Recognition using statistical distribution features. In: Proceedings of the IEEE Conference on Information Processing, pp. 586–591 (2015)
7. Vaidya, M., Joshi, Y.V: Handwritten numeral identification system using pixel level distribution features. In: Proceedings of the 2nd International Conference on Information and Communication Technology for Intelligent Systems, vol. 2, pp. 307–315 (2017)
8. Pal, U., Chaudhuri, B.B.: Automatic recognition of unconstrained off-line Bangla handwritten numerals. In: Advances in Multimodal Interfaces—ICMI 2000, pp. 371–378. Springer, Berlin, Heidelberg (2000)
9. Tripathy, N., Panda, M., Pal, U.: System for Oriya handwritten numeral recognition. In: Proceedings of the Imaging International Society for Optics and Photonics, pp. 174–181 (2003)
10. Rajput G., Hangarge, M.: Recognition of isolated handwritten Kannada numerals based on image fusion method. Pattern Recogn. Mach. Intell. 153–160 (2007)
11. Fan, Z., Queiroz, R.: Maximum likelihood estimation of JPEG quantization table in the identification of bitmap compression history. In: Proceedings of the IEEE International Conference on of Image Processing 2000, vol. 1, pp. 948–951 (2000)
12. Ahmed, N., Natarajan, T., Rao, K.R.: Discrete cosine transform. IEEE Trans. Comput. **C-25**, 90–93 (1974)
13. Plamondon, R., Srihari, S.N.: On-line and off-line hand-writing recognition: a comprehensive survey. IEEE Trans. Pattern Anal. Mach. Intell. **22**(1), 63–84 (2000)
14. Bhattacharya, U., Chaudhuri, B.B.: Handwritten numeral databases of Indian scripts and multistage recognition of mixed numerals. IEEE Trans. Pattern Recogn. Mach. Intell. **31**(3), 444–457 (2009)
15. Garain, U., Chaudhuri, B.B.: Segmentation of touching characters in printed devnagari and bangla scripts using fuzzy multifactorial analysis. IEEE Trans. Syst. Man Cybern. Part C Appl. Rev. vol. 32, no. 4, pp. 449–459, 2002
16. Han, J., Pei, J., Kamber, M.: Data Mining: Concepts and Techniques. Elsevier (2011)
17. Shaw, B., Parui, S.K., Shridhar, M.: Offline handwritten Devanagari word recognition: a segmentation based approach. In: Proceedings of the International Conference on Pattern Recognition (ICPR), Tampa, Florida, USA, pp. 1–4 (2008)
18. Lu, Z., Chi, Z., Siu, W.: Extraction and optimization of B-spline PBD templates for recognition of connected handwritten digit strings. IEEE Trans. Pattern Anal. Mach. Intell. **24**(1), 132–139 (2002)
19. Otsu, N.: A threshold selection method from gray-level histograms. IEEE Trans. Syst. Man Cybern. **9**(1), 62–66 (1979)

20. Aburas A., Rehiel, S.A.: JPEG for Arabic handwritten character recognition: add a dimension of application. In: Advances in Robotics, Automation and Control, InTech, pp. 21–32 (2008)
21. Birajdar, G., Subhedar, M.: Use of JPEG algorithm in handwritten Devanagari numeral recognition. Int. J. Distrib. Parallel Syst. 2(4), 152–160 (2011)
22. Reddy, G.S., Sharma, P., Prasanna, S.R.M., Mahanta, C., Sharma, L.N.: Combined online and offline assamese handwritten numeral recognizer. In: Proceedings of the IEEE National Conference on Communications, pp. 1–5 (2012)
23. Mishra, T.K., Majhi, B., Panda. S.: A comparative analysis of image transformations for handwritten Odia numeral recognition. In: Proceedings of the IEEE International Conference on Advances in Computing, Communications and Informatics (2013)
24. Lwin, T.N., Soe, T.: Comparison of handwriting characters accuracy using different feature extraction methods. Int. J. Sci. Eng. Technol. Res. 3(6), 1027–1032 (2014)
25. Sethy, A., Patra, P.K.: Off-line Odia handwritten numeral recognition using neural network: a comparative analysis. In: Proceedings of the IEEE International Conference on Computing, Communication and Automation, pp. 1099–1103 (2016)

Tracking System for Driving Assistance with the Faster R-CNN

Kai Yang, Chuang Zhang and Ming Wu

Abstract The vehicle detection and tracking in driving assistance system are ordinarily achieved by the optical or radar technology. In this work, we explore video processing for driving assistance system. An object's detection and tracking system based on the Faster R-CNN and Camshift algorithm is proposed, and Kalman filtering algorithm is used to predict the position of objects. Our system differs with other object-tracking algorithms based on deep learning, which directly detects each frame and has no tracking part, such as YOLO and SSD. We also introduce an evaluation criterion for driving assistance system according to the dangerous level of the surrounding cars. Through the experiments on the videos recorded by drive recorder, we show that our approach can achieve better performance in complex scenes than YOLO or SSD and can satisfy the real-time requirement by processing 30 frames per second.

Keywords Object detection and tracking · Driving assistance · Faster R-CNN Camshift

1 Introduction

With the development of technology, the car is becoming more and more important in our life. Those existing driving assistance system, mainly using the Tesla radar driving assistance system or the remote sensing system based on LiDAR technology. In recent years, the rapid development of deep learning, as well as in-depth study

K. Yang · C. Zhang (✉) · M. Wu
Pattern Recognition and Intelligent System Lab, School of Information and Communication Engineering, Beijing University of Posts and Telecommunications, Beijing, China
e-mail: zhangchuang@bupt.edu.cn

K. Yang
e-mail: dangercane@gmail.com

M. Wu
e-mail: wuming@bupt.edu.cn

© Springer Nature Singapore Pte Ltd. 2019
S. K. Bhatia et al. (eds.), *Advances in Computer Communication and Computational Sciences*, Advances in Intelligent Systems and Computing 760, https://doi.org/10.1007/978-981-13-0344-9_31

of artificial intelligence, makes vehicles through software, or some algorithms to perceive the external environment becomes possible. Therefore, using deep learning to make the driving assistance system safer has significant research significance.

In the field of vehicle detection and tracking, Kostia Robert has proposed a new research direction based on hierarchical feature detection as a detection method [1]. With the further maturation of image processing and pattern recognition, some research algorithms based on image texture feature are proposed. M. Boumediene, A. Ouriri, and Keche use the vehicle's edge features (horizontal and vertical edges) to carry out the vehicle detection algorithm [2]. The main purpose of the study was to use it on a vehicle-assisted driving system, where they focused on applying Hough transforms to uninterrupted video streams. In the target tracking, Anton Milan proposed a RNN network based on the online multi-target tracking system [3]. It is different from the previous and is involved a variety of complex models. It provides a complete end-to-end learning based on depth learning and solves a variety of problems in multi-target tracking, such as the number of unknown time-varying targets, the combination of continuous state estimates, and discrete data. Jiapeng Wu, Zhaoxuan Yang, Jun Wu proposed a method using the number of virtual coil brightness and color to detect the vehicle algorithm [4]. However, the multi-objective feature fusion, no matter from which direction, cannot be a complete description of a vehicle. There are too many problems to be considered in the target tracking such as the speed of recognition. It is unnecessary to complete the estimation of the continuous state if each frame can complete the object detection.

Based on the reasons above, we consider starting from the deep learning to solve the problem of object detection and tracking for driving assistance. We use Faster R-CNN for object detection, it can achieve very high accuracy in car detection and still has a good speed. Combined with the Camshift algorithm, with the output of Faster R-CNN as input, which can complete the real-time tracking requirements of drive recorder video. Then, we use Kalman filtering algorithm to improve tracking system. Finally, we propose a state-of-the-art, real-time object detection and tracking system for driving assistance.

2 Object Detection and Tracking System

The traditional target tracking algorithm finds the characteristics of moving objects through the frame changes, then detects and locates the object. The traditional tracking method is widely used, but it cannot distinguish the moving background. Therefore, we need to consider a new method to detect and tracking for driving assistance system.

Deep learning in object detection greatly improves the detection result. It can maintain good efficiency. Currently, method for tracking object mostly uses video as a continuous image which directly detects each frame. It has two shortcomings, first is the speed. The network must reduce the accuracy of object detection in order to balance the speed. The second is the result of each frame which is not exactly the

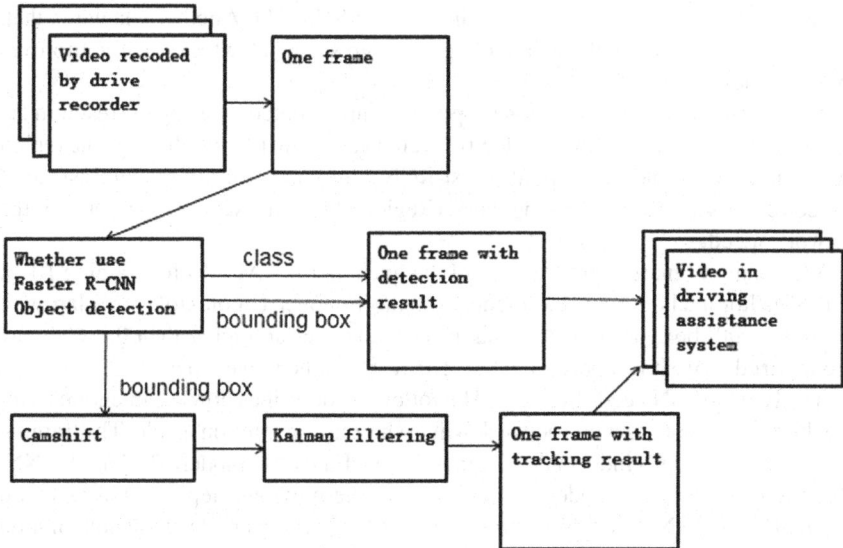

Fig. 1 Tracking system based on the Camshift + Kalman filtering algorithm

same. It will make tracking instability. Therefore, the Camshift tracking algorithm can effectively improve the tracking efficiency. We do not need to detect for each frame. We can only detect one frame, and the other frame uses the detection result as input through tracking algorithm, which can offer an easy tradeoff between speed and accuracy.

The object detection network is Faster R-CNN. Although the algorithm efficiency is not good, less than SSD and YOLO, but the object detection accuracy is the highest. For the complementarity of the algorithm, use tracking system based on the Camshift + Kalman filtering algorithm. This method is simple and efficient and is very suitable for tracking the vehicle object on the road. The system structure is shown in Fig. 1.

2.1 Faster R-CNN

The concept of R-CNN uses the concept of CNN into the field of object detection [5]. It greatly improves the object detection. The next research, including Fast R-CNN, Faster R-CNN, has been effectively optimized. R-CNN propose a method called regional proposal (RP), using selective search. On the other hand, R-CNN extracts features in depth and performs feature integration at high levels to make the results more accurate. Using ImageNet ILSVC 2012 and ASCAL VOC 2007. In the Fast R-CNN, the entire picture is convoluted, does the RP on the feature map, then is unified through the FC layer, finally through the SVM [6]. Another change is that, the SVM is combined with the bounding box regression [7], which greatly improves the efficiency.

In the Faster R-CNN, the region proposal is implemented by using the result in the convolution layer, instead of the selective search algorithm [8]. Finally, lead the region proposal network integrate with the Fast-RCNN into one network model. Instead of the selective search, the RPN network (Region Proposal Networks) improves the network speed.

We assign a binary label class to all the anchors (N * M * k) for training RPN. If the anchor/anchors with the highest Intersectionover-Union (IoU) overlap with a ground-truth box, or an anchor that has an IoU overlap higher than 0.7 with any ground-truth box. It is a positive label. The rest, as a negative label.

The Fast R-CNN needs training. The following describes how we integrate training this two network, get one network which sharing convolution result. The training steps is, RPN requires independent training, using ImageNet model. The Fast-RCNN, also use the ImageNet model to train, and get the previous step's output as input, get the Fast-RCNN network's output, take out the parameters, lead the convolution layer learning rate to zero, and then train a new RPN. Fixed two networks' shared convolution layer, then, train the Fast-RCNN remaining untrained part, get a new network without fine-tuning. Finally, fine-tuning Fast-RCNN network, get a RPN + Fast-RCNN network.

2.2 Tracking Algorithm Based on Camshift + Kalman

Camshift algorithm is a method, which leads image color histogram transform into color probability distribution, according to the previous frame information, adaptive adjustment of the size and position of the search box, complete the target tracking algorithm in video. Camshift is very effective in deformation target, or part of blocking. On the other hand, the Camshift algorithm is relatively simple, the time complexity is good, and it is especially suitable for driving assistance.

In order to enhance the stability of the tracking system. Object tracking meet problem when the background color is more complicated or the object color is similar with background. Kalman filtering is added when the target is lost. The core of Kalman filtering is iterated along the function gradient direction, and then converges to a range according to the number of iterations and the error threshold. This range is the position of the object in the next frame [9]. The purpose of optimizing the Camshift tracking algorithm is using past positions to predict the future location of the object [10]. The Kalman filtering algorithm can predict the position of the object by using the past positions. To overcome the shortcomings of similar background interference.

Camshift combined with Kalman filtering algorithm, the specific process is as follows: First, the object in the previous frame, using Camshift algorithm, according to the color probability distribution, predict the object in this frame. Then set the threshold; if the center and centroid's distance is too large or the new search box cannot be predicted, the Kalman filtering algorithm is used to perform the gradient estimation, whether the object is existing or not. If the two decisions are consistent, the position estimated by the Camshift algorithm is used as the observation and updated the state of the Kalman filtering. If not, the result of the Kalman filtering algorithm is executed.

However, there are two serious cases, the first is object blocking, as driving assistance system, when the occurrence of such problem exists, this is not a big threat, cannot tracking it is ok, so ignored. We need to deal with the second case, when the object color similar to the background, because the road situation in a very short period, such as fps is 20, the gap between each frame is only 50 ms, the object movement should be slow, and size changes should be gradual. If it appears this,

$$\text{Dect\_rect} > \alpha \times \text{dect\_last\_rect}, \alpha \text{ is a constant coefficient} \qquad (1)$$

Can be identified that the Camshift algorithm meets a problem. At this time according to the previous settings, should still use Kalman filtering algorithm to update the object state.

3 Evaluation Standard

In order to compare with other methods, we need an evaluation standard. We have consulted a number of existing evaluation standard. Most of them are based on accuracy which is not suitable for us. So we propose our own evaluation standard, which has two part, for detection cars and for tracking cars. We use this evaluation standard to compare several methods, including our own method. Using the score to visually demonstrate the effectiveness of the methods.

3.1 Evaluation Standard for Detection Cars

The car on the road needs real-time, stable, and accurate tracking system. It means on the one hand, a detection network requires a high degree of accuracy; on the other hand, it must ensure algorithm efficiency. Except these two requirements, adapt to changing environment, such as cloudy days and the accuracy of the algorithm in multi-target situations also need us to consider. While we were driving on the road, many car running beside us, also many car parking on the roadside. Whether can we identify as many cars as possible, is the most important problem we need to solve. And it is the most efficient way to show an object tracking system performance. Based on the analysis above, we define the following indicators.

For the target recognition, the size of the picture is fixed to 1080p, which is the general drive recorder camera configuration. Comparing time cost per frame and the number of effective tracking objects. The network we compare is YOLO and SSD [11].

3.2 *Evaluation Standard for Tracking Cars*

For the multi-target tracking of driving assistance, the main purpose is warning system, prevent the car from road accidents. That is, the nearest target needs highest priority. To sum up, on the basis of ensuring the speed of the tracking system, comparing two aspects, one is the sudden emergence of the target. The longer the target does not capture, the greater the loss increase. The second is, the closer the target is, the greater the importance of tracking, and the greater the loss increase without capture, so the loss e can be defined as follows:

$$e = \sum_{0}^{t} \sum_{obj_1}^{obj_n} P_i * L_{cost}(i) \tag{2}$$

Here, t represents the time, P_i means whether the object is detected, object detected set 0, did not set 1. L_{cost} means loss at different locations; the closer the system is, the greater the value.

4 Experiments

In this chapter, we will use experiment to verify the object detection and tracking effect of the driving assistance system we proposed. We use PYTHON and OpenCV libraries to write the algorithm on the computer with GTX1080. The test set we used are collected from the drive recorder, including the daytime with high and low brightness. The position of the video, including main road and side road, where main road is always in heavy traffic. It can effectively reflect the detection and tracking effect in the real driving assistance system. The experiment will be carried out from the two aspects mentioned above, object detection and target tracking. The object detection evaluation will use the detected number and the average time cost per frame. The target tracking will be mainly evaluated by the loss e. In this experiment, we will compare the YOLO, SSD, and our own method proposed in this paper to see the effect of the algorithm [12]. We use the YOLO and SSD for alignment because these methods are highly expensive in tracking and have a stable r.

Fig. 2 Faster R-CNN can detect object accurate and in real-time

Table 1 Detection frameworks on PASCAL VOC 2007

	Train	mAP	FPS
Faster R-CNN	2007 + 2012	76.1	7
YOLO	2007 + 2012	62	42
SSD	2007 + 2012	74.2	19

4.1 Object Detection

For the selection of the objection detection method, the previous section has been discussed, not only need the accuracy, so that when target be blocked, or suddenly appear in the video, it can detect as soon as possible. The speed requirements are also necessary. The effect of objection detection is shown in Fig. 2.

Here only show the complex case of the object detection results, the object is car, my complex situation here can be seen from the figure, it is at a crossroads where many vehicles on it, and on the roadside, there are a lot of car parked. Moreover, at this moment, it is in the winter evening the weather is cloudy and the brightness is very low. Which can be seen, for complex environments, low brightness the multi-object detection system can get a good result. Due to this result, there is no need to test other environment which is easy to detect. Then compared with YOLO and SSD, using mAP (mean average precision) on VOC2007, the results shown in Table 1.

It can be seen from the above table, in speed, although the Faster R-CNN has no advantage, but in the driving assistance object detection and tracking system, within every second, only a few frames needs detection, other frame can be based on the objection detection results, access to the target tracking algorithm. Therefore, in terms of speed, all in real-time can reach 30 frames per second. From the object detection classification effect, compared with the other two algorithms, Faster R-CNN has a better effect and can better complete the complex environment for multi-target tracking task.

Table 2 Speed and effect in different detect tracking ratio

γ	FPS	Loss e
1	11.4	0.125
3	21.6	0.131
4	27.9	0.157
5	32.3	0.172
6	35.3	0.224
10	51.9	0.371
20	73.5	0.679

Fig. 3 Driving assistance system we proposed can detect and track object accuratly and in real-time

4.2 Target Tracking

As discussed above, since the target tracking algorithm used here is relatively simple and the algorithm is efficient. The time of tracking one frame is very short. Therefore, the ratio of the target tracking frame with the object detection frame is important. The test set video is recorded by a drive recorder, and labeled by us. The results are shown in Table 2.

When the ratio is small, due to the tracking algorithm on the time constraints, resulting in tracking speed is not high. When the ratio exceeds a certain value, the loss function rises faster, resulting in tracking effect decreased. So here, we choose a balance of tracking rate and tracking effect, set the ratio 5 or 6. Because drive recorders are mainly 30FPS. In the next test, the ratio is five. The result is shown in Fig. 3, which also shows only the results of a complex scene, did not show the simple environment, the reason is the same as above:

Table 3 Tracking loss in different environment

Object number	Brightness	F+CK	SSD	YOLO
1–3	High	0.006	0.006	0.009
1–3	Low	0.008	0.007	0.013
4–10+	High	0.169	0.214	0.467
4–10+	Low	0.175	0.221	0.556

From the figure, it can ensure real-time tracking. When the driving assistance system in a relatively low brightness, the road is complex, including roadside parking, cars running on the road, it still able to maintain a good effect. All the nearby cars are effectively detected and tracked.

In this paper, we will compare three algorithms of multi-target tracking in the same environment. Compared with the object detection and tracking algorithm based on Faster R-CNN and Camshift + Kalman filtering. SSD and YOLO are two fast and outstanding object detection algorithm. The number of pictures can be detected per second can reach more than 30, so the three algorithms can complete meet requirements of real-time tracking. Drive recorder records the test video set. The results are shown in Table 3.

Here, we can see that the three methods almost the same when the environment is relatively simple, the loss function is very small. It means all the important objects are tracked. When the brightness is not good, the loss function has risen, still in a small range, which is due to the deep learning network extract the deep level of the object feature. Finally, the loss function has a large increase when the road conditions are complex, whether the brightness is high or low. There are many objects which need tracking. Because of driving assistance system is mainly serves car warning, therefore, the loss function can be a very good way to see the algorithm effect, for the need of driving assistance. From the results, the object detection and tracking algorithm based on Faster R-CNN and Camshift + Kalman filtering can maintain a good tracking effect in complex cases. Although some objects, which are in a distance or blocked by other object, still cannot be detected or tracked. However, for the entire driving assistance system, too far from the car is not important, can be ignored.

5 Conclusion

Our method is different from the traditional tracking method and also different from YOLO or SSD. We aimed to improve the accuracy of the object detection and ensure real-time tracking. We propose a target tracking method based on the Faster R-CNN and Camshift + Kalman filtering. After the experiments, it shows that it can effectively improve the recognition rate of the cars in driving assistance system. At the same time, the speed is as good as YOLO and other high-efficiency algorithm.

We proposed a new way for driving assistance which uses the risk of surrounding car to compare the object detection and tracking method, and still need to continue to explore and improve the model; we will continue to carry out research on this issue.

References

1. Robert, K.: Video-based traffic monitoring at day and night vehicle features detection tracking. In: International IEEE Conference on Intelligent Transportation Systems, pp. 1–6. IEEE (2009)
2. Alshaqaqi, B., Baquhaizel, A.S., Ouis, M.E.A., et al.: Driver drowsiness detection system. In: International Workshop on Systems, Signal Processing and Their Applications, pp. 151–155. IEEE (2013)
3. Milan, A., Rezatofighi, S.H., Dick, A., et al.: Online multi-target tracking using recurrent neural networks (2016)
4. Jiapeng, W.U., Yang, Z., Jun, W.U., et al.: Virtual line group based video vehicle detection algorithm utilizing both luminance and chrominance. In: IEEE Conference on Industrial Electronics and Applications, pp. 2854–2858. IEEE (2007)
5. Girshick, R., Donahue, J., Darrell, T., et al.: Rich feature hierarchies for accurate object detection and semantic segmentation. Comput. Sci. 580–587 (2014)
6. He, K., Zhang, X., Ren, S., et al.: Spatial pyramid pooling in deep convolutional networks for visual recognition. IEEE Trans. Pattern Anal. Mach. Intell. **37**(9), 1904–1916 (2015)
7. Girshick, R.: Fast R-CNN. Comput. Sci. (2015)
8. Ren, S., He, K., Girshick, R., et al.: Faster R-CNN: towards real-time object detection with region proposal networks. IEEE Trans. Pattern Anal. Mach. Intell. (99), 1–1 (2015)
9. Fang, Q., Mei, X.C., Zhang, Y.P.: Design of Kalman filter for maneuvering target track. Radar Sci. Technol. (2006)
10. Qu, J.B., Qu, C.R., Wang, S.J.: Multi-objective forecast track based on Kalman-Camshift algorithm. Adv. Mater. Res. **211–212**, 137–141 (2011)
11. Liu, W., Anguelov, D., Erhan, D., et al.: SSD: Single Shot MultiBox Detector (2016)
12. Redmon, J., Divvala, S., Girshick, R., et al.: You only look once: unified, real-time object detection, pp. 779–788 (2016)

Tropical Cyclone Hazardous Area Forecasting Using Self-adaptive Climatology and Persistence Model

Akara Prayote and Arthit Buranasing

Abstract A tropical cyclone is one of the natural disaster which is the most economical and human losses in the world and tends to be more damaging and more frequent in the future due to climate change and human behaviors. To mitigate catastrophic phenomenon, a modern natural disaster management model (MNDM) is designed and the most important phase in MNDM is emphasis on process before the catastrophic phenomenon or preparing tropical cyclone track forecasting, intensity forecasting, and risk area identification. Although various tropical cyclone track and intensity forecasting techniques have been developing and improving for several years, the errors of the forecasting model still remain. Moreover, risk area assessment and uncertainty of the major model which are the most important phase in MNDM are excluded. To address these problems, this paper proposes an integrated short-range tropical cyclone hazardous area forecasting system which includes both track and hazardous area forecasting in system by using only 12 features which were extracted from satellite images with improvement of the traditional statistical methods. The performance of the model is satisfactory; the average error from the experimental results of R_{34}, R_{50}, and R_{64} forecasting with unknown tropical cyclone data between years 2013–2015 on Mercator projection map is lower than traditional techniques by 28.99%, 22.81%, and 24.38%, respectively.

Keywords Risk area assessment · Natural disasters management · Tropical cyclone track forecasting · Remote sensing · Decision support system

A. Prayote (✉) · A. Buranasing
Department of Computer and Information Science, King Mongkut's
University of Technology North Bangkok, Bangkok, Thailand
e-mail: akara.p@sci.kmutnb.ac.th

A. Buranasing
e-mail: arthit.bur@hotmail.com

© Springer Nature Singapore Pte Ltd. 2019
S. K. Bhatia et al. (eds.), *Advances in Computer Communication
and Computational Sciences*, Advances in Intelligent Systems
and Computing 760, https://doi.org/10.1007/978-981-13-0344-9_32

371

1 Introduction

Today, the weather and climate on the earth is rapidly changing and causing of various severe natural disasters such as various kinds of storms, volcano eruptions, earthquakes, tsunamis, floods, droughts, fires [1]. According to the report of ECMWF [2], there were many severe natural disasters around the world and these were the main causes of numerous death and injury to humans, damage or loss of properties, national economic and ecological losses, etc. Especially, the tropical cyclone (TC) is one the most catastrophic phenomenon of natural disasters on the earth because it often causes damaging winds, torrential rainfall, flooding, etc. [3].

According to the report of TMD [4]. Since 1951–2010, there are many severe TC hit in Thailand, for example, Typhoon Harriet, Typhoon Gay, Typhoon Linda and these cause of death or injury of humans, damage or loss of valuable good, and national economic. To mitigate catastrophic phenomenon, a MNDM model is designed [5]. The model consists of four phase approach, i.e., mitigation, preparedness, response, and recovery. Nevertheless, the most important phase in MNDM is the process before the catastrophic phenomenon due to the efficiency losses deduction is risk area assessment [6], i.e., preparing TC track/intensity forecasting and risk area identification.

So far, there are many techniques for TC forecasting but these can be grouped into three main classes of forecasting models. First, statistical models. Second, dynamical models (numerical weather prediction models—NWP model), and third models that use elements of both approaches are called statistical-dynamical models [7]. In Thailand, there are two primary techniques in operation. First, the statistical method which is a conventional method used in TMD [8]. Another method is dynamical model which runs on a WRF software (Weather Research and Forecasting [9]). The WRF model requires various meteorological features of which Thailand lacks. As the result, TC track forecasting still have high errors [10].

TC track and intensity forecasting require various meteorological features for prediction. To provide meteorological data to the NWP model is a high volume in various measure equipment investment/maintenance and weather observation. Although recently, Buranasing and Prayote [11, 12] developed short-range (6–24 h) economical track–intensity techniques but the model is excluded risk area assessment which will be effected by TC. On the other hand, this paper suggests an economical alternative solution for TC track, intensity, and risk area forecasting which gives satisfactory up to 24 h forecasting by using only satellite images data for analysis.

This paper is organized into the following sections: Experimental Area and Historical Tropical Cyclone Data in Sect. 2. Tropical Cyclone Identification-Analysis and Track Forecasting in Sect. 3. Tropical Cyclone Hazardous Area Self-Adaptive Forecasting in Sect. 4. Performance of Tropical Cyclone Hazardous Area Self-Adaptive Forecasting in Sect. 5. Geological Hazardous Area Graphic Display in Sect. 6, and a conclusion and remark are drawn in Sect. 7.

Fig. 1 Example of TC
historical data from JTWC

WP, 27, 2015111718, , BEST, 0,
52N, 1583E, 45, 989, TS, 34,
NEQ, 55, 65, 65, 55, 1009, 170,
45, 0, 0, W, 0, , 0, 0, IN-
FA, M,

WP, 27, 2015111800, , BEST, 0,
55N, 1570E, 55, 982, TS, 34,
NEQ, 75, 65, 50, 75, 1009, 112,
45, 0, 0, W, 0, , 0, 0, IN-
FA, M,

WP, 27, 2015111800, , BEST, 0,
55N, 1570E, 55, 982, TS, 50,
NEQ, 35, 35, 35, 35, 1009, 112,
45, 0, 0, W, 0, , 0, 0, IN-
FA, M,

Fig. 2 MTSAT-1R/MTSAT-
2 satellite
properties

Channel	Wavelength (micrometer)
VS	0.55 - 0.90
IR1	10.3 - 11.3
IR2	11.5 - 12.5
IR3	6.5 - 7.0
IR4	3.5 - 4.0

2 Experimental Area and Historical Tropical Cyclone Data

The experiment area in this paper is between latitudes 70°N to 20°S and longitudes 70°E to 160°E which Thailand is covered and all of these tropical cyclones which moving into Thailand are considered. The model used two data types for experiment, i.e., (a) historical tropical cyclone data or best track data were derived from Joint Typhoon Warning Center (JTWC) [13] which includes time of analysis, storm intensity, latitude/longitude of the storm, central pressure, maximum sustained wind speed (MSW), radius of maximum wind (RMW) at R_{34}, R_{50}, and R_{64}, etc., which is shown in Fig. 1. Historical tropical cyclone data is reported every 6 h following the World Meteorology Organization (WMO) regulation.

(b) Japan Meteorological Agency's Satellite is a geostationary meteorological satellites which operated at coordinate N70–S20 and E70–E160 since 1978, producing data that helps to prevent and mitigate weather-related disasters based on monitoring of typhoons and other weather conditions in the Asia-Oceania region [14]. JMA's Satellite has five channels whose detail is shown in Fig. 2, and all channels will scan image every hour with the resolution of 1800×1800 pixels which is shown in Fig. 3.

Fig. 3 Example of IR1
Image by MTSAT-2 on
November 7, 2013

3 Tropical Cyclone Identification-Analysis and Track Forecasting

Satellite image IR1 [15, 16] data from JMA's Satellite in 6 h interval is used for analyzing the maximum sustain wind (intensity) and extracts the location of the tropical cyclone. To get the storm position, the model for detection and location identification [17, 18] is applied. But the simple technique for detection and location identification is the research of Wong Ka Yan, et al. and more detail at [19] which was applied in this work for detection and extraction location of tropical cyclone.

However, most of the automatic tropical cyclone location identification leads to large error during formative and decaying phase of tropical cyclone due to the absence of robust pattern in the images or the spiral/eyes of tropical cyclone are not present in the cloud pattern. The error of location identification often occurs during formative and decaying phase because at these phases this is still a challenging research issue. As the result, tropical cyclone forecasting in Sect. 6 will show only result experiment data from historical data files for testing accuracy of TC hazardous area forecasting model because only data of the maximum tropical cyclone level from image extraction in best track is quite small data set.

In intensity analysis, methodology is demonstrated in two parts as follows. (a) First phase is an image extraction and reconstruction phase, satellite image data from JMA's Satellite in 6 h interval which were extracted center of the storm in images were mapped and reconstructed tropical cyclone image with size 600×600 pixels with center of image are position at $x = 300$, $y = 300$ which is TC center and classified into three levels; Tropical Depression (TD), Tropical Cyclone (TC), and Typhoon (TY) and, (b) Second, each image from first phase were analyzed by Dvorak techniques and more detail at [20], analyzed images were recorded as wind speed (km/h).

Climatology and Persistence (CLIPER) [21] is a tropical cyclone track forecasting technique which includes 13 predictors and is able to forecast 6–72 h by using multiple regression techniques. However, traditional CLIPER (T-CLIPER) is emphasis only track forecasting and only based on historical data equation and gives an unsatisfied result when forecast more than 12 h. Therefore, improvement of the model (Self-Adaptive CLIPER or SA-CLIPER) by Buranasing and Prayote [11, 12] is applied in this work.

4 Tropical Cyclone Hazardous Area Self-adaptive Forecasting

Tropical cyclone hazardous area forecasting in this paper used statistical methods based on climatology and persistence by using multiple regression techniques. However, traditional technique is only based on historical data equation and gives an unsatisfied result when forecast more than 12 h. Therefore, improvement of the SA-CLIPER model called Integrated Self-Adaptive Climatology and Persistence or ISA-CLIPER in this paper selected 12 predictors which were extracted and analyzed from satellite images from Sect. 3 as follows.

First, calculate next radius of maximum wind (RMW) at R_i from statistical-based equation (R_i-SBE) as follows.

$$FR_i = \beta_0 + \beta_1 x_1 + \beta_2 x_2 + \cdots + \beta_i x_i + \cdots + \beta_{12} x_{12} \tag{1}$$

where FR_i is next radius of maximum wind (RMW) at R_i. Where β_0 to β_{12} can be calculated as follows.

$$n\beta_0 + \beta_1 \sum_{i=1}^{n} x_{i,1} + \beta_2 \sum_{i=1}^{n} x_{i,2} + \cdots + \beta_{12} \sum_{i=1}^{n} x_{i,12} = \sum_{i=1}^{n} y_i \tag{2}$$

$$\beta_0 \sum_{i=1}^{n} x_{i,1} + \beta_1 \sum_{i=1}^{n} x_{i,1} x_{i,1} + \beta_2 \sum_{i=1}^{n} x_{i,1} x_{i,2} + \cdots + \beta_{12} \sum_{i=1}^{n} x_{i,1} x_{i,12} = \sum_{i=1}^{n} x_{i,1} y_i \tag{3}$$

$$\beta_0 \sum_{i=1}^{n} x_{i,2} + \beta_1 \sum_{i=1}^{n} x_{i,2} x_{i,1} + \beta_2 \sum_{i=1}^{n} x_{i,2} x_{i,2} + \cdots + \beta_{12} \sum_{i=1}^{n} x_{i,2} x_{i,12} = \sum_{i=1}^{n} x_{i,2} y_i \tag{4}$$

$$\beta_0 \sum_{i=1}^{n} x_{i,3} + \beta_1 \sum_{i=1}^{n} x_{i,3} x_{i,1} + \beta_2 \sum_{i=1}^{n} x_{i,3} x_{i,2} + \cdots + \beta_{12} \sum_{i=1}^{n} x_{i,3} x_{i,12} = \sum_{i=1}^{n} x_{i,3} y_i \tag{5}$$

$$\beta_0 \sum_{i=1}^{n} x_{i,4} + \beta_1 \sum_{i=1}^{n} x_{i,4} x_{i,1} + \beta_2 \sum_{i=1}^{n} x_{i,4} x_{i,2} + \cdots + \beta_{12} \sum_{i=1}^{n} x_{i,4} x_{i,12} = \sum_{i=1}^{n} x_{i,4} y_i \tag{6}$$

$$\beta_0 \sum_{i=1}^{n} x_{i,5} + \beta_1 \sum_{i=1}^{n} x_{i,5} x_{i,1} + \beta_2 \sum_{i=1}^{n} x_{i,5} x_{i,2} + \cdots + \beta_{12} \sum_{i=1}^{n} x_{i,5} x_{i,12} = \sum_{i=1}^{n} x_{i,5} y_i \tag{7}$$

$$\beta_0 \sum_{i=1}^{n} x_{i,6} + \beta_1 \sum_{i=1}^{n} x_{i,6} x_{i,1} + \beta_2 \sum_{i=1}^{n} x_{i,6} x_{i,2} + \cdots + \beta_{12} \sum_{i=1}^{n} x_{i,6} x_{i,12} = \sum_{i=1}^{n} x_{i,6} y_i \tag{8}$$

$$\beta_0 \sum_{i=1}^{n} x_{i,7} + \beta_1 \sum_{i=1}^{n} x_{i,7} x_{i,1} + \beta_2 \sum_{i=1}^{n} x_{i,7} x_{i,2} + \ldots + \beta_{12} \sum_{i=1}^{n} x_{i,7} x_{i,12} = \sum_{i=1}^{n} x_{i,7} y_i \tag{9}$$

$$\beta_0 \sum_{i=1}^{n} x_{i,8} + \beta_1 \sum_{i=1}^{n} x_{i,8} x_{i,1} + \beta_2 \sum_{i=1}^{n} x_{i,8} x_{i,2} + \cdots + \beta_{12} \sum_{i=1}^{n} x_{i,8} x_{i,12} = \sum_{i=1}^{n} x_{i,8} y_i \tag{10}$$

$$\beta_0 \sum_{i=1}^{n} x_{i,9} + \beta_1 \sum_{i=1}^{n} x_{i,9}x_{i,1} + \beta_2 \sum_{i=1}^{n} x_{i,9}x_{i,2} + \cdots + \beta_{12} \sum_{i=1}^{n} x_{i,9}x_{i,12} = \sum_{i=1}^{n} x_{i,9}y_i \quad (11)$$

$$\beta_0 \sum_{i=1}^{n} x_{i,10} + \beta_1 \sum_{i=1}^{n} x_{i,10}x_{i,1} + \beta_2 \sum_{i=1}^{n} x_{i,10}x_{i,2} + \cdots + \beta_{12} \sum_{i=1}^{n} x_{i,10}x_{i,12} = \sum_{i=1}^{n} x_{i,10}y_i$$

$$(12)$$

$$\beta_0 \sum_{i=1}^{n} x_{i,11} + \beta_1 \sum_{i=1}^{n} x_{i,11}x_{i,1} + \beta_2 \sum_{i=1}^{n} x_{i,11}x_{i,2} + \cdots + \beta_{12} \sum_{i=1}^{n} x_{i,11}x_{i,12} = \sum_{i=1}^{n} x_{i,11}y_i$$

$$(13)$$

$$\beta_0 \sum_{i=1}^{n} x_{i,12} + \beta_1 \sum_{i=1}^{n} x_{i,12}x_{i,1} + \beta_2 \sum_{i=1}^{n} x_{i,12}x_{i,2} + \cdots + \beta_{12} \sum_{i=1}^{n} x_{i,12}x_{i,12} = \sum_{i=1}^{n} x_{i,12}y_i$$

$$(14)$$

From all equations above (2)–(14), β_0 to β_{12} are able to solve equations by using matrices method and n is all tropical cyclone in database. Note that, x_1 to x_{12} in (1)–(14) will be replaced by list of predictor variables in Table 1. Finally, calculate error elimination or adjustment equation (AE) as follows.

$$\varepsilon_i = \left[\sum_{i=1}^{t} FR_i - R_i \right] \Big/ t \quad (15)$$

$$FR_i = FR_i - \varepsilon_i \quad (16)$$

where ε_i is an average error of R_i in latest t time windows hours. In addition, t is 24 h in this paper. From Eq. (16), FR_i is next radius of maximum wind (RMW) at R_i of tropical cyclone which are errors eliminated by self-adjustment methodology. Note that R_i is separated calculation where each i is 34 nm, 50 nm, 64 nm which means radius at wind speed 34 nm, 50 nm, 64 nm, respectively, following World Meteorology Organization (WMO).

In summary, all methodologies described in this paper can be rewritten into a sequence of tropical cyclone hazardous area forecasting models, shown in Fig. 4. SBE Eqs. (1)–(14) are used for a training class in track (T-SBE) and radius hazard (R_i-SBE) as shown in Fig. 4(1.1), 4(1.2), respectively. Figure 4(2) is the satellite image from a geostationary satellite. Figure 4(3) is the pre-processing phases, Fig. 4(3.1) is the TC detection and Fig. 4(3.2) is the TC location identification. Figure 4(3.3) is the intensity analysis. Figure 4(4.1), 4(4.2), and 4(4.3) are statistical-based equations (SBE) calculated by using track (T-SBE) and radius hazard (R_i-SBE) in which β_0 to β_{12} are calculated in the first block of the model but replaced x_1 to x_{12} by unknown data for forecasting. Then, adjust track (T-SBE) and radius hazard (R_i-SBE) by using the T-Adjustment Equation and R_i-Adjustment Equation using Eqs. (15)–(16) as shown in Fig. 4(5.1), 4(5.2), and 4(5.3). Figure 4(6) is the result of the next values

Table 1 List of predictor hazardous area forecasting variables

Predictors	Description
Julian_Date	Julian date
Initial_LAT	Initial latitude
Initial_LONG	Initial longitude
Current_LAT	Current latitude
Current_LONG	Current longitude
P12h_LAT	Latitude over past 12 h
P12h_LONG	Longitude over past 12 h
P24h_LAT	Latitude over past 24 h
P24h_LONG	Longitude over past 24 h
Current_MSW	Current Maximum Sustain Wind
AVG12h_SPEED	Avg. Speed over past 12 h
AVG24h_SPEED	Avg. Speed over past 24 h

of latitude, longitude, and radius hazard forecasting. Furthermore, the model can draw a graphic display which is described in Sect. 6 and shown in Fig. 4(7).

5 Performance of Tropical Cyclone Hazardous Area Self-adaptive Forecasting

The performance of model was evaluated in Tables 2, 3 and 4. The experiment of tropical cyclone radius of maximum wind (RMW) at R_i forecasting model was divided into two classes, one is training class which the result is absented in this paper due to objective of training class is only create statistical-based equation and another is testing class. In training class, the model used all historical tropical cyclone data between years 2003–2012 (10 years or 77%) to create statistical-based equation and testing class used historical tropical cyclone data between years 2013–2015 (3 years or 23% with over 85 tropical cyclones) which are unknown data to testing the ISA-CLIPER model and compared with T-CLIPER. However, the experiments were tested only data from historical files data for accuracy testing due to data from extracted images is quite small data set, which is lack of image identification processing phases.

In Tables 2, 3, and 4 show the experiment results of T-CLIPER and ISA-CLIPER forecasting with unknown tropical cyclone data between years 2013–2015. In 6 h-forecasting experiment, T-CLIPER give an average R_{34}, R_{50}, and R_{64} error are 18.16 nm, 8.21 nm, 4.33 nm, respectively, on Mercator projection map and ISA-CLIPER are 10.16 nm, 5.00 nm, 24 nm or ISA-CLIPER is lower than T-CLIPER about 44.05%, 39.09%, and 44.57%. In 12 hour-forecasting experiment, T-CLIPER give an average R_{34}, R_{50}, and R_{64} error are 17.61 nm, 8.08 nm, 4.31 nm, respectively, on Mercator projection map and ISA-CLIPER are 13.40 nm, 6.46 nm, 3.41 nm, or

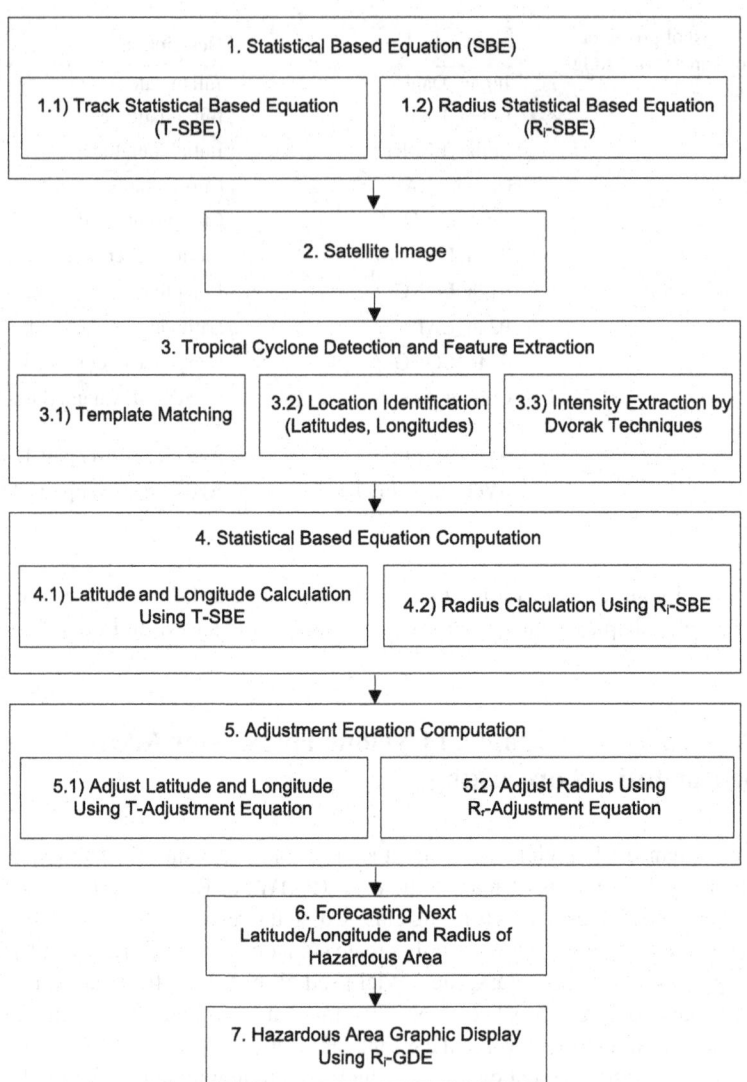

Fig. 4 ISA-CLIPER Forecasting Model

ISA-CLIPER is lower than T-CLIPER about 23.90%, 20.04%, and 20.88%. Also, in 24 h-forecasting experiment, T-CLIPER give an average R_{34}, R_{50}, and R_{64} error are 16.18 nm, 6.66 nm, 3.11 nm, respectively, on Mercator projection map and ISA-CLIPER are 13.10 nm, 6.04 nm, 3.24 nm or ISA-CLIPER is lower than T-CLIPER about 19.03%, 9.30%, and 7.69%. Note that nm unit stands for Nautical Miles in Table 3.

Table 2 Performance of ISA-CLIPER model in 6 h (nm)

Model	Forecasting								
	2013 (33 TC)			2014 (23 TC)			2015 (29 TC)		
	R34	R50	R64	R34	R50	R64	R34	R50	R64
T-CLIPER	16.01	7.95	4.82	16.50	7.58	4.15	21.99	9.11	4.03
ISA-CLIPER	7.91	4.03	1.83	9.20	4.35	2.43	13.37	6.64	2.95

Table 3 Performance of ISA-CLIPER model in 12 h (nm)

Model	Forecasting								
	2013 (33 TC)			2014 (23 TC)			2015 (29 TC)		
	R34	R50	R64	R34	R50	R64	R34	R50	R64
T-CLIPER	15.23	7.62	4.76	16.21	7.60	4.16	21.39	9.02	4.03
ISA-CLIPER	10.02	4.90	2.74	13.06	6.54	3.52	17.12	7.94	3.99

Table 4 Performance of ISA-CLIPER model in 24 h (nm)

Model	Forecasting								
	2013 (33 TC)			2014 (23 TC)			2015 (29 TC)		
	R34	R50	R64	R34	R50	R64	R34	R50	R64
T-CLIPER	13.21	6.43	2.96	14.55	7.15	3.43	20.80	10.39	4.16
ISA-CLIPER	9.44	4.64	2.70	12.08	6.75	3.30	17.80	6.73	3.74

6 Geological Hazardous Area Graphic Display

Tropical cyclone hazardous area forecasting model in Fig. 4 can display hazardous area for tropical cyclone radius of maximum wind (RMW) at R_i (R_{34}, R_{50}, and R_{64}) by calculating R_i of Graphic Display Equations (R_i-GDE) as follows. First, calculate interpolation from track forecasting between each pair of longitude as follows.

$$L_i(LONG_j) = a_0 + a_1 LONG_j \qquad (17)$$

$$L_{i+1}(LONG_{j+1}) = a_0 + a_1 LONG_{j+1} \qquad (18)$$

where $L_i(LONG_j)$ is current latitude of i forecasting at longitude of j forecasting. $L_{i+1}(LONG_{j+1})$ is next latitude of i forecasting at longitude of j forecasting. a_0 and a_1 are coefficient that is able to solve equations by using matrices method. Second, after a_0 and a_1 are calculated. Then, calculate interpolation between $L_i(LONG_j)$ and $L_{i+1}(LONG_{j+1})$ as follows

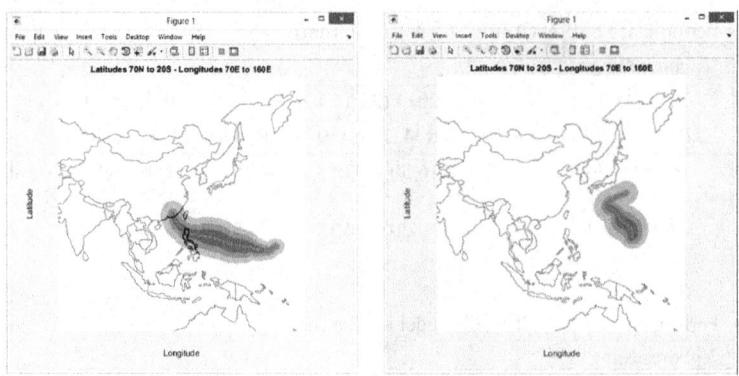

Fig. 5 ISA-CLIPER Graphic forecasting for Typhoon HAIYAN (left) and FRANCISCO (right)

$$L_{i+x}(LONG_{j+x}) = a_0 + a_1 LONG_{j+x} \tag{19}$$

where $LONG_i < LONG_{j+x} < LONG_{i+1}$ and x can calculated as follows

$$x = |LONG_{j+1} - LONG_j| \tag{20}$$

Then, hazardous area radius can be calculated as follows.

$$HR_i = LONG_{i+x} + FR_i \tag{21}$$

where HR_i is hazardous radius at longitude x and FR_i is radius of maximum wind (RMW) forecasting at R_i in the past at time i (Fig. 5).

7 Conclusion and Remark

Although tropical cyclone (TC) track and intensity forecasting techniques have been improving for several decades. However, some uncertainty still remains [22]. Moreover, risk area assessment and uncertainty of the major model which is the most important phase in modern natural disaster management is excluded. End users of TC forecasts, such as risk managers and public agencies, need both reliable track forecasts and an estimation of forecast uncertainty. To address these problems, this paper proposes an ISA-CLIPER which is a self-adaptive model using only 12 features which were extracted from satellite images with improvement of the traditional statistical methods. The model is able to forecast both track/intensity and risk area assessment is integrated into the model by based on using strike zone diagram theory [23] which is a lines around TC's center position and indicate the 34-knot, 50-knot, and 64-knot wind radii. The performance of the model is satisfactory; the average error from experimental results of R_{30}, R_{50}, and R_{64} forecasting with unknown tropical cyclone data between years 2013–2015 on Mercator projection map is lower than traditional techniques by 28.99%, 22.81%, and 24.38%, respectively.

References

1. Buranasing, A.: Storm intensity prediction using an artificial neural network. In: International Conference on Southeast Asian Weather and Climate 2013 (ICSA-WC2013), Chiang Mai, Thailand
2. Miller, M.: Global tropical cyclone forecasting for days, weeks and a season ahead. The European Centre for Medium-Range Weather Forecasts (ECMWF)
3. Buranasing, A., Prayote, A.: Storm intensity estimation using symbolic aggregate approximation and artificial neural network. In: 2014 International Computer Science and Engineering Conference (ICSEC 2014: International Track), Khon Kaen, Thailand
4. Climatological Center, Meteorological Development (TMD): Tropical Cyclones in Thailand Historical Data 1951–2010. Thai Meteorological Department
5. Coppola, D.P.: Introduction to International Disaster Management. Elsevier's Science & Technology
6. Pruksapong, M.: Disaster Risk Reduction and Development. Asian Disaster Preparedness Center (ADPC), 7 June 2013
7. Pasch, R., Clark, J.S., et al.: Technical Summary of the National Hurricane Center Track and Intensity Models. National Oceanic and Atmospheric Administration (NOAA)
8. Jampanya, P.: Tropical storm forecasting technique in Thailand. ACTS Workshop, Taiwan, 6–7 June 2013
9. The Weather Research & Forecasting Model. Weather Forecasting. http://www.wrf-model.org. Accessed 14 July 2015
10. Yavinchan, S.: Tropical Storm Track Prediction for Thailand by a High Resolution Numerical Weather Prediction Model. The Meteorological Department, Weather Forecast Bureau, Bangkok, Thailand
11. Buranasing, A., Prayote, A.: Application of remote sensing for tropical cyclone track forecasting based on statistical methods. In: ICSEC 2015—The 19th International Computer Science and Engineering Conference, Chiang Mai, Thailand
12. Buranasing, A., Prayote, A.: Tropical cyclone track and intensity forecasting using remotely-sensed images. In: ICITEE 2015—The 7th International Conference on Information Technology and Electrical Engineering, Chiang Mai, Thailand
13. Joint Typhoon Warning Center (JTWC). http://www.usno.navy.mil/JTWC/. Accessed 14 July 2015
14. Meteorological Satellites - Japan Meteorological Agency (JMA). http://www.jma.go.jp/jma/jma-eng/satellite/index.html. Accessed 14 July 2015
15. Mather, P.M.: Computer Processing of Remotely-Sensed Images, 4th edn. Wiley (2011)
16. Kaňák, J.: Overview of the IR channels and their applications. EUMeTrain, 14 June 2011
17. Jaiswal, N., Kishtawal, C.M.: Automatic determination of center of tropical cyclone in satellite-generated IR images. IEEE Geosci. Remote Sens. Lett. (2011)
18. Jaiswal, N., Kishtawal, C.M.: Objective detection of center of tropical cyclone in remotely sensed infrared images. IEEE J. Sel. Top. Appl. Earth Obs. Remote Sens. (2013)
19. Yan, W.K., Lap, Y.C., Wah, L.P., Wan, T.W.: Automatic template matching method for tropical cyclone eye fix. In: The 17th International Conference on Pattern Recognition (ICPR'04)

20. Olander, T.L., Velden, C.S.: The advanced Dvorak technique: continued development of an objective scheme to estimate tropical cyclone intensity using geostationary infrared satellite imagery. Am. Meteorol. Soc. (2007)
21. Xu, Y., Neumann, C.J.: A Statistical Model for the prediction of Western North Pacific Tropical Cyclone Motion
22. Dupont, T., Plu, M., Caroff, P., Faure, G.: Verification of ensemble-based uncertainty circles around tropical cyclone track forecasts. Weather Forecast. Am. Meteorol. Soc. (2011)
23. Dupont, T., Plu, M., Caroff, P., Faure, G.: Operational implementation of ensemble-based dynamical uncertainty circles around tropical cyclone track forecasts. In: The 30th Conference on Hurricanes and Tropical Meteorology, Apr 2012

Flexible EM-Type Algorithms for Spatial Clustering

Wen-Liang Hung

Abstract This paper studies a problem of grouping the spatial data, which are important data in regional science. We propose two analytically simple EM-type algorithms for spatial clustering. One is the suppressed NEM, and the other is spatial EM algorithm. Practical algorithms to implement the proposed methods are further discussed. The empirical performance of the proposed methods is evaluated through numerical studies including two real data analysis, which demonstrates promising results of the proposed clustering algorithms.

Keywords EM algorithm · Error rate · Gaussian mixture · MR image · Spatial clustering

1 Introduction

Spatial clustering is an important tool for spatial data analysis. The goal of spatial clustering is to arrange spatial data into some meaningful sets, called spatial clusters, such that spatial data points in the same cluster are as similar as possible and data in the different clusters are as different as possible. Existing spatial clustering algorithms can be classified into two categories [9]. One is gained from the spatial point pattern statistical analysis field, such as spatial autocorrelation analysis and spatial scan methods [9]. The other is developed in the field of data mining, such as model-based clustering, partitioning algorithms, and hierarchical algorithms [11]. The most well-known model-based clustering is EM algorithm [13]. Applying the idea of EM algorithm, Ambroise et al. [2] proposed a neighborhood EM (NEM) algorithm that uncovers the cluster structure of the spatial data sets. Hu and Sung [8] proposed a hybrid EM (HEM) algorithm to combine the EM and NEM algorithms. However, these two algorithms have two limitations: slow convergence and the confusion between noise.

W.-L. Hung (✉)
Center for Teacher Education, National Tsing Hua University, Hsin-Chu, Taiwan
e-mail: wlhung0625@gmail.com

© Springer Nature Singapore Pte Ltd. 2019
S. K. Bhatia et al. (eds.), *Advances in Computer Communication and Computational Sciences*, Advances in Intelligent Systems and Computing 760, https://doi.org/10.1007/978-981-13-0344-9_33

383

The problem in this article is "Given a set of spatial data, how do we propose an effective clustering algorithm to group these data into similar groups?". By combining the ideas of Ambroise and Govaert [1], Ambroise et al. [2], Chuang et al. [3], and Fan et al. [5], we proposed two extension EM algorithms. One is suppressed NEM algorithm, and the other is spatial EM algorithm. The outline of the paper is as follows. We provide an overview of the EM and NEM algorithms in Sect. 2. The proposed methods are illustrated in Sect. 3. Numerical results are presented in Sect. 4. Finally, conclusions and future work are discussed in Sect. 5.

2 EM and Neighborhood EM Algorithms

2.1 EM Algorithm

A powerful and flexible statistical method for clustering is the finite mixture model. Specifically, a random vector X is distributed as a finite mixture model if its probability distribution function (pdf) can be given as

$$f(x|\psi) = \sum_{k=1}^{K} \pi_k f_k(x|\theta_k),$$

where $\psi = (\pi, \theta)$, $\theta = (\theta_1, \ldots, \theta_K)$ and $\pi = (\pi_1, \ldots, \pi_K)$, subject to $\sum_{k=1}^{K} \pi_k = 1$ $(0 < \pi_k < 1)$, is a vector of mixing proportions, $f_k(x; \theta_k)$ represents the pdf of the kth mixing component.

In this article, let $f_k(x|\theta_k)$ be the d-variate Gaussian model with mean vector μ_k and covariance Σ_k; i.e.,

$$f_k(x|\theta_k) = (2\pi)^{-d/2} |\Sigma_k|^{-1/2} \exp\left(\frac{-1}{2}(x - \mu_k)' \Sigma_k^{-1}(x - \mu_k)\right).$$

To obtain the parameter estimates of the finite mixture model, a feasible used approach is the expectation-maximization (EM) algorithm [4, 12].

2.2 Neighbor EM Algorithm

Ambroise and Govaert [1] proposed a NEM algorithm for spatial data. They also illustrated that NEM converges. The loglikelihood function of the sample $\{x_1, \ldots, x_n\}$ is given by

$$L(\psi) = \sum_{i=1}^{n} \ln f(x_i|\psi).$$

By introducing the

$$c_{ik} = \frac{\pi_k f_k(x_i|\theta_k)}{\sum_{s=1}^{K} \pi_s f_s(x_i|\theta_s)}, \ i = 1, \ldots, n, k = 1, \ldots, K.$$

and using the constraint $\sum_{k=1}^{K} c_{ik} = 1$, the loglikelihood function $L(\psi)$ is

$$L(\psi) = \sum_{i=1}^{n} \sum_{k=1}^{K} c_{ik} \ln \pi_k f_k(x_i|\theta_k) - \sum_{i=1}^{n} \sum_{k=1}^{K} c_{ik} \ln c_{ik}.$$

Using Hathaway's idea [7], the matrix $\mathbf{c}$ must satisfy the following condition:

$$\mathbf{c} = \{c_{ik} | \ 0 \le c_{ik} \le 1, \ \sum_{k=1}^{K} c_{ik} = 1, \ \sum_{i=1}^{n} c_{ik} > 0, \ 1 \le i \le n, \ 1 \le k \le K\}.$$

Considering the spatial dependence of objects, Ambrooise and Govaert [1] formalized the spatial structure via a matrix $W = (w_{ij})$ defined as follows:

$$w_{ij} = \begin{cases} 1 & \text{if } x_i \text{ and } x_j \text{ are neighbors and } i \ne j, \\ 0 & \text{otherwise.} \end{cases}$$

Thus, Ambrooise and Govaert [1] added the penalizing term

$$\frac{1}{2} \sum_{i=1}^{n} \sum_{j=1}^{n} \sum_{k=1}^{K} w_{ij} c_{ik} c_{jk},$$

to the loglikelihood function $L(\psi)$. That is, the new objective function would be

$$\tilde{L}(\psi, \mathbf{c}) = L(\psi) + \frac{\beta}{2} \sum_{i=1}^{n} \sum_{j=1}^{n} \sum_{k=1}^{K} c_{ik} c_{jk} w_{ij},$$

where $\beta \in [0.5, 1]$. Under some constraints, to maximize the objective function $\tilde{L}(\psi, \mathbf{c})$, we obtain the equation of c_{ik} as follows:

$$c_{ik} = \frac{\pi_k f_k(x_i|\theta_k) \exp\left(\beta \sum_{j=1}^{n} w_{ij} c_{jk}\right)}{\sum_{s=1}^{K} \pi_s f_s(x_i|\theta_s) \exp\left(\beta \sum_{j=1}^{n} w_{ij} c_{js}\right)}. \tag{1}$$

$$\pi_k = \frac{1}{n} \sum_{i=1}^{n} c_{ik}.$$

Now, we also consider the d-variate Gaussian mixture model. Then the equations for those parameters are as follows:

$$\mu_k = \frac{\sum_{i=1}^{n} c_{ik} x_i}{\sum_{i=1}^{n} c_{ik}}$$

$$\Sigma_k = \frac{\sum_{i=1}^{n} c_{ik} (x_i - \mu_k)(x_i - \mu_k)'}{\sum_{i=1}^{n} c_{ik}}$$

Thus, the NEM algorithm can be seen in Ambrooise and Govaert [1].

3 The Proposed Algorithms

3.1 Suppressed NEM Algorithm

With large data sets under consideration, the NEM algorithm has a slow convergence rate. Hu and Sung [8] proposed a hybrid EM algorithm to speed up the convergence rate. To alleviate this problem, we proposed the suppressed NEM (S-NEM) algorithm using the idea of Fan et al. [5], which increases the largest membership degree and decreases the second largest membership degree. The main difference from NEM is to revise memberships c_{ik} given in Eq. (1):

Let the membership c_{ip} be the ith data point belonging the pth cluster and be the largest among K clusters. (Note: If there are two or more largest memberships, then randomly choose one.) Then, the memberships are revised by

$$c_{ip} = 1 - \alpha \sum_{k \neq p} c_{ik} = 1 - \alpha + \alpha c_{ip} \text{ and } c_{ik} = \alpha c_{ik}, \ k \neq p \tag{2}$$

where $0 \leq \alpha \leq 1$.

Based on Eq. (2), the S-NEM algorithm seems to prefer the largest membership and inhibits the others. Note that when $\alpha = 0$, the S-NEM algorithm is reduced to the hard NEM and when $\alpha = 1$, this algorithm is the same as the NEM. As an appropriate parameter α is selected, we see that S-NEM may reflect both the fast convergence speed of the hard NEM and the superior partition performance of the NEM.

3.2 Spatial EM Algorithm

We consider an image which neighboring pixels are highly correlated. It means that these neighboring pixels have similar feature values, and there are high probability that they belong to the same cluster. According to Sect. 2.2, the spatial information is important in clustering. To use the useful information, a spatial function may be given by [3]

$$h_{ik} = \sum_{j \in NB(x_i)} \hat{z}_{jk},$$

where $NB(x_i)$ denotes a square window centered on pixel x_i in the spatial domain. In this subsection, 3×3 window was used in clustering. Based on the definition of h_{ik}, h_{ik} can be regarded as the degree of membership of pixel x_i belonging to kth cluster. Hence, h_{ik} is incorporated into $\hat{z}_{ik}$ as follows:

$$\hat{z}_{ik}^* = \frac{\hat{z}_{ik}^p h_{ik}^q}{\sum_{c=1}^{K} \hat{z}_{ic}^p h_{ic}^q},$$

where p and q are parameters to control the relative importance of both functions and

$$\hat{z}_{ik} = E(z_{ik}|x_i) = \frac{\pi_k f_k(x_i|\theta_k)}{\sum_{s=1}^{K} \pi_s f_s(x_i|\theta_s)}.$$

$\text{sEM}_{p,q}$ represents the spatial EM with parameter p and q. Note that $\text{sEM}_{1,0}$ is identical to the EM algorithm. Therefore, the spatial EM algorithm is similar to EM algorithm.

4 Numerical Results

To assess the performance of the proposed S-NEM and spatial EM algorithm, we carried out on two real datasets.

Experiment 1. We will first assess the performance of the proposed S-NEM algorithm for the real house price dataset [6] available at LeSage. For simplicity's sake, we use the house price of 506 towns in Boston area. Since we have no information about this dataset, we select the middle-point between 0 and 1, i.e., let $\alpha = 1/2$ in S-NEM algorithm. According to the histogram shown in Fig. 4b [8], there are two clusters in this house price dataset. One is high price and the other is low price. Their definition is:

$$x_i = \begin{cases} \text{high,} & \text{if } x_i \geq 22.533; \\ \text{low,} & \text{otherwise,} \end{cases}$$

where 22.53 is the mean of house values of 506 towns and $i = 1, \ldots, 506$. Next, we examine the clustering performance of EM, NEM, HEM, and the proposed S-NEM ($\alpha = 1/2$) algorithms on this dataset. According to Hu and Sung [8], we set $\beta = 1$ in NEM and HEM algorithms. According to Table 1, the error produced by

Table 1 The error rates of EM, NEM, HEM, and the proposed S-NEM ($\alpha = 1/2$) algorithms

Algorithms	Error rate (%)
EM	51.19
NEM	42.49
HEM	40.32
S-NEM	33.99

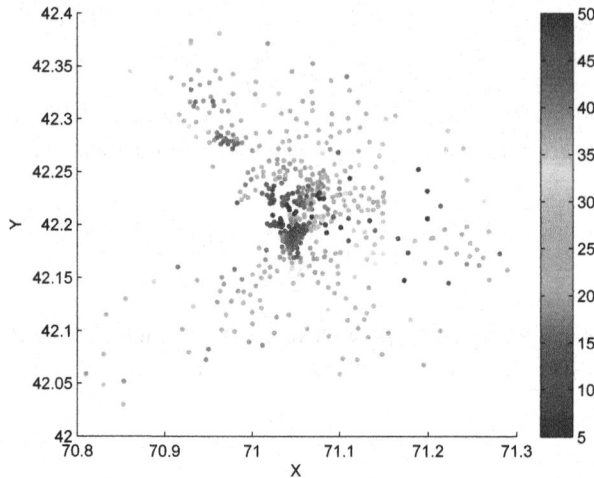

Fig. 1 The true house values of 506 towns in Boston area

S-NEM is the lowest (33.99%). However, EM, NEM, and HEM are significantly higher at 40.32%. This implies that the clustering performance of the proposed S-NEM algorithm outperforms EM, NEM, and HEM algorithms in terms of the error rate (Figs. 1, 2, 3, 4, and 5)

Experiment 2. Here we analyze the MR image [14]. To compare the segmentation results of the EM, $sEM_{1,1}$, $sEM_{0,2}$, and S-NEM ($\alpha = 1/2$) algorithms, we compute the score S_{ik} (see [10, 15]), where i means that the ith algorithm and kth is the kth reference or segmented image. The large value of $S_{ik} \in [0, 1]$ indicates a better segmentation result. From Table 2, the $sEM_{0,2}$ technique produces the best segmentation result and the segmentation results of $sEM_{1,1}$ and S-NEM ($\alpha = 1/2$) are acceptable.

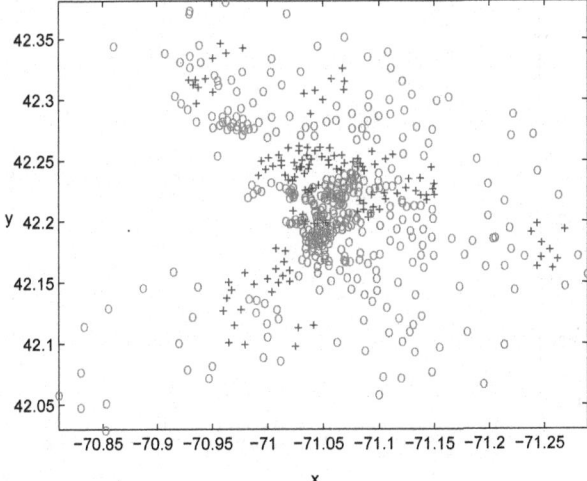

Fig. 2 The clustering result of EM algorithm. The blue "+" represents the high price, and the green "o" represents the low price

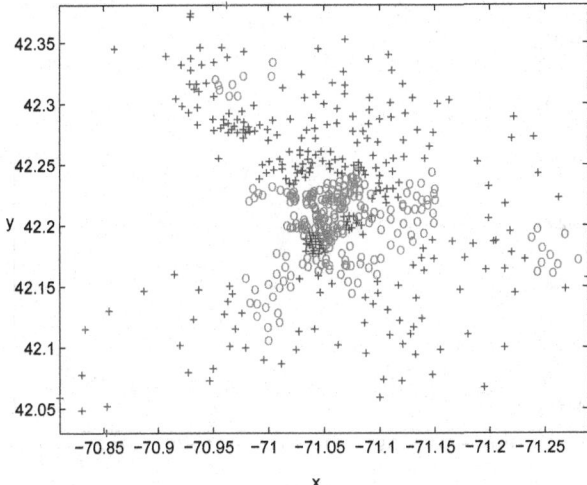

Fig. 3 The clustering result of NEM algorithm. The blue "+" represents the high price, and the green "o" represents the low price

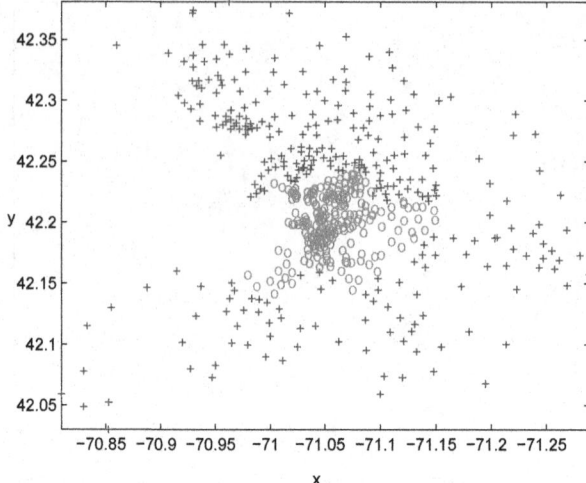

Fig. 4 The clustering result of HEM algorithm. The blue "+" represents the high price, and the green "o" represents the low price

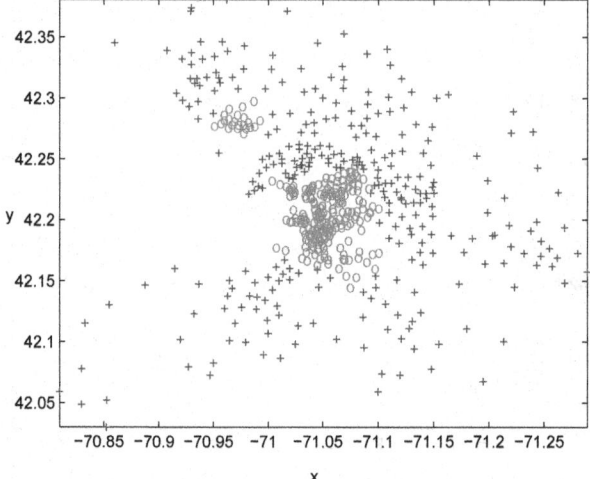

Fig. 5 The clustering result of S-NEM ($\alpha = 1/2$) algorithm. The blue "+" represents the high price, and the green "o" represents the low price

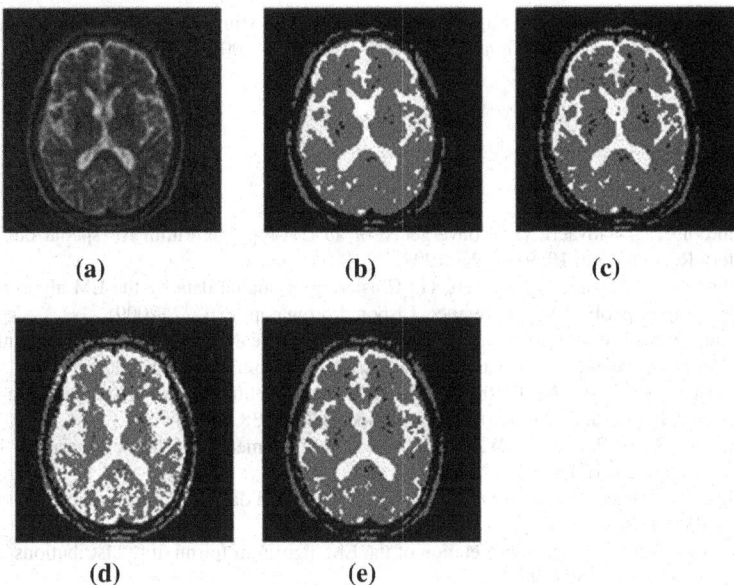

(a)   (b)   (c)

(d)   (e)

Fig. 6 **a** Original image; **b** the segmentation result by EM algorithm; **c** the segmentation result by sEM$_{1,1}$ algorithm; **d** the segmentation result by sEM$_{0,2}$ algorithm; **e** the segmentation result by S-EM ($\alpha = 1/2$) algorithm

Table 2 Comparison scores S_{ik} of the ith algorithm and the kth class for the images in Fig. 6a

ith algorithm	kth class		
	class 1 (%)	class 2 (%)	class 3 (%)
EM	12.08	61.38	97.57
sEM$_{1,1}$	87.94	99.43	99.68
sEM$_{0,2}$	93.45	99.47	99.43
S-NEM($\alpha = 1/2$)	87.36	99.43	99.68

5 Conclusions

In this article, we have proposed two flexible EM algorithms to perform clustering of spatial data. The proposed S-NEM algorithm may reflect both the fast convergence and the superior partition. Moreover, we have proposed practical algorithms for implementing the proposed methods. Through numerical experiments involving real data analysis, we have conclude the proposed methods are capable of efficiently clustering spatial data. In the future, there are several related topics that we intend to pursuit. First, we plan to show the convergence of the proposed S-NEM and spatial EM algorithms. We are also interested in investigating "How to select appropriate values of p and q to improve the segmentation results?".

Acknowledgements This work was supported in part by the Ministry of Science and Technology of the Republic of China under Grant MOST 105-2118-M-007-007.

References

1. Ambrooise, C., Govaert, G.: Convergence of an EM-type algorithm for spatial clustering. Pattern Recogn. Lett. **19**, 919–927 (1998)
2. Ambrooise, C., Dang, M., Govaert, G.: Clustering of spatial data by the EM algorithm. In: Proceeding of geoENV96 Conference, Lisbon, Portugal, pp. 20–22 (2009)
3. Chuang, K.S., Tzeng, H.L., Wu, J., Chen, T.J.: Fuzzy c-means clustering with spatial information for image segmentation. Comput. Med. Imaging Graph. **30**, 9–15 (2006)
4. Dempster, A.P., Laird, N.M., Rubin, D.B.: Maximum likelihood from incomplete data via the EM algorithm (with discussion). J. R. Stat. Soc. B **39**, 1–38 (1977)
5. Fan, J.L., Zhen, W.Z., Xie, W.X.: Suppressed fuzzy c-means clustering algorithm. Pattern Recogn. Lett. **24**, 1607–1612 (2003)
6. Gilley, O.W., Pace, R.K.: On the Harrison and Rubinfeld data. J. Environ. Econ. Manag. **31**, 403–405 (1996)
7. Hathaway, R.J.: Another interpretation of the EM algorithm for mixture distributions. J. Stat. Probab. Lett. **4**, 53–56 (1986)
8. Hu, T., Sung, S.Y.: A hybrid EM apprach to spatial clustering. Comput. Stat. Data Anal. **50**, 1188–1205 (2006)
9. Lu, Y.: Spatial clustering detection and analysis. In: International Encyclopedia of Human Geography 1, pp. 317–324. Elsevier, Oxford (2009)
10. Masulli, F., Schenone, A.: A fuzzy clustering based segmentation system as support to diagnosis in medical imaging. Artif. Intell. Med. **16**, 129–147 (1999)
11. Miller, H., Han, J.: Geographic Data Mining and Knowledge Discovery, 2nd edn. CRC, New York (2009)
12. McLachlan, G.J., Peel, D.: Finite Mixture Models. Wiley (2001)
13. Xu, R., Wunsh II, D.: Clustering. Wiley (2009)
14. Yang, M.S., Tsai, H.S.: A Gaussian kernel-based fuzzy c-means algorithm with a spatial bias correction. Pattern Recogn. Lett. **29**, 1713–1725 (2008)
15. Zhang, D.Q., Chen, S.C.: A novel kernelized fuzzy C-means algorithm with application in medical image segmentation. Artif. Intell. Med. **32**, 37–50 (2004)

Intelligent Fouling Detection System Using Haar-Like Cascade Classifier with Neural Networks

Cheng Siong Chin, JianTing Si, A. S. Clare and Maode Ma

Abstract Numerous solutions were introduced to combat issues related to marine fouling. But recent studies have shown that the current approach is unable to detect types of fouling on the ships automatically. A fouling detection system using machine learning is therefore proposed. The proposed system incorporates Haar cascade detection with camera module in a waterproof enclosure to detect fouling in real time. The images of barnacles were taken from Singapore's coastline to train the Haar cascade classifier. The experimental results show that the system can detect the type of barnacle with a reasonable accuracy. The study also includes a fouling repository for users to contribute the fouling images.

Keywords Fouling · Object recognition · Barnacles · Database server · Haar-like cascade · Neural networks

1 Introduction

The colonization of fouling organisms on marine structures had been a perceptual phenomenon in the marine industry. Barnacles and algae tenaciously grip onto marine structures such as the ship's hull, columns of offshore structures, and marine debris. Fouling activity can take place almost immediately when a vessel is stationary. These organisms form a layer of increased surface roughness that would lead to a turbulent flow layer along the surface of the hull, creating a viscous effect and thereby influencing the design an ideal hull form. By reducing frictional resistance generated in seawater, a shipowner can save additional fuel consumption since frictional

C. S. Chin (✉) · J. Si · A. S. Clare
Faculty of Science, Agriculture, and Engineering,
Newcastle University, Newcastle upon Tyne NE1 7RU, UK
e-mail: cheng.chin@ncl.ac.uk

M. Ma
School of Electrical & Electronic Engineering, Nanyang Technological University,
Nanyang Avenue, Singapore 639798, Singapore

© Springer Nature Singapore Pte Ltd. 2019
S. K. Bhatia et al. (eds.), *Advances in Computer Communication and Computational Sciences*, Advances in Intelligent Systems and Computing 760, https://doi.org/10.1007/978-981-13-0344-9_34

Table 1 Fouling rank

Rank	Descriptions	Visual estimate of fouling cover (%)
0	No visible fouling	0
1	Slime fouling only	0–1
2	Light fouling	1–5
3	Considerable fouling	6–15
4	Extensive fouling	16–40
5	Very heavy fouling	41–100

resistance constitutes up to 90% of the total drag force. Slow speed vessels with a deep draft tend to attract the habitation of fouling organisms faster [1, 2] although it reduces the fuel consumption. Thus, monitoring of fouling conditions is essential. Wigforss [3] conducted a study to analyze fuel efficiency of 44 general cargo ships using automatic identification system (AIS). Half of the fleet was loaded with 7700 deadweight tonnage, while the other half was loaded with 12,700 deadweight tonnage. The result shows that the speed range for the optimal fuel consumption was around 7.8–9.7 knots. The proper loading of cargo and fouled hull condition are therefore essential to improve the fuel efficiency. In addition, fouling ranking developed by Floerl et al. [4] enables a quantification of fouling. This method was developed by capturing five random images from the surface of the yacht and visually judging the percentage of fouling. Table 1 shows the fouling ranking developed.

An experiment was also conducted on a Japanese destroyer to determine the consequences on fouling and ship speed. The result showed that more fouling tonnage was accumulated when longer durations were spent at sea than traveling at a lower speed [5]. Numerous precautions had been used to prevent the occurrence of fouling. For example, the antifouling (AF) systems have been used. Millett and Anderson [6] used environmental-friendly foul release (FR) coatings to reduce the fuel consumption on a 33-knot aluminum catamaran by 12%. However, the condition of the hull for different FR coating cannot be monitored until a vessel is at dry dock for cleaning. By using image-processing methods from Ismail et al. [7], they observed the growth rate of barnacles throughout the year by counting the number of pixels for barnacles on a captured image. Wang et al. [8] used self-organizing map (SOM) neural network for classification of ship-hull fouling conditions on 300 image samples. The fouling condition on the ship's hull such as percent of fouling and implementation of the fouling detection system was not shown. A surveillance system was then designed by Goh [9] using computer vision using the hue–saturation–value (HSV) to identify the types of fouling located on the ship's hull. The method experimented on few fouling images that gave limited accuracy. The Haar classifier has been commonly used in the face recognition [10–13] and other applications [14, 15]. The application on fouling recognition has not been applied to ship's hull. There is no automated or intelligent approach to identifying the types of fouling and the percent of fouling. Hence, this paper focuses on the method to determine the types of fouling and the severity of fouling on the ship's hull using Haar classifier with neural network approach. This method provides a semi-automated detection and identification of the fouling for the shipowners and operators.

2 Proposed System Architecture

The design of the system architecture has been through many iterations with the aim to create a system that is portable and low cost for the automated fouling detection. The proposed system consists of a mobile mounting device designed on the bulwark of the offshore support vessel (OSV) named PACC offshore services holdings (POSH) constant with variable extension rods from the gunwale to the design waterline to observe fouling conditions. The Raspberry Pi microcontroller and a 5 megapixels camera module will be used. A Raspberry Pi camera enclosed in a watertight enclosure to capture the image of the fouling will be installed. The system is designed to perform the following tasks. There are, namely (a) to capture a picture of fouling; (b) to highlight the area of fouling and compute the fouling percentage and level of fouling severity as in Table 1, and (c) to allow the user to detect and analyze fouling image via a graphical user interface (GUI).

2.1 Software Components

The use of Raspberry Pi serves as a low-cost approach for using an algorithm to detect and identify the fouling. Python and OpenCV are the two main software modules installed in the microcontroller. The fouling information required by a shipowner can be obtained from the GUI through a wireless Secure Shell (SSH) connection via a wireless-enabled device (Tables 2 and 3).

Table 2 Software components

Software	Purpose
Solidworks 2013	System design
Raspbian OS	Operating system
TightVNC Viewer	Remote control
Adobe Photoshop CS 5	Image adjustment
EasyPHP 3.0	Web server, MySQL
Notepad++	Programming tool
Python 2.7	Programming tool
WinSCP	File transfer

Table 3 Hardware components

Hardware	Quantity	Purpose
Winch	1	Hoisting/lifting
Suction cup	2	Support device
Rod	5	Extension
Prototype design	1	Primary structure
Raspberry Pi with charger	1	Microcontroller
Camera module	1	Capture device
Wi-fi USB dongle	1	Provide SSH
8 GB micro-SD card	1	Storage

2.2 System Mechanical Design

The prototype was designed to hold at the bulwark of a vessel. A variation in the length of the prototype is necessary to cater for different sizes of the ship. Two different orientations were provided for the actual implementation of the system. The proposed prototype design in Fig. 1 is intended for mounting on a mid-ship section of a hull or the hull with minimal curvature. In this paper, the first orientation in Fig. 1 was selected for prototype testing. The proposed orientation of the prototype is broken down into three sections, namely Sections A, B, and C. Section A consist of a winch and a gunwale clamp mount that allows different bulwark thickness. The winch at Section A has a load capacity of 100 kg.

A total length of 5-m rods will be implemented on the OSV according to its maximum draft. The rod clamp shown in Fig. 2 is secured to a part of the rod attached to the hook. However, the suction cup was used to hold the rods in place and to restrict movement when the ship is moving. The portable system design requires only one man to operate. The preinstallation work includes accurate alignment of suction cups where the handles of the suction cups must be perpendicular to the waterline. The installation of the gunwale clamp onto the bulwark plate must be in line with the position of the suction cups. Figure 3 shows the suction cups capable of securing itself onto a metal plate using a vacuum. It is also designed to accommodate hoisting and lowering of the assembly.

3 Program Methodology

The live video will convert the picture frames into a perspective view. The program takes into consideration that the keel has a curvature that caused the image to be taken at a wrong angle. The captured image is then compared using Haar cascade algorithm as shown in Fig. 4. The outcome of the program will determine the percent of barnacles in a given surface area. The data obtained will be saved in each session.

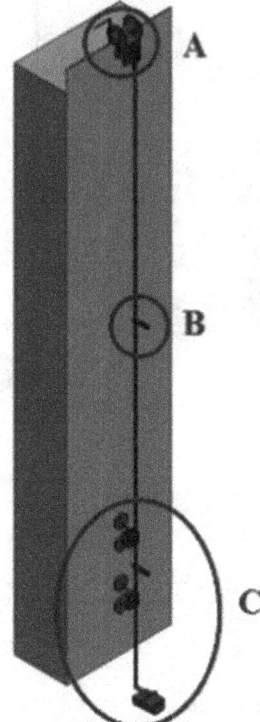

Fig. 1 Layout of proposed system on surface of ship's hull

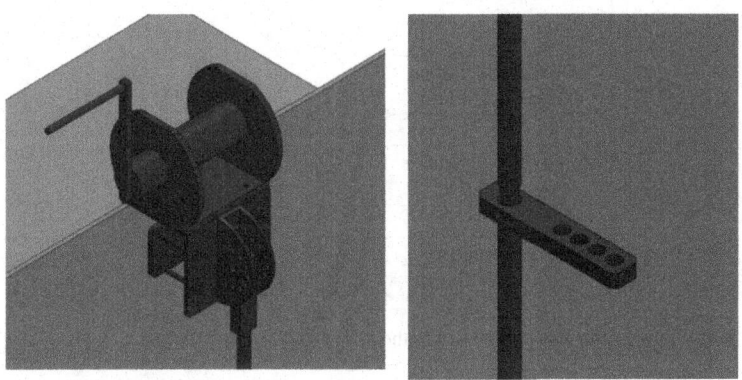

Fig. 2 Section A (left) with winch and B (right) in Fig. 1

Fig. 3 Section C in Fig. 1

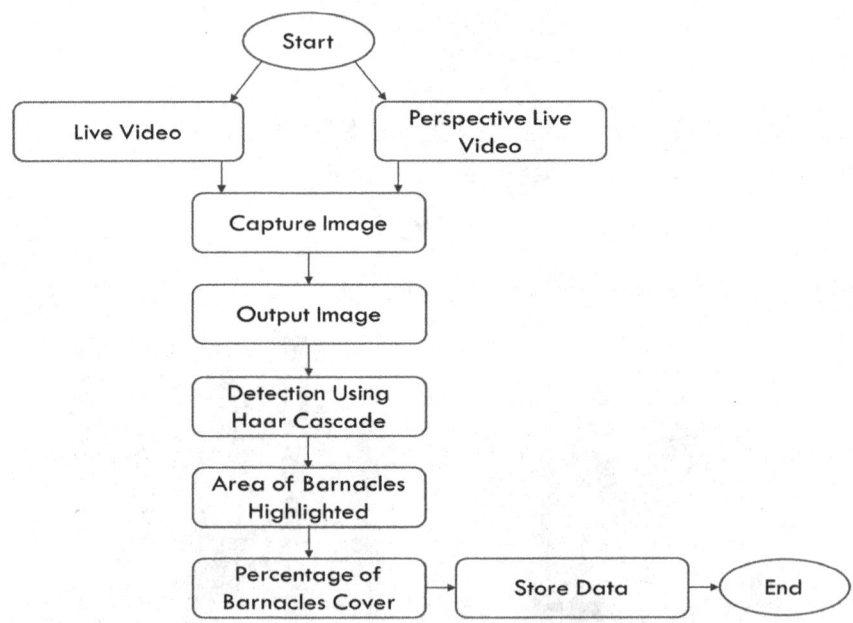

Fig. 4 Software program flowchart using Python

3.1 Haar Cascade Object Recognition and Artificial Neural Networks (ANNs)

There are around 500 images of acorn barnacles, mostly Chthamalus taken from the coastline of Singapore using a high-quality DSLR camera. As shown in Fig. 5,

Fig. 5 Forty images of barnacles selected for training

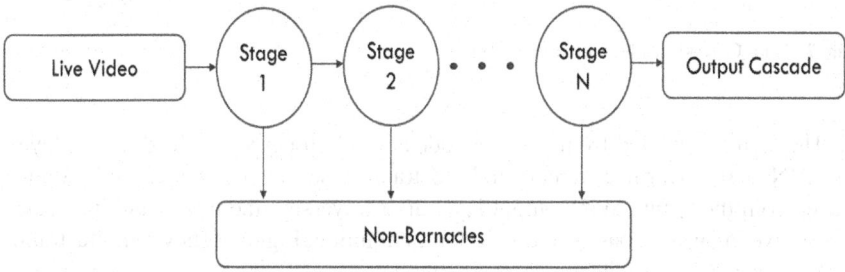

Fig. 6 Cascade classifier structure

forty best images were manually selected for classifier's training. As observed, some parts of the images did not contain any fouling. Haar cascade uses machine learning algorithms for classification of objects based on the massive input images provided. It learns from the input patterns and makes generalizations based on given images. There are around 4000 negative images which were used from the internet sources. The negative images are larger than positive images to allow superimposing of these pictures. The superimposing creates a greater number of positive images by overlaying each cropped image of barnacles on top of each negative image. A total of 25 vector files were generated. Merging of these vector files will produce a final vector file for Haar cascade training. The cascade.xml will be trained and used to compare input images taken from the Raspberry Pi camera module (as the target). The process of the training is illustrated in Fig. 6.

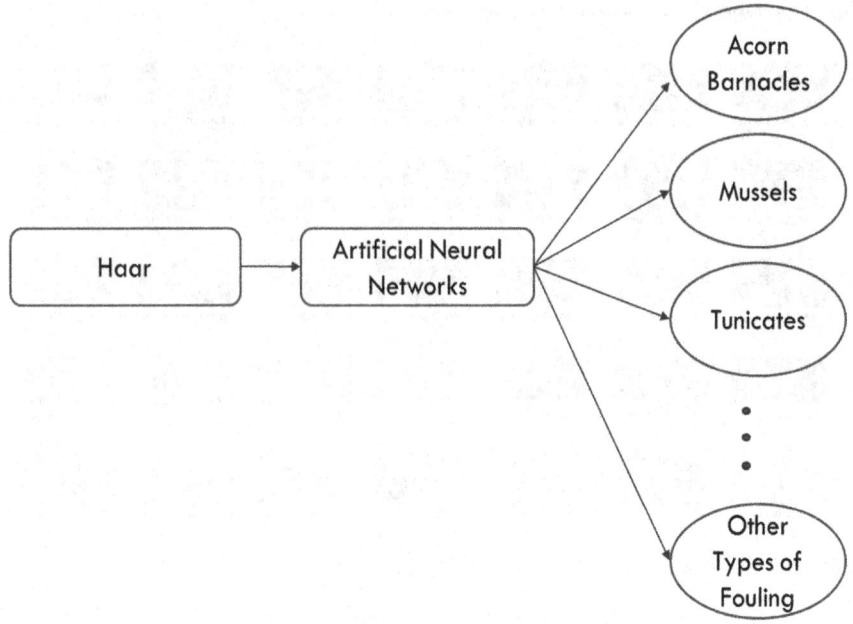

Fig. 7 Haar Cascade with Neural Networks

The results obtained from Haar cascade are used for ANNs. The different layers in ANN will perform different kinds of transformations on their inputs. Signals travel from the inputs to the output layer after traversing the layers multiple times. A massive image database is required for the training. Figure 7 shows an illustration of using the Haar cascade and neural networks.

3.2 Fouling Database

A Web interface is designed for the user to contribute different types of fouling images. It was created by PHP 5.0, and the Web interface was hosted on the Apache web server supported by MySQL for storage. The database server stores and organizes content in a Web browser used by an end user. The Hypertext Preprocessor (PHP) is an interpreter embedded within a Web page to interpret the PHP file when accessing a page. Figure 8 shows the flow in setting up the Web page to allow users to contribute the fouling organisms.

The users can contribute fouling organisms to the database server by choosing image files for uploading. The Web page was supported by a drag-and-drop plug-in which simplifies the process. The information (such as name, type, and category of the fouling) needs to be specified by the contributor before they can upload to the

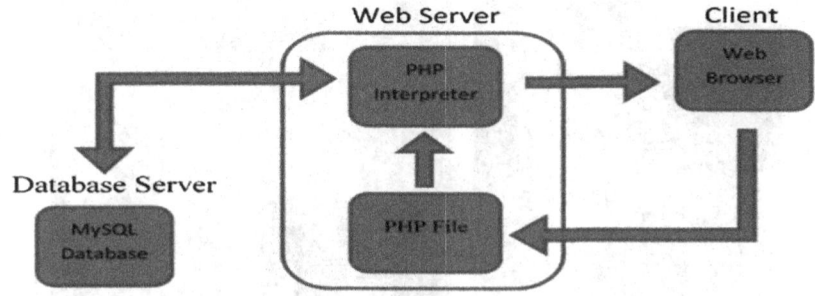

Fig. 8 Web server-to-database server relationship

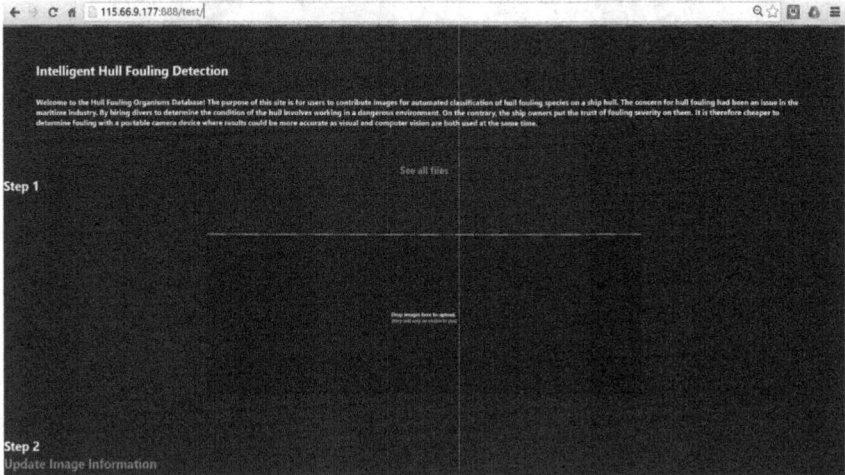

Fig. 9 Fouling database Web page for contributors

server. The contributors can browse and download all fouling images stored in the database as shown in Fig. 9.

4 Experimental Results

The prototype was fabricated using aluminum with black anodized coating which is shown in Fig. 10. A few seals were placed in the camera enclosure to ensure watertight capability. The experiment was conducted at a 1.8-m-depth swimming pool. An underwater test was carried out by mounting the clamp on a goalpost as a substitute for the gunwale of the OSV. There was no leakage observed between the enclosure cover and the enclosure in the water. The Wi-fi connectivity in water was set up as shown in Fig. 11. The Wi-fi was ensured using an RG174 coaxial

Fig. 10 Experimental setup in pool (left) and close view of proposed system

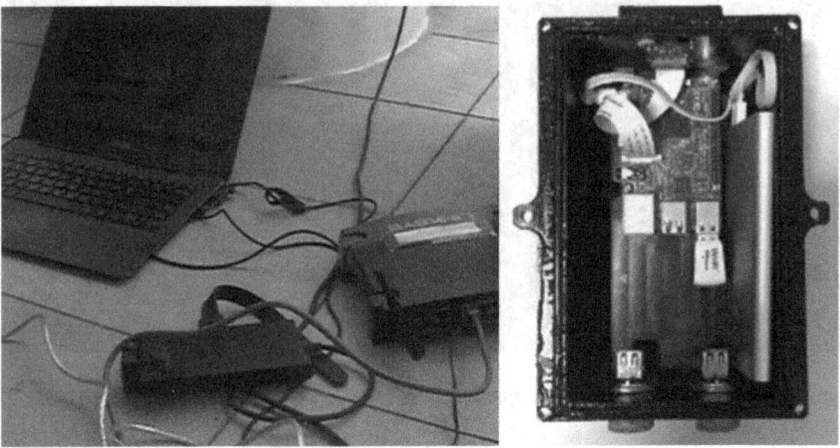

Fig. 11 Wi-fi connectivity setup (left) and enclosure components layout (right)

cable placed between the enclosure and the router. The Raspberry Pi can be either powered by a portable charger inside the enclosure or through a waterproof GTC USB connector as shown in Fig. 11.

A few images of barnacles were used to test the algorithm. The algorithm converts the images to gray scale before comparing the image from the Haar cascade. Figure 11 shows the results obtained from the GUI with a fouling percentage of 75% when tested underwater in Fig. 12. The barnacles such as Semibalanus balanoides seen in the image were identified. Further improvement in the results will require more training on the fouling images.

Fig. 12 Underwater test results (identified Semibalanus balanoides indicated in square boxes)

4.1 Fouling Percentage Test

Fouling percentage in the GUI was also computed using (1). Ten images filled with barnacles were tested to determine the mean area of the barnacles. Figure 13 shows one set of the test results obtained from the GUI. The mean area of fouling was found to be around 20.8%. As given in Table 4, seven test results produced a fouling level of 4 (extensive fouling), 1 of the result has a fouling level of 2 (slight fouling), and remaining tests have a fouling level of 3 (considerable fouling).

$$\text{Fouling Area} = \frac{\text{Area of Fouling}}{\text{Total Area of Image}} \times 100\% \tag{1}$$

However, the direct comparison of the accuracy of other algorithms is unable to perform since the region of interest (ROI) was trained differently and can only be improved if more images of barnacles will use. Furthermore, it will take a longer time for the training. However, the accuracy of the system can be determined by manual counting of barnacles in an image as compared to the output image in (2).

$$\text{Detection Accuracy} = \frac{\text{No. of barnacles detected by algorithm}}{\text{Total no. of barnacles detected visually}} \times 100\% \tag{2}$$

As shown in Table 4, ten tests were performed, and the mean accuracy of the validation is around 50.51% with a standard deviation of 10.44%. Some false detection can be observed in some results due to the poor quality of the camera used on the Raspberry PI.

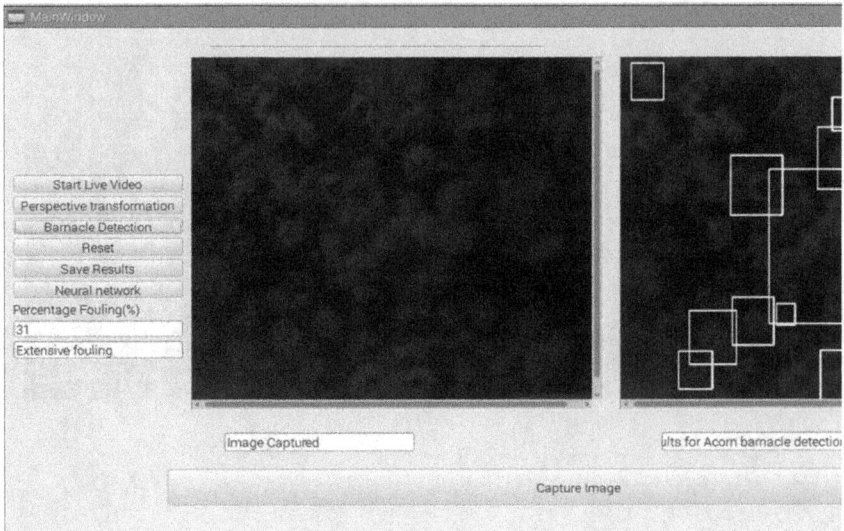

Fig. 13 An example of test results displayed for users

Table 4 Percentage of fouling test results

Test	Manual count	Haar count	Fouling by Haar (%)	Accuracy (%)
1	46	31	17	67.39
2	43	15	23	34.88
3	20	11	17	55.00
4	5	3	5	60.00
5	40	24	31	60.00
6	41	20	38	48.78
7	17	9	10	52.94
8	34	15	12	44.12
9	34	15	25	44.11
10	37	14	30	37.84

Although, a similar application of Haar-like cascade recognition method by Zhao et al. [16] was performed on five different types of the license plate of various sizes. The study showed a success rate of 89.5% with some false detection due to brand logos. This comparison of a result proved that using Haar-like cascade algorithms could produce a high accuracy of detection rate if large numbers of high-quality positive and negative samples are trained. In this paper, the barnacles can differ substantially from another as compared to the letter and numbers used on the license plates. In addition, the design of the system requires the angle of the detection to be specified and manually adjusted may cause inaccuracy in the training results.

5 Conclusion

An intelligent fouling detection using Haar-like cascade classifier was presented. By monitoring the fouling condition on the ship's hull, the shipowners can plan docking schedules and routes to reduce fouling activity. It enables a significant amount of savings on maintenance and docking time. The results achieved by the detection system would also allow better research of new antifouling coatings for the ship's hull. With the fouling database coupled with Haar cascade algorithm, improvements to the accuracy of the fouling detection can be performed automatically in a shorter time frame. Future work will include the use of more machine learning techniques such as unsupervised and extreme machine learning in classifying the types of fouling on the ship's hull.

Acknowledgements The author would like to express the deepest appreciation to Mr. Junyi Goh and Mr. Wong Jun Jie Jeremy graduated from Newcastle University for sharing their graduation reports and logbooks on fouling. The author would like to thank the Global Excellence Fund Award (ID: ISF#51) for sponsoring and supporting the project since November 2015.

References

1. Schultz, M.: Effects of coating roughness and biofouling on ship resistance and powering. Biofouling **23**(5), 331–341 (2007)
2. Schultz, M.P., Bendick, J.A., Holm, E.R., Hertel, W.M.: Economic impact of biofouling on a naval surface ship. Biofouling **27**(1), 87–98 (2011)
3. Wigforss, J.: Benchmarks and Measures for Better Fuel Efficiency. Chalmers University of Technology, Undergraduate (2012)
4. Floerl, O., Inglis, G., Hayden, B.: A risk-based predictive tool to prevent accidental introductions of nonindigenous marine species. Environ. Manage. **35**(6), 765–778 (2005)
5. Woods Hole Oceanographic Institution [WHOI]: Marine fouling and its prevention. United States Naval Institute, Annapolis, MD (1952)
6. Millett, J., Anderson, C.D.: Fighting fast ferry fouling. In: Fast'97. Conference Papers, Australia, vol. 1, pp. 493–495 (1997)
7. Ismail, S.B., Salleh, Z., Yusop, M.Y.M., Fakhruradzi, F.H.: Monitoring of barnacle growth on the underwater hull of an FRP boat using image processing. Procedia Comput. Sci. **23**, 146–151 (2013)
8. Wang, P.F., Lieberman, S., Ho, L.: Unsupervised learning neural network for classification of ship-hull fouling conditions. In: IEEE International Joint Conference on Neural Network Proceedings, pp. 4601–4604 (2006)
9. Goh, J.Y.: Intelligent surveillance system for hull fouling. Undergraduate Final Year Project Report. University of Newcastle Upon Tyne (2015)
10. Yu, W., Xiu, J., Liu, C., Yang, Z.: A depth cascade face detection algorithm based on AdaBoost. In: IEEE International Conference on Network Infrastructure and Digital Content (IC-NIDC), pp. 103–107 (2016)
11. Killioğlu, M., Taşkiran, M., Kahraman, N.: Anti-spoofing in face recognition with liveness detection using pupil tracking. In: 15th International Symposium on Applied Machine Intelligence and Informatics (SAMI), pp. 87–92 (2017)
12. Delgado, J.I.F., Santos, L. G. M., Lozano, R., Hernandez, I. G., Mercado, D.A.: Embedded control using monocular vision: face tracking. In: International Conference on Unmanned Aircraft Systems (ICUAS), pp. 1285–1291 (2017)

13. Guennouni, S., Ahaitouf, A., Mansouri, A.: Face detection: Comparing Haar-like combined with cascade classifiers and edge orientation matching. In: International Conference on Wireless Technologies, Embedded and Intelligent Systems (WITS), pp. 1–4 (2017)
14. Xiang, X., Lv, N., Zhai, M., Saddik, A.E.: Real-time parking occupancy detection for gas stations based on Haar-AdaBoosting and CNN. IEEE Sens. J. **17**(19), 6360–6367 (2017)
15. Wang, L., Zhang, Z.: Automatic detection of wind turbine blade surface cracks based on UAV-taken images. IEEE Trans. Industr. Electron. **64**(9), 7293–7303 (2017)
16. Zhao, Y., Gu, J., Liu, C., Han, S., Gao, Y., Hu, Q.: License Plate Location Based on Haar-Like Cascade Classifiers and Edges, pp. 102–105 (2010)

Image Mosaic Intelligent Vehicle Around View Monitoring System Based on Ring Fusion Method

Yinglin Zhang, Xinyu Zhang, Hongbo Gao and Yuchao Liu

Abstract In order to solve the problem that the edge correction effect of fish-eye image is poor, we proposed a radial de-distortion method. The mapping relationship between the distortion point and the aberration point in the image coordinate system is established by the division model based on the radial distortion. We also setup ideal pattern to extract the target position of correction. Aiming at the problem of low degree of automation and low accuracy in the process of mosaic, the integrated corner detection method is used to automatically extract the coordinate position of the corner points in the calibration pattern. Aiming at the problem that the brightness of each region is inconsistent and the transition effect is not good, the ring fusion method is designed to fuse the image. The experimental results show that the proposed algorithm has the average pixel error of 1.792, the accuracy and efficiency of corner extraction is 100%, which effectively improve the automation degree of the algorithm.

Keywords Around view monitoring system · Radial distortion
Corner detection · Image mosaic · Image fusion

Y. Zhang
State Key Laboratory of Advanced Design and Manufacturing for Vehicle Body,
Hunan University, Changsha 410000, China
e-mail: zyl5251159@163.com

Y. Zhang · X. Zhang
Information Technology Center, Tsinghua University, Beijing 100083, China
e-mail: xyzhang@tsinghua.edu.cn

H. Gao (✉)
State Key Laboratory of Automotive Safety and Energy, Tsinghua University,
Beijing 100083, China
e-mail: ghb48@mail.tsinghua.edu.cn

Y. Liu
Institute of Electronic Engineering of China, Beijing 100039, China
e-mail: yuchao_liu@163.com

© Springer Nature Singapore Pte Ltd. 2019
S. K. Bhatia et al. (eds.), *Advances in Computer Communication
and Computational Sciences*, Advances in Intelligent Systems
and Computing 760, https://doi.org/10.1007/978-981-13-0344-9_35

1 Introduction

In the past decades, because of the rapid growth of road transportation and private cars, road traffic safety has become an important problem in society [1]. The statistics result of the national traffic accident show that the proportion of accidents caused by driver's vision limitation, the delay of reaction, judgment error, and improper operation accounted for up to 40% of the total accidents [2]. In order to solve the above problem, the advanced driving assistance system (ADAS) received more and more attention [3, 4]. In which, around view monitoring system (AVM) is used to provide the driver with the 360-degree video image information around the body, in the parking and crowded city traffic conditions, to help users to better judge the road traffic conditions around the vehicle, to avoid collision with pedestrians and vehicles around, to make driving process more safe and convenient [5]. The existing around view system has many problems such as unaligned mosaic, poor transition effect, and inconsistent region brightness. These problems are mainly caused by defects in two key technologies: fish-eye image correction and image mosaic.

The radial distortion correction method based on convex optimization is proposed with the monotonicity constraint of radial distortion function [6]. The distortion of the image is processed by simultaneously calibrating the camera's radial distortion function and other internal parameters [7]. The literature shows that even the 3-order radial distortion model is not sufficient to adequately correct the fish-eye image [8]. It is difficult to completely eliminate all kinds of distortion of fish-eye image, but we can remove partial distortion according to the specific situation and meet the target requirements. It presents a content-based method, using the conformal constraints and linear constrained function and smoothing equations to ensure maximum correction effect of significant object, but this method needs to manually extract the significant objects in the scene [9]. The proposed local correction method does not depend on the distortion model, using the planarity of target plane (chessboard) and orthogonality constraint to minimize interpolation distortion and residual distortion, but this method cannot guarantee the global optimal correction effect [10].

The overlap points of two correction image are extracted by matching the SFIT features with scale and direction invariance [11]. The feature is usually used to extract overlapping corner points including Harris [12], Canny [13], Moravec [14], and other operators. In the around view monitoring system, we realize image registration by using specific calibration pattern. The overlapping corners are already known, and there are no needs to use feature to find them. The key is to detect corners in calibration pattern. The literature detects the corner points of checkerboard patterns by quadrilateral join rules [15, 16]. The initial corner set is obtained by the improved Hessian corner detector, and the initial concentration of false points is eliminated by the intensity and geometrical features of the chessboard pattern [17]. But the above two methods are mainly designed for the chessboard pattern used in camera calibration, which is not applicable to the requirement of the calibration pattern used in the mosaic process.

The optimal seam is searched by minimizing the mean square variance of pixel in the region, and the adjacent interpolation method is used to smooth the mosaic effect, but this method is not suitable for the scene with too large difference of exposure [18]. Fuse image by weighting several images captured by the same camera with different parameters from the same angle of view, but the method can only adjust the brightness and cannot reduce the color difference [19]. Seamless mosaic is achieved by tone compensation near the optimal seam, but the mosaic seam of around view system is fixed. So, the AVM image cannot be fully fused by this method [20, 21].

This paper makes full use of the advantages of radial distortion model and the objective function based on the radial distortion assumption, and through the ideal pattern set the target of optimal location, to ensure the stability of model calculation and the visual effect of salient objects. Then, aiming at the calibration pattern used in the mosaic process of around view monitoring system, an automatic corner extraction method is designed. According to the characteristic of AVM image, such as the large exposure difference of different view angle, the fixed position of the mosaic seam, annular closed mosaic results, we proposed ring fusion method for AVM image fusion.

The rest of this paper is organized as follows: Sect. 2 introduces the extraction of corner position from the camera image of intelligent vehicle. Section 3 introduces the images stitching. Section 4 introduces algorithm flow. Section 5 illustrates its effectiveness via experiment results. Section 6 concludes this paper.

2 Extract Corner Position

2.1 Fish-Eye Image Correction

The optimization process of the distortion model parameters is shown in Fig. 1. Step 1, input the position of distortional corner P_{id}, target position P_{pu}, initialization parameters $[\lambda_1, \lambda_2]$, $[C_x, C_y]$, γ. Step 2, CoD_{t-1}, Λ, P_{id} are substituted into the distortion model to compute P_{iu}, and use P_{pu}, P_{iu} to compute the Homogeneity matrix H, and then use the matrix H to unify the coordinate system of CoD_{t-1}, P_{pu}, P_{iu}. After that, we get CoD_H, P_{pu}, P_{iuH} to optimize the position of center of distortion by cost function 1. Then we obtain the optimization result CoD_t. Step 3, CoD_t is used to compute P_{iu} again with Λ, P_{id}, and use P_{pu}, P_{iu} to compute the homogeneity matrix H, and then use the matrix H to unify the coordinate system of CoD_t, P_{pu}, P_{iu}. After that, we get CoD_H, P_{pu}, P_{iuH} to optimize the distortion coefficient by cost function 2.

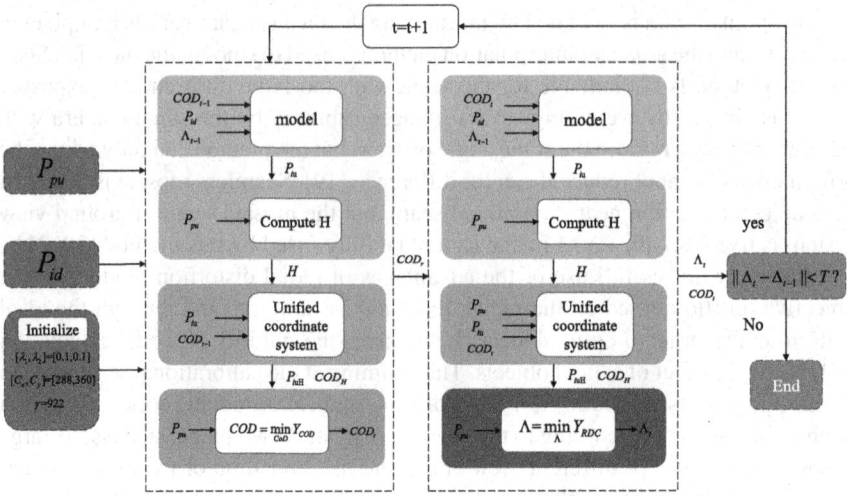

Fig. 1 Algorithm flow of distortion correction

2.2 Extract the Corner Position of Calibration Scene

Through the integrated corner detection method, the coordinate positions of feature points of the mosaic calibration pattern in correction image are extracted and stored for the next registration matrix calculation.

Algorithm Steps

1. Input distortion calibration image from each camera;
2. Use the Rufli corner detection method to detect the corners in the chessboard array;
3. Use the relative position relationship between the black box and the checkerboard array to obtain the ROI area of the big black box.
4. Preprocess the ROI region by adaptive binarization and a closed operation;
5. Obtain the contour of big black box by contour fitting method, where we use the following method to move redundant contour.

   (1) Limit the minimum area of the contour. The threshold is calculated as shown in Eq. (1).

$$\text{b\_area}_{min} = \begin{cases} 1.5 * \text{cb\_area}_{avg} \ (\textit{front and rear view}) \\ 0.1 * \text{cb\_area}_{avg} \ (\textit{left and right view}) \end{cases} \tag{1}$$

where b_area$_{min}$ denotes the threshold of minimum area of big black box, cb_area$_{avg}$ denote the average area of small box in chessboard array.

(2) Limit contour shape. The shape restriction rules mainly include aspect ratio, symmetry, and the area ratio between contour and envelope rectangular, as shown in Eq. (2).

$$\begin{cases} d_1 \geq 0.15 * p \cap d_2 \geq 0.15 * p \\ 8 * d_3 \geq d_4 \cap 8 * d_4 \geq d_3 \\ area_{contour} > 0.3 * area_{rect} \end{cases} \tag{2}$$

where d_1, d_2 denotes the diagonal length of the quadrilateral, d_3, d_4 denotes the length of adjacent side, p denotes quadrilateral circumference, $area_{contour}$ denotes the area of contour, and $area_{rect}$ denotes the area of envelope rectangle.

6. Use the SUSAN method to locate the exact position of the vertex near the initial position obtained by above step;
7. Substituting the corner position extracted from the distorted picture into the distortion correction model, to calculate its coordinate position in the corrected image.

2.3 Compute the Target Position in Bird's-Eye View Output Image Plane

Set up the calibration scene required for splicing, as shown in Fig. 2. The target position in the bird-view output image corresponding to position of the calibration pattern corner in the world coordinate system can be calculated. For example, we describe how to calculate the target position of point 1 (as shown in Fig. 2). First, we calculate the position of point 1 in the world coordinate system, as shown in Eq. (3).

$$\begin{cases} x_1 = -\frac{1}{2}W - w_w - w_{b1} \\ y_1 = -\frac{1}{2}L - w_w - w_{b1} \end{cases} \tag{3}$$

where the original point of world coordinate system is located in the center of the calibration scene, as shown in Fig. 2. (x_1, y_1) denotes the position of point 1, W denotes the vehicle width, L denotes the vehicle length, w_w is the white edge width, w_{b1} is the width of big black box. And then we can use the position in world coordinate system to calculate the position in output image coordinate system, as shown in Eq. (4).

$$\begin{cases} scale = W_{img} / W_{real} \\ u_1 = scale * x_1 + W_{img}/2 \\ v_1 = scale * y_2 + L_{img}/2 \end{cases} \tag{4}$$

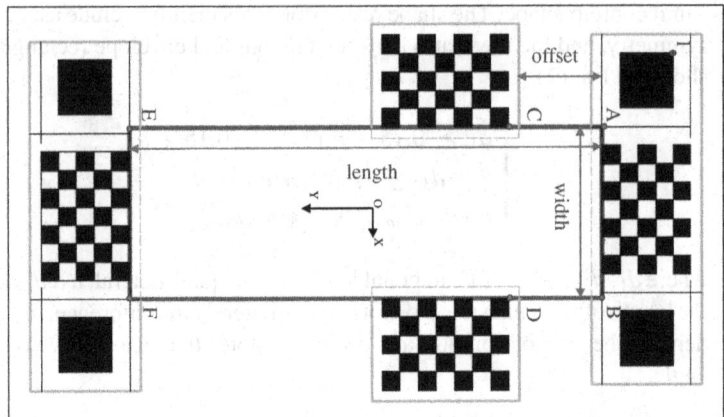

Fig. 2 Illustration of calibration scene

where *scale* denotes the scaling factor from world coordinate system to image coordinate system. W_{img} denotes the width of bird-view output image. W_{real} denotes the width of real scene. (u_1, v_1) denotes the position in bird-view output image coordinate system. (x_1, y_1) denotes the position in the world coordinate system.

3   Image Mosaic

3.1   *Image Registration and Coordinate System Unification*

We can obtain the correction image of each fish-eye camera with good visual effect by LUT. However, the final goal of the AVM is to generate a 360° bird's-eye view image. Therefore, it is necessary to rotate, zoom and perspective transform the correction images, and unify the coordinate system of them, to generate a 360° bird's-eye view image, as shown in Fig. 3.

By the homography matrix, we can project correction images in the bird's-eye view coordinate system. The transform matrix can be calculated after obtaining the position of each corner point in the corrected image coordinate system and its target position in the top view image coordinate system. The form of the homography matrix is shown in Eq. (5), and the calculation formula is shown in Eq. (6).

$$H = \begin{bmatrix} a_{11} & a_{12} & a_{13} \\ a_{21} & a_{22} & a_{23} \\ a_{31} & a_{32} & a_{33} \end{bmatrix} \tag{5}$$

$$P_b = H * P_u \tag{6}$$

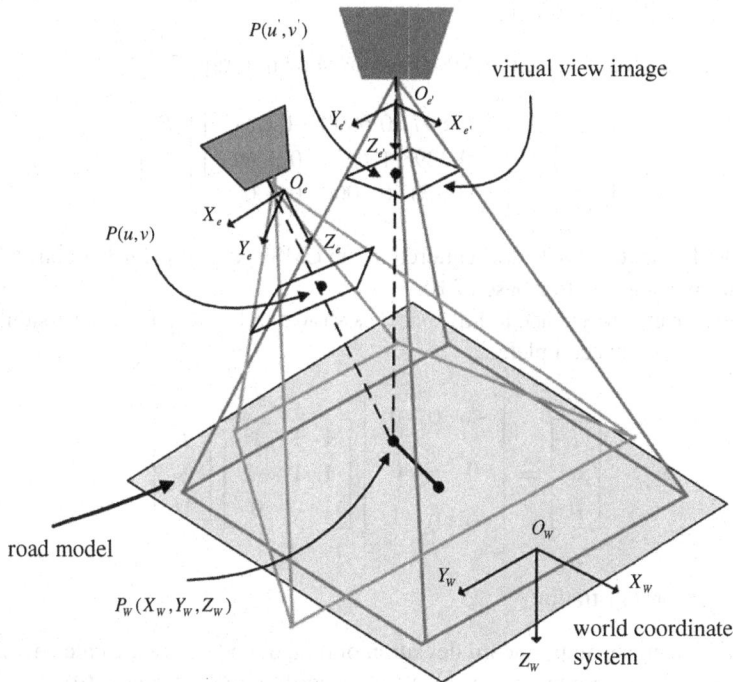

Fig. 3 Illustration of image mosaic

where $P_u = [x_u, y_u, 1]^T$ denotes the corner position in the corrected image coordinate, system. $P_b = [x_b, y_b, 1]^T$ denotes the target position in the top view image coordinate system.

3.2 Image Blend

As the content and light is different on 4-view angle of the same scene, the image obtained from four cameras is different in the brightness, saturation, and color. So, the effect of mosaic directly is not good. In order to realize the seamless mosaic, a ring fusion method is proposed. The method includes equalization preprocessing, ring color matching, and interpolation processing.

(1) Equalization preprocessing

The imadjust color adjustment function in MATLAB is used to equalize each view image.

(2) Ring color matching

Step 1:   Spatial transformation

Firstly, from RGB space to LMS space, as shown in Eq. (7).

$$\begin{bmatrix} L \\ M \\ S \end{bmatrix} = \begin{bmatrix} 0.3897 & 0.6890 & -0.0787 \\ -0.2298 & 1.1834 & 0.0464 \\ 0.0000 & 0.0000 & 0.0000 \end{bmatrix} \begin{bmatrix} R \\ G \\ B \end{bmatrix} \tag{7}$$

Secondly, since the data are scattered in the LMS space, it is further converted to a logarithmic space with a base of 10.

Finally, from LMS space to $l\alpha\beta$ space, as shown in Eq. (8). The conversion from RGB to $l\alpha\beta$ space is completed.

$$\begin{bmatrix} l \\ \alpha \\ \beta \end{bmatrix} = \begin{bmatrix} \frac{1}{\sqrt{3}} & 0 & 0 \\ 0 & \frac{1}{\sqrt{6}} & 0 \\ 0 & 0 & \frac{1}{\sqrt{2}} \end{bmatrix} \begin{bmatrix} 1 & 1 & 1 \\ 1 & 1 & -2 \\ 1 & -1 & 0 \end{bmatrix} \begin{bmatrix} L \\ S \\ M \end{bmatrix} \tag{8}$$

Step 2:   Color registration

Firstly, the mean and standard deviation of on l, α, and β axes are calculated. The color registration parameters are calculated according to the formula (9).

$$\begin{cases} t_{21}^{l} = \sigma_{v1}^{l}/\sigma_{v2}^{l} \\ t_{21}^{\alpha} = \sigma_{v1}^{\alpha}/\sigma_{v2}^{\alpha} \\ t_{21}^{\beta} = \sigma_{v1}^{\beta}/\sigma_{v2}^{\beta} \end{cases} \tag{9}$$

where t_{21}^{l} denotes registration parameters from the viewing angle 2 to the viewing angle 1 on the channel l, the rest is similar. σ_{v1}^{l} denotes the variance of viewing angle 1 on the channel l, and the rest is similar. According to the above calculation of the registration parameters, let image 2 be registered with image 1.

Step 3:   Global optimization

By using the ring color scheme, the global optimum is guaranteed:

$$V_2 \rightarrow V_1, V_3 \rightarrow V_2, V_4 \rightarrow V_3, V_1 \rightarrow V_4$$

(3) **Interpolation processing**

After color registration, the visual effects of the panorama mosaic image have been greatly improved. But around the mosaic seam, the effect is still not enough. Therefore, we use the bilinear interpolation technology around mosaic seam to ensure smooth transition.

4   Algorithm Flow

The algorithm flow of Image Mosaic is shown in Fig. 4. Firstly, we input scene image and correct the distortion of image by using LUT. Secondly, we extract the corner position in correction image by using integrated corner detection method and compute the target position in bird's-eye view image. After that, the homography matrix H is calculated. Finally, we project correction images of 4 view angle on the bird's-eye view image plane by using matrix H and blend them by using ring fusion method.

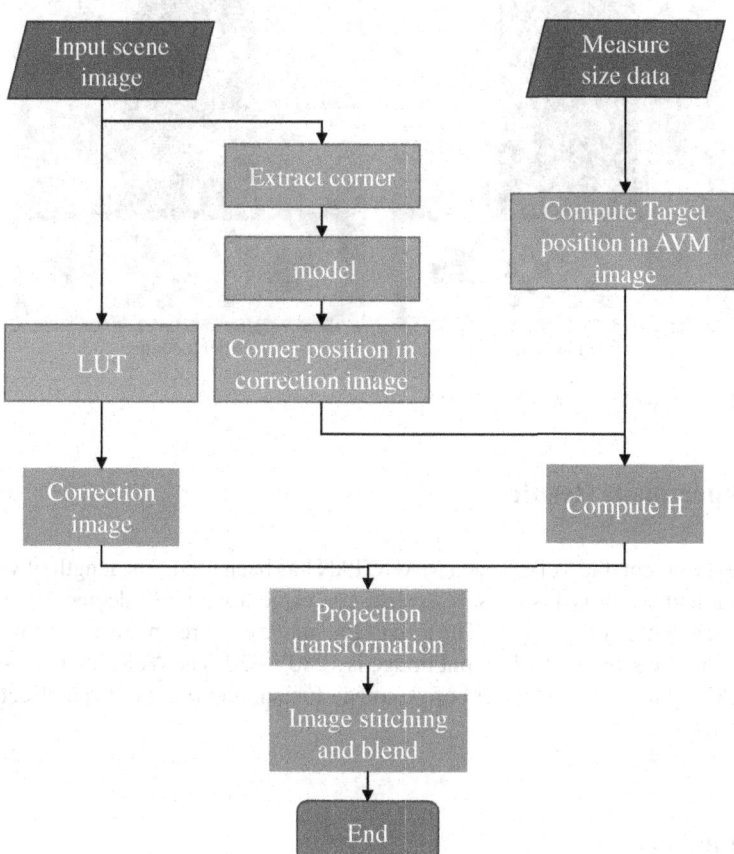

Fig. 4 Flowchart of algorithm

(a)   **(b)**

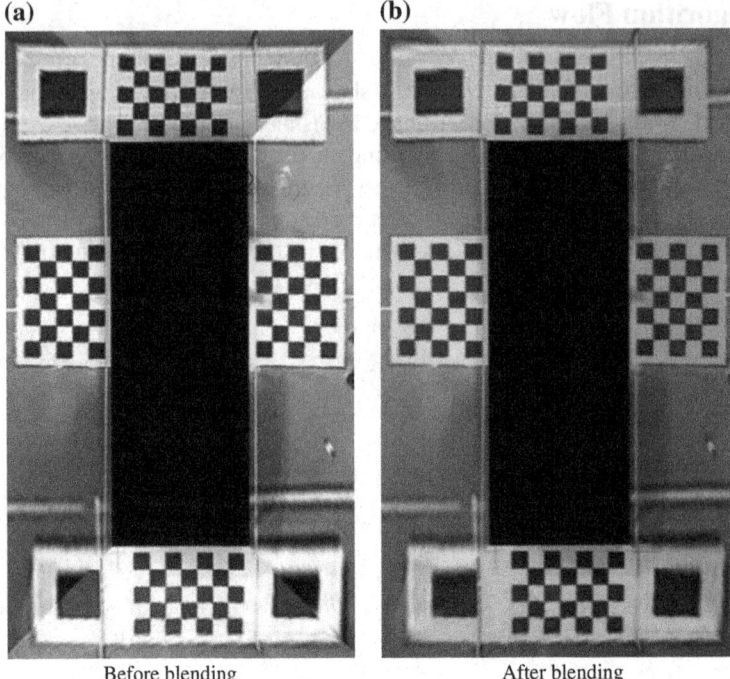

Before blending   After blending

Fig. 5 Stitched bird-view image of AVM

5 Experiment Result

In our experiment, the Volkswagen MAGOTAN has been used. The length of vehicle is 4.8 m, and width is 1.8 m. We use fish-eye camera with 180 degree large view angle, focal length of 2.32 mm. The size of the image captured by fish-eye camera is 720 * 576. The size of AVM output image is 1280 * 720. The AVM image is shown in Fig. 5. Figure 5a is the effect before image fusion, and Fig. 5b is the effect after image fusion.

6 Conclusion

The radial de-distortion method proposed in this paper fully guarantees the effect of fish-eye image edge correction. The integrated corner detection proposed in the process of mosaic automatically extracts the corner position of the calibration pattern, improving the degree of automation of the algorithm. The performance of ring fusion method we proposed is very good, especially around the mosaic seam. However, some problems still need to be improved. In the future, we can use the scene image to directly calibrate the camera to improve the degree of algorithm automation further. At the same time, it is also possible to combine the around view monitoring system with the lane departure warning algorithm and the dynamic object detection algorithm to provide more auxiliary driving function for the driver at a lower cost and improve the road traffic safety.

Acknowledgements This work was supported by Junior Fellowships for Advanced Innovation Think-tank Program of China Association for Science and Technology under Grant No. DXB-ZKQN-2017-035, Project funded by China Postdoctoral Science Foundation under Grant No. 2017M620765, the National Key Research and Development Program of China under Grant No. 2016YFB0100903, Beijing Municipal Science and Technology Commission special major under Grant No. D171100005017002 and No. D171100005117002, and National Natural Science Foundation of China under Grant No. U1664263.

References

1. Guo, C., Meguro, J., Kojima, Y., et al.: A multimodal ADAS system for unmarked urban scenarios based on road context understanding. J. IEEE Trans. Intell. Transp. Syst. **16**(4), 1690–1704 (2015)
2. Ministry of Public Security Traffic Administration People's Republic of China Road Traffic accident statistic annual report Jiangsu Province Wuxi: Ministry of Public Security Traffic Management Science Research Institute (2011)
3. Su, C.L., Lee, C.J., Li, M.S., et al.: 3D AVM system for automotive applications. In: International Conference on Information, Communications and Signal Processing, pp. 1–5. IEEE (2015)
4. Gao, H.B., Zhang, X.Y., Zhang, T.L., Liu, Y.C., Li, D.Y.: Research of intelligent vehicle variable granularity evaluation based on cloud model. J. Acta Electronica Sinica **44**(2), 365–374 (2016)
5. Santhanam, S.M., Balisavira, V., Roh, S.H., et al.: Lens distortion correction and geometrical alignment for Around View Monitoring system. In: International Symposium on Consumer Electronics, pp. 1–2 (2014)
6. Ying, X., Mei, X., Yang, S., et al.: Radial distortion correction from a single image of a planar calibration pattern using convex optimization. In: IEEE International Conference on Image Processing, pp. 3440–3443. IEEE (2015)
7. Hartley, R., Kang, S.B.: Parameter-free radial distortion correction with center of distortion estimation. J. IEEE Trans. Pattern Anal. Mach. Intell. **29**(8), 1309–1321 (2007)
8. Shah, S., Aggarwal, J.K.: Intrinsic parameter calibration procedure for a (high-distortion) fisheye lens camera with distortion model and accuracy estimation*. J. Pattern Recogn. **29**(11), 1775–1788 (1996)
9. Carroll, R., Agrawal, M., Agarwala, A.: Optimizing content-preserving projections for wide-angle images. J. ACM Trans. Graph. **28**(3), 1–9 (2009)
10. Marcon, M., Sarti, A., Tubaro, S.: Piecewise distortion correction for fisheye lenses. In: IEEE International Conference on Image Processing, pp. 4057–4061. IEEE (2015)
11. Jiang, Z., Wu, J., Cui, D., et al.: Mosaic method for distorted image based on SIFT feature matching. In: International Conference on Computing and Networking Technology, pp. 107–110. IEEE (2013)
12. Sipiran, I., Bustos, B.: Harris 3D: a robust extension of the Harris operator for interest point detection on 3D meshes. J. Vis. Comput. **27**(11), 963 (2011)
13. Huo, Y.K., Wei, G., Zhang, Y.D., et al.: An adaptive threshold for the Canny Operator of edge detection. In: International Conference on Image Analysis and Signal Processing, pp. 371–374. IEEE (2010)

14. Jiang, L., Liu, J., Li, D., et al.: 3D point sets matching method based on Moravec vertical interest operator. In: Proceedings of the 2011 2nd International Congress on Computer Applications and Computational Science, pp. 53–59. Springer, Berlin, Heidelberg (2012)
15. Zhang, X.Y., Gao, H.B., Guo, M., Li, G.P., Liu, Y.C., Li, D.Y.: A study on key technologies of unmanned driving. J. CAAI Trans. Intell. Technol. 1(1), 4–43 (2016)
16. Rufli, M., Scaramuzza, D., Siegwart, R.: Automatic detection of checkerboards on blurred and distorted images. In: IEEE International Conference on Intelligent Robots and Systems, pp. 3121–3126. IEEE (2008)
17. Liu, Y., Liu, S., Cao, Y., et al.: Automatic chessboard corner detection method. J. Image Process. IET 10(1), 16–23 (2016)
18. Pulli, K., Tico, M., Xiong, Y.: Mobile panoramic imaging system, pp. 108–115 (2010)
19. Tang, Y., Shin, J.: Image mosaic with efficient brightness fusion and automatic content awareness. In: International Conference on Signal Processing and Multimedia Applications, pp. 60–66. IEEE (2016)
20. Cha, J.H., Jeon, Y.S., Moon, Y.S., et al.: Seamless and fast panoramic image mosaic. In: IEEE International Conference on Consumer Electronics, pp. 29–30. IEEE (2012)
21. Gao, H.B., Zhang, X.Y., Liu, Y.C., Li, D.Y.: Uncertainty control of autonomous vehicles based on cloud Model. J. Sci. Programm. 24(12), (2016)

Part III
Advanced Communications

Innovative Security Techniques to Prevent Attacks on Wireless Payment on Mobile Android OS

Maurizio Cavallari, Francesco Tornieri and Marco de Marco

Abstract Mobile technologies are increasingly pervading a substantial portion of everyday life. In particular, the economic sector of consumers and private sales has shown a very high rate of utilization of mobile applications. Mobile payments are no exception, and the economic development relies more and more on mobile technologies. Bank institutions and financial firms are privileged targets for cyber attacks and organized crime, exploiting vulnerabilities of smart mobile devices in particular for host card emulation wireless payments. The research analysis was based on mobile platform Android and identified ten original (novel) controls that can avoid possible attacks on payment transactions and/or privacy. The study has practical implications as practitioners and organizations, like banks, shall control the risks associated with the taxonomy of tamper IDs. Organizational implications can be regarded as the need for banks to look into software development for mobile applications.

Keywords Smartphones · Android · Mobile · Payments · Tampers
Attacks · Security

1 Introduction

Mobile technologies are increasingly pervading a substantial portion of everyday life. The economic sector of consumers and private sales has shown, in particular, a very high rate of utilization of mobile applications. Mobile payments are no exception, and economic development resides more and more in mobile technologies [1]. Banks and

M. Cavallari (✉) · F. Tornieri · M. de Marco
Università Internazionale Telematica UniNettuno, Rome, Italy
e-mail: maurizio.cavallari@uninettunouniversity.it; maurizio.cavallari@uninettunouniversity.net

F. Tornieri
e-mail: francesco.tornieri@uninettunouniversity.it

M. de Marco
e-mail: marco.demarco@uninettunouniversity.it

© Springer Nature Singapore Pte Ltd. 2019

S. K. Bhatia et al. (eds.), *Advances in Computer Communication
and Computational Sciences*, Advances in Intelligent Systems
and Computing 760, https://doi.org/10.1007/978-981-13-0344-9_36

financial firms are privileged targets for cyber attacks and organized crime, tampering and compromising smart mobile devices [2].

Smart mobile devices can act as effective computing devices, and this characteristic made them very much utilized for mobile commerce and money transfer. Security experts and security firms such as Finland-based F-Secure are stating that the widespread of smart devices will drive malware creators and attackers to target smartphones [3].

The mentioned study, of more than 1,200 Android malware application samples [2], found that malicious programs *"are rapidly evolving and existing anti-malware solutions are seriously lagging behind"* cit. [3]; the sentence is confirmed, even highlighting the worst scenario, in the F-Secure report of November 2015 [3].

Other studies show the magnitude of intrusion and data exfiltration from mobile devices, giving space to attacks to payment transactions and even possible cyber attacks on a large scale [4].

As smartphones have access to the entry points of vendor's networks and bank's operational systems, the risk is shared from the personal devices and could step up to a systemic-level escalating to the vendors of goods, card processors, and issuing banks, among others intermediaries.

Also automation and robotics (A&R) is a field where smartphones entered with pervasive presence and critical tasks are also dependent on smartphones and professional skills. In fact, not only A&R requires skilled and trained employees in order to effective deployment of processes and procedures, but also issues about smartphone security shall be considered [5].

Mass markets are being reached by mobile technologies as demonstrated in a recent confirmatory case study, where paths are found in order to outline the market penetrability of smartphones and their ability to reach large consumer masses. The technology level is demonstrated to have substantial possibilities of interaction and operation of devices among different platforms, such as the case of smartphones [6]. Research into customers' acceptance of mobile payment applications suggests that several factors are to be considered to understand the acceptance of mobile payment services; those factors are: hardware and software compatibility, individual mobility, and subjective norm (i.e., rules and laws) [5, 7].

For the aforementioned reasons, research into smartphone host card emulation (HCE), the breaking technology adopted to facilitate mobile payments, is particularly important to protect customers of mobile banking [8].

HCE is the frontier technology used to process payments, so-called in proximity, allowing a smoother and easy payment process. Card issuers, bankers, processors, and business owners are all subject to PCI DSS regulation [9].

Networked aspects of everyday life have gained common currency in society. This observable reality depicts a society that is profoundly intermingled with information technologies and communication networks, embracing a wide variety of aspects of social life [10]. Research into e-Government, points out that new trends in the public sector are changing the way governments are organizing and offering their services to citizens, rapidly moving onto smartphone applications [11].

2 Literature Review

Among the global community of scholars and practitioners, the expectations are high that cyber threat intelligence (including payment systems' disruption and information stealing) will be the main threat that will drive organizations and network infrastructure to improve security. Security recommendations and best practices are mandatory on pushing any threat as far as possible from the target [12]. The mentioned study shows that new mobile malware increased in 2015 to almost 2.5 million new malware, while the total known mobile malware accounts for more than 12 million at the end of 2015. Android mobile OS remains the most attacked environment as on the 12 million total mobile malware [12]. Data analyses from 2016 reports show same trends and risks [13].

Internet banking is becoming very popular, and in mobile banking and payment applications, banks are finding new means to reduce costs and raise margins, like in [14]. This pertains also to relatively new markets like China and Asia. The security issues surrounding Internet banking and mobile payments should represent the focus of both banks and customers. Much emphasis has been laid on advanced security technology and the human factor in security, and recent researches show a low customer awareness of Internet banking security such as the research in China [15]. The findings of this research evaluated current levels of customer awareness of Internet banking security as scoring rather low. However, other research results show that beside perceived cost and perceived risk, which are generally regarded as two important adoption drives, the other three factors that influence the use of the third-party mobile payment are perceived benefit, perceived convenience, and herd mentality convenience [16]. The authors argued that a qualitative approach should give space to quantitative methods.

The dilemma of the qualitative versus quantitative approach to information security risk has been much debated in the scientific literature [17]. Some authors state that standard methodologies reveal lack of rigor, suggesting that research in the field shall introduce time constraint and moral hazard as factors influencing the researcher risk assessment, both in quantitative information security risk assessment models and in qualitative models. These research findings present practical limits because it is very difficult to predict in the long term the financial losses coming from insufficient security policies, lack of communication, and training about security procedures [18].

Other directions of research into performance and risk point to human resources and human factors [19, 20].

Interesting research into organizational culture and the so-called risk culture argues that just spreading within an organization the fashion of "risk culture" will not resolve the real problems of high risks [21]. The author states that in order to stimulate the proper awareness of risk throughout the organization, the baseline culture itself should be changed. Pointing to the real and actual organization culture means that each firm is individual and there is not a generalized model that would fit, about risk. The study argues that different organizations would require different models, and different business units within the same organization would require different approaches. Results demonstrate that it is not possible to impose an organization's culture, but it should be first understood in order to be able to possibly change it and to eventually ensure that it remains changed [22]. The research [21] proposes a risk culture framework based on a (partially) successful effort to change banking culture in the UK (i.e., Treating Customers Fairly—the TCF initiative) developed by the UK financial regulator.

Research into mobile payment demonstrated that mobile payments had been a strong driver of socioeconomic enhancement of general markets and in e-Government gaining a comprehensive perspective of social meaning [23, 24].

Advanced technologies, economic environments, and ubiquitous diffusion of smart devices have pushed smartphone payment app utilization to a high edge. The mentioned research states that effective facilitators or, on the contrary side, barriers to the mobile payment diffusion, are: security, interoperability, and ease of use.

Innovative approaches to mobile payments and mobile credit cards using host card emulation/near-field communication (HCE/NFC) have been investigated by researchers [25] considering human factors and referring to well-established theory [26]. The novelty of the approach proposed by Tan, Ooi, Chong, and Hew broadens the traditional technology acceptance model (TAM), implying four add-ons constructs. The findings showed that the only non-significant factors are financial risks and that gender had no moderating effect. The mentioned study confirmed the validity of TAM theory for HCE/NFC mobile payments realm [25].

A recent research based on a case study [6] examines openness to evaluate the mobile payment platform's market potential within a threefold approach: technology, user aspects, and provider levels.

The study searched responses to the following question: "*to what extent can opening (or closing) each level increase or decrease a platform's market potential?*" cit. [6].

The technology level pertains to the cross-operability of application services among various mobile technologies.

The user level segments the customer base with respect to different technological platforms.

The provider level points to the strategic commitment of businesses that provide an application operational environment for mobile payments.

Based on both, analytical and theoretical arguments, the mentioned research proposes structured analysis about the possible opens of a variety of markets, demonstrating that there is a strong demand of cross-platforms operability of mobile payment environments, including the novelty of HCE/NFC as a key factor.

3 Host Card Emulation Basics

In order to avoid the need for Secure Element (SE) that resides on mobile phone SIM cards, host card emulation (HCE) came into play. The HCE technology allows the virtualization of a smart card by emulating via software the same functionalities, so that the presence of the hardware component (i.e., SIM card) is no longer needed and connectivity and payments are made possible independently from hardware components [27].

HCE is a technology based on near-field communication (NFC) that provides connectivity in two-way and short-range mode: When two NFC devices (the initiator and the target) are matched (within a radius of about 4 cm), a peer-to-peer connection between the two is created and both can send and receive information. The main cases for using this communication mechanism are in the financial world: NFC-enabled devices can be used in contactless payment systems in the mobile environment.

NFC is widely used on terminals equipped with version 4.0 (Ice Cream Sandwich), and since the release of Android 4.4 (KitKat), Google subsequently introduced support for secure transactions based on NFC via HCE. In systems based on iOS (Apple), the NFC chip is preinstalled as in iPhone 6 and iPad Air 2 (the chip was missing in earlier devices), but HCE payment transactions are possible only with Apple Pay® system; Windows Phones do not present, at present time, "root" methods probably due to their little diffusion and are not considered a profitable target. Thus, the current market of HCE mobile payment systems is based on solutions designed for Android OS.

The HCE solution permits the creation of an exact, virtual representation of a smart card using only software components, effectively eliminating the need for SE hardware in the device. HCE is defined as the ability to exchange information via NFC between a terminal (enabled for the exchange of information with an NFC card) and a mobile application configured to emulate the functionality of an NFC card. The novelty lies in the fact that HCE enables the NFC protocol to be routed directly through the operating system of the device rather than being first processed by SE hardware.

HCE can therefore be defined as open architecture on which it is possible to develop application solutions that have the possibility of emulating the payment card and that interface directly with the NFC controller, without the need to resort to SE hardware available on the device. When a smartphone is placed near the reader, the NFC controller calls an Android (HCE service) previously associated with identifiers reserved for payment networks that identify payment applications.

Having eliminated the need for SE hardware in the smartphone, card data to be emulated is stored by the issuer's authorization center in the "cloud": The payment application's task is to recover the information needed to make the payment from the cloud. This shift, however, is not to be considered only as a modification of the data flow (data movement) but also as the renouncing of the use of the secure hardware cryptographic capability of the SE in storing and processing such data in a secure session with the terminal; the mobile device becomes a transducer that connects the POS terminal with the image of the card stored in the network.

The current process between the POS and SE is divided into two parts by delegating the task (local) of the SE to applications on the issuer card images (remote). This solution also simplifies the chain, reducing the role of third parties and enabling freedom from mobile network operators and, contrary to what happens with the classic NFC, from any kind of conditioning by SIM or SD card, with a consequent reduction in fees. At the same time, the current network consisting of POS terminals requires no changes for using contactless cards or NFC-enabled smartphones with SE hardware.

4 Host Card Emulation in Practice

The host card emulation (HCE) architecture in Android revolves around the concept of "HCE services" which run in the background without any user interface.

This is convenient since many HCE applications (like loyalty cards) do not require the user to launch any application to use the service, but simply place the device near to an NFC reader to start the service correctly and execute the transaction in the background. When the user places the smartphone near to an NFC reader, the Android operating system must be able to identify which HCE service is going to be contacted. To solve this problem, and to define a means of selecting the wallet application required, the concept of application identification (AID) is employed. These AIDs are nothing more than a sequence of 16 bytes that are well known and publicly recorded by the major payment networks (such as Visa and Mastercard) and used by Android to determine which service the NFC reader is trying to contact.

Major credit card issuers/managers/processors like Mastercard® and Visa® use the following application identifiers (AIDs):

- Mastercard, "A0000000041010"
- Visa, "A0000000031010".

It should be noted that the Android implementation of HCE was created to allow simultaneous support to other card emulation methods, including the use of SE hardware (the so-called standard NFC). This simultaneous presence is based on the mechanism of "AID routing," according to which the NFC controller maintains a routing table consisting of a list of rules, each of which contain an AID and a destination (which may be either the CPU running Android applications or an SE).

The first message sent by the NFC terminal contains the "SELECT AID" command, which is intercepted by the NFC controller, which then controls if the specified AID belongs to at least one of its routing rules. If there is a match, the destination in the routing rule selected is identified as the recipient of the subsequent NFC communication.

The main difference between the two routes then resides in the recipients of the communications. If the target is the SE, everything proceeds in the traditional method

Table 1 Cardholder's data, both normal and sensitive, contained on credit cards and transmitted/exchanged while pursuing a payment transaction

Card holders' ("normal") data	Sensitive authentication data
Primary account number (PAN)	Comprehensive data on magnetic strip (or equivalent present on chip-card)
Card holder's name	CAV2/CVC2/CVV2/CID
Expiry date	PIN/PIN Blocks
Service code	...

of contactless payments via NFC. If the communication is HCE, the controller routes the communication directly to the operating system. Android considers as payment applications all those HCE services that have declared, in their manifest, the use of AIDs in the "payment" category. Furthermore, Android contains a "Tap and Pay" menu in its settings, which lists all these payment features and apps. Within the app menu, users can choose the default application to be used for payments, whenever the smartphone comes into pairing with a payment terminal, i.e., point of sale (PoS).

Currently, payment solutions based on HCE are supported by all those terminals enabled with NFC-connections and which use Android 4.4 (the benchmark platform for the entire industry) as the operating system, given the spread of Android at the expense of the Windows Phone (NB: The NFC chip is only a recent addition to Apple devices).

In its implementation of HCE, Google (now owner of the system) states that Android 4.4 supports many protocols and standards adopted by the most common types of NFC readers, for which contactless payment applications have currently been developed. The new operating system implements card emulation based on the NFC-Forum ISO-DEP specifications (in turn based on ISO/IEC 14443-4) and can handle Application Protocol Data Units (APDUs) defined by ISO/IEC 7816-4.

5 International Regulation and Payments

The reference standard for payment channels, mail order, telephone, and e-commerce is called the "Payment Card Industry Data Security Standard," i.e., PCI DSS. Since companies are constantly at risk of losing sensitive data relating to holders of credit or debit cards, resulting in the possibility of incurring fines, lawsuits, and bad publicity, PCI DSS regulatory compliance is one of the main objectives of companies that handle, store, transmit, or process credit card data [9].

PCI DSS standards apply wherever data is archived, retrieved, computed, or transmitted. Data streams account of personal cardholder information as well as "Sensitive Authentication Data," i.e., SAD, as indicated in Table 1.

If the name of the cardholder, the service code, and/or expiry date are archived, computed, or transmitted along with the Primary Account Number (PAN) or are otherwise included in the cardholder information, such data shall be secured complying with regulations and constraints of PCI DSS except for Requirements 3.3 and 3.4 [9] which are applicable only to the PAN (which corresponds to what is most commonly understood as "credit card number").

The PCI DSS represents a compulsory baseline of controls that can be expanded by laws and regulations at local, regional, and sector levels. Moreover, the legislative and regulatory statements may demand the securitization of personal information or other identifiable elements (e.g., card holder name).

Privacy laws and Consumer Data Protection and European Directives about information security are only few examples of legislative regulation that impacts on mobile payment transactions. Compliance with the following PCI DSS requirements is required:

- **Requirement 3: Protect stored credit card cardholder data**. Various protection solutions are required to protect cardholder sensitive information. Those solutions include encryption, data-truncation, information-masking, and hash-coding. If, by chance, an attacker bypasses frontline security mechanisms and accesses internal data, he/she would find only encrypted information. The robustness of encryption algorithms to be utilized is so that, lacking the appropriate decryption keys, the attacker would not be able to utilize data. To limit possible risks, other effective methods for the protection of stored data should be considered (e.g., truncating cardholder information).
- **Requirement 4: Encrypt cardholder data transmitted over open, public networks**. Sensitive information must be encrypted during transmission over networks that unauthorized users can access.
- **Requirement 6: Develop and maintain secure systems and applications**. Unauthorized users exploit vulnerabilities to gain privileged access to systems. Many of these vulnerabilities are fixed by vendor-supplied security patches, which must be installed by the entities that manage the systems. All critical systems must have the latest software patches.
- **Requirement 8: Assign a unique ID to anyone with access to a computer**. Assign a unique ID to all users who have access, to ensure that each user is uniquely accountable for their actions. The mentioned requirements are applicable to all processors, also to PoS. Requirements 8.1, 8.2, and 8.5.8 to 8.5.15 are not required, however, for accounts within a payment application from a PoS that can process single card numbers, so to make single transaction easier and ergonomic (e.g., cash transactions).
- **Requirement 10: Track and monitor all access to network resources and cardholder data**. This requirement from PCI DSS demands the capabilities of pertaining systems to strictly control access and to keep records of users' operations and activities. Those elements are required to prevent, to detect, and to mitigate the negative effects of a possible data breach. All processing environment shall maintain logs for revision and analysis if data compromises arise. Logs and activity-recall mechanisms, and in general the possibility to back track user activities, are key pillars in avoiding, detecting, and mitigating the negative impact of data tampering

or exfiltration, specifically regarding sensitive data or simply trigger all stages of problem determination or alerting should problems arise.

The requirements of PCI DSS are mandatory in order to obtain certification needed to operate in mobile payments [9].

6 Smartphone Security

The method devised by the authors and proposed in this paper is meant to highlight new and innovative mitigation techniques to prevent and reduce tampering of mobile devices. Tamper techniques applied by smartphone users are designed to exploit vulnerabilities in mobile systems to gain administrative privileges (the so-called user root) and so avoid all the restrictions imposed by the vendor (i.e., Google). The user's goal is to exploit the potential of the system at will and install pirated apps without having to pay the fee to the company that developed the software.

The analysis was based on the more widespread mobile platform Android (iOS as explained earlier is not feasible for investigation; Windows Phone platform was excluded from this phase because of its limited diffusion and the inability, at present, to carry out "rooting" of the system). Ten tamper checks for Android were then identified.

The level of risk can be assessed with respect to the proposal of "Open Source Security Testing Methodology Manual" [28], as follows.

Overall Risk = Probability of the event * Impact of the compromising event
where

- The "probability of the event" can be defined as the degree of real perspective that the unwanted (malicious) event, or attack would occur.
- The "impact of the compromising event" can be described as the potential damage resulting from the actual occurrence of the unwanted event.
- The "Overall Risk" is then the product of the álea (i.e., the uncertainty of the realization of that unwanted event), and the concrete damage that the specific event would generate.

The big picture can be appreciated by the matrix below (Fig. 1—Risk Matrix).

Once a set of controls, described below, had been defined, an impact value for each control was used to limit any false positives (to be considered in view of the stringent controls in place); once the maximum value of "impact" (I) had been established as equal to 10, a minimum value (5) above which the device will be considered "compromised" was identified, therefore: TAMPERED DEVICE: $I \geq 5$.

The identified risks may belong to the following categories: high risk, where $I = 5$; medium risk, where $I = \sum I_i$ $i = 1...n$. Lower risks had not been considered.

For the mentioned environment (i.e., Android mOS), the relevant controls are defined as follows.

(CTRL001) Standard checks carried out using open source "RootTools" library integrate the following controls.

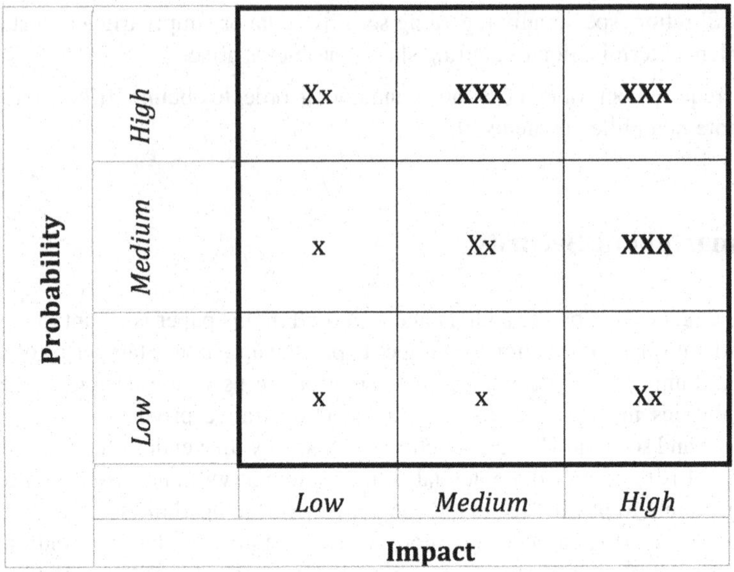

Fig. 1 Risk matrix. The level of overall risk is configured from estimated possibility of a negative event intertwined with the actual damage the occurrence would generate

Table 2 Taxonomy and classification of controls with corresponding impact values

Tamper ID	Risk	Impact value (I)
CTRL001	HIGH	5
CTRL002	MEDIUM	2
CTRL003	MEDIUM	2
CTRL004	HIGH	5
CTRL005	HIGH	5
CTRL006	MEDIUM	2
CTRL007	HIGH	5
CTRL008	MEDIUM	2
CTRL009	HIGH	5
CTRL010	MEDIUM	2

a. Check "busybox"
b. Check "su"
c. Check "root command"

(CTRL002) Verification of system default attribute: In a non-tampered-with device, the "ro.build.tags" variable must contain the "release-keys" value, whereas a "rooted" device will have the "test-keys" value.

(CTRL003) Verification for certificates issued by vendors considered "trust" necessary for over-the-air (OTA) updates (file presence "otacerts.zip") (Table 2).

(CTRL004) Check to identify those developers mainly involved in the release of rooting or hiding software. The identification of these will be made based on the id associated with the developer by the "Play Store" (which enabled the minimization of false positives).

a. chainfire
b. chainsdd
c. kinguser
d. clockworkmod
e. geohot
f. jcase
g. beaups
h. root_de_la_vega
i. devadvance

(CTRL005) Check for malicious applications installed by the user for rooting the device or to hide tampering. The following list shows the most widely spread malicious applications.

a. com.noshufou.android.su
b. com.thirdparty.superuser
c. eu.chainfire.supersu
d. com.koushikdutta.superuser
e. com.zachspong.temprootremovejb
f. com.ramdroid.appquarantine
g. kinguser
h. OTA RootKeeper
i. Superuser
j. sa.root.toolkit
k. framaroot
l. WeakSauce
m. Towelroot
n. Xposed Installer
o. Stump
p. RootCloak
q. Cygwin

(CTRL006) Check for non-official, third-party Android operating system (ROM) (non-tampered devices will use the Android operating system chosen by the manufacturer). The following list shows the most widely used non-official (dangerous) ROM.

a. Cyanogenmod (Cyanogen)
b. JellyBam (Jelly Bam)
c. AOKP
d. Paranoidandroid (Paranoid)
e. MIUI

f. Slim ICS
g. Liquid Smooth
h. SuperNexus
i. SlimKat

(CTRL007) The tampering process of the device needs the binary "su" (i.e., SuperUser) required to obtain administrator privileges; then, it checks for the following binaries.

a. /system/bin/su
b. /system/xbin/su
c. /sbin/su
d. /system/su
e. /system/bin/.ext/.su
f. /system/usr/we-need-root/su-backup
g. /system/xbin/mu.

Note. If the "su" command is found, it is advisable to execute it and to check, using the "id" command, for the current "root" permissions (administrator). For example: uid = 0(root) gid = 0(root).

(CTRL008) In a compromised system, the file system portions that are not normally accessible because of permission restrictions can be accessed. A check of the folder read permission was carried out (if granted, the system is compromised).

a. /data.
   A check of write permission for the following folders was also carried out (if granted, the system is compromised).
b. /data
c. /
d. /system
e. /system/bin
f. /system/sbin
g. /system/xbin
h. /vendor/bin
i. /sys
j. /sbin
k. /etc.
l. /proc
m. /dev.

(CTRL009) Tampering with an Android device is normally related to the installation of BusyBox, a tool used to obtain a number of Unix binaries normally used by advanced users to execute commands on systems. The following commands were tested in the system under analysis.

a. busybox whoami
b. busybox df

(CTRL010) A folder can be kept "hidden" by using execution-only permissions (a technique used by some malware). A "hidden" folder may have been created ad hoc by the user to hide code or executable code that would, in normal checks, reveal tampering. Checks can be carried out as follows.

a. "--x--x--x"

7　Conclusions

A research team from Technische Universität Darmstadt and Fraunhofer SIT [4], along with McAfee Labs [12, 13], investigated in March 2015 two million mobile applications from three of the major backend as a service providers (i.e., Cloud-Mine, Amazon AWS, and Facebook Parse). The research team found 56 million sets of unencrypted data (and nor otherwise protected). Large cloud databases were accessed utilizing original, untampered mobile apps, exposing highly sensitive personal information, such as names, e-mails, passwords, pictures, money transactions, and health records. All the (legitimately) accessed information that was exposed by mobile applications could have been easily used by malicious attackers to perpetrate identity theft, spamming, malware distribution, and the like [12].

Our research demonstrates that Android smartphones show an explosion to vulnerabilities and attacks due to their mobile operating system (mOS) characteristics, as also reported by other empirical observations [29]. The results confirm other findings from prior research into external storage compromising and installing malware through storage data cards (SD Cards) [30].

The novelty of the research undertaken is a brand-new taxonomy of tamper IDs (see Table 2) that are to be considered in terms of internal control and restriction of permissions while developing and testing a mobile application. The taxonomy applies to Android systems as HCE transactions on iOS are restricted—to date—to Apple Pay® system; Windows Phones do not represent a target.

The present study has practical as well as theoretical implications.

Practical implications as practitioners and organizations, like banks, shall control and demand, from mobile payment applications developers, the utmost attention to cope with the risks and impact associated with the taxonomy of tamper IDs [31]. Awareness of the tampers taxonomy and their impact for banks and other organizations while dealing with internal mobile software development teams, as well as with external software developers, can make the difference between a robust mobile payment application capable of resisting attacks, even terrorist or cyber attacks, and an application that fails under tampering attempts and puts money, resources, and privacy at high risk [32].

Organizational implications of the present study can be regarded as the need for banks and other financial intermediaries to look into software development for mobile applications [33] and to have a special kind of internal compliance to evaluate and test—periodically—payment applications. Nowadays, there is no awareness of such risk and everything falls under the "assurances" made by developers [34, 35]. External developers are the most frequent case for mobile payment applications (HCE or otherwise) [36, 37]. Hence, the need for internal controls (within the committing organizations) and a set of compliance rules that take into account our research results are envisaged and would be particularly appropriate to prevent risks and cyber attacks on money and privacy [38, 39].

8   Envisaged Future Research Directions

Brand-new research findings that fit well with present paper results are published by Brown et al. [40]. Research into similarities with human immune system permitted the proposal of a multiple-detector set artificial immune system (mAIS) [40] serving the discovery of Android smartphone malware.

Other recent results that link perfectly with present paper findings can be observed in Leeds and Atkinson [41] where the researchers experimented machine learning techniques to discriminate malicious Android apps from benign ones. The level of accuracy ranged between 75 and 80%, leaving a huge scope for further investigation in that direction. An early detection of malware by artificial intelligence machine learning could represent a breakthrough into Android mOS security.

Controls over massive app stores (e.g., Google Play) and, in particular, automatic controls over Android applications, both by code analysis and inference from neural networks [41–43], are a direction to look to, as up-to-date research indicates that existing control mechanisms are not sufficient to protect from malicious software [44]. Trust and consumer behaviors when downloading Android apps from massive Application Markets are also aspects to be considered for future research [45].

Acknowledgements A preliminary, abridged, version of the present study was presented at the Conference of the Lebanese Association for Information Systems, Lebanese Chapter of AIS: "ICT for greater development impact," Beirut (Lebanon) June 2–4, 2016.

References

1. Vasquez, S., Simmonds, J.: Mobile application monitoring. In: Proceedings—International Conference of the Chilean Computer Science Society, SCCC, art. no. 7814430, pp. 30–32 (2017)
2. Deloach, J., Caragea, D., Ou, X.: Android malware detection with weak ground truth data. In: Proceedings—2016 IEEE International Conference on Big Data, Big Data 2016, art. no. 7841008, pp. 3457–3464 (2017)
3. F-Secure: Threat Report (2015). https://www.f-secure.com/documents/996508/1030743/Threat_Report_2015.pdf

4. Rasthofer, S., Bodden, E., Castillo, C., Hinchliffe, A., Huber, S.: We know what you did this summer: Android banking Trojan exposing its sins in the cloud (2015). https://www.virusbull etin.com/uploads/pdf/conference_slides/2015/Huber-etal-VB2015.pdf
5. Casalino, N., De Marco, M., Rossignoli, C.: Extensiveness of manufacturing and organizational processes: an empirical study on workers employed in the European SMEs. In: 2nd International KES Conference on Smart Education and Smart e-Learning, SEEL 2015. Smart Innovation, Systems and Technologies, vol. 41, pp. 469–479. Sorrento (2015)
6. Ondrus, J., Gannamaneni, A., Lyytinen, K.J.: The impact of openness on the market potential of multi-sided platforms: a case study of mobile payment platforms (September 2015). J. Inf. Technol. **30**(3), 260–275 (2015)
7. Huang, H., Zheng, C., Zeng, J., Zhou, W., Zhu, S., Liu, P., Chari, S., Zhang, C.: Android malware development on public malware scanning platforms: a large-scale data-driven study. In: Proceedings—2016 IEEE International Conference on Big Data, Big Data 2016, art. no. 7840712, pp. 1090–1099 (2017)
8. Li, L., Li, D., Bissyande, T.F., Klein, J., Le Traon, Y., Lo, D., Cavallaro, L.: Understanding Android App piggybacking: a systematic study of malicious code grafting. IEEE Trans. Inf. Forensics Secur. **12**(6), art. no. 7828100, 1269–1284 (2017)
9. PCI DSS (2006–2016). https://www.pcisecuritystandards.org/document_library
10. vom Brocke, J., Becker, J., De Marco, M.: The networked society. Bus. Inf. Syst. Eng. **58**(3), 159–160 (2016)
11. Zardini, A., Rossignoli, C., Mola, L., De Marco, M.: Developing municipal e-Government in Italy: the city of Alfa case. Lecture Notes in Business Information Processing, vol. 169 LNBIP, pp. 124–137 (2014)
12. McAfee: McAfee Threat Reports November 2015 (2015). http://www.mcafee.com/us/resourc es/reports/rp-quarterly-threats-nov-2015.pdf
13. McAfee: McAfee Threat Reports September 2016 (2016). https://www.mcafee.com/us/resour ces/reports/rp-quarterly-threats-sep-2016.pdf
14. Leeds, M., Atkison, T.: Preliminary results of applying machine learning algorithms to android malware detection. Proceedings—2016 International Conference on Computational Science and Computational Intelligence, CSCI 2016, art. no. 7881497, pp. 1070–1073 (2017)
15. Zhu, R.: Customer awareness of Internet Banking security in China. In: WHICEB 2015 Proceedings. Paper 2 (2015). http://aisel.aisnet.org/whiceb2015/2
16. Zhao, K., Xi, Z.: Analysis on affecting factors of the users' adoption of third-party mobile payment. In: WHICEB 2015, Proceedings. Paper 76 (2015). http://aisel.aisnet.org/whiceb201 5/76
17. Cavallari, M.: The role of extraordinary creativity in response to digital threats. In: D'Atri et al. (ed.) Information Technology and Innovation Trends in Organizations, 1st edn., pp. 479–486. XVI, Physica Verlag Heidelberg (2011)
18. Schilling, A.: A framework for secure IT operations in an uncertain and changing environment. Comput. Oper. Res. **85**, 1339–1351 (2017)
19. Casalino, N., Cavallari, M., De Marco, M., Ferrara, M., Gatti, M., Rossignoli, C.: Performance management and innovative human resource training through flexible production systems aimed at enhancing the competitiveness of SMEs IJKM, IUP J. Knowl. Manag. **XIII**(4), 29–42 (2015)
20. Iannotta, M., Gatti, M., D'Ascenzo, F.: The diffusion of ICT across Italian corporate universities: An exploratory study. Lecture Notes in Information Systems and Organisation, vol. 19, pp. 37–47 (2016)
21. McConnell, P.J.: A risk culture framework for systemically important banks. J. Risk Gov. **3**(1)
22. Cavallari, M.: A grand master and an exceptional mind. Eur. J. Inf. Syst. **14**(5), 463–464 (2005)

23. Cavallari, M.: Information systems security and end-user consciousness—a strategic matter. In: Management of the Interconnected World—ItAIS: The Italian Association for Information Systems, pp. 251–258 (2010)
24. Sorrentino, M., De Marco, M.: Implementing e- government in hard times. When the past is wildly at variance with the future. Inf. Polity 18(4), 331–342 (2013)
25. Tan, G.W.-H., Ooi, K.-B., Chong, S.-C., Hew, T.-S.: NFC mobile credit card: The next frontier of mobile payment? Telemat. Inform. 31(2), 292–307 (2012)
26. Bellini, F., D'Ascenzo, F., Dulskaia, I., Savastano, M.: Digital service platform for networked enterprises collaboration: a case study of the NEMESYS project. Lecture Notes in Business Information Processing, vol. 247, pp. 313–326 (2016)
27. Cavallari, M., Adami, L., Tornieri, F.: Organisational aspects and anatomy of an attack on NFC/HCE mobile payment systems. In: David, R., Carlos, B., Daniel, M. (eds.) ICEIS 2015—17th International Conference on Enterprise Information Systems, Proceedings, vol. 685–700, pp. 27–30. Springer, Barcellona (2015)
28. OSSTMM Open Source Security Testing Methodology Manual (2016). http://www.isecom.org/mirror/OSSTMM.3.pdf
29. Chebyshev, V.: Mobile attacks! Kasperski Labs (2014). http://www.securelist.com/en/blog/805/Mobile_attacks
30. Do, Q., Martini, B., Choo, K.-K.R.: Enforcing file system permissions on Android external storage. In: Proceedings of 13th IEEE International Conference on Trust, Security and Privacy in Computing and Communications (TrustCom 2014). IEEE Computer Society Press (2014)
31. Feizollah, A., Anuar, N.B., Salleh, R., Suarez-Tangil, G., Furnell, S.: AndroDialysis analysis of Android intent effectiveness in malware detection. Comput. Secur. 65, 121–134 (2017)
32. F-Secure: Threat Report 2014 H2 (2014). https://www.f-secure.com/documents/996508/1030743/Threat_Report_H2_2014
33. Imgraben, J., Engelbrecht, A., Choo, K.-K.R.: Always connected, but are smart mobile users getting more security savvy? A survey of smart mobile device users. Behav. Inf. Technol. 33(12), 1347–1360 (2014). Taylor & Francis, Inc. Bristol, PA, USA
34. Schierz, P.G., Schilke, O., Wirtz, B.: Understanding consumer acceptance of mobile payment services: an empirical analysis. Electron. Commer. Res. Appl. 9(3), 209–216 (2010)
35. Casalino, N., Cavallari, M., De Marco, M., Gatti, M., Taranto, G.: Defining a model for effective e-government services and an inter-organizational cooperation in public sector. In: ICEIS 2014—Proceedings of the 16th International Conference on Enterprise Information Systems, vol. 2, pp. 400–408 (2015)
36. Spagnoletti, P., Resca, A., Lee, G.: A design theory for digital platforms supporting online communities: a multiple case study. J. Inf. Technol. 30 (2015)
37. Cavallari, M.: Analysis of evidences about the relationship between organisational flexibility and information systems security. In: Information Systems: Crossroads for Organization, Management, Accounting and Engineering: ItAIS: The Italian Association for Information Systems, pp. 439–447 (2013)
38. Ferrari, A., Rossignoli, C., Zardini, A.: Enabling factors for SaaS business intelligence adoption: a theoretical framework proposal. In: D'Atri, A., Ferrara, M., George, J.F., Spagnoletti, P. (eds.) Information Technology and Innovation Trends in Organizations, pp. 355–361. Springer, Berlin (2011)
39. Dameri, R.P., Sabroux, C.R., Saad, I.: Driving IS value creation by knowledge capturing: theoretical aspects and empirical evidences. In: D'Atri, A., Ferrara, M., George, J.F., Spagnoletti, P. (eds.) Information Technology and Innovation Trends in Organizations, pp. 73–81. Springer, Berlin (2011)
40. Brown, J., Anwar, M., Dozier, G.: An artificial immunity approach to malware detection in a mobile platform. Eurasip J. Inf. Secur. 2017(1), art. no. 7 (2017)

41. Leeds, M., Atkison, T.: Preliminary results of applying machine learning algorithms to Android malware detection. In: Proceedings—2016 International Conference on Computational Science and Computational Intelligence, CSCI 2016, art. no. 7881497, pp. 1070–1073 (2017)
42. Wang, K., Song, T., Liang, A.: Mmda: Metadata based malware detection on android. In: Proceedings—12th International Conference on Computational Intelligence and Security, CIS 2016, art. no. 7820536, pp. 598–602 (2017)
43. Oulehla, M., Oplatkova, Z.K., Malanik, D.: Detection of mobile botnets using neural networks. In: FTC 2016—Proceedings of Future Technologies Conference, art. no. 7821774, pp. 1324–1326 (2017)
44. Choliy, A., Li, F., Gao, T.: Obfuscating function call topography to test structural malware detection against evasion attacks. In: International Conference on Computing, Networking and Communications, ICNC 2017, art. no. 7876235, pp. 808–813 (2017)
45. Za, S., Marzo, F., De Marco, M., Cavallari, M.: Agent based simulation of trust dynamics in dependence networks. In: Nóvoa, H., Drăgoicea, M. (eds.) Exploring Services Science. IESS 1.5. Lecture Notes in Business Information Processing, vol. 201, 243–252. Springer, Heidelberg (2015)

Role of Virtualization Techniques in Cloud Computing Environment

Geeta and Shiva Prakash

Abstract In Internet, cloud computing plays an important role to share information and data. Virtualization is an important technique in the cloud environment to share data and information. It is also important computing environment to enables academic IT resources or industry through on-demand dynamically allocation. The main aim of this research paper is to explore the basic knowledge terms of the virtualization and how virtualization works in cloud system. We will explain about how to maintain the virtualization with optimized resources such as storage, network, application, server, and client in cloud computing. We will compare different open-source-based hypervisors or virtual monitor machines (VMM) that are in use today, and we will discuss several issues of virtualization which will be very helpful to the researchers for further study.

Keywords Hypervisor · Cloud computing · Virtual machine
RedHat enterprise virtualization (RHEV)

1 Introduction

Nowadays, cloud computing is a latest on-demand pay as you go technology that allows customers to access large amount of data, information, and various other computing resources. There is a term virtualization (virtual machine), it is a one type of techniques that helps to enhance the efficiency and availability of cloud computing. It is possible to work on various applications and operating systems simultaneously over the single server with the help of virtualization; hence, virtualization enhances the flexibility and utility of hardware. It decreases the utilization of hardware, costs,

Geeta (✉)
Dr. APJ. Abdul Kalam Technical University Lucknow, Lucknow, India
e-mail: geetasingh02@gmail.com

S. Prakash
M.M.M. University of Technology, Gorakhpur, India
e-mail: shiva.plko@gmail.com

© Springer Nature Singapore Pte Ltd. 2019
S. K. Bhatia et al. (eds.), *Advances in Computer Communication and Computational Sciences*, Advances in Intelligent Systems and Computing 760, https://doi.org/10.1007/978-981-13-0344-9_37

Fig. 1 Type 1 hypervisor

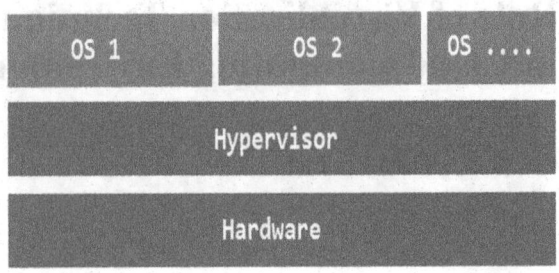

saves energy, and helps to run several applications and several OS on the single server at one time. Virtualization improves the efficiency, flexibility, and utilization of underlying hardware of computer systems.

The basic idea of the virtualization is to be controlling or managing the workload of computing to make it more effective, efficient, economical, and scalable. Virtualization [1–4] is a method or technique which allows to create abstract layer of cloud computing resources and hides the complexity of software and hardware of the systems. Virtualization commonly implemented with hypervisor [5, 6] technology, which is a software or firmware elements that can virtualizes system resources. Virtualization provides several benefits including saving energy and time, minimizing overall system risk, and decreasing costs. It provides capability to increase productivity, prevent data loss, manage resources, and provide remote access securely.

(I) **Hypervisor or Virtual Machine Monitor**: Virtualization is implemented with the help of software called as hypervisor or a virtualization manager. Hypervisor is also called as virtual machine monitor (VMM) [7]. Hypervisor lies between hardware and guest OS. The hypervisor controls the memory, network connectivity [8], processor, and other parts by allowing various operating systems to run on the single machine without help of source code. The guest operating system cannot manage works in cloud computing environment and works only within its virtual machine. However, it is available for a guest operating systems to manage VM acts if a cloud service is IaaS [9]. A hypervisor can be divided into two types based on the various levels of implementation.

(i) **Bare Metal/Native Hypervisor/Type 1 hypervisor**
(ii) **Hosted Hypervisor/Type 2 hypervisor**

Type 1 hypervisors sitting direct on the hardware and communicate between virtual machine and hardware. Type 1 hypervisor is also known as a "bare metal" or "native" or "embedded" hypervisor in vendor literature. There is no requirement of host [10] OS in type 1 hypervisor because they run direct on the hardware (physical machine) as shown in Fig. 1. Citrix XenServer, Microsoft Windows Server 2012 Hyper-V VMware vSphere/ESXi, open-source Kernel-based Virtual Machine (KVM) and RedHat Enterprise Virtualization (RHEV) are examples of bare metal or type 1 hypervisors.

Fig. 2 Type 2 hypervisor

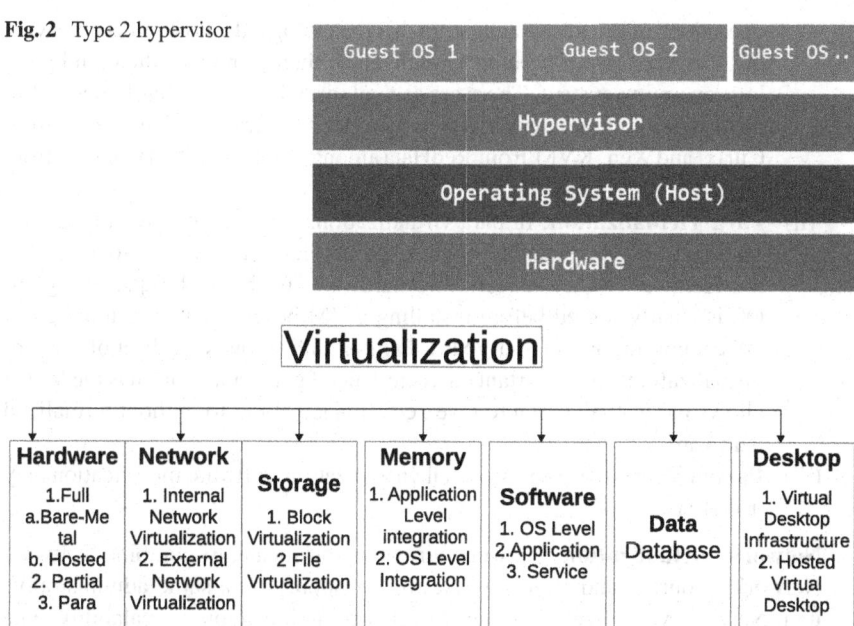

Fig. 3 Types of virtualization in cloud computing

Type 2 hypervisor is sitting on the OS and then manages virtual machine and support other guest operating systems shown in Fig. 2. There is an extra layer between physical machine and virtual machine in the hosted hypervisor. Type 2 hypervisor is dependent on host OS. Microsoft Virtual PC, VirtualBox and Vmware Workstation and Oracle Virtual Box are examples of type 2 hypervisor. The host or guest machines (or domain) are used to describe various roles in the hypervisor. Host OS (or host machine or domain) holds a hypervisor which is used to control VMs, and in guest OS (or guest machine or domain) each VM lying on host machine in a secure and isolated environment with its logical domain.

(II) **Types of Virtualization in Cloud Computing**: There are different types of virtualization in cloud computing. Some important types of virtualization are shown in Fig. 3.

a. **Server/Hardware Virtualization**: It is very important virtualization technique as it gives benefits of hardware utilization and application uptime. In this technology, several small servers (physical server) combined into single big physical server, so the processor can be utilized more efficiently and effectively.
   Hardware virtualization is of the following types:

   (i) **Full Virtualization**: In full virtualization, VM can run on any OS supported by the underlying hardware. Full virtualization is the common virtualization technology in which the virtual processors reproduce the CPU operations of the host machine. In this technology, the host OS is not aware of other versions of it being virtualized so there is no modification by the host operating system. There are several virtualization technologies of full virtualization these are VMware's family of hypervisors, XenServer from Citrix, and Xen, KVM from RedHat (among others), and Virtual Box from Oracle.

(ii) **Para Virtualization**: In para virtualization, unmodified software runs on modified OS as a separate system and provides an interface to VMs that are identical to their underlining hardware. In this technique, the guest OS is clearly ported before installing a VM because a non-tailored guest OS cannot run on top of a VMM. Para virtualization is a subset of a server virtualization. An important characteristic of par virtualization is the VMM allows par virtualization to give performance closer to without virtualized hardware.

(iii) **Partial Virtualization**: In partial virtualization, software modification may need to run.

b. **Network Virtualization**: Network virtualization means to combine Software Network resources and Hardware Network resources in a same administrative unit. Network virtualization allows to network optimization of scalability, data transfer rates, security, flexibility, and reliability. It also automates several network administrative tasks. This type of virtualization is particularly useful for extremely large, rapid, and unpredictable increase of usage. The result of network virtualization provides improved network efficiency and productivity.
Two important types of network virtualization are as follows

(i) **External**: In external network virtualization, it combines several networks into a virtual unit.

(ii) **Internal**: It provides a network-like functionality to a single system. An internal virtual network consists of one system using VMs whose network interfaces are configured over at least one physical NIC. These network interfaces are also known as virtual NICs (virtual network interface cards).

c. **Storage Virtualization**: In storage virtualization, several network storage resources are available as a single storage device managed for more efficient and easier from central console. The abstraction of logical storage created from physical storage. Three types of data storage are available in storage virtualization; they are SAN (Storage Area Network), NAS (Network Attached Storage), and DAS (Direct Attached Storage). SAN is a device shared with several servers over a high-speed network. NAS is a shared storage method which connects through network. The NAS is used for sharing file, storing backup, and sharing device among the machines. DAS is the typical approach of data storage where storage drives are directly attached to server machine.

d. **Memory Virtualization**: Memory virtualization [11] includes sharing the physical memory of the system and dynamically allocating it to virtual machines. This includes a method to separate memory from server to provide a networked, distributed, or shared function. It improves the performance of the cloud system by giving larger memory capacity without addition to the main memory. Due to this reason, a part of disk drive works as an extension of the main memory.

e. **Software Virtualization**: This provides the capability to a computer system to create and run one or more virtual environments. Software virtualization is used to facilitate whole computer systems to enable a guest operating system to run. Software virtualizations are of three types: (i) operating system, (ii) service virtualization, and (iii) application virtualization.

f. **Data Virtualization**: In data virtualization, we can easily manipulate data without any technical details, and we know where it is physically located and how it is formatted. It decreases workload and data errors.

g. **Desktop Virtualization**: Desktop virtualization [12] gives the work secure, convenience, and easy. As someone can access remotely, we are able to do work from any location or anywhere and any computer. It gives a lot of flexibility for users to work from home or on the go. Desktop virtualization also protects confidential data by keeping it safe on central servers. Desktop virtualization is of two types "server side" and "client side" desktop virtualization [13]. For first type virtualization, the applications of end-user are executed remotely, on a central server, and streamed toward the endpoint via a Remote Display Protocol or other presentation and access virtualization technology. In second type virtualization, the applications are executed at the endpoint, which is the user location, and presented locally on the user's computer.

In next Sect. 2, we have compared some important open-source hypervisors. In Sect. 3, we have explained different levels of virtualization. In Sect. 4, we have discussed related work about virtualization with their comparison, and we have discussed it in Sect. 5, and in final Sect. 6, we have concluded it.

2 Comparison of Different Open-Source-Based Hypervisor Virtualizations

In Table 1, we present the virtualization [14] methods from open-source providers like RedHat, Citrix systems, OpenVZ, Oracle, Proxmox, and Linux-Vserver. Here we compare several hypervisor models with several virtualization techniques. We have discussed some virtualization techniques in previous sections; virtualization creates virtual resources with the help of physical resources as network, storage components, and operating systems.

We have compared different open-source-based hypervisors and virtual monitor machines (VMM) in the above Table 1 that are in use today which will be very helpful to researchers for further study.

Table 1 Comparison of important open-source-based hypervisor virtualization

S. no	Hypervisor name	Hypervisor model	Company	Virtualization types			
				Full	Para	Hardware assisted	OS
1	Xen	Bare metal	Citrix Systems, Inc.	No	Yes	Yes	No
2	KVM	Hosted	RedHat	Yes	No	No	No
3	VirtualBox	Hosted	Oracle	No	Yes	Yes	No
4	Linux-V Server	Hosted	Linux-V Server	No	No	No	Yes
5	OpenVZ Linux	Hosted	Open VZ	No	No	No	Yes
6	ProxmoxVE	Bare metal	ProxmoxVE	Yes	No	No	Yes
7	QEMU	Bare metal	RedHat	No	Yes	Yes	No
8	Bochs	Bare metal	Linux	Yes	No	Yes	Yes

3 Levels of Virtualization

In virtualization, several customer applications, handled by their own OS (guest operating system) independent of host OS, can run on a single hardware. This is because of adding additional layer or software, known as virtualization layer. This software creates an abstraction of virtual machines by introducing a virtualization layer at different levels of a computer system shown in Fig. 4. The essential function of this software layer is to virtualize the physical hardware of a host machine into virtual resources to be used by the virtual machines, exclusively. It is implemented at different operational levels.

(i) Instruction Set Architecture (ISA) Level
(ii) Hardware Abstraction (HAL) Level
(iii) OS Level
(iv) Library Support Level
(v) User Application Level.

(i) **Instruction Set Architecture (ISA) Level**

It is a processor virtualization method that enables imitation of the Instruction Set Architecture (ISA) of one processor on a different processor. It allows the running or emulating of the instruction set architectures of different processors among each other delivered as a virtualization layer.

(ii) **Hardware Abstraction Level**

This virtualization level is performed right on top of the bare hardware of computer and generates a virtual hardware for a virtual machine. With the help of virtualization, this process controls the underlying hardware. The basic idea of this level is to virtualize the computer resources, i.e., I/O devices, memory,

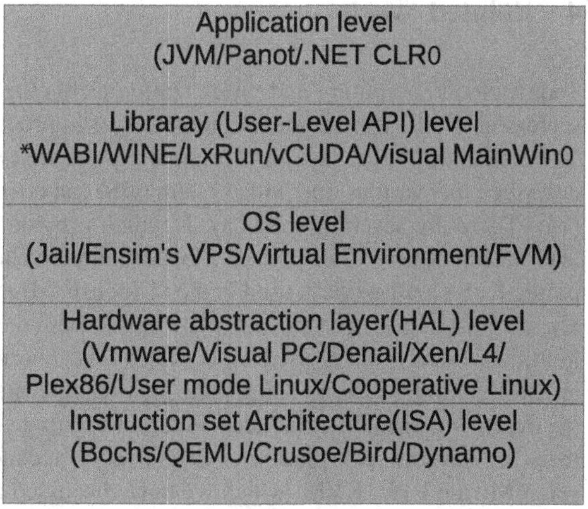

Fig. 4 Virtualization
abstraction level

| Application level
(JVM/Panot/.NET CLR0 |
|---|
| Libraray (User-Level API) level
*WABI/WINE/LxRun/vCUDA/Visual MainWin0 |
| OS level
(Jail/Ensim's VPS/Virtual Environment/FVM) |
| Hardware abstraction layer(HAL) level
(Vmware/Visual PC/Denail/Xen/L4/
Plex86/User mode Linux/Cooperative Linux) |
| Instruction set Architecture(ISA) level
(Bochs/QEMU/Crusoe/Bird/Dynamo) |

and processors. The main idea is to improve the rate of hardware utilization by several users simultaneously. This idea implemented in 1960s in IBM VM/370 and recently, the Xen hypervisor applied to virtualize x86-based machines to run Linux or other guest operating system applications.

(iii) **Operating System Level**

In OS-level virtualization, an abstraction layer is sitting between user applications and traditional OS. This level creates separated containers on any physical server and OS instances to utilize the software and hardware in the data centers. These containers act like the real servers. Basically, this level is helped in creating virtual host environments to allocate hardware resources among many mutually distrusting users.

(iv) **Library Support Level**

Virtualization with library interfaces is possible by controlling the communication link between applications and the other parts of a computer system through API hooks. The software tool WINE has implemented this approach to support Windows applications on top of UNIX hosts. Another example is the vCUDA which allows applications executing within VMs to leverage GPU hardware acceleration.

(v) **User Application Level**

This level is also called as process-level virtualization as we know that an application often runs as a process on traditional OS. One of the popular methods is deploying high level language (HLL) virtual machines. A program written in HLL and compiled for this virtual machine will be enabled to run on it. Examples of this class of virtual machines are Java virtual machine (JVM) and The Microsoft .NET CLR.

4 Related Work

Today, cloud computing and virtualization technologies are used in most of the IT sectors as well as in business enterprises to improve their growth and efficiency. The concept of virtualization is originally pioneered by IBM in 1960s where it has described that various operating system (OS) can be installed in the single computer [15]. There are several public clouds which provide IaaS with the help of virtualization techniques. The Amazon Elastic Compute Cloud (EC2) is the most popular public IaaS cloud services used in the IT industry. Eucalyptus [16] and Nimbus [17] are the most popular private IaaS platforms in both the industrial and scientific communities. Eucalyptus is used for a private and hybrid cloud environment to build an elastic compute cloud using their own resources and Nimbus, originating from the deploying virtual workspaces on top of existing Grid infrastructure. Nowadays, there are several open platforms virtualization technologies such as Xen hypervisor, VMware ESX, KVM, which we have discussed above available in the market. Several researchers have proposed and implemented related work with the help of various virtualization techniques. Here we are discussing some important virtualization methods. In [18] Muditha Perera has discussed that virtualization is making a virtual image of network resources, storage devices, servers, or operating systems so that all these can be used on different PC at the same time at anywhere.

In [19] Chaudhary and Mance have discussed computer virtualization and its benefits to universities and have discussed several advantages related to cost and power consumption to provide the benefit to universities. The authors have given some examples to improve computer laboratories of universities and decrease over all IT budget with the help of virtualization techniques. They also described the IT infrastructure required to implement such innovative tools. They highlighted two benefits decrease energy consumptions and condensed costs. Microsoft, Vmware, and Citrix were discussed as possible solutions and how the size and structure of the various universities would need various settings for the virtualization deployment to increase benefits. In [20] Kaungo has described the role of virtualization for increase efficient energy and resource utilization in cloud system.

A VM is a combination of attributes and parameters, including the VM image, OS kernel, network interfaces, CPU capacity, and memory [21]. The authors have discussed about the performance of the system with virtualization. In [22], virtualization technology enhancements of cloud computing based on virtual resources. In [23], Watfa has presented a structural framework of virtualization technology based on real campus deployment and provided Virtual Desktop Infrastructure (VDI) for academic members to convenient access virtual applications. Here, using virtualization technologies the result showed that academic staff and students had good experiences and have increased accessibility to their academic materials and course documents. They presented structure of a framework includes attributes and significant measures of virtualization on academic staff and on students. These attributes are accessibility, usability, privacy, performance, and applications.

Riskhan and Muhammed in [24] have proposed a framework EPOES and implementing new techniques with the help of cloud computing and system level virtualization technology. Isolation, Bare metal virtualization and virtual machine templates are some important techniques to implement the proposed framework. It gives the opportunity to improve the performance of online education system by sharing physical resources.

A table of different virtualization methods in cloud computing used to improve the performance of the OE systems.

In this Table 2, we are describing some important latest virtualization methods with their description and advantages to improve the performance of online education in cloud computing environment.

5 Discussion

There are several issues of virtualization that threaten its capabilities. There are some analyzed drawbacks of virtualization.

Data Standardization: Consumers find a problem to move their data from one cloud system to the other cloud system because of dissimilarity in data format and representation. It indicates apathy way of service providers to give consumers address and attention for their concerns. One best solution is to set up with a standard procedure in managing, presenting, creating, deploying, and storing data in the cloud adopted by all participating providers.

Computing Security: The hypervisor controls and manages the whole virtualization task in a cloud computing. The hypervisor is a software or an OS that has a work of the traffic cop to assure all things happen in according orderly manner. The whole cloud computing system is in a give-and-take manner and an attacker benefits to access it. But unlike a traditional OS that is flexible in its dealing, i.e., it controls all the application software, hardware components and utilities. However, there is no perfect system and skewing hypervisor(s) to aspect in the cloud computing would create problem for attackers to manage or gain of the whole virtual machines.

Storage Concerns: In a cloud computing system, the cloud permits for removal and addition of VM and if a specific virtual machine leaves the cloud, and its data is not cleared from primary memory of the server and same space is reallocated to another virtual machine, there would be a greater chance of steal data usual in the public cloud computing system. One best method for VM users to solve this problem is to responsibly delete or free all appropriate data on their assigned cloud computing system before abdicating the space.

Monitoring Network: There are a dynamic configuring, monitoring, and setting in the cloud environment because of its virtual nature, cloud system leaves a lot to be require. However, there are some software that are used to diminish and fix the problem, these companies are IBM, Vmware, CA, and Hewlett Packard their software monitors and manages virtual systems and dynamic systems. Virtualization is an effective approach which should be tuned-up to better improvement the ever-growing benefits of cloud environment.

Table 2 Latest virtualization methods to improve cloud computing performance

S. no.	Title/Author	Description/techniques	Advantages
1	Enhancing the Performance of Current Online Education System—A Study of Cloud Computing and Virtualization/Riskhan (2015) [24]	1. The author has proposed a framework to Enhance the Performance of Online Education System (EPOES) to find virtual machine templates the isolation and bare metal virtualization 2. Implemented the proposed framework with the help of system-level virtualization	1. Increase efficiency with the help of VM template 2. It operates the server in less time
2	An Educational Virtualization Infrastructure/Mohamed K. Watfa (2016) [23]	1. Presented structure of a framework includes attributes and significant measures of virtualization on academic staff and on students 2. These attributes are accessibility, usability, privacy, performance, and applications	1. Improve efficiency 2. Save cost in the IT infrastructure development and maintenance
3	Efficient Resource Management in Cloud Computing Using Virtualization/Kananga P. (2016) [20]	1. The authors explained that virtualization provides the efficient resource utilization as well as it permits energy efficient computing. 2. Virtualization may increase this utilization from 70 to 80% 3. The model is based on virtualization of data center	1. Server consolidation 2. Resizing physical infrastructure at runtime 3. Low-power requirements
4	Virtualization: Providing Better Computing To Universities/Chawdhary (2016) [19]	1. The authors discussed about benefits of virtualization for the universities to access laboratories and IT department in minimum budget 2. They described about VMware, Citrix, and Microsoft as solutions how differ virtualization 3. Virtualization also allows the university to provide the latest operating systems and applications with maintaining compatibility of older operating systems and applications	1. Reduced cost budget to maintain university laboratories using virtualization 2. Increase reliability

6 Conclusion

In this research paper, we have focused on virtualization in cloud computing. Cloud computing plays an important role in Internet to share data and information and virtualization is a popular technique to share data and information in cloud computing. Here we have discussed different types of virtualization and have compared some popular open-source hypervisors and virtual monitor machines (VMM) in the table that are in use today. We have discussed five virtualization levels which improve scalability besides making the cloud computing solutions cost effective, and these virtualization technologies improves the overall efficiency of cloud computing and we have discussed some important previous related work and have compared some of them with their techniques and advantages which will help to researchers to do work in this area. At last we have discussed some drawbacks of virtualization with some solutions which will be very helpful to the researcher for further research study or implementation.

References

1. Loganayagi, B., Sujatha, S.: Creating virtual platform for cloud computing. In: IEEE International Conference on Computational Intelligence and Computing Research (ICCIC 2010), pp. 1–4 (2010)
2. Karen, S., Murugiah, S., Paul, H.: Guide to Security for Full Virtualization Technologies, pp. 1–35. Special Publication National Institute of Standards and Technology (NIST) (2011)
3. Barham, P., Dragovic, B., Fraser, K., Hand, S., Harris, T., Ho, A., Neugebauer, R., Pratt, I., Warfield, A.: Xen and the art of virtualization:. In: Proceedings of the 19th ACM Symposium on OS Principles. SOSP, pp. 164–177 (2003)
4. Dawei, S., Guiran, C., Qiang, G., Chuan, W., Xingwei, W.: A dependability model to enhance security of cloud environment using system-level virtualization techniques. In: 1st International Conference on Pervasive Computing. Signal Processing and Applications (PCSPA), pp. 305–310 (2010)
5. Joanna, R., Alexander, T.: Bluepilling the Xen Hypervisor. In: Xen owning Trilogy par-III, Black Hat USA, 2008, pp. 1–85 (2008)
6. Samuel, T.K., Chen, P.M., Yi-min, W., Chad, V., Helen, W., Jacob, R., Lorch, S.: Implementing malware with Vms. In: IEEE Symposium on Security and Privacy, pp. 314–327 (2006)
7. Rosenblum, M., Garfinkel, T.: Virtual machine monitors: current technology and future trends. IEEE Comput. Sci. **38**(5), 39–47 (2005)
8. Popek, G.J., Goldberg, R.P.: Formal requirements for virtualizable third generation architectures. Commun. ACM **17**(7), 412–421 (1974)
9. Farzad, S.: Intrusion detection techniques performance in cloud environments. In: Proceedings Conference on Computer Design and Engineering. Kuala Lumpur, Malaysia, pp. 398–402 (2011)
10. Obasuyi, G.C., Arif, S.: Security challenges of virtualization hypervisors in virtualized hardware environment. Int. J. Commun. Netw. Syst. Sci. **08**(07), 260–273 (2015)
11. Chen, W., Lu, H., Shen, L., Wang, Z., Xiao, N., Chen, D.: A novel hardware assisted full virtualization technique. In: A Novel Hardware Assisted Full Virtualization Technique (IEEE, 2008), pp. 1292–1297 (2008)
12. Microsoft 2008 desktop virtualization strategy. http://download.microsoft.com/download/6/F/8/6F8EF4-26BD-48EABF45BFF00A3B5990/Micros%20Client%20Virtualization%20Strategy%20White%20Paper_final.pdf
13. Spruijt, R.: Desktop virtualization and the power of App-V and Windows. http://www.brianmadden.com/blogs/rubenspruijt/archive/2010/02/22/desktopvirtualizationand-the-power-of-windows-7.asp
14. Sarna, D.E.Y.: Implementing and Developing Cloud Computing Applications. Taylor and Francis Group, LLC, (2011)
15. Macro, A., Valerio, B., Alberto, C., Marco, C., Ernesto, D., Fulvio, F., Joel, T.H., Davide, R.: Learning computer networking on open paravirtual laboratories. In: IEEE Trans. Educ. **50**(4), 302–311 (2007)

16. Daniel, N., Rich, W., Chris, G., Graziano, O., Sunil, S., Lamia, Y., Dmitrii, Z.: The eucalyptus open-source cloud-computing system. In: Proceedings of Cloud Computing and Its Applications. pp. 1–16 (2014)
17. Katarzyna, K., Ian, F., Timothy, F., Xuehai, Z., Daniel, G.: Virtual workspaces in the grid. Lecture Notes in Computer Science, vol. 3648, pp. 421–431 (2005)
18. Muditha, P., Keppitiyagama, C.: A performance comparison of hypervisors. In: A Performance Comparison of Hypervisors, IEEE, 2011, edn., pp. 120–120 (2011)
19. Chawdhry, A., Mance, C.: Virtualization: providing better computing to universities. In: Information Systems Educators Conferenc (ISECON) Nashville Tennessee, USA, vol. 27, no. 1401, pp. 1–7 (2010)
20. Kanungo, P.: Efficient resource management in cloud computing using virtualization. Int. J. Adv. Res. Comput. Commun. Eng. 5(4), 650–652 (2016)
21. Ramakrishnan, L., Canon, R.S., Muriki, K., Sakrejda, I., Wright, N.J.: Evaluating interconnect and virtualization performance for high performance computing. In: ACM 2nd International Workshop on Performance Modeling, Benchmarking & Simulation of High Performance Computing Systems, pp. 1–2 (2011)
22. G5Networks–Technology Partner (2013). http://www.g5networks.net/virtualization.html
23. Watfa, M.K., Udoh, V.A., Al Abdulsalam, S.M.: An educational virtualization infrastructure. In: ICST Institute for Computer Sciences, Social Informatics and Telecommunications Engineering 2016, pp. 12–21 (2016)
24. Riskhan, B., Muhammed, R.: Enhancing the performance of current online education system—a study of cloud computing and virtualization. J. Comput. Commun. 3, 43–51 (2015)

Load Balancing and Clustering Scheme for Real-Time VoIP Applications

Shubhani Aggarwal, Nitish Mahajan, Sakshi Kaushal and Harish Kumar

Abstract Voice over IP (VoIP) is a methodology for transmitting data, video, voice, chat services, and messages over the Internet. Session Initiation Protocol (SIP) is an application layer protocol that is used to initiate, establish, modify, and terminate the session. Load balancing is the mechanism for improving the performance of a distributed and parallel system by redistribution of load between the servers. The main purpose of load balancing and clustering is to handle large number of VoIP calls on different processors. This paper presents the design, implementation, and performance evaluation of a load balancer for cluster-based SIP servers. We proposed a new measurement scheme based on OpenSIPS server to a group of machines with varying configuration and analyzed various parameters like jitter, packet loss, response time, throughput, and hence, it is possible to provide better Quality of Service (QoS) in network.

Keywords VoIP · SIP · PBX · FreeSWITCH · Load balancing and clustering
QoS

1 Introduction

The Internet telephony provides communication services (voice, SMS, voice-messaging) over the Internet, instead of using Public Switched Telephone Network (PSTN). VoIP is the communication method that transmits voice, data, and multimedia content over IP networks [1]. VoIP network uses International Telecommunication Union (ITU) standard codecs, such as G.729; used for compressing the packets and G.711; used to transmit the uncompressed packets. When compression occurs, voice quality may reduce, but it also sinks the bandwidth requirements [2]. SIP is a protocol that is used to initiate, establish, modify, and terminate the session in an

S. Aggarwal (✉) · N. Mahajan · S. Kaushal · H. Kumar
Computer Science and Engineering, University Institute of Engineering and Technology, Panjab University, Chandigarh 160014, India
e-mail: Shubhaniaggarwal529@gmail.com

© Springer Nature Singapore Pte Ltd. 2019
S. K. Bhatia et al. (eds.), *Advances in Computer Communication and Computational Sciences*, Advances in Intelligent Systems and Computing 760, https://doi.org/10.1007/978-981-13-0344-9_38

application layer [3]. An IP-PBX is IP-Private Branch Exchange server that swaps the calls between the VoIP users on the local lines while permitting all users to share a few number of external telephone lines [4]. FreeSWITCH is an open-source IP-based platform. It is capable of acting like a PBX (often an office telephone system), a Soft-Switch (software that operates like legacy hardware-based telephone company switches), or anything in between [5–7]. FreeSWITCH can handle the calls in two modes: peer-to peer (P2P) mode in which it handles registration and signaling load only, i.e., SIP packets flow through the PBX or FreeSWITCH but the voice in the form of RTP packets travels directly between endpoints or clients. In other mode, i.e., back-to-back (BTB) configuration, all the SIP and RTP packets flow through the server, i.e., FreeSWITCH acts as an intermediary between its clients. BTB provides the PBX much better control over the voice or RTP packets as compared to P2P.

Load balancing is a way to distribute operation, i.e., voice and multimedia onto the multiple computers [8]. A clustering is a group of servers that runs the same program in cooperation. The basic function of a cluster is to distribute the load onto distinct servers and for providing failover for each other [9]. In this paper, we have implemented load balancing and clustering technique with OpenSIPS server for providing better QoS in VoIP network.

The rest of this paper is organized as follows. Section 2 presents the basic concepts of load balancing and clustering and also describes existing work related to this technique. Proposed algorithm is presented in Sect. 3. Section 4 presented the simulation and results. Conclusion and future scope are presented in Sect. 5.

2 Literature Review

This section focuses on the study of the related work carried out by various researchers. The quality of VoIP-based real-time applications is usually measured with Quality of Service (QoS). For high quality of VoIP calls, consider various factors like packet loss, delay, jitter, and Mean Opinion Score (MOS). The performance of the VoIP application also depends on the processor with different codecs that varies in speech quality, complexity, and bandwidth [10]. In VoIP signaling, when SIP servers are short of resources for handling all the SIP messages that they received, then overload occurs in SIP networks. A scheme has been proposed that defined the nature of SIP servers' and also identifies a loss-based overload method for SIP [11]. One of the features of Media Access Control (MAC) layer is designed for differentiating VoIP services among traffic categories with different multimedia requirements to provide QoS [12]. An IP-based Telecommunication System (IPTS) is a highly stable and well-arranged LAN connection. Instead of using central IP-PBX, IPTS structure has inexpensive and changeable features that can be changed, upgraded, or deleted without manipulating the function of IPTS system [13]. In this [14], researchers evaluated the performance of various VoIP codecs over the network and the performance parameters are jitter, delay, packet loss, and MOS which are used for evaluating the effectiveness of VoIP codecs and provide QoS. The basic goal of load balancing is to divide a load of work onto several computers, so that the program can process a higher work load [15]. Various algorithms are Call-Join-Shortest-Queue (CJSQ), round robin, Transaction-Least-Work-Left (TLWL), and Transaction-Join-Shortest-Queue (TJSQ). They represented their analysis w.r.t call response time and throughput performance for SIP servers [16, 17]. Depending upon the composition, TLWL improves both response time and throughput and achieves best performance. By connecting various characteristics, TLWL removes the differences of different transactions in processing costs, resulting, improvements of throughput up to 24% and improvements of response time up to 2% [18]. In this, researchers has been used various algorithms and provide significantly better response time by distributing requests across the cluster-based SIP server. TLWL-1.75 provides 25% better throughput than a standard hash-based algorithm and 14% better throughput than a dynamic round robin algorithm. TJSQ provides nearly the same level of performance. CJSQ performs poorly since it does not distinguish transactions from calls and does not consider variable call hold times [19].

Singh and Schulzrinne [20] compare the various failover and load sharing approaches based on SIP call routing and registration. For registration, an identity table includes an identity entry for some servers and each entry comprising a Fully Qualified Domain Name (FQDN) and load balancing information. A valid entry is stored and indicates a persistent connection between the client and the selected second server [21]. For load balancing, two algorithms are considering: CPU Load Monitoring (CLM) and Control Through Thresholds (CTT). A CTT mechanism leverages the concepts for managing the traffic on thresholds, and CLM mechanism destroys the parameters for processing measurement of load to manage the overload periods [22]. In this paper, we proposed an algorithm based on load balancing and clustering in OpenSIPS and analyzed various parameters which provide better QoS.

3 Proposed Work

In this section, an algorithm for load balancing and clustering is presented. Based on this, QoS parameters like jitter, packet loss, throughput, response time, and Call Manager parameters like RAM and CPU utilization are considered while load balancing and clustering.

The detailed algorithm is shown in Fig. 1. The proposed algorithm shows how load balancing and clustering can be done in OpenSIPS. OpenSIPS is a multi-functional SIP server that is capable of keep tracking all the load information of all destinations. It will consider the destination which is less loaded and has largest available slots to handle input load. This method is for incoming SIP request or incoming traffic. The incoming calls can be managed by load balancer module and decided how they route to different set of destinations.

Further, based on this algorithm, we have conducted different performance tests that are presented in next section.

Steps to perform Load Balancing and Clustering

Step I- Authentication of incoming requests.
    Check whether capacity is present or not in Step II.
Step II- Total capacity reached yet or not for the cluster.
    If capacity is present Goto Step III
    Otherwise Step V
Step III-Choose the server with largest number of slots.
    Do Step IV and Step V
Step IV- Direct the call forward the server.
Step V- If the particular server is not available, direct the calls to the server which is next to
    the previous server in terms of available slots.
Step VI- Repeat above steps till we have servers available.
    else send error message

Fig. 1 Proposed algorithm

4 Results and Analysis

In this section, emulated results and analysis of the proposed algorithm for load balancing and clustering is presented. Performance tests have been defined and executed that allow us to measure resource usage and Quality of Service accurately for VoIP calls in P2P mode with codecs G.711 and G.729.

4.1 Emulation Parameters for Call Load Testing

StarTrinity is used as an emulation framework for generating different VoIP calls. StarTrinity is an emulator used for load testing and performance of SIP servers [23]. It also monitors the VoIP quality of live IP network servers and displays QoS parameters like delay, jitter, MOS. It is able to monitor and simulate thousands of simultaneously outgoing and incoming SIP calls and build real-time reports. For emulation study, blocking factor set to be zero and Call Holding Time (CHT) of 1 and 2 min. Consider four PCs as clients such that two PCs behave as call originators (senders) and two PCs are used to receive the incoming calls (receiver). Total 2000 users are registered on Call Manager. An input to Call Manager (SIP Server) can be given in form Calls per Second (CPS). The performance testing is done by considering two different systems as Call Manager with following configuration as shown in Table 1.

For all tests, two important parameters are taken into consideration, i.e., CPU utilization and RAM utilization. CPU utilization means usage of CPU for processing resources, i.e., amount of work handled by CPU at a particular time. It depends on the type and amount of load being processed. It must not exceed 80% of its utilization. Similarly, RAM usage also increases as the load on the system increases. It must be less than 60%. If CPU and RAM utilization increases beyond its value, then stability of the system is affected and the services are not handled as efficiently as they should.

Table 1 Specification of processors

Specifications	i7/Elite 8300	Raspberry Pi/V1.2 2015 version
Clock speed	3.6 GHz * 8	1.2 GHz
No. of CPU cores	8	4 (QUAD)
RAM	4 GB/1600 MHz speed	1 GB/SDRAM
NIC	100 mbps	10/100 mbps
Hard disk	500 GB	Micro SD
Operating system	Debian 8.1	Debian 8.1

Table 2 Emulation results for P2P calls in i7 with G.711 codec

CPS	CPU utilization (%)	RAM utilization (%)	Max concurrent calls	Failed calls (%)
10	10	33.9	755	0
15	20	40.2	908	0
20	25	45	1511	0
25	32	52	1681	3.8
30	38 (unstable)	60	3101	[a]Network congestion

[a]Network Congestion means server performance degrades due to overloading

Table 3 Emulation results for P2P calls in i7 with G.729 codec

CPS	CPU utilization (%)	RAM utilization (%)	Max concurrent calls	Failed calls (%)
10	10	15	804	0.1
20	20	39.4	1294	0.2
25	30	48.2	1594	0.86
26	32 (unstable)	60.1	1706	1.97
30	42.5 (unstable)	62.4	2417	5.68

Emulation has been performed 10 times, and average values have been taken for analysis. For emulation study, consider only one real-time applications in VoIP network, i.e., P2P. In P2P calling, server handles the registration and signaling load in the form of SIP packets while the RTP packets flow directly between the clients.

4.1.1 Result Analysis for Call Load

Using Call Manager utilization parameters, i.e, CPU and RAM, the performance of the Call Manager has been evaluated and analyzed how many maximum concurrent calls can be handled by a Call Manager. Results are presented for P2P calls as given in Tables 2, 3, 4, and 5.

Table 4 Emulation results for P2P calls in Raspberry Pi with G.711 codec

CPS	CPU utilization (%)	RAM utilization (%)	Max concurrent calls	Failed calls (%)
1	9	17.08	61	0
5	45	24	312	0.06
7	66.5	34.7	486	2.88
8	70.75	36.7	500	3.62
9	78.2	37.8	501	10.1
10	82	39	521	16.96

Table 5 Emulation results for P2P calls in Raspberry Pi with G.729 codec

CPS	CPU utilization (%)	RAM utilization (%)	Max concurrent calls	Failed calls (%)
1	9.25	16.2	74	0
4	44.75	26.8	314	0.16
5	52	27.4	357	0.33
9	70	34.5	509	16.41
10	84.5	36.5	528	29.8

From results, we analyzed, for P2P calls, both codecs G.711 and G.729 have a similar impact on the server as RTP data is handled by the client ends.

With i7, the system is able to handle up to 25 CPS or maximum of 1600 concurrent calls for both codecs. Beyond this value, StarTrinity displays the message of hardware limitation.

Raspberry Pi is able to handle maximum of 500 concurrent calls for given call holding time. Beyond that, generated calls fail and StarTrinity displays the server error (503 Service Unavailable).

4.2 Result Analysis for Load Balancing and Clustering

In this section, emulation for load balancing and clustering is done in OpenSIPS [24, 25]. For starting load balancing and clustering in OpenSIPS, firstly, load the module, i.e., load.balancer.so and set the module parameters as per the requirements. Secondly, generate the routing logic which can control and manage all the SIP requests.

To generate maximum call load in OpenSIPS, set a cluster of FreeSwitch's (FS's) (FS's are i7 processor and Raspberry Pi) that balancing the load and given input is in the form of maximum concurrent calls that each FS server can handle which are evaluated from the proposed algorithm.

For load balancing and clustering emulation study, blocking factor set to be zero and Call Holding Time (CHT) of 15 s. Total 2000 users are registered on Call Manager. An input to OpenSIPS SIP server can be given in form of CPS by StarTrinity SIP

Fig. 2 Packet loss with calls per second

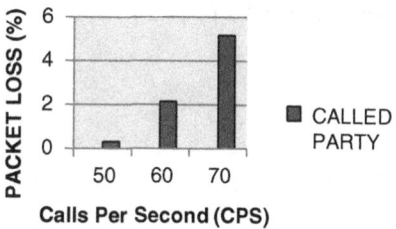

Tester and maximum concurrent calls to each FS server in OpenSIPS that each FS server can handle. For emulation tests, important parameters are taken into consideration, *i.e.*, CPU utilization, RAM utilization, packet loss, response time, throughput, and jitter. Emulation has been performed ten times with emulation run time 5 min and average values have been taken for analysis. For result analysis, two processors only can be used, i.e., i7 and low-end Raspberry Pi. FS1 and FS2 are i7 processors that can handle of maximum concurrent calls of 1500, and FS3 is low-end Raspberry Pi that can handle up to 500 maximum concurrent calls are evaluated. For testing VoIP call load in OpenSIPS with P2P mode, results are analyzed with different QoS parameters like packet loss, jitter, Call Manager Utilization parameters like CPU and RAM, throughput and response time are in following subsections.

4.2.1 Packet Loss

Packet loss occurs when one or more packets of data traveling across a network fail to reach their destination. It is caused by congestion and is measured as percentage of packet lost w.r.t. sent. It should not be more than 3%.

Figure 2 shows the result of packet loss over calls per second. In this, it can be seen that number of packets lost only from called party because the packet lost from the caller party is zero. As a number of CPS increases, number of packets lost from called party side also increases. At one time, it can beyond its range value which is not acceptable.

4.2.2 Jitter

Jitter is a measure of variation in time of interval for consecutive packets delivery. The variation can be caused by serialization, contention effect, and queuing of IP networks. The packets enter into the jitter buffer at a variable rate and are taken out at a constant rate for decoding. The ideal value of jitter should not be more than 40 ms.

$$Avg\,Jitter = \frac{\sum_i^n |D_i|}{n} \tag{1}$$

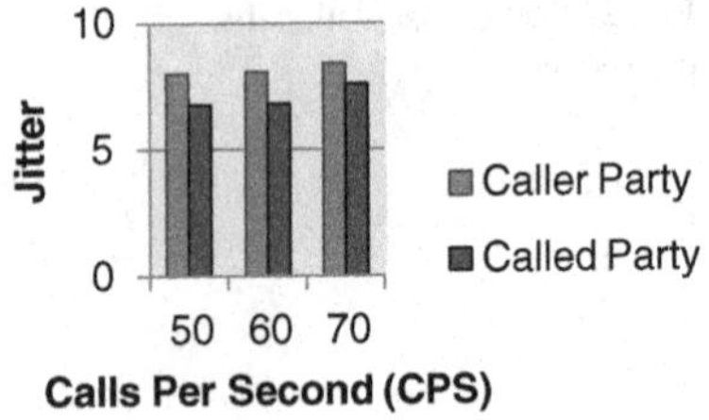

Fig. 3 Jitter with calls per second

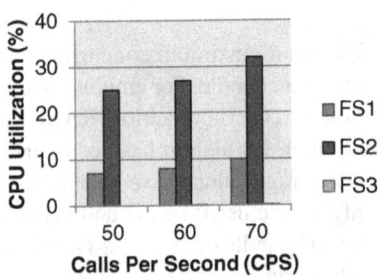

Fig. 4 CPU utilization with calls per second

Figure 3 shows the results of average jitter over calls per second. As number of CPS increases, average jitter also increases. As the load on network increases, variation in delay also increases because of fixed jitter buffer. The server needs more time for processing a large number of packets after signaling. As the number of user increases, there would be the large number of RTP packets and their processing in a network can introduce jitter.

4.2.3 CPU and RAM

CPU utilization means usage of CPU for processing the resources, i.e., an amount of work handled by CPU at a particular time. It depends on the type and amount of load being processed. It must not exceed 80%.

Figure 4 shows the result of CPU utilization over calls per second of all the FS's server. In this, it can be seen that as the number of CPS increases, CPU utilization on each server can also increase. Each FS server can handle call load according to their specifications.

Similarly, RAM usage also increases as the load on the system increases. It must be less than 60%.

Figure 5 shows the result of RAM utilization over calls per second of all the FS's server. In this, it can be seen that as the number of CPS increases, RAM usage on each server can also increase.

Fig. 5 RAM utilization with calls per second

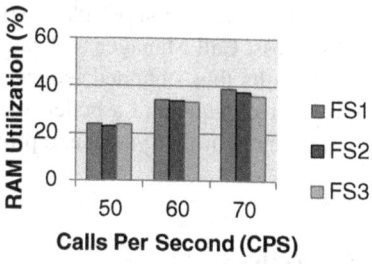

Table 6 Comparison with existing technique

Technique	Throughput (%)	Response time (ms)
Jiang [18]	85.10	800.625
Proposed work	99.18	152.80

4.2.4 Throughput and Response Time

Throughput is the maximum rate at which the data can be processed. In communication networks, such as Ethernet, network throughput is the rate of successful message transmission over a communication channel. The peak throughput is the maximum throughput which can be sustained while successfully handling more than 99.99% of all requests.

Response time is the time between sending INVITE request and receiving 200 OK requests. Its value should not be more that 250 ms [26].

Table 6 shows that proposed technique gives better results than existing technique. It can be seen that throughput value of proposed technique is approx 99% as compared to existing technique, i.e., 85% which is less than peak throughput.

Similarly, response time also gives better results. Response time is the minimum time between the requests sent and received. By comparing it with existing technique, proposed technique gives output as 152.80 ms response time that is less than 800.625 ms. It is within acceptable range of 0–250 ms.

In the end, from emulation work, we can say that load balancing and clustering in OpenSIPS is used for handling maximum call load and it gives better results in terms of throughput and response time. This would help VoIP service providers to provide better QoS for real-time VoIP applications.

5 Conclusion and Future Scope

In this paper, a method has been proposed for load balancing and clustering in VoIP network. Firstly, by using experimental results, evaluation of maximum concurrent calls that a particular Call Manager can handle is done. With the help of maximum concurrent calls, a setup is designed for load balancing and clustering. For testing the maximum load, various emulation runs and analysis of QoS parameters like jitter and packet loss, Call Manager Utilization parameters like CPU and RAM were carried out. Results thus obtained indicate that the proposed technique is an addition on existing technique and a better load balancing is achievable that leads to capacity enhancement of the system and provides QoS.

References

1. Idrees,F., Khan, U.A.: A generic technique for voice over Internet protocol (VoIP) traffic detection. Int. J. Comput. Sci. Netw. Secur. 8(2) (2008)
2. Varshney, U., Snow, A., McGivern, M., Howard, C.: Voice over IP magazine. Commun. ACM 45(1), 89–96 (2002)
3. Rosenberg, J., Schulzrinne, H., Camarillo, G., Johnston, A., Peterson, J., Sparks, R., Handley, M., Schooler, E.: SIP: Session Initiation Protocol, RFC 3261, June 2002
4. Sulkin, A.: PBX Systems for IP Telephony: Migrating Enterprise Communications. McGraw-Hill Professional, Book (2001)
5. FreeSWITCH Home Page (2016). http://www.freeswitch.org/accessed
6. FreeSwitch: http://www.freeswitch.org. Accessed Apr 2013
7. Asterisk.org: About Asterisk PBX. http://www.asterisk.org/get-started. Accessed Apr 2013
8. http://tutorials.jenkov.com/software-architecture/load-balancing.html. Acceseed Oct 2014
9. https://f5.com/resources/white-papers/load-balancing-101-nuts-and-bolts. Accessed May 2017
10. Goode, B.: Voice over internet protocol (VoIP). Proc. IEEE 90(9), 1495–1517 (2002)
11. Gurbani, V., Schulzrinne, H., Hilt, V.: Session Initiation Protocol (SIP) Overload Control, RFC 7339, Sept 2014
12. Cicconetti, C., Lenzini, L., Mingozzi, E., Eklund, C.: Quality of service support in IEEE 802.16 networks. IEEE network. IEEE Commun. Soc. 20(2), 50–55 (2006)
13. Thompson, C.A., Latchman, H.A., Angelacos, N., Pareek, B.K.: A distributed IP-based telecommunication system using SIP, Dec 2013. arXiv:1312.2625
14. Tariq, M.I., Azad, M.A., Beuran, R., Shinoda, Y.: "Performance analysis of VoIP codecs over BE WiMAX network. Int. J. Comput. Electr. Eng. Singapore 5(3), 345 (2013)
15. http://tutorials.jenkov.com/software-architecture/load-balancing.html. Accessed Oct 2014
16. Akbar, A., Basha, S.M., Sattar, S.A.: A comparative study on load balancing algorithms for SIP servers. Information Systems Design and Intelligent Applications, Feb 2016, pp. 79–88. Springer, India
17. Anandhan, K., Prabu, V.D., Kumar, C.: A novel approach for load balancer in SIP clusters. Netw. Commun. Eng. 5(4), 207–213 (2013)
18. Jiang, H., Iyengar, A., Nahum, E., Segmuller, W., Tantawi, A.N., Wright, C.P.: Design, implementation, and performance of a load balancer for SIP server clusters. IEEE/ACM Trans. Netw. (TON), USA, 20(4), 1190–1202 (2012)
19. Jiang, H., Iyengar, A., Nahum, E., Segmuller, W., Tantawi, A., Wright, C.P.: Load balancing for SIP server clusters. In: IEEE International Conference on Computer Communications (INFOCOM-2009), Rio de Janeiro, Brazil, April 2009, pp. 2286–2294
20. Singh, K., Schulzrinne, H.: Failover, load sharing and server architecture in SIP telephony. J. Comput. Commun. 30(5), 927–942 (2007)
21. Bharrat, S.J, Asveren, T., Hart, J.: Load balancing among VoIP server groups, inventors, Sonus Networks, Inc., assignee, United States patent application US 12/771, 618, Nov 2011

22. Montagna, S., Pignolo, M.: Performance evaluation of load control techniques in sip signaling servers. In: Third International Conference on International Society of Public Law Systems (ICONS 08), 2008, Cancun, Mexico, Mexico, April 2008, pp. 51–56
23. www.startrinity.com
24. http://www.opensips.org/Documentation/Manuals. Accessed Mar 2017
25. http://www.opensips.org/Documentation/Modules-2-3. Accessed Mar 2017
26. Bruno, A., Jordan, S.: CCDA 640-864 Official Cert Guide, 9 June 2011. Cisco Press

Detection of Wormhole Attack in Static Wireless Sensor Networks

Manish Patel, Akshai Aggarwal and Nirbhay Chaubey

Abstract Wireless sensor networks are vulnerable to many more attacks. Wormhole attack is very dangerous to wireless sensor networks because it is a gateway to many more attacks such as black hole, gray hole, Sybil, jellyfish, denial of service. Without knowing the protocols used in the network, an attacker launches a wormhole attack by placing two malicious nodes in two different parts of the network which are far away from each other. In this way, an attacker tries to disturb the routing process. An attacker can drop the packets, modify the packets, and analyze the traffic. In this paper, we have proposed a wormhole detection method based on neighborhood information and alternate path length calculation. Simulation results show that our approach has good detection accuracy with less storage requirements.

Keywords Security · Vulnerable · Jellyfish · Denial of service

1 Introduction

Wireless sensor network has unique characteristics compared to other wireless ad hoc networks such as sensor nodes are resource limited, nodes are densely deployed, nodes use broadcast communication instead of point to point, nodes are prone to failure and their topology often changes [1, 2]. They are remotely managed. Security is very crucial for wireless sensor networks. In [3], Karlof and Wagner classify a number of attacks in wireless sensor networks. Among all the possible attacks, wormhole attack is very dangerous because it opens gate to many more attacks. To launch the attack, attacker does not need any cryptographic break. Two malicious

M. Patel (✉) · A. Aggarwal · N. Chaubey
Smt. S R Patel Engineering College, Gujarat Technological University, Ahmedabad, India
e-mail: it43manish@gmail.com

A. Aggarwal
e-mail: akshai.aggarwal@gmail.com

N. Chaubey
e-mail: nirbhay@ieee.org

© Springer Nature Singapore Pte Ltd. 2019
S. K. Bhatia et al. (eds.), *Advances in Computer Communication and Computational Sciences*, Advances in Intelligent Systems and Computing 760, https://doi.org/10.1007/978-981-13-0344-9_39

Fig. 1 Wormhole attack
with two malicious nodes

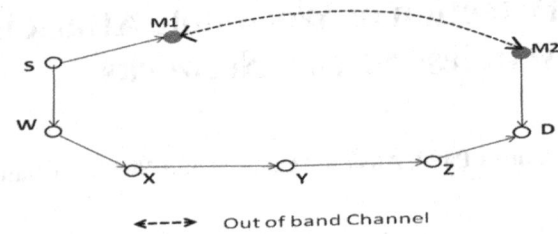

<---> Out of band Channel

nodes are connected by high-speed low latency tunnel. One malicious node captures traffic from one area, forwards it to another malicious node located far away from it, and disturbs the routing process. This paper presents wormhole detection mechanism based on neighborhood information and alternate path length calculation.

The rest of the paper is organized as follows: Sect. 2 presents description of the wormhole. Section 3 presents various existing approaches for wormhole attack detection. Proposed detection algorithm is discussed in Sect. 4. Section 5 provides simulation results. Conclusion and future work are presented in Sect. 6.

2 Description of Wormhole Attack

Two malicious nodes located far away from each other create a high-speed low latency tunnel between them. One malicious node captures traffic from one area of the network and tunnels it to another malicious node. The received traffic is replied in another area of the network. Due to this, routing is disturbed. After launching the attack, an attacker can modify the packet data, drop the packets, analyze the traffic, and open the door for many more attacks. As shown in Fig. 1, source node S broadcasts the packets, which is captured by malicious node M1. Malicious node M1 tunnels it to another malicious node M2 through out of band high-speed channel. Destination node D receives packets from malicious node M2. Destination node D receives them as if it was receiving them directly from the source node S. Packets also reach to the destination node through the normal route S-W-X-Y-Z-D later than those conveyed through the high-speed tunnel, and they are dropped because the normal path contains more number of hops.

3 Existing Approaches for Wormhole Detection

Wormhole attack can be identified by using location information. It requires GPS or directional antenna which increases the network cost. In [4–9], authors have presented location-based approaches for wormhole attack detection. Detection techniques based on neighborhood information [7, 10, 11] require additional storage for storing neighborhood information. These approaches do not work efficiently for mobile wireless sensor networks because neighbor list frequently changes.

A wormhole path contains shorter hop count compared to a normal path. Hop count and distance-based approaches are presented in [12–17]. Time-based approaches for wormhole detection are presented in [18–21]. If the average time per hop count is greater than that for the normal route, it indicates the existence of a wormhole. Network connectivity-based approaches for wormhole detection are presented in [22–24]. A malicious node results in incorrect connectivity information.

4 Detection Methodology

4.1 Assumptions

Our first assumption is all the sensor nodes are static. Second assumption is at the time of deployment malicious nodes are not present during some initial interval in the network. During initial interval, all legitimate sensor nodes safely establish their neighbor information.

4.2 Adversary Model

Two malicious nodes launch high-speed tunnel. They both are located far away from each other. From one area, one malicious node captures the traffic and forwards the traffic to another malicious node located in different areas. Due to this, routing is totally disturbed. After launching the wormhole attack, malicious nodes can analyze the traffic, drop the packets, and also modify the packets.

4.3 Proposed Protocol

Initially during some time interval, malicious nodes are not participated in the network. To establish neighbor information, each node sends hello message to its neighbors. When the node receives the message, it immediately replies back. The sender node adds the receiving node to its neighbor list. In this way, every node establishes one-hop neighborhood list. After establishing the one-hop neighborhood list, each node sends its first-hop neighbor list to its neighbors. In this way, each node establishes two-hop neighborhood lists.

After some time, if any node overhears packets from some new nodes, then the new neighbors are added into suspicious list. Suppose node A overhears packets

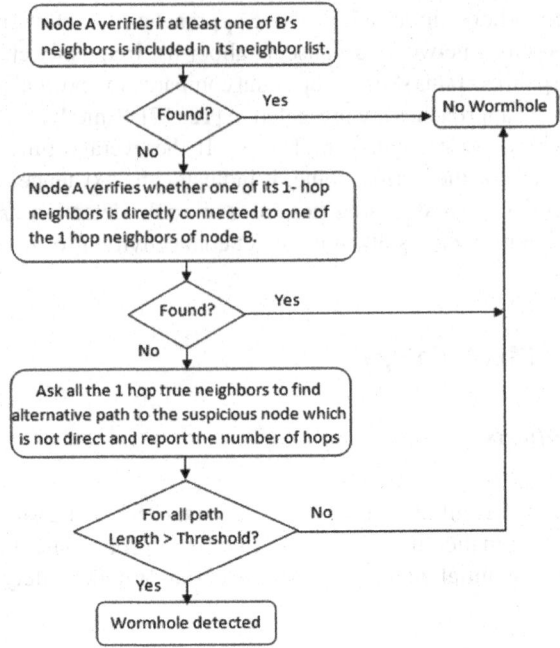

Fig. 2 Flow diagram of proposed methodology

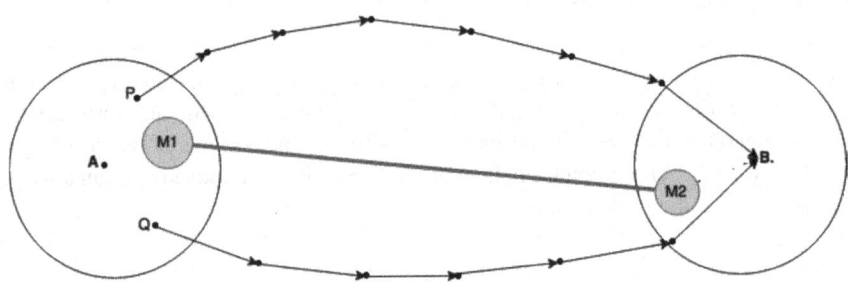

Fig. 3 Alternate path length calculation

from new node B; then, node B is added into suspicious neighbor list of node A and node A executes the procedure as shown in Fig. 2.

As shown in Fig. 3, all the trusted neighbors of node A find the shortest path to suspected node B which is not direct and it avoids the one-hop neighbors of node A. The direct path from node A to B is also not included. For all path, if the length is greater than the predefined threshold, then the link A -> B is declared as fake link and wormhole attack is detected.

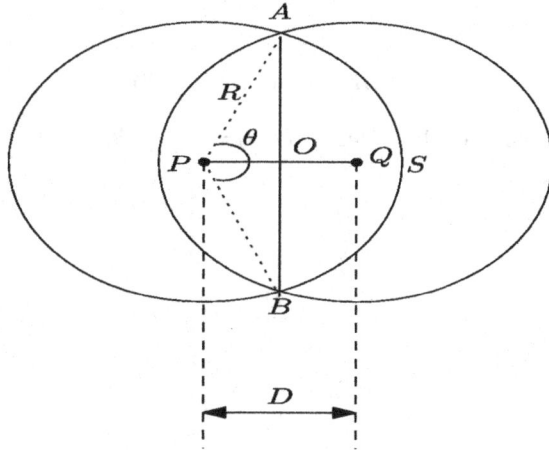

Fig. 4 Common area shared by two neighbor nodes

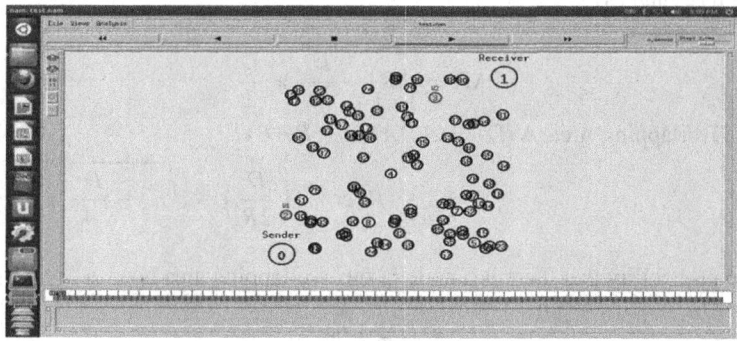

Fig. 5 Wormhole attack in dense network

4.4 Mathematical Analysis

Mathematically, we have shown that two genuine neighbor nodes always share common one-hop neighbors (Fig. 4).

In Fig. 5, the distance between nodes P and Q is D and the radius is R. We have calculated the probability of occurrence of the other node in the overlapped transmission zone.

$$\text{Sector Area(PASB)} = \frac{1}{2} * R * S$$

$$= \frac{\theta}{2} * R^2 \tag{1}$$

$$\frac{\theta}{2} = \cos^{-1}\frac{D}{2R} \tag{2}$$

From (1) and (2),

$$PASB = R^2 . \cos^{-1}(D/2R) \tag{3}$$

$$\text{Area of the triangle } (PAB) = \frac{1}{2} * AB * PO$$

$$= \frac{1}{2} * OA * D/2 \tag{4}$$

$$R^2 = PO^2 + OA^2$$

$$= \frac{D^2}{4} + OA^2$$

$$\therefore OA^2 = R^2 - \frac{D^2}{4}$$

$$\therefore OA = \sqrt{R^2 - \frac{D^2}{4}} \tag{5}$$

From (4) and (5),

$$PAB = \sqrt{R^2 - \frac{D^2}{4}} * \frac{D}{2} \tag{6}$$

$$\text{Overlapping area } A\,(D) = 2(PASB - PAB)$$

$$= 2 * (R^2 \cos^{-1}\left(\frac{D}{2R}\right) - \sqrt{R^2 - \frac{D^2}{4}} * \frac{D}{2} \tag{7}$$

The probability that there is a node in the overlapping area is

$$= e^{-\delta . A(d)} \tag{8}$$

The maximum distance between nodes P and Q is R,

$$\therefore P(D) = \frac{1}{R^2} * \frac{\partial D^2}{\partial D} = \frac{2D}{R^2} \tag{9}$$

From (8) and (9), the probability that none of the nodes exists in the common area is

$$P = \int_0^R \frac{2D}{R^2} * e^{-\delta . A(d)} * dD$$

$$= e^{-1.18 * \delta * R^2}$$

If we take density = 100 nodes per km^2 and R = 250 m, then p < 0.1%.

Table 1 Simulation parameters

No. of nodes	14, 25, 50, 100
Area	1000 * 1000 m^2
Routing protocol	AODV
Attacker	1 pair
Propagation	Two-ray ground

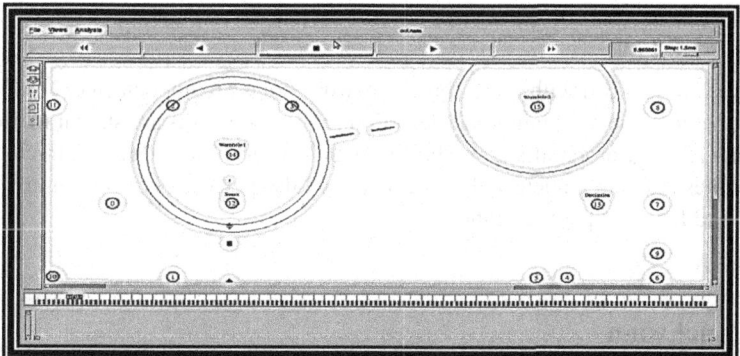

Fig. 6 Wormhole attack in sparse network

5 Result and Analysis

The average number of neighbors is represented by N_{AV}. The total number of nodes is represented by N_T. The size of ID is represented by S_{ID}. The storage cost required for storing the neighbor information is $S_{ID}N_{AV}$. To store the neighbors' neighbor list, the storage cost required is $S_{ID}N_{AV}N_{AV}$. For each node, the total storage cost required is $\{S_{ID}N_{AV} + S_{ID}N_{AV}N_{AV}\}$.

If the ID size = 4 bytes and avg. no. of neighbors of a node = 10, then the storage cost is 440 bytes for each node. Proposed protocol uses very less memory, and so it is applicable to resource-constrained wireless sensor networks. We have simulated wormhole attack in both dense and sparse networks. The simulation parameters are given in Table 1.

In normal scenario, packet delivery ratio is 99.88, in attack scenario it decreases to 36%, and after applying proposed approach it is 98.10. Similarly, in normal scenario throughput is 86 kbps, in attack scenario it decreases to 34 kbps, and after applying proposed approach it is 84.70 kbps. In the presence of attack, packet delivery ratio and throughput sharply decrease. After applying the proposed protocol, both packet delivery ratio and throughput have significant improvement (Fig. 6).

False positive is totally reduced. False negative occurs when wormhole launched for short distance. Detection accuracy is nearly 100% when wormhole launched for long distance. For reported path length, the threshold value λ taken is 3. We can detect short wormholes with $\lambda = 1$, but it will increase false positives. The false

Table 2 Accuracy analysis (in %)

No. of nodes	Pworm (2015) [23]	RTT-based MDS (2016) [21]	Proposed approach
14	0.80	0.93	0.97
25	0.82	0.95	0.98
50	0.86	0.96	0.99
100	0.91	0.98	0.99

positives are reduced with $\lambda = 5$, but short wormholes are not detected. $\lambda = 3$ is the best suitable value to obtain good detection ratio. The results are shown in Table 2. Pworm [23] is a probabilistic method in which when little traffic is attracted then wormholes may not be detected. In [21], a few false positives occur with less no. of nodes and for short path wormhole.

6 Conclusion

The research related to wormhole attack in WSNs has been followed with much interest in recent years. To launch the attack, attacker does not need any cryptographic breaks and no need to know the protocols use in the network. Most of the methods employ hardware which increases the manufacturing cost of a sensor node. Existing detection algorithms are resource hungry and have limitations in accuracy. Secure neighbor discovery in mobile wireless sensor network is a challenging research issue.

Acknowledgements The authors are highly thankful to Smt. S R Patel Engineering College, Gujarat Technological University, for providing the opportunity to conduct this research work.

References

1. Aboelaze, M., Aloul, F.: Current and future trends in sensor networks: a survey. In: Second IFIP International Conference on Wireless and Optical Communications Networks (2005)
2. Padmavathi, G., Shanmugapriya, D.: A survey of attacks, security mechanisms and challenges in wireless sensor networks. Int. J. Comput. Sci. Inf. Secur. (IJCSIS) 4, 117–125 (2009)
3. Karlof, C., Wagner, D.: Secure routing in wireless sensor networks: attacks and countermeasures. Ad Hoc Netw. J. (Spec. Issue Sens. Netw. Appl. Protoc.), 113–127 (2003)
4. Hu, Y.C., Perrig, A., Johnson, D.B.: Packet leashes: a defense against wormhole attacks in wireless networks. IEEE Comput. Commun. Soc. IEEE 3, 1976–1986 (2003)
5. Hu, Y.C., Perrig, A., Johnson, D.B.: Wormhole attacks in wireless networks. IEEE J. Sel. Areas Commun. 24(2), 370–380 (2006)
6. Poovendran, R., Lazos, L.: A graph theoretic framework for preventing the wormhole attack in wireless ad hoc networks. Wirel. Netw. 13, 27–59 (2007)
7. Zhang, Y., Liu, W., Lou, W., Fang, Y.: Location-based compromise—tolerant security mechanisms for wireless sensor networks. IEEE J. Sel. Areas Commun. 24(2) (2006)

8. Lazos, L., Poovendran, R.: SeRLoc: robust localization for wireless sensor networks. ACM Trans. Sens. Netw., 73–100 (2005)
9. Lazos, L., Poovendran, R.: HiRLoc: high-resolution robust localization for wireless sensor networks. IEEE J. Sel. Areas Commun. 24(2), 233–246 (2006)
10. Lu, X., Dong, D., Liao, X.: MDS-based wormhole detection using local topology in wireless sensor networks. Int. J. Distrib. Sens. Netw. 2012(Article ID 145702), 9 pp.
11. Buttyan, L., Dora, L., Vajda, I.: Statistical wormhole detection in sensor networks. SAS 2005, pp. 128–141. Springer
12. Xu, Y., Chen, G., Ford, J., Makedon, F.: Detecting wormhole attacks in wireless sensor networks. In: International Conference on Critical Infrastructure Protection, ICCIP 2007, pp. 267–279
13. Liu, J., Chen, H., Zhen, Z., Sha, M.: Intrusion detection algorithm for the wormhole attack in ad hoc network. In: International Conference on Computer Science and Informational Technology, pp. 147–154 (2013)
14. Chen, H., Lou, W., Sun, X., Wang, Z.: A secure localization approach against wormhole attacks using distance consistency. EURASIP J. Wirel. Commun. Netw. 2010, 11 pp.
15. Shokri, R., Poturalski, M.: A practical secure neighbor verification protocol for wireless sensor networks. ACM, WiSec'09, March 16–18, 2009, Zurich, Switzerland
16. Xu, Y., Ouyang, Y., Le, Z., Ford, J., Makedon, F.: Analysis of range-free anchor-free localization in a WSN under wormhole attack. ACM, MSWiM'07, Oct 22–26, 2007, Chania, Greece
17. Yadav, J., Kumar, M.: Detection of wormhole attack in wireless sensor networks. In: International Conference on ICT for Sustainable Development, pp. 243–250 (2016)
18. Singh, R., Singh, J., Singh, R.: WRHT: A Hybrid Technique for Detection of Wormhole Attack in Wireless Sensor Networks. Mob. Inf. Syst. 2016(Article ID 8354930), 13
19. Khabbazian, M., Mercier, H., Bhargava, V.K.: Severity analysis and countermeasure for the wormhole attack in wireless ad hoc networks. IEEE Trans. Wirel. Commun. 8(2), 736–745 (2009)
20. Qazi, S., Raad, R., Mu, Y., Susilo, W.: Securing DSR against wormhole attacks in multirate ad hoc networks. J. Netw. Comput. Appl., 582–593 (2013)
21. Mukherjee, S., Chattopadhyay, M., Chattopadhyay, S., Kar, P.: Wormhole detection based on ordinal MDS using RTT in wireless sensor network. J. Comput. Netw. Commun. 2016(Article ID 3405264), 15 pp.
22. Wang, W., Bhargava, B.: Visualization of wormholes in sensor networks. In: WiSe'04, Proceeding of the 2004 ACM Workshop on Wireless Security, pp. 51–60. ACM Press (2004)
23. Lu, L., Hussain, M.J., Luo, Q., Han, Z.: Pworm: passive and real-time wormhole detection scheme for WSNs. Int. J. Distrib. Sens. Netw. 2015(Article ID 356382), 16 pp.
24. Karapistoli, E., Sarigiannidis, P., Economides, A.A.: Visual-assisted wormhole attack detection for wireless sensor networks. In: International Conference on Security and Privacy in Communication Systems, pp. 222–238 (2015)

SDN-Based Optimization Model of Virtual Machine Live Migration Over Layer 2 Networks

Hongyan Cui, Bo Zhang, Yuyang Chen, Tao Yu, Zongguo Xia and Yunjie Liu

Abstract Virtualization can provide significant benefits for service user by cutting the cost as well as enabling easy access. However, because the MAC table capacity of each host is limited, the traditional layer 2 networks have been unable to meet the needs of the numbers of virtual machines in the center. VXLAN has become an important layer 2 over layer 3 network transmission tunneling technology, a good solution to the data center internal virtual machine isolation and cross-data center transmission problems, but VXLAN does not propose a complete solution for virtual machine migration in the data center. Therefore, we design and implement the network model based on SDN, which optimizes the current VXLAN technology. This model separates the network plane from migration control plane, which can simplify the process of VXLAN and the complex network in data center. Moreover, we propose a detailed process for virtual machine migration supported by OpenFlow protocol. Some design in our model can really reduce the complexity of VXLAN technology and make it convenient for the users of virtual machine.

H. Cui (✉) · B. Zhang (✉) · Y. Chen (✉) · Y. Liu
State Key Laboratory of Networking and Switching Technology, Beijing University
of Posts and Telecommunications, Beijing 100876, China
e-mail: cuihy@bupt.edu.cn

B. Zhang
e-mail: zhangbo_china@live.cn

Y. Chen
e-mail: chenyuyang@bupt.edu.cn

H. Cui · B. Zhang · Y. Chen · Y. Liu
Beijing Laboratory of Advanced Information Networks, Beijing, China

H. Cui · B. Zhang · Y. Chen · Y. Liu
Key Laboratory of Network System Architecture and Convergence, Beijing, China

T. Yu
Institute of Network Science and Cyberspace Tsinghua University, Tsinghua, China

Z. Xia
School Official University of Massachusetts Boston, Boston, USA

© Springer Nature Singapore Pte Ltd. 2019
S. K. Bhatia et al. (eds.), *Advances in Computer Communication
and Computational Sciences*, Advances in Intelligent Systems
and Computing 760, https://doi.org/10.1007/978-981-13-0344-9_40

Keywords Virtual machine live migration · VXLAN · SDN · OpenFlow

1 Introduction

With the sharp increase in data services and the appearance of cloud data center services such as IAAS, PAAS, and SAAS, an increasing number of enterprise users and
even individual users prefer to use cloud computing and cloud storage to cut cost,
bringing cloud computing an explosive increase to cloud computing. For cost reasons,
virtualization technology has been applied to cloud computing to realize multiplex
function [1]. But with the dramatic increase in data centers and the corresponding
increase in servers, it is increasingly hard to maintain and manage virtual private
machine, causing problems like the low utilization of servers and high energy consumption. To solve the above problems, the notion of resource dynamic combination
is proposed, in which virtual machine live migration is one of the core technologies.

Though the problem of network congestion in memory copy and migration of
virtual machine live migration is studied by many [2, 3], the domain scope which
can be realized by virtual machine live migration catches little attention. We consider
that the conflict between the continuous increase in cloud computing users as well
as virtual machine and the relatively small domain scope of virtual machine live
migration is escalating. Therefore, a prudent model should be proposed to solve the
problem without damaging its security. What deserves particular note is that virtual
machine live migration is not only to solve the problems of network latency in the
process of copy of the CPU and storage information, and data transmission, but also
to enlarge the scale and realm of the migration. In the current virtual machine live
migration technology, we use layer 2 networks as the main network structure. Due
to the limitation of the numbers of virtual machine among one layer 2 network, the
migration scope is strictly limited.

Cooperating with leading network-connectivity and electronics venders (including Cisco systems, Juniper), VMware created VXLAN, which is a common method
on network and top-level "floating" virtual domain of virtual infrastructure. We can
create mountains of virtual domains in existing networks by using standard Ethernet
technology, and they are completely isolated from each other and from the underlying networks. VXLAN can solve the problems of the limited table capacity of MAC
address, expanding the scope of virtual machine live migration from a certain extent
in turn. However, it is still certainly inefficient under the condition of the layer 2
networks. Therefore, we try to design a network model to break the limitation of
virtual machine live migration in layer 2 networks.

In this paper, we have analyzed limitation which causing the scale and scope of
live migration of virtual machine in data center under traditional network structure.
In addition, we study the scenarios and problems that VXLAN is target in expanding
the live migration range of virtual machines and the limitations and shortcomings of
VXLAN technology in solving this problem. We focus on its limitation and design
a SDN-based optimization model of virtual machine live migration over layer 2

networks. The main reason why we focus on SDN technology is that SDN controller can be a great supervisor to catch the global network condition and respond the network request. On the other side, SDN provides easier method to dispatch network source and network transportation rules. Meanwhile, we also considered the user access problem after live migration of the virtual machine over layer 2 networks and try to solve this problem through OpenFlow.

The rest of the paper is organized as follow: In Sect. 1, we present some related work done before as the foundation of our research. In Sect. 2, we discuss the process of virtual machine live migration and the problems faced by live migration under the traditional network structure. In Sect. 3, we will explain the SDN-based optimization model of virtual machine live migration over layer 2 networks put forward by us and clarify what extent it addresses the scale of dynamic migration of virtual machines. In Sect. 4, we make a conclusion of our design.

2 Related Works

Now, in solving the VLAN restrictions on cloud data center networks, there are mainly two tunneling strategies based on layer 2 over layer 3 network: NVGRE and VXLAN.

Garg et al. proposed NVGRE [4] which builds a layer 2 over layer 3 tunneling mechanism by means of Generic Routing Encapsulation (GRE). Mahalingam et al. proposed VXLAN [5] which also achieves layer 2 over layer 3 network virtualization. Both VXLAN and NVGRE can realize data center extensions very well. Nick proposed the OpenFlow protocol [6]. As a southward agreement of SDN, OpenFlow implements the function of SDN very well. It provides a very effective solution to the various problems in cloud data center network. In [7], Bochra Boughzala proposes the method of using SDN to solve the strategy of cross-domain virtual machine migration. However, they still use the traditional layer 2 over layer 3 communication means, but they only use the advantages of flexible configuration of SDN to complete the migration of virtual machines, and do not realize the expansion of data center scale. Ajila in [8] proposed using packet redirection and dynamic DNS method to ensure the achievement of functional connectivity while allowing the virtual machine IP address change.

3 The Process of Migration and the Network Challenge

Virtual Machine Live Migration

The concept of virtual machine migration is proposed to achieve physical machine load balancing to save costs and resources [9]. Due to the increase in the number of users of cloud computing platform, the division and management of physical machine's resource becomes very complicated, and the amount of tasks realized by each user on virtual machine is very easy to cause excessive load or inefficient use of physical resource. So we need to migrate to the virtual machines on physical machines to ensure a relatively good use of resources. Virtual machine migration is divided into two ways, static migration and live migration. In a static migration project, the source virtual machine will be aborted to facilitate the use of the storage device to copy the data to the destination virtual machine, where the user's service is interrupted during the migration process. While the live migration, the source virtual machine will remain running state. Here, we only discuss live migration.

Virtual machine live migration can be divided into two stages:

- Preparation stage: The migration detector checks the state of the physical machines continuously and notifies the migration decision maker to start the migration process when the load of a certain physical machine is detected to be exceeding the threshold. The migration decision maker selects the source virtual machine with a set criterion and determines the destination virtual machine by the interface-aware component.
- Migration stage: The complete operation status of CPU, RAM, and devices of the source virtual machine is transmitted to the destination virtual machine via the network.

One of the key factors in the migration phase is the network, and the network topology and bandwidth or other parameters will affect the live migration of the virtual machine. In the effect of the network, we attach the most attention to two indicators, the migration speed and migration scale. There have been a lot of researchers studying migration speed improvement, and this is not the focus of our research. We will focus on the expansion of the migration scale.

The Problems Faced by the Scale of Migration Under the Traditional Network Structure

- Layer 2 networks: For layer 2 networks, different virtual machines are distinguished only by MAC address. In the communication between two virtual machines, the ARP protocol is used for addressing. ARP protocol is broadcast addressing in the same layer 2 networks, and the host of the address matching will record the source MAC address and the forwarding path in the MAC table and respond.

  However, the MAC table capacity of each host is limited, and when the number of virtual machines in the layer 2 networks increases rapidly, the MAC table will overflow, resulting in the failure to record new MAC addresses and forwarding.

  At the same time, in the layer 2 networks, in order to prevent flooding and broadcast storm, the STP protocol is adopted to eliminate the loop in the topology structure. Due to the convergence performance of STP and other reasons, in general, the network size of STP will not exceed 100 switches. At the same time, because STP needs to block redundant devices and links, it also reduces the bandwidth utilization of network resources. Therefore, in the actual network planning, from the forwarding performance, utilization, reliability, and other considerations, it will control the STP network range as much as possible.

The above two points limit the number of virtual machines in the layer 2 networks, and further limit the scale of the live migration of the virtual machines in the layer 2 networks.

- VLAN network: The introduction of VLAN technology has expanded the scale of layer 2 networks to a certain extent. The VLAN network uses VLAN id to make further partition and isolation of the layer 2 networks. The virtual machines under the same VLAN id can communicate with each other through the ARP protocol, while the virtual machines under different VLAN ids are isolated.

This is equivalent to that a big layer 2 network is divided into many separate groups, the number of virtual machines in each group can reach the original size, and the virtual machine capacity within the same data center after the integration is enhanced.

However, in VLAN technology, the number of VLAN ID available is only 4094. Facing the explosive growth of the number of virtual machines, this number is still somewhat inadequate. The limitation of the layer 2 networks using only MAC addresses as unique identifiers has not been eliminated.

4 Our Design

Overview

The network model of virtual machine live migration over layer 2 networks based on SDN designed by us is to use SDN controller for the optimization as to the point that broadcast is needed when data packets are forwarded in VXLAN Tunneling End Point (VTEP) for existing VXLAN technology. In this way, the broadcast required by ARP addressing is effectively changed to unicast when the connection between VTEP is establishing, so that the efficiency of network connection establishment is improved.

We illustrate our model in Fig. 1. Considering cost and other reasons, the internal structure of existing data centers needs to be compatible. Our model does not affect the network within the data center. Similarly, under the cost permit, the layer 2 networks in the data center can also be improved by using the SDN controller to improve efficiency, but this is not the focus of our model discussion. The SDN controller is separated from the control components such as virtual machine monitor (VMM), physical machine monitor (PMM), migration decision-making component, which are used for preparation and judgment in virtual machine migration. The control information is transmitted between them to achieve the transmission of judgment information required by virtual machine migration through the network, and the configuration information of the network is provided through the virtual machine migration management module.

To some extent, our model achieves some degree of separation of virtual machine migration network plane from migration control plane. In addition, our model analyzes and resolves the user access after virtual machine migration and also makes

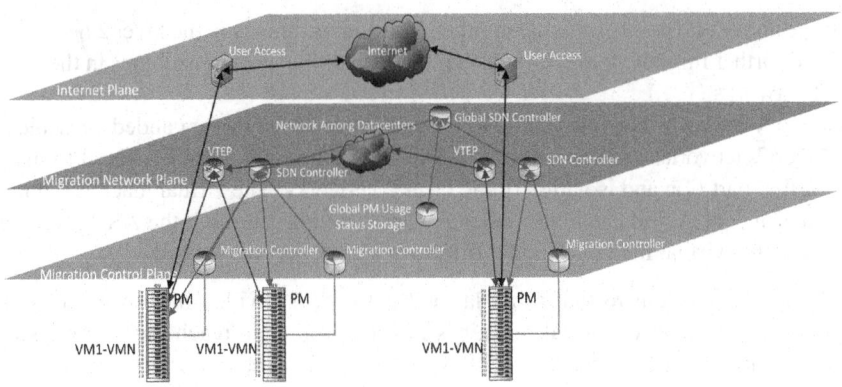

Fig. 1 Overview of the optimization model

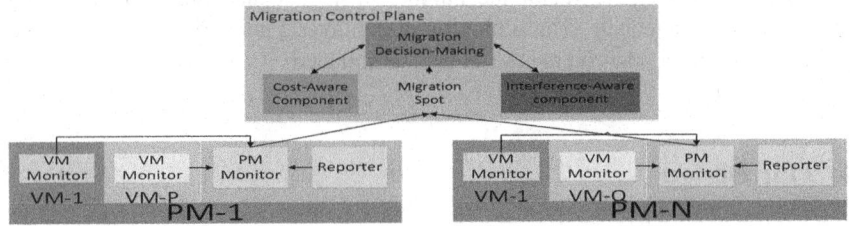

Fig. 2 Migration control module

use of the collaboration between SDN controller and virtual machine migration controller to ensure the user can still be directed to the virtual machine after the migration in the access of original IP address and port.

The separation of virtual machine migration network plane from migration control plane

Zhang W et al. proposed in [10] that in the virtual machine management module, each virtual machine contains a monitor-VM which is responsible for collecting the system parameters of the virtual machine. On each physical machine, there is a monitor-PM that collects the system parameters of the physical machine like in Fig. 2. Migration spot detector continuously monitors the use status of resources from the PM and transfers the data to a storage space of the SDN master controller via the connected SDN controller for summary of global PM resources use status. When PM resources use, such as CPU or disk usage, exceeds the threshold set by the policy, the virtual machine dynamic migration begins. That is, the migration spot detection component determines when to notify the migration decision-making component to start migration. In Fig. 3, we propose a model separate the migration plane and network plane, making the process clearer.

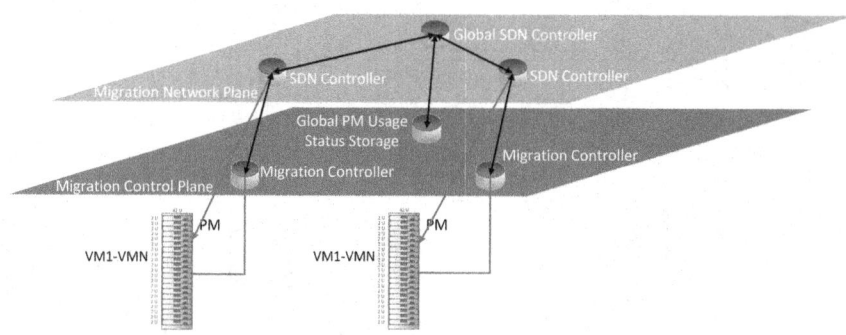

Fig. 3 Separation of VM migration network plane and migration control plane

Use of SDN to optimize VXLAN

First of all, in implementation, we consider that each switch within the data center supports the OpenFlow protocol and is able to communicate with the controller. VTEP can be configured using virtual switches (such as Open vSwitch). Considering the performance of the controller and the flow of data in data interacting with the switch, we can use one controller to control for each of the layer 2 networks under the VTEP, including the VTEP. Then, the whole network is managed by the method of multi-controller cooperation. For controller collaboration, we employ a master control scheme to control several subcontrollers.

When a physical connection is made between the switch and the controller, the two parties establish the secure channel through the OpenFlow protocol, and the controller can obtain all kinds of information about the whole network.

According to the above analysis of transfer control management module, after virtual machine migration begins, the controller can divide the migration process into the same layer 2 networks and cross layer 2 networks through the network global information without the need for broadcast search.

In the same layer 2 network (Fig. 4):

SDN's flexible control of the network and the characteristics of OpenFlow to abstract network traffic as a flow make the process very simple, requiring only the controller to issue the flow table to the switch and establish a path in the layer 2 networks to establish communication between the two virtual machines, so the migration can begin. After the migration, the flow table can be deleted to achieve the isolation between the VMs. However, based on the traditional VXLAN, two-tier broadcast in the same layer 2 networks increases the burden on the internal network of the data center.

Cross layer 2 networks:

When the controller knows that migration requires a cross layer 2 networks, a flow table is built to transmit data frames to the VTEP to encapsulate the packet.

In VXLAN technology, VTEP needs to manage MAC of virtual machines and corresponding table between VXLAN ID and VTEP IP, and we might as well call it

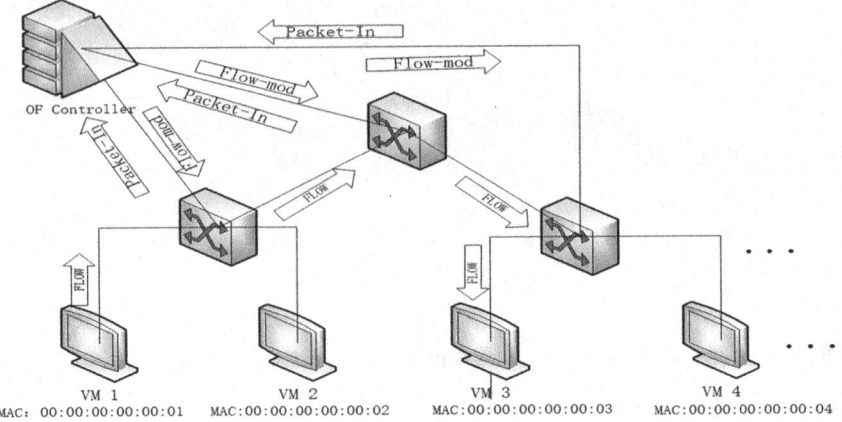

Fig. 4 Overview of network condition in the same layer 2 networks

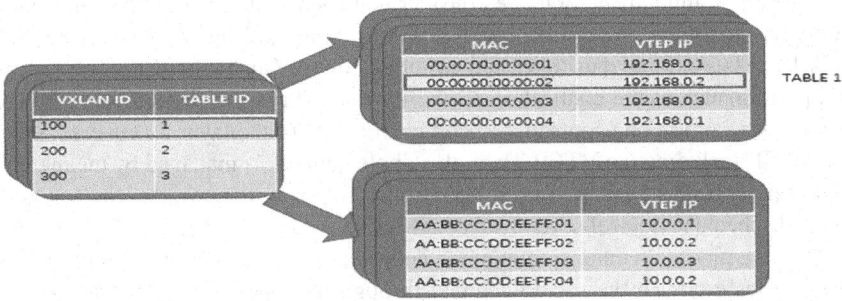

Fig. 5 Process to find VTEP IP in VTEP table

VTEP TABLE. This table is managed by the controller, transmitted from VTEP to the subcontroller, and then transmitted from the subcontrol to the master controller.

In the lookup table in master controller, in order to reduce the complexity of lookup table, we can use the thought of the multi-level flow table proposed in OpenFlow1.1 protocol. First, we check the VXLAN ID in a set of tables; each VXLAN ID is corresponding to one table that is corresponding between MAC and VTEP IP; in this table, we can find the MAC of virtual machines of the same VXLAN ID and the corresponding VTEP IP.

As shown in Fig. 5, if the VXLAN ID of the VM needing migration is 100, then directly go to the table whose TABLE ID is 100 for query, and then query the corresponding VTEP IP whose MAC address is 00:00:00:00:00:02.

After this, the subcontroller already knows the MAC of the destination virtual machine and the corresponding VTEP IP. At this time, let VTEP encapsulate the package, and the package then shall be transmitted to the specified VTEP for decapsulation, which shall ultimately realize the migration of virtual machines. This process actually turns the originally required multi-cast into a unicast process (Fig. 6).

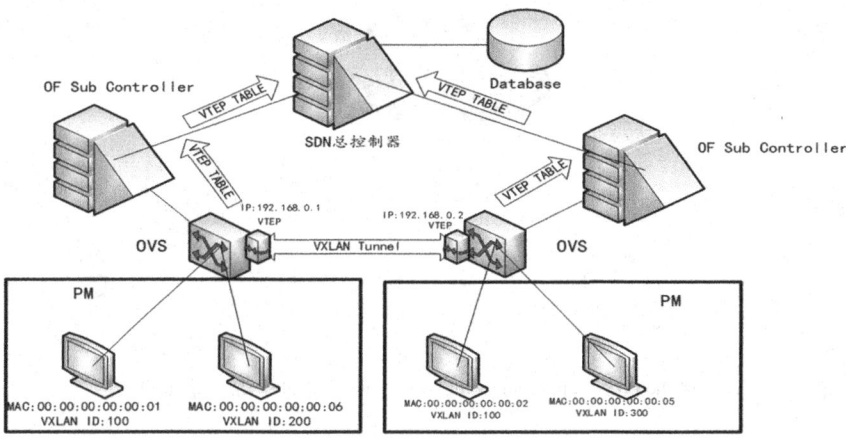

Fig. 6 Overview of the network condition cross layer 2 networks

After the migration

When the virtual machine migration is completed, the controller needs to inform the jump server before the migration of the IP and port of the jump server after the migration. When the user visits the original virtual machine again, it will first visit the original jump server and obtain the IP and port number of the new jump server, so that the access to the virtual machine after the migration can be achieved through the original IP and port number. This process is invisible to the user. Therefore, a migration table is required to be generated at the master controller, which can be achieved through the port-status message in the OpenFlow protocol. When change in the switch port occurs, this message will be sent to the subcontroller to inform the controller of the change; after the subcontroller transmits the message to the master controller, the master controller can obtain location information before and after the migration according to the received message. Considering the scope of the data center and the capacity of the master controller, we design distributed storage in each of the jump server. The stored tables are updated according to the data sent by the controller.

As shown in Fig. 7, the VM2 under jump server 1 completes the migration of the VM3 under the jump server 2, its access IP and port number under jump server machine 1 is 192.168.0.1:2001. The user accesses the original jump server through this IP, and then, the jump server queries whether there is migration information in the storage, and if not, the virtual machine is accessed directly. If there is migration information in the storage, the virtual machine is re-accessed according to the jump server IP (10.0.0.1:3000) after the migration which is gained by table query. It is also necessary here to use OpenFlow for fine-grained operations on data streams, that is, to change the destination IP address in the IP packet. At this point, the user is unaware that the virtual machine has completed the migration.

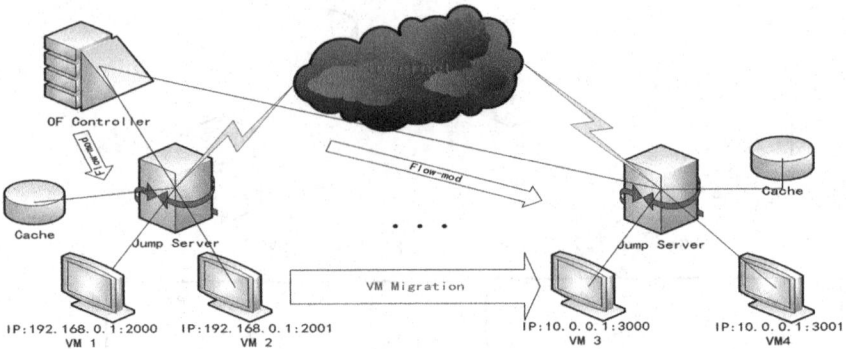

Fig. 7 Overview of the process to connect the virtual machine after the migration

5 Conclusion

In this chapter, we focus on expanding the network range of virtual machine live migration and describe the design and implementation process of the SDN-based optimization model of virtual machine live migration over layer 2 networks.

1. Through the deployment and configuration of the SDN controller in the network, we implement our model on the virtual machine migration network. With the separation and collaboration between virtual machine migration control management plane and network plane, we can optimize and simplify the whole process of virtual machine migration.
2. For the user access problem after the virtual machine live migration across layer 2 networks, we propose an IP address redirection scheme based on OpenFlow to ensure the invisibility of the migration process for the user.

However, in this chapter, we do not give detailed data to illustrate the effectiveness of our proposed model and the comparison with existing schemes. In addition, we do not consider the cost. Our design requires network components to support the OpenFlow protocol and needs to set up a large number of storage devices, and even network links, so the transformation of existing networks requires a great deal of resource input. In the future, we plan to carry out experiments on the designed model to verify its actual effect and compare the experimental data with the existing schemes for performance comparison. At the same time, we will consider the cost problem and reduce costs as much as possible.

References

1. Han,Z., Tan, H., Chen, G., Wang, R., Chen, Y., Lau, F.C.M.: Dynamic virtual machine management via approximate Markov decision process. IEEE INFOCOM 2016—The 35th Annual IEEE International Conference on Computer Communications, San Francisco, CA, 2016, pp. 1–9
2. Zhang, J., Lu, X., Panda, D.K.: High-performance virtual machine migration framework for MPI applications on SR-IOV enabled InfiniBand clusters. In: 2017 IEEE International Parallel and Distributed Processing Symposium (IPDPS), Orlando, FL, pp. 143–152 (2017)
3. Zhang, F., Fu, X., Yahyapour, R.: CBase: a new paradigm for fast virtual machine migration across data centers. In: 2017 17th IEEE/ACM International Symposium on Cluster, Cloud and Grid Computing (CCGRID), Madrid, 2017, pp. 284–293
4. Wang, Y.-S., Garg, P.: NVGRE: Network Virtualization Using GenericRouting Encapsulation. RFC 7637, Oct. 2015
5. Mahalingam, M., et al.: VXLAN: a framework for overlaying virtualized layer 2 networks over layer 3 networks. Internet http://draft.draft-mahalingam-dutt-dcops-vxlan-08.txt, IETF, Feb 2014
6. McKeown, N., Anderson, T., Balakrishnan, H., Parulkar, G., Peterson, L., Rexford, J., Shenker, S., Turner, J.: OpenFlow: enabling innovation incampus networks. ACM SIGCOMM Comput. Commun. Rev. 38(2), 69–74 (2008)
7. Boughzala, B., Ali, R.B., Lemay, M., Lemieux, Y., Cherkaoui, O.: OpenFlow supporting inter-domain virtual machine migration. In: Proceedings of the 8th International Conference on WOCN, pp. 1–7 (2011)
8. Ajila, S.A., Iyamu, O.: Efficient live wide area VM migrationwith IP address change using type II hypervisor. In: IEEE International Conference on Information Reuse & Integration, pp. 372–379 (2013)
9. Rawat, S., Tyagi, R., Kumar, P.: An investigative study on challenges of live migration. In: 2016 5th International Conference on Reliability, Infocom Technologies and Optimization (Trends and Future Directions) (ICRITO), Noida, pp. 129–134 (2016)
10. Zhang, W., Ahu, M., Mei, Y., et al.: LVMCI: efficient and effective VM live migration selection scheme in virtualized data centers. In: IEEE 8th International Conference on Parallel and Distributed Systems (ICPDS), pp 368–375 (2012)

Waveform Design of Linear Frequency Modulation Pulse Signal Based on Ambiguity Function

Bin Wang

Abstract The rapid development of science and technology brings exigent demands for radar performance. Intelligent radar should have the ability to be flexible and agile and transmit different waveforms according to different working environment. In this paper, in order to obtain more flexible waveform, we propose a novel waveform design method of linear frequency modulation pulse signal after analyzing principles of ambiguity function. In simulations, time–frequency characteristics and ambiguity characteristics are provided. Simulations results demonstrate the proposed method which has good range and velocity resolution. Finally, the full paper is summarized.

Keywords Intelligent radar · Linear frequency modulation pulse signal
Ambiguity function

1 Introduction

With the development of modern science and technology, excessive use of electromagnetic spectrum will make electromagnetic environment become more complex. There is a huge demand for intelligent sensors based on sophisticated technology, and the radar system is developing toward the intelligent direction.

In order to meet the needs of practical application, more and more modern intelligent radar systems are proposed. Cognitive radar is a new concept of intelligent radar systems. Simon Haykin has proposed it as a technological solution for performance optimization in limited and constrained environments. In cognitive radar, the radar can study the environment continuously through prior knowledge with the working condition and transmit waveforms adaptively to the environment [1].

B. Wang (✉)
EOSA Institute, Northeastern University at
Qinhuangdao, Taishanrd. 143, Qinhuangdao 066004, China
e-mail: wangbinneu@qq.com

© Springer Nature Singapore Pte Ltd. 2019
S. K. Bhatia et al. (eds.), *Advances in Computer Communication
and Computational Sciences*, Advances in Intelligent Systems
and Computing 760, https://doi.org/10.1007/978-981-13-0344-9_41

Waveform design is becoming a hot topic in radar research, and many researchers have done much work in related field. In [2], the authors research on correlation function problem and propose a new form for sine-phased binary offset carrier tracking, which is confirmed from numerical results. In [3], the authors research on signal coding problem and propose a three-state signal coding scheme, which is suitable for driving Class-S amplifiers with higher efficiency. In [4], in order to reduce the peak to average power ratio, the authors propose the $\beta-\alpha$ pulse shaping filter, which has a better performance. In [5], the authors research on the synthesis problem of ambiguity function and propose two algorithms which will minimize the squared error. In [6], the authors introduce delay-Doppler ambiguity function in multiple-input multiple-output radar, which is related to the clutter's power. In [7], from the radar point of view, the authors research the Iridium L band downlink signals based on ambiguity function. In [8], in the presence of clutter, the authors obtain ambiguity functions for radar. Matched illumination transmit signals are used for the detection of range-spread targets. However, the waveform design methods mentioned above lack flexibility. In [9], it is shown that when looking upwind or downwind, local spectrum intensity is strongly linked to mean Doppler shift. In [10], a new model is proposed, which is comprised by functions of radar frequency, polarization, sea state, and grazing angle.

In this paper, after analysis of characteristics of ambiguity function, we propose a novel waveform design method of linear frequency modulation pulse signal, which can obtain more flexible waveform.

2 Ambiguity Function of Point Target Echo

Ambiguity function not only can easily describe the radar signal form and some related features of the corresponding matching filter, but also is very useful in the analysis of the target resolution, side lobe performance, and Doppler coupling and other related problems.

Set two-point targets M_1 and M_2 that need to be resolved and have the same detection angle. Figure 1 shows the relative position of the two-point targets on the two-dimensional plane of the delay-Doppler shift. The delay of the received signal of the detection target M_1 is t_d, and the Doppler frequency shift is f_d. The delay of the received signal of the detection target M_2 is $t_d + \tau_d$, and the Doppler frequency shift is $f_d + \xi_d$. Therefore, the echo reception signal of the detection target M_1 can be expressed as

$$
\begin{aligned}
s_1(t) &= u(t - t_d)e^{j2\pi(f_0+f_d)(t-t_d)} \\
&= u(t - t_d)e^{j2\pi f_0(t-t_d)}e^{j2\pi f_d(t-t_d)}
\end{aligned}
\tag{1}
$$

where the complex envelope of the echo signal of the detection target M_1 can be expressed as

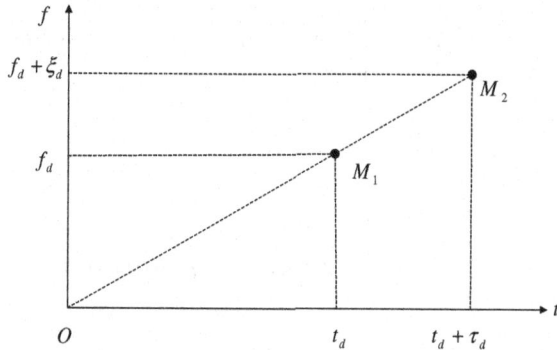

Fig. 1 Time–frequency two-dimensional coordinates of two-point targets

$$u_1(t) = u(t - t_d)e^{j2\pi f_d(t-t_d)} \tag{2}$$

Similarly, the complex envelope of the echo signal of the detection target M_2 can be expressed as

$$u_2(t) = u(t - t_d - \tau_d)e^{j2\pi(f_d+\xi_d)(t-t_d-\tau_d)} \tag{3}$$

It can be seen from Fig. 1 that the larger the relative delay τ_d and relative Doppler frequency shift ξ_d is, the easier to distinguish the two-point targets. In order to distinguish the two-point targets, we need to use mathematical tools. The commonly used method is to use mean square error criterion as the best resolution criterion. Mean square value of the difference of complex envelope of two-point target echo signal can be expressed as

$$
\begin{aligned}
\varepsilon^2_{\tau_d,\xi_d} &= \int_{-\infty}^{\infty} |u_1(t) - u_2(t)|^2 dt \\
&= \int_{-\infty}^{\infty} [u_1(t) - u_2(t)][u_1(t) - u_2(t)]^* dt \\
&= \int_{-\infty}^{\infty} u(t - t_d)u^*(t - t_d)dt + \int_{-\infty}^{\infty} u(t - t_d - \tau_d)u^*(t - t_d - \tau_d)dt \\
&\quad - 2\mathrm{Re}\left\{ \int_{-\infty}^{\infty} u^*(t - t_d)u(t - t_d - \tau_d)e^{j2\pi[\xi_d(t-t_d-\tau_d)-f_d\tau_d]}dt \right\}
\end{aligned}
\tag{4}
$$

where $u^*(t)$ is the conjugate function of $u(t)$. The first and second terms are twice the energy of the echo reception signal of the detection target M_1 and M_2, and they are denoted as $2E_1$ and $2E_2$. Then let $t' = t - t_d - \tau_d$, we can get

$$\varepsilon^2_{\tau_d,\xi_d} = 2\left\{ E_1 + E_2 - \mathrm{Re}\left[e^{-j2\pi f_d \tau_d} \int\limits_{-\infty}^{\infty} u(t')u^*(t' + \tau_d)e^{j2\pi \xi_d t'}\, dt' \right] \right\}$$

$$= 2\{E_1 + E_2 - \mathrm{Re}[e^{-j2\pi f_d \tau_d}\chi(\tau_d, \xi_d)]\} \tag{5}$$

where

$$\chi(\tau_d, \xi_d) = \int\limits_{-\infty}^{\infty} u(t')u^*(t' + \tau_d)e^{j2\pi \xi_d t'}\, dt' \tag{6}$$

Replace the integral variable t' with t, and we can get

$$\chi(\tau_d, \xi_d) = \int\limits_{-\infty}^{\infty} u(t)u^*(t + \tau_d)e^{j2\pi \xi_d t}\, dt \tag{7}$$

The function $\chi(\tau_d, \xi_d)$ can be defined as ambiguity function of the radar signal. It is a composite autocorrelation function of the time and frequency of the complex envelope of the two-point target echo.

$$\mathrm{Re}[e^{j2\pi f_d \tau_d}\chi(\tau_d, \xi_d)] \leq |\chi(\tau_d, \xi_d)| \tag{8}$$

So, we can get

$$\varepsilon^2_{\tau_d,\xi_d} \geq 2[E_1 + E_2 - |\chi(\tau_d, \xi_d)|] \tag{9}$$

Radar resolution of the target is generally after processing and detection of the radar system on the echo signal. That is to say to use echo signal's modulus value to distinguish. So, the modulus $|\chi(\tau_d, \xi_d)|$ of ambiguity function represents a measure of the distance and velocity resolution of two similar detection targets. In the processing of radar distinguishing target, the module $|\chi(\tau_d, \xi_d)|$ will increase with the decrease of τ_d and ξ_d, and the mean square error $\varepsilon^2_{\tau_d,\xi_d}$ will increase which means it is easy to distinguish the detection targets.

For a radar measuring a target, Doppler frequency and range are very essential. Besides, two orthogonal space angles are also important. However, if a radar wants to measure targets in angle coordinates, that is almost impossible. According to the theory of radar detection, we usually make an independent consideration to angle resolution. So, their resolution properties are different. We usually design waveform to achieve the performance of either good range resolution or good Doppler. However, the two performances are contradictory.

3 AF-Based Linear Frequency Modulation Pulse Signal

Complex envelope of LFM pulse signal can be expressed as

$$
u(t) = \begin{cases} Ae^{j\pi Kt^2} & (0 \le t \le \tau) \\ 0 & \text{(others)} \end{cases}
\tag{10}
$$

where chirp rate $K = B/\tau$, and B is frequency modulation bandwidth.
LFM pulse signal can be expressed as

$$
s(t) = A\,rect(t/\tau)\cos(2\pi f_0 t + \pi Kt^2)
\tag{11}
$$

where $rect(t/\tau)$ is rectangular function

$$
rect(t/\tau) = \begin{cases} 1 & (0 \le t \le \tau) \\ 0 & \text{(others)} \end{cases}
\tag{12}
$$

According to the definition of ambiguity function, we can get

$$
\chi(\tau_d, \xi_d) = \int_{-\infty}^{\infty} u(t)u^*(t+\tau_d)e^{j2\pi\xi_d t}dt = \int_{\alpha}^{\beta} e^{j2\pi(\xi_d - K\tau_d)t}e^{-j\pi K\tau_d^2}dt
\tag{13}
$$

When $|\tau_d| > \tau$, so

$$
\chi(\tau_d, \xi_d) = 0
\tag{14}
$$

When $-\tau < \tau_d \le 0$, according to the existence interval of two signals, integral lower limit $\alpha = -\tau_d$, and integral upper limit $\beta = \tau$.

$$
\chi(\tau_d, \xi_d) = e^{j\pi[(\xi_d - K\tau_d)(\tau - \tau_d) - K\tau_d^2]}\left[\frac{\sin\pi(\xi_d - K\tau_d)(\tau + \tau_d)}{\pi(\xi_d - K\tau_d)}\right]
\tag{15}
$$

When $0 < \tau_d \le \tau$, according to the existence interval of two signals, integral lower limit $\alpha = 0$, and integral upper limit $\beta = \tau - \tau_d$.

$$
\chi(\tau_d, \xi_d) = e^{j\pi[(\xi_d - K\tau_d)(\tau - \tau_d) - K\tau_d^2]}\left[\frac{\sin\pi(\xi_d - K\tau_d)(\tau - \tau_d)}{\pi(\xi_d - K\tau_d)}\right]
\tag{16}
$$

Merge the above three formulas, and we can get

$$\chi(\tau_d, \xi_d) = \begin{cases} e^{j\pi\left[(\xi_d - K\tau_d)(\tau - \tau_d) - K\tau_d^2\right]}\left[\frac{\sin[\pi(\xi_d - K\tau_d)(\tau - |\tau_d|)]}{\pi(\xi_d - K\tau_d)(\tau - |\tau_d|)}(\tau - |\tau_d|)\right] & (|\tau_d| \leq \tau) \\ 0 & (|\tau_d| > \tau) \end{cases}$$

$$(17)$$

The modulus is

$$|\chi(\tau_d, \xi_d)| = \begin{cases} \left|\frac{\sin[\pi(\xi_d - K\tau_d)(\tau - |\tau_d|)]}{\pi(\xi_d - K\tau_d)(\tau - |\tau_d|)}(\tau - |\tau_d|)\right| & (|\tau_d| \leq \tau) \\ 0 & (|\tau_d| > \tau) \end{cases}$$

$$(18)$$

4 Simulations

The simulation parameters are set as follows. Normalize amplitude A, signal duration $T = 10\,\mu s$, carrier frequency $f_0 = 0\,Hz$, and signal bandwidth $B = 30\,MHz$. Figure 2 is characteristics of linear frequency modulation pulse signal. Carrier frequency changes into $f_0 = 10\,MHz$, and Fig. 3 is characteristics of linear frequency modulation pulse signal. Comparing Fig. 2 with Fig. 3, the amplitude–frequency characteristic spectrum of the signal is moved to 10 MHz. According to the above two experimental conditions, time-bandwidth product is $BT = 300$.

When $T = 5\,\mu s$, $f_0 = 0\,Hz$, $B = 30\,MHz$, that is to say $BT = 150$. Figure 4 is characteristics of linear frequency modulation pulse signal. When $T = 30\,\mu s$, $BT = 900$, Fig. 5 is characteristics of linear frequency modulation pulse signal.

It can be concluded that when time-bandwidth product BT grows larger, amplitude–frequency characteristic curve of LFM pulse signal will approach rectangle. Its spectral width does not change with time.

When signal bandwidth $B = 2.5\,Hz$, signal time-width $T = 2\,s$, Fig. 6 is ambiguity graph of linear frequency modulation pulse signal. Figure 7 is ambiguity degree graph of linear frequency modulation pulse signal.

Modify experimental parameters, signal bandwidth $B = 1.5\,Hz$, signal time-width $T = 1.5\,s$. Figure 8 is ambiguity graph of linear frequency modulation pulse signal. Figure 9 is ambiguity degree graph of linear frequency modulation pulse signal.

Cutting the fuzzy graph, when $\xi_d = 0$, we can get

$$|\chi(\tau_d, 0)| = \begin{cases} \left|\frac{\sin[(-K\tau_d)(\tau - |\tau_d|)]}{\pi(-K\tau_d)(\tau - |\tau_d|)}(\tau - |\tau_d|)\right| & (|\tau_d| \leq \tau) \\ 0 & (|\tau_d| > \tau) \end{cases}$$

$$(19)$$

When $\tau_d = 0$,

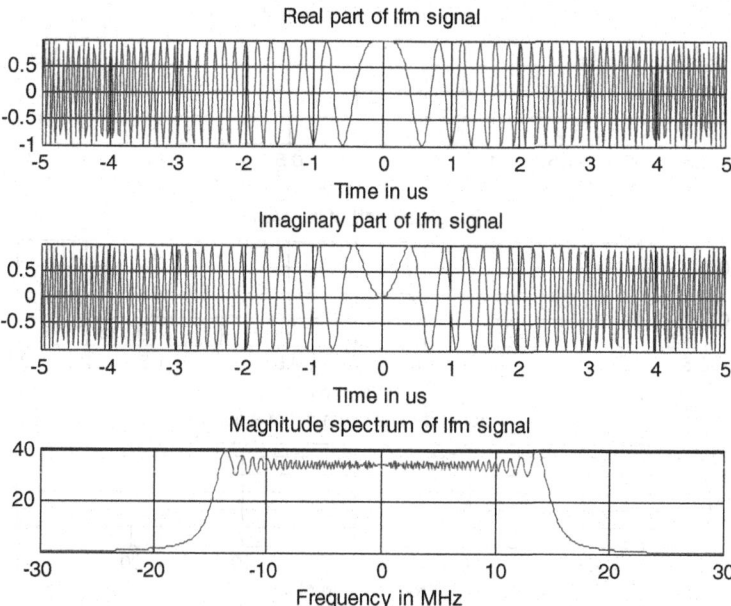

Fig. 2 Characteristics of linear frequency modulation pulse signal (30 MHz)

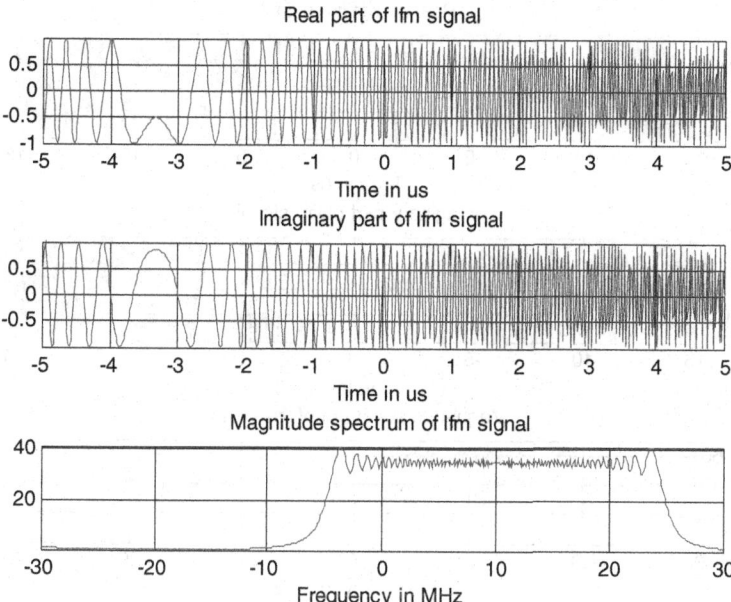

Fig. 3 Characteristics of linear frequency modulation pulse signal (10 MHz)

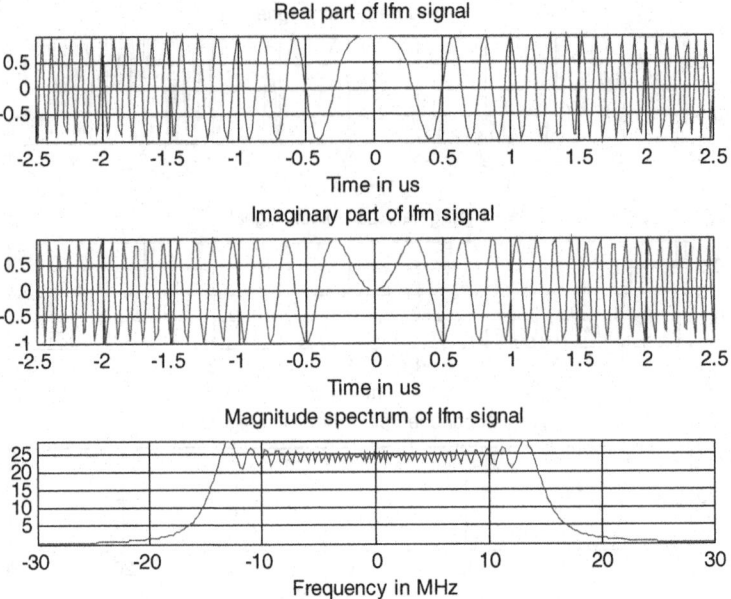

Fig. 4 Characteristics of linear frequency modulation pulse signal (BT = 150)

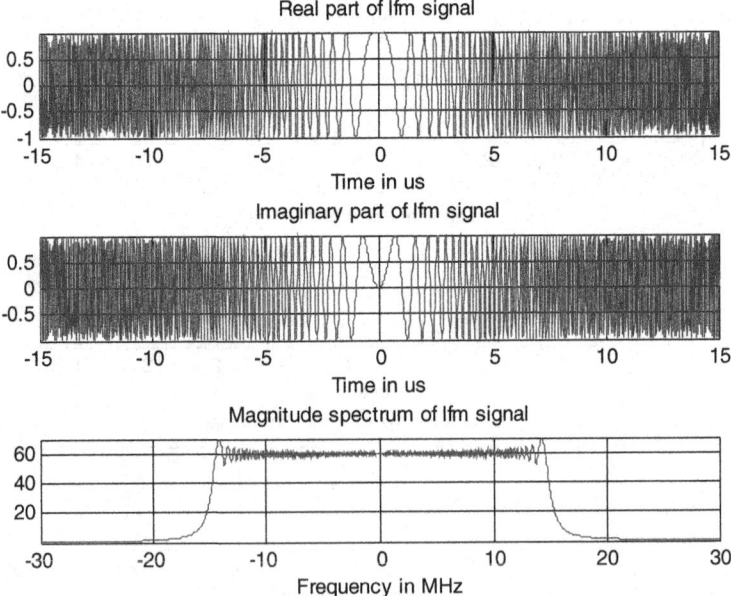

Fig. 5 Characteristics of linear frequency modulation pulse signal (BT = 900)

Fig. 6 Ambiguity graph of linear frequency modulation pulse signal

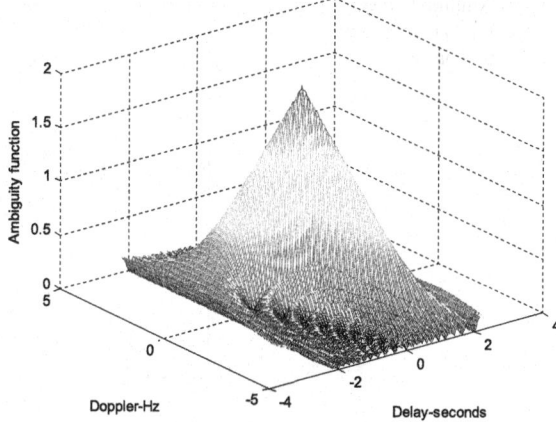

Fig. 7 Ambiguity degree graph of linear frequency modulation pulse signal

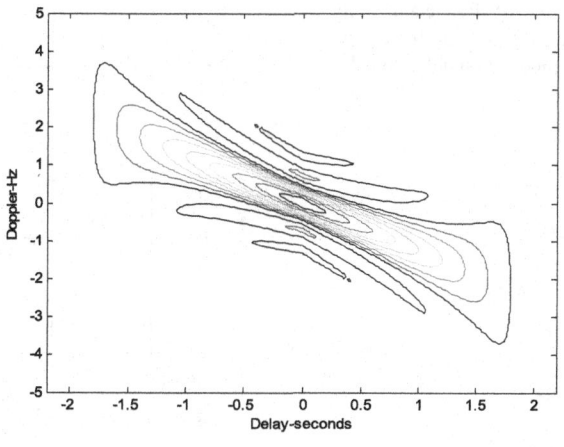

Fig. 8 Ambiguity graph of linear frequency modulation pulse signal

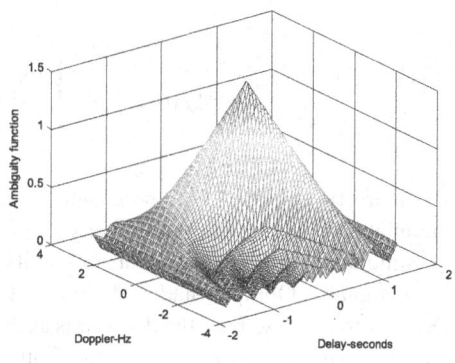

Fig. 9 Ambiguity degree graph of linear frequency modulation pulse signal

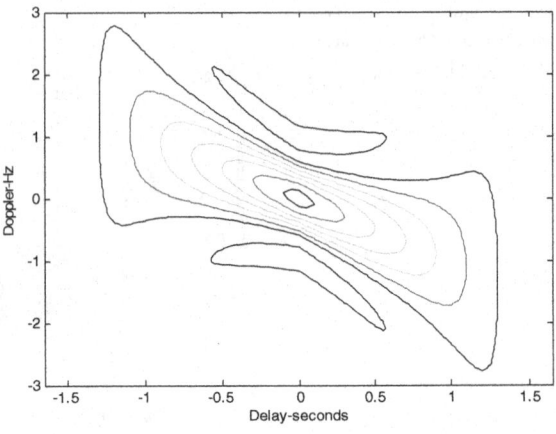

Fig. 10 Range ambiguity graph of linear frequency modulation pulse signal

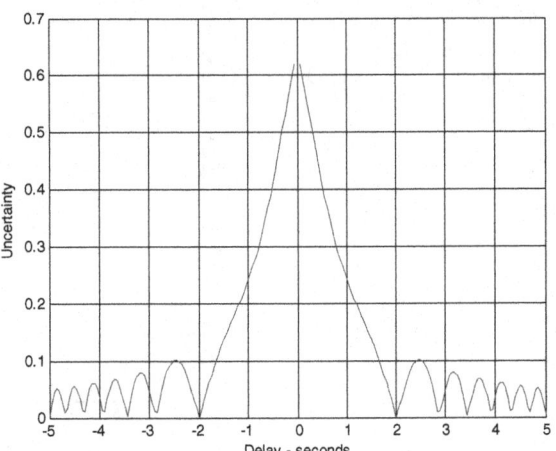

$$|\chi(0,\xi_d)| = \begin{cases} \left|\frac{\sin(\pi\xi_d\tau)}{\pi\xi_d\tau}\tau\right| & (0 \le \tau) \\ 0 & (0 > \tau) \end{cases} \qquad (20)$$

Figure 10 and Fig. 11 are range and velocity ambiguity graph of linear frequency modulation pulse signal, respectively.

Simulation results show that ambiguity graph of linear frequency modulation pulse signal is Oblique blade. The blade differs from the axis of the delay and the Doppler frequency. Take the delay axis as the reference axis; chirp rate K determines the angle and direction of the blade from the shaft. Linear frequency modulation pulse signal can control range resolution and velocity resolution through changing parameters. So, the waveform designed is more flexible and agile, and it is suitable for intelligent radar. A target which has positive Doppler is more closer than its true range.

Fig. 11 Velocity ambiguity graph of linear frequency modulation pulse signal

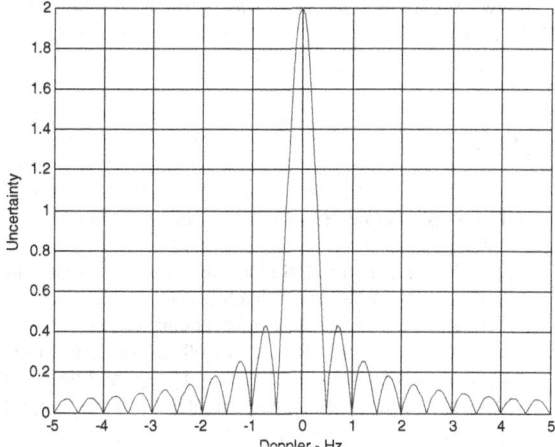

In ambiguity graph, the small the frequency modulation bandwidth B and time-width is, the greater the width of the intersection between the inclined ellipse and the delay axis, the larger the width of the intersection with the Doppler frequency axis is, then range resolution and velocity resolution become low.

Therefore, it is necessary to choose the linear frequency modulation pulse signal with large time bandwidth and large bandwidth, the large time-width can make the radar detection farther, and the large bandwidth has high range resolution. The range and velocity ambiguity graph of linear frequency modulation pulse signal is similar to that of Sinc function, so the signal has a good range and velocity resolution. It also represents speed and range.

From the different simulation results with two sets of different parameters, we can see that good performance is achieved in different working environment. So, the designed waveform is flexible, and it is suitable for intelligent radar with the ability to be flexible and agile.

5 Conclusions

With the increasing complexity of radar working environment, more and more requirements for the intelligence of radar are put forward. Intelligent radar can transmit different waveform according to different environment. In this paper, based on the principle of ambiguity function of point target echo, a novel waveform design method of linear frequency modulation pulse signal is proposed. Simulation results of time–frequency characteristics and ambiguity characteristics demonstrate that the proposed method has good range and velocity resolution. How to design waveform accurately is an important problem in the research field of intelligent radar.

Acknowledgements This work was supported by the National Natural Science Foundation of China (No. 61403067).

References

1. Haykin, S.: Cognitive radar: a way of the future. IEEE Signal Process. Mag. **23**(1), 30–40 (2006)
2. Chae, K., Liu, H., Yoon, S.: Unambiguous BOC signal tracking based on partial correlations. In: IEEE 80th Vehicular Technology Conference, pp. 1–5. IEEE Press (2014)
3. Morgan, D.R.: A three-state signal coding scheme for high efficiency Class-S amplifiers. IEEE Trans. Circuits Syst. I: Regul. Pap. **60**(7), 1681–1691 (2013)
4. Alexandru, N.D., Diaconu F.: Thorough spectral investigation of the β-α pulse shaping filter. In: International Symposium on Signals, Circuits and Systems, pp. 1–4. IEEE Press, New York (2015)
5. Zhang, J., Shi, C., Qiu, X., Wu, Y.: Shaping radar ambiguity function by L-Phase unimodular sequence. IEEE Sens. J. **16**(14), 5648–5659 (2016)
6. Abramovich, Y.I., Frazer, G.J.: Bounds on the volume and height distributions for the MIMO radar ambiguity function. IEEE Signal Process. Lett. **15**, 505–508 (2008)
7. Lyu, X., Stove, A., Gashinova, M., Cherniakov, M.: Ambiguity function of iridium signal for radar application. Electron. Lett. **52**(19), 1631–1633 (2016)
8. Santra, A., Srinivasan, R., Jadia, K., Alleon, G.: Ambiguity functions, processing gains, and Cramer-Rao bounds for matched illumination radar signals. IEEE Trans. Aerosp. Electron. Syst. **51**(3), 2225–2235 (2015)
9. Watts, S., Rosenberg, L., Bocquet, S., Ritchie, M.: Doppler spectra of medium grazing angle sea clutter; part 1: characterization. IET Radar, Sonar Navig. **10**(1), 24–31 (2016)
10. Gregers-Hansen, V., Mital, R.: An improved empirical model for radar sea clutter reflectivity. IEEE Trans. Aerosp. Electron. Syst. **48**(4), 3512–3524 (2012)

Global Positioning System: A New Opportunity in Human Life Logging System

Dinislam Bakdaulet, J. J. Yermekbayeva, S. T. Suleimenova
and K. S. Kulniyazova

Abstract We live in the modern world, where technologies evolve very fast. However, very few people know that many of the technologies widely used nowadays firstly were developed for military purposes. One of such technologies, positioning system (global positioning system (GPS)), was chosen as the main topic of this research. Nevertheless, in indoor environments where GPS signals are blocked by walls and ceilings, extra equipment may be required for such purposes. Thus, special attention was paid to indoor positioning in this study. The model was verified by comparing it with the accelerometer values of function filtered by discrete Fourier transform. At last, the advised step positioning technique is determined through real set test data. The given results validate the practical and efficiency of the proposed method.

Keywords Path estimation system · Android device · GPS · Navigation · DFT algorithm

1 Introduction

In recent years, people have become interested in navigation systems also known as path estimation systems. Most of them are used to help people to easily reach an unknown place. Such kind of systems is widely studied to automate driving a car. One more part of our lives where path estimation system can be used is life-log. For life

D. Bakdaulet (✉) · J. J. Yermekbayeva · S. T. Suleimenova · K. S. Kulniyazova
L.N. Gumilyov Eurasian National University, Astana, Kazakhstan
e-mail: bakdauletdinislam@gmail.com

J. J. Yermekbayeva
e-mail: erjanar@gmail.com

S. T. Suleimenova
e-mail: s.t.suleimenova@gmail.com

K. S. Kulniyazova
e-mail: kulniyazova_ks@enu.kz

© Springer Nature Singapore Pte Ltd. 2019
S. K. Bhatia et al. (eds.), *Advances in Computer Communication and Computational Sciences*, Advances in Intelligent Systems and Computing 760, https://doi.org/10.1007/978-981-13-0344-9_42

support, it is useful to look at population walking, as instance, to detect employee's traffic in factories or to be informed where person had gone. The proposed system for indoor path estimation certainly shares two paths: Android application which collects data from acceleration and orientation sensors of the device.

Server which receives collected data from the device analyzes it and performs all calculations to estimate path of the user. In fact, [1, 2] analyzed the drawbacks of this tool as well.

The aim is to study available positioning systems and develop the path estimation system based on Android device with its sensors and a server.

For first, the step-based localization application was projected and tested. Algorithms present a set of operations to be held. This outcome is new and should lead to critical values for inland navigation in the room. The system will link on a number of factors of the DFT [3, 4].

The paper consists of introduction, four chapters, and the conclusion. The introduction describes relevance of the chosen area and its purposes and specifies the objectives. The first chapter gives information about positioning systems, their history, and positioning techniques. The second chapter describes the technologies that were used in implementation of the path estimation system. The third chapter examines problems and shows implementations of main functions. The last chapter demonstrates the results of the system testing. Conclusion contains summary of the project.

2 Overview of GPS Systems

Currently, estimating the people's position has become available due to such technologies as wireless networks and global positioning system (GPS). Almost all of the new mobile devices are equipped with GPS sensors with use of which position of the device can be spotted. Those systems cannot be applied in defended area because any wave is shut down by sides and division. For instance, the wireless PS has major point of entry similar by their functionality to satellites in GPS. Those points will monitor user's position. Almost all positioning systems act similar to GPS. Nevertheless, there is another type of positioning called inertial navigation system (INS). The main difference of this type of positioning is that INS does not require extra equipment and only uses motion and rotation sensors which can be found together in modern mobile devices. Such system including special and time data in all weather criteria, if on highway visible to four or more satellites and which is openly available to anyone who has a receiving machine. Such system including special information, which is accessible to anyone who has a receiving machine. Much attention is paid to satellite monitoring of transport and in monitoring movements and vibrations of plates, active leisure, and geotagging. The mixing of devices have the capability, which is the smallest of misalignment error than three discrete models, which are subsequently packaged and united.

Fig. 1 Organization of common chain

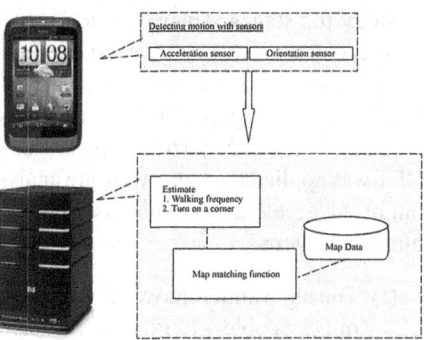

3   Methods and Structure of System

The previous chapter looked at the problems in integrated navigation systems and then assessed their advantages. Stages in this area make it complex to measure the exact XYZ for designing internal maps. The purpose of this research was to elaborate a system that would be considered a walk of people in a room or place where it is not easy to use GPS. Below are some examples of the application of the system: monitoring the movement and location of employees in factories or offices, monitoring the movement of older citizens and the size of work in care institutions, etc. To achieve the top objective, a system stage is briefly described below: a tracking in real time is not required. The sender must remove the sidewalks every hour or every day. The system shows the hallways or box where the client is located, and there is no need to specify the exact XYZ coordinates. Detectors for this system should be accessible to everyone, for example, for sensors on Android devices, where basic structure presented at (Fig. 1).

According this picture, the top block has two detecting modules: two types of sensors. The main unit contains three sub-blocks: "Estimate", which organizes walking and turning processes. The next stage is matching map function and lastly set of map data. The data stream is processed/passes instantly and exactly. The diagram below shows how system for assessing pedestrian lines is examined. Our main task is to define the special map as the optimal distances and directions information.

4   Frequency Estimation of Walking with Accelerometer Support

With the advent of technological advancement, the DSP field has undergone enormous changes. It is argued that in addition to the traditional core processing which are sensors, network and connecting, filtering [5, 6] should be incorporated into the official position mapping. To evaluate the distance between two nodes, it is necessary to study the stages. The extraction steps evaluated the acceleration amplitude. For two acceleration waves, a case exceeding a simple threshold degree was performed.

However, the acceleration fluctuations immediately become the error of counting the displacement steps. The basic process of selecting peaks are shown on the Fig. 2

The values of 1.5 and 6 are only noise. Since the threshold method was not efficient, DFT was applied for more accurate analysis of the accelerations. The frequency goes out of the acceleration, which is converted to the frequency domain to calculate the number of steps.

(I)   The algorithm removes the acceleration wave 4 s, when any choice, and determines people—to go or not to publish using BalanceIndex.

(II) The extracted runaway is multiplied by the Hann window.

(III) At a frequency of 0.5–3 Hz, it differs from the previously selected frequency of less than 1 Hz by the walking frequency. It is not a good state when the balance exceeds 0.998 and the change in acceleration is small.

Fig. 2 Diagram of accelerometer values using simple threshold method

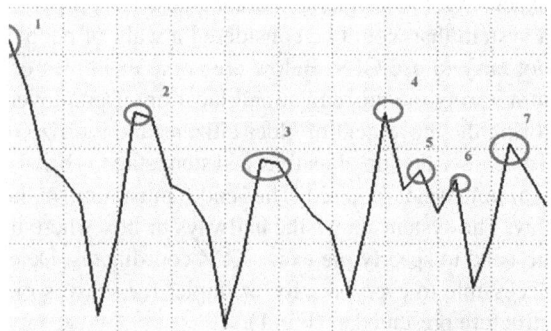

Fig. 3 Instruction algorithm

```
Implementation
/** Discrete Fourier Transform
        @param inreal input real part
        @param inimag input imaginary part
    */
    public static double[] dft(double[] inreal, double[] inimag) { int n = inreal.length;
        double dataOut[] = new double[n]; for (int k = 0; k < n;
        k++) {
                double sumreal = 0; double
                sumimag = 0;
                for (int t = 0; t < n; t++) { // For each input element sumreal +=
                        inreal[t]*Math.cos(2*Math.PI * t * k / n) +
inimag[t]*Math.sin(2*Math.PI * t * k / n);
                        sumimag += -inreal[t]*Math.sin(2*Math.PI * t * k / n)
                        |+ inimag[t]*Math.cos(2*Math.PI * t * k / n);
                }
//according to stackoverflow
                dataOut[k] = Math.sqrt((sumreal*sumreal+sumimag*sumimag));
        }
        return dataOut;//Arrays.copyOfRange(dataOut, 2, n/2);
    }
```

Function Hannah is usually used as window functions for digital signal processing. Often choose a subset of the number of samples to perform the Fourier transform or other computation.

If the Hann window is used for sampling the signal to convert the frequency region, it is difficult to resume in the time domain without adding distortions.

In this problem, in simple words DFT algorithm helps to filter data [7–9] of accelerometer and obtain the most constant change in acceleration with big amplitude. The realization DFT algorithm is shown in Fig. 3.

5 Case Study

We turned on the estimation by orientation sensor, which compares and tests the average of previous steps' azimuths. In case, where the difference is more than 55° that can be achieved by turning next step, turn occurs, and from now, take next step as the first in new sequence. Multiple measurement of accelerometer and comparing non-filtered data shows a work of Hann window [10].

Figure 4 shows that the non-filtered accelerometer values for 20 steps. The Hann window for every 4 s of research motion is shown in Figs. 4 and 8. The oscillation/traffic is observed in motion speed on time and steps count, with the subsequent return to steady state (Figs. 5, 6 and 7).

The main purpose of the three experiments was to identify the reason for the motion speed.

Figure 9 shows the transform time-domain plot to frequency-domain plot. This figure describes the walking frequency in the frequency domain; thus, we can see a general picture of the process. The peak or maximum value reaches the value 200.

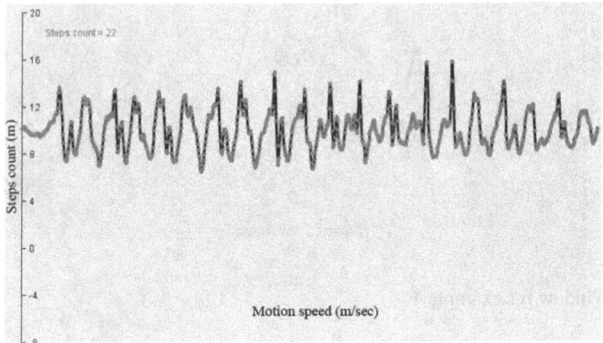

Fig. 4 Accelerometer values for example 1

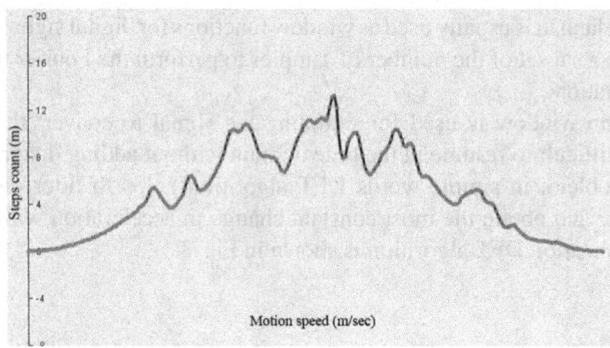

Fig. 5 Hann Window for example 2

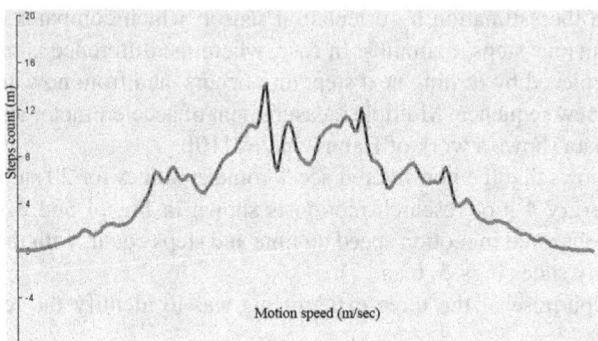

Fig. 6 Hann Window for example 3

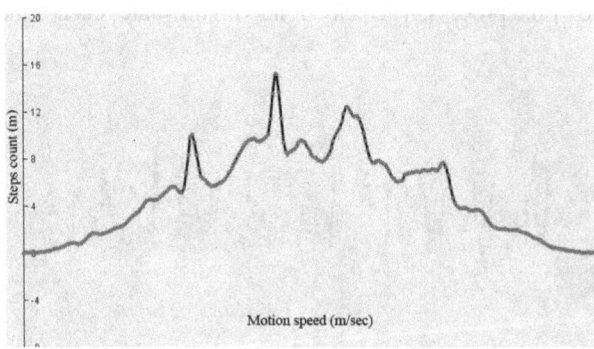

Fig. 7 Hann Window for example 4

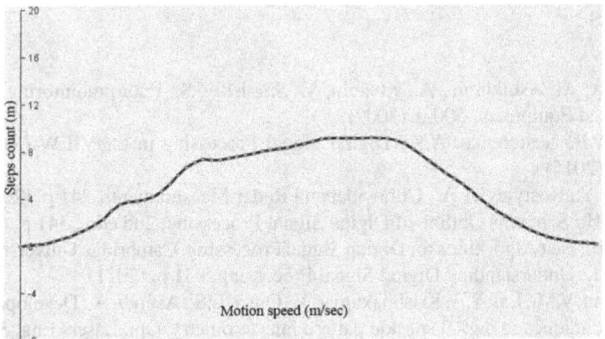

Fig. 8 Hann Window for example 5

Fig. 9 Function filtered by
discrete Fourier transform

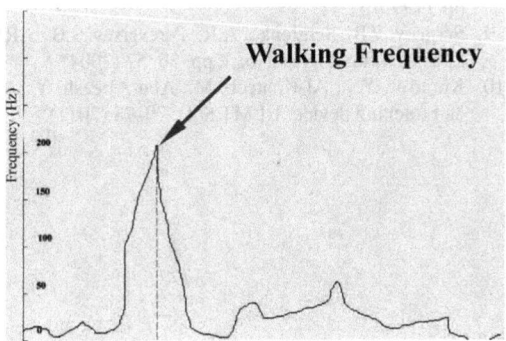

6 Conclusion

We conducted a study on the development of a positioning system. Research study
draws the development of the positioning system. Different ways have been investigated trying to count. At the end, the foundation was laid for a new method of estimating pedestrian walkway. This method involves the analysis of accelerations and
azimuthal orientations, balanced using sensors on the Android device. To evaluate
walking, frequency of walking is estimated by analyzing the accelerations. Rota-
tions are evaluated by analyzing the azimuthal orientation. In accordance with the
frequency of walk and turns, combined with pre-made vector map we can estimate
the walking path with great accuracy.

Finally, it has shown much stronger forecast stability, which indicates that it
can achieve more satisfactory generalization potential. The major importance of
these results will motivate and combine novelty to seek way on typical application
techniques and to implement vector map by which walking path will be corrected
with the use of round-robin search.

References

1. Rembovsky, A., Ashikhmin, A., Kozmin, V., Smolskiy, S.: Radio monitoring. In: Problems, Methods and Equipment, 500 p. (2009)
2. Fedosov, V.P., Nesterenko, A.K.: Digital Signal Processing in LabVIEW (Russian Edition) (Russian) (2015)
3. Astanin, L.Y., Kostylev, A.A.: Ultrawideband Radar Measurements, 241 p. (2009)
4. Hayes, M.H.: Schaums Outline of Digital Signal Processing, 2nd edn., 341 p. (2009)
5. Blahut, R.E.: Fast Algorithms for Digital Signal Processing. Cambridge University Press (2010)
6. Lyons, R.G.: Understanding Digital Signal Processing, 931 p. (2011)
7. Murukeshan, V.M., Lai, Y.F., Krishnakumar, V., Onga, L.S., Asundi, A.: Development of Matlab filtering techniques in digital speckle pattern interferometry. Opt. Lasers Eng. **39**(4), 441–448 (2003)
8. Zhodzizhskiy, M.I., Mazepa, R.B., Ovsiannikov, E.P.: Digital Radio Receiving Systems: Reference Book. (in Russian). Under edition of Zhodzizhskiy, M.I. Moscow, Radio i sviaz, 208 pp. (1990)
9. Sergeev, V.B., Sergienko, A.R., Pereversev. S.B.: ARK-D1TP Panoramic Measuring Receiver. Special technologies, No. 3, pp. 50–57 (2004)
10. Khraisat, Y.S., Al-Khateeb, M., Abu-Alreesh, Y., Ayyash, A., Lahlouh, O.: GPS navigation and tracking device. IJI MT **5**(4), 70–78 (2011)

Measuring Quality of C2C E-Commerce Websites in Lao PDR

Bounthanome Vilaysuk, Lachana Ramingwong, Sakgasit Ramingwong,
Narissara Eiamkanitchart and Kenneth Cosh

Abstract E-commerce gives chances to businesses of all sizes to compete globally. It opens the doors to new product, services, and creates opportunities. Though e-commerce was made possible 25 years ago, it is relatively new in Lao PDR. Merely 12.8% of firms owns website. Lao government's recent development in infrastructure and policy, in conjunction with cooperative agreement with developed countries, presents a promising future for e-commerce. One key success factor for e-commerce is service quality. A high-quality website draws customer and leads to purchase. This study proposes an improvement to ES-QUAL, a framework for electronic service quality, to address missing quality dimensions and adaption for quality measurement of C2C websites in Lao PDR.

Keywords E-commerce website · C2C · E-service quality assessment
Questionnaire

1 Introduction

Internet transforms the world into a place with no border where people wirelessly communicate, work, learn, watch movies, shop, etc. A number of digital services have grown tremendously ever since the Internet was used worldwide.

B. Vilaysuk (✉) · L. Ramingwong · S. Ramingwong · N. Eiamkanitchart · K. Cosh
Faculty of Engineering, Department of Computer Engineering,
Chiang Mai University, Chiang Mai, Thailand
e-mail: bounthanome@hotmail.co.th

L. Ramingwong
e-mail: lachana@eng.cmu.ac.th

S. Ramingwong
e-mail: sakgasit@eng.cmu.ac.th

N. Eiamkanitchart
e-mail: narisara@eng.cmu.ac.th

K. Cosh
e-mail: drkencosh@gmail.com

© Springer Nature Singapore Pte Ltd. 2019
S. K. Bhatia et al. (eds.), *Advances in Computer Communication
and Computational Sciences*, Advances in Intelligent Systems
and Computing 760, https://doi.org/10.1007/978-981-13-0344-9_43

E-commerce raises an opportunity for small businesses to grow faster and a chance to compete with other local and global businesses. However, e-commerce adoption in developing countries is still much far behind developed countries [8, 22]. Known problems include infrastructure, policy, and regulations are difficult to control, and therefore, are excluded from this paper.

Despite increasing popularity in mobile platform, websites are still perceived as a more convenient way than native mobile application when it comes to online shopping [21]. Lao PDR is one of the countries ranked in the bottom in terms of economic and infrastructure [10, 19]. The number of cellular subscription in Laos was much lower than nearby countries such as Cambodia, Vietnam, Thailand, and Malaysia. Yet the number of internet users in 2016 (21.9% penetration) was tripled compared to 6 years ago. Given the fact that the number of broadband internet subscription is rapidly increasing [27], there is a great potential for Laos to grow its e-commerce division and digital business prospects [15].

Moreover, differences in nature of mobile website and desktop website makes the two being equally important in delivering user experience. Major applications like Facebook, YouTube, Digg, and Flickr have selected to implement separate sites for mobile and desktop [26]. This suggests the importance of desktop website in spite of increasing mobile usage. This paper emphasizes on desktop commerce websites in Laos.

Lao PDR was ranked in the last place on B2C e-commerce index in United Nations conference on trade and development [24]. No similar index is found for C2C (Customer-to-Customer e-commerce) websites in this country. An encouraging news, reported by Lao news agency which is a movement unit, is that the Lao government and the Republic of Korea have approved the collaboration in expanding electronic business opportunities to compete with other countries [14]. E-commerce is still relatively new to Lao businesspersons.

The key success factor for e-commerce is its service quality. A high-quality website draws customer and leads to purchase, word of mouth, loyalty, and repurchase. Various service quality scales are studied in this paper. Limitations are identified. An improved questionnaire, based on the well-known service quality scale, is developed and assessed. The results are visualized for communication purpose.

This paper studies assessment of C2C e-commerce websites in Laos. Section 2 presents related work and scales to assess quality of websites. Section 3 discusses research methodology used in this research. Section 3.1 evaluates the questionnaire. Section 4 discusses the results and finally Sect. 5 gives conclusion.

2 Related Work

2.1 E-Business in Laos

E-business in Lao PDR has been gradually expanded over the past few years [28]. The moves toward e-business implementation started when Laos laid out the e-government action plan in 2006 [5]. In July 2017, the Republic of Korea held a workshop in Vientiane, Lao PDR's capital city, to convey Korean e-commerce lessons to the local [16]. A series of actions from government sectors toward increasing e-business readiness shows that Lao PDR is making progress. Since the e-government plan was initiated, financial institution, university, and ministry have implemented e-business projects. Consequently, interactive websites have been initiated in order to facilitate application of the government's online databases. This project has yet fully implemented.

2.2 E-Commerce in Laos

The 2016 survey from the World Bank revealed that only 12.8% of 368 firms in Lao PDR has their own website while the average is 44.3% for all countries [27]. An article from Laotian Times, based on Amazon Alexa ranking, revealed the top 10 Highest Traffic Lao Websites [15] and also the 11th–25th rank. There was only one e-commerce website, and it was ranked 4th. The second one was on 11th rank. The 20th rank was an online-classified website which was an ally to e-commerce. Other high ranking included news and entertainment, job portal, finance, and government. There were not any e-commerce websites in the last year's top 10 ranks. This gives an impression that e-commerce is still new to Laotians, but with progress. This is consistent with the cooperative agreement in e-commerce between Lao PDR and the Republic of Korea [14, 16]

2.3 Website Quality

Customer satisfaction was found to be the significant factor in e-commerce [6]. It positively affects trust, spending [17], and loyalty [11]. On the other hand, perceived service quality has a substantial impact to customer satisfaction [11]. A study by Hsu and colleges confirms that website quality greatly influences customer satisfaction [9]. Other factors that help shaping customer satisfaction are word of mouth (WOM) [11] and electric word of mouth (eWOM) [13]. Based on the recognition that WOM/eWOM is positively related to customer satisfaction, it can be implied that WOM/eWOM induces customer satisfaction in other customers to some extent [25], which in turn produces WOM/eWOM and, ultimately, a cycle. Therefore, it can be safe to claim that providing with appropriate infrastructure, success of e-commerce implementation and expansion equally depend on website quality (internal effort), in addition to the government (or external effort).

A high-quality website enhances the prospect of maintaining customers [12]. System quality (multimedia capability, search facility, responsiveness), information quality (information accuracy, information relevance), service quality (empathy, trust) and attractiveness were identified as the key factors which customers perceived as e-commerce's website quality factors.

2.4 Service Quality

Various service quality scales were proposed. Scales developed in the past years include SERVQUAL, SITEQUAL, ISRQ, WebQual, eTailQ, and a few others [1]. One scale may be appropriate for one online industry but not the others. Blut and co-workers reported that Website design and security were the attributes found in all nine models studied, indicating them as crucial elements in e-service quality. Another study [4] uncovered that the modified version of SERVQUAL yielded higher predictive validity than the original version which was developed in the USA. This can be implied that service quality is affected by culture [23].

2.5 Website Design and Quality

Website design has a direct effect to the perceived quality of website. A website with an unfriendly visual design creates negative perception, and, therefore, is likely to influence users to abandon the shopping cart [7]. Evaluation of websites can lead to improvement in website design as well as quality of e-service.

There are different views to website quality evaluation. ISO 9260 includes a quality model that distinguishes three different perspectives to quality. WebQEM offered a tool for measuring web application quality based on over predefined 90 attributes [18]. This approach, however, requires quality requirements specification document to be created beforehand, therefore expert-centered approach. Consequently, results obtained from such measurement do not reflect the consumer point of view as it is not a user-centered approach.

ES-QUAL was used for measuring service quality offered by online shopping websites [20]. Yet ES-QUAL did not include information and service quality aspects which are considered a vital factor to influences purchase decision. In addition, the instrument did not distinctly address quality of information and service. Sitequal, WebQual, and eTailQ were developed for measuring website interface, [1]. However, as WebQual did not take the entire online purchasing process into account, some aspects were neglected when the service quality was measured [3]. Although eTailQ was developed for measuring quality of online shopping experience, it omitted fulfillment aspects of customer service [2]. Similar to WebQual, SITEQUAL did not capture all aspects of online purchase process.

3 Research Methodology

3.1 Questionnaire Design and Validation

In response to the limitations discussed above, a new questionnaire, E-Commerce Information and Service Quality (EC-IS-QUAL), is developed based on the first part of ES-QUAL that measures service quality. New questions are added to include information and service quality aspects. The original wordings are slightly modified to make them suitable for data collection from participants from Lao PDR.

In order to evaluate for validity and reliability, the questionnaire was sent to five experts in the field for content and face validities. Index of Item-Objective Congruence (IOC) of each question is calculated. Every question scored an IOC more than 0.5. Thus, it is considered appropriate. Additionally, a reliability test was conducted on 30 website developers to provide a measure of the internal consistency of the new questionnaire. The resulted Alpha is greater than 0.9, thus acceptable.

The original instrument, ES-QUAL, and the proposed EC-IS-QUAL are compared for IOC and Alpha. The results show that the new questionnaire yielded significantly higher scores.

3.2 Questions

Questions in EC-IS-QUAL are divided into 13 dimensions including (1) information quality, (2) design, (3) organization, (4) aesthetic, (5) correctness, (6) data collection, (7) reliability, (8) security and authentication, (9) responsiveness, (10) goods and services, (11) reputation, (12) warranty, and (13) customer care.

3.3 Data Collection

The data was collected from 402 customers who transacted on three C2C websites. They are divided into three groups according to the number of websites used in this research. Two websites are well-known C2C websites in Laos, referred as A and B. C is a well-known global C2C website.

All three website's designs are significantly different. One follows a traditional website design with main items being texts and pictures. Another one incorporates more multimedia items. The last one's design is presented in a clean, modern, and elegant way. The quality of these websites was measured and then compared.

4 Results and Discussion

4.1 Initial Findings

The participants included students, civil servants, entrepreneurs, and employees. They were divided into three groups, 134 participants in each group. 73% of them earned less than 242 USD a month (approximately 2,005,398 Laotian Kip). Most of them used Internet from home. Interestingly, almost half of the population stated that they used Internet for shopping and mobile was the most common platform for purchases. Merely 26% admitted they used PC to purchase. Surprisingly, credit cards and direct payments to sellers were the common methods for majority of participants. This can be implied that, though, there are many customers who trusted the system enough to give their credit card numbers online, the similar number of participants still prefers direct payment to sellers. This reflects a contrast as well as marking the rise of e-commerce in Lao PDR.

Three websites were used in this study. Participants were divided into three separate groups of 134 each. Each website (A, B, C) was used by one group of participants. The first group (56% women, 44% men) used website A. The second group (33.6% women, 66.4% men) used website B. The third group (47% women, 53% men) used website C. The majority of participants' age ranges from 20 to 39 years old.

4.2 Comparison of Website Quality

The summary of responses from the questionnaire is shown on Table 1. Means and standard deviations for each question are shown in the table. The responses are rated from 1 (strongly disagree) to 5 (strongly agree). The average scores for website A, B, and C are 3.38, 4.03, and 4.6, respectively. The pattern found correlates to the characteristics of the websites mentioned earlier. Website C is expected to outperform websites A and B, and the results confirm that also website B outperforms website A as it contained more types of media and was organized in such a user-centered way.

The highest score on website C was on information quality dimension. Information quality is very important aspect of website quality. The website with a very good score on information quality dimension is likely to score well in other dimensions. The scores obtained from websites A and B show a similar pattern.

T-test and F-test for each question were calculated, but no significant difference was found.

Table 1 Comparison of three websites

Dimensions	Website A		Website B		Website C	
	Mean	SD	Mean	SD	Mean	SD
1. Information quality	3.50	0.409	4.01	0.466	4.64	0.359
2. Website properties	3.45	0.337	3.94	0.525	4.49	0.469
3. Information organization	3.38	0.458	4.00	0.544	4.61	0.419
4. Aesthetics	3.30	0.402	3.96	0.534	4.59	0.464
5. Correctness	3.30	0.490	4.02	0.680	4.53	0.454
6. Data collection	3.32	0.406	4.02	0.509	4.52	0.411
7. Security and authentication	3.39	0.272	4.09	0.485	4.59	0.375
8. Reliability	3.53	0.341	4.10	0.443	4.60	0.303
9. Responsiveness	3.26	0.431	4.07	0.599	4.50	0.406
10. Goods and services	3.42	0.424	4.01	0.577	4.62	0.434
11. Reputation	3.37	0.467	4.04	0.578	4.55	0.458
12. Warranty	3.38	0.505	4.13	1.38	4.56	0.463
13. Customer care	3.35	0.485	4.04	0.565	4.52	0.431
Overall	**3.38**	**0.252**	**4.03**	**0.408**	**4.56**	**0.304**

4.3 Discussion

Figure 1 presents the average scores for each quality dimensions including overall scores. The solid lines on top of the score bars were connected to show the trend for each website. Each quality dimensions of the selected Laos C2C websites is discussed below.

Information quality: Though website A's information quality is lower than the others, participants of the questionnaire were positive about the reliability of the information provided, and they would be happy to transact with the website. This can be implied that reliable information is likely to convert visitor to buyer.

Website properties: The website that provides the right amount of multimedia items and tools that can be personalized is viewed as a higher quality. A well-designed website with more multimedia items is perceived as a high-quality one.

Navigation and Information organization: The website with a more complete and comprehensive navigation system appears to be more superior. It allows users to

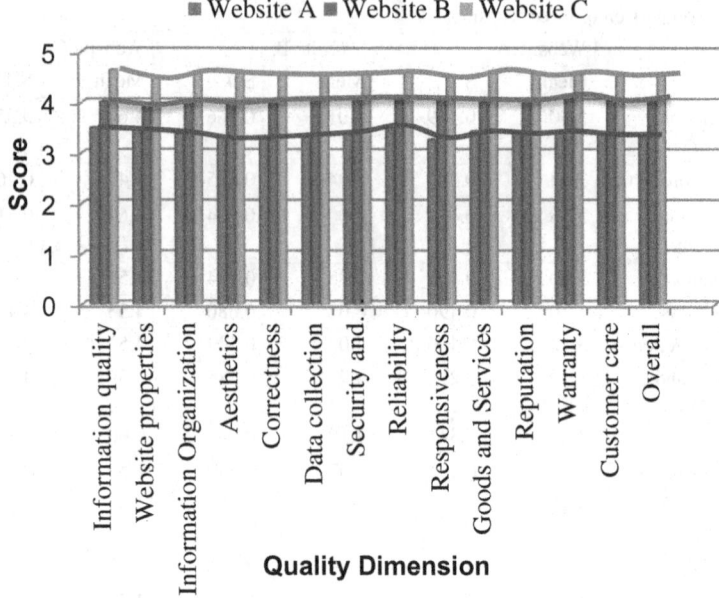

Fig. 1 Scores on all quality dimensions of the three websites

efficiently access information they want easier. Moreover, information should be sorted and grouped to allow faster information retrieval, hence increase satisfaction.

Aesthetics: Colors comfortable to consumer eyes should be used on the websites. While using multimedia on the websites tends to improve their quality, having too many animations and colors might not produce a positive impact. Furthermore, an easy to remember and attractive logo or metaphor has a positive impact to quality.

Correctness: Information perceived as less accurate or inaccurate affects quality of the websites. Each website, therefore, must provide the most accurate and up-to-date information on products, services, buyers, and sellers.

Data collection: Keeping and maintaining visitor data is perceived as an important factor to quality as visitors are aware of the usefulness of the data such as keeping history of transaction and signing in.

Security: Identity confirmation, user authentication, data encryption are factors that let visitors feel secure. Websites should also implement widely used payment systems, which support onsite payment, and track shipment to give users the confidence.

Reliability: The websites visitor perceives that a reliable website should have their own domain name and run on a reliable host, not a free one. Factors affecting the reliability include page design and presentation, contact information, reviews. The websites that accept credit card payments are regarded as more reliable.

Responsiveness: Consumers seem to prefer communication with online sellers to be similar to sellers in traditional businesses. Thus, the websites should be able to answer questions and responded to customers quickly.

Goods and services: Clear product information helps consumers in decision-making process. High-quality Web sites are expected to provide services to multiple customers at the same time.

Reputation: The websites must process all payments and shipment correctly and quickly. In addition, they must have a system in place to prevent frauds and protect customers. Consumers consider the websites with awards or equivalences more credible. A large number of visitors are one of the factors visitors consider when determining website quality.

Warranty: This is one of the most important factors which must be made visible on listing information. Visitors to website B rate this dimension higher than the others.

Customer care: Websites offer different channels to communicate to users about promotions and offers, and also encourage them to give feedback tend to be perceived as of higher quality.

5 Conclusion

EC-IS-QUAL is developed to fulfill the gap of lacking of information quality dimensions in ES-QUAL. The questionnaire consists of 13 dimensions. The questionnaire was validated before putting into use. Then, it was administered to a sample size of 402 respondents on three websites. Interesting insights are drawn from the results.

Although the businesses are conducted online, sellers and buyers never meet in person, unlike in traditional businesses, buyers still expect to communicate with online sellers at similar pace to sellers in brick-and-mortar business. The scores of the responsiveness dimension of the two websites are less than the other dimensions. This indicates a room for development for these websites.

Information quality rating is rated high on two of the websites and the highest on the other. This is highest quality facet among all dimensions. Some dimensions are rated in an opposite direction to information quality such as website properties and responsiveness. This suggests that the quality of information alone could be enough to influence purchases. On the other hand, information quality seems to be related to several other dimensions of the websites. A further study is required to validate the correlation.

References

1. Blut, M., Chowdhry, N., Mittal, V., Brock, C.: E-service quality: a meta-analytic review. J. Retail. **91**(4), 679–700 (2015)
2. Boshoff, C.: A psychometric assessment of ES-QUAL: a scale to measure electronic service quality. J. Electron. Commer. Res. **8**(1), 101 (2007)
3. Bressolles, G., Nantel, J.: Electronic service quality: a comparison of three measurement scales. In: Proceedings of the 33th EMAC Conference, Murcia, Spain, May 2004
4. Carrillat, F.A., Jaramillo, F., Mulki, J.P.: The validity of the SERVQUAL and SERVPERF scales: A meta-analytic view of 17 years of research across five continents. Int. J. Serv. Ind. Manag. **18**(5), 472–490 (2007)
5. Charmonman, S., Mongkhonvanit, P.: Information technology preparation to enter ASEAN community. Int. J. Comput. Internet Manag. **22**(2), 1–6 (2014)
6. Choshin, M., Ghaffari, A.: An investigation of the impact of effective factors on the success of e-commerce in small-and medium-sized companies. Comput. Hum. Behav. **66**, 67–74 (2017)
7. Hasan, B.: Perceived irritation in online shopping: the impact of website design characteristics. Comput. Hum. Behav. **54**, 224–230 (2016)
8. Hoque, M.R., Boateng, R.: Adoption of B2B e-commerce in developing countries: evidence from ready made garment (RMG) industry in Bangladesh. Pac. Asia J. Assoc. Inf. Syst. **9**(1) (2017)
9. Hsu, C.L., Chang, K.C., Chen, M.C.: The impact of website quality on customer satisfaction and purchase intention: perceived playfulness and perceived flow as mediators. Inf. Syst. e-Bus. Manag. **10**(4), 549–570 (2012)
10. Internet World Stats: Internet usage, broadband and telecommunications report. http://www.internetworldstats.com/asia/la.htm (2016). Accessed 15 Aug 2017
11. Kassim, N., Asiah Abdullah, N.: The effect of perceived service quality dimensions on customer satisfaction, trust, and loyalty in e-commerce settings: a cross cultural analysis. Asia Pac. J. Mar. Logist. **22**(3), 351–371 (2010)
12. Khalil, H.: The Role of the Quality of a Website in Consumer Perception (2017)
13. King, R.A., Racherla, P., Bush, V.D.: What we know and don't know about online word-of-mouth: a review and synthesis of the literature. J. Interact. Mark. **28**(3), 167–183 (2014)
14. Lao News Agency: Laos, S. Korea Agree to Expanding E-Commerce. http://kpl.gov.la/En/Detail.aspx?id=25955 (2017). Accessed 15 Aug 2017
15. Laotian Times: Korea Eyes Laos and Cambodia for E-commerce Opportunities. https://laotiantimes.com/2017/07/04/korea-eyes-laos-cambodia-e-commerce-opportunities/ (2017). Accessed 15 Aug 2017
16. Lee, R.: ASEAN Workshop boosts Lao PDR's e-commerce sector. In: The Korea Times. http://www.koreatimes.co.kr/www/nation/2017/07/176_232644.html (2017). Accessed 15 Aug 2017
17. Nisar, T.M., Prabhakar, G.: What factors determine e-satisfaction and consumer spending in e-commerce retailing? J. Retail. Consum. Serv. **39**, 135–144 (2017)
18. Olsina, L., Rossi, G.: Measuring web application quality with WebQEM. IEEE Multimed. **9**(4), 20–29 (2002)
19. Palvia, S.C.J., Sharma, S.S.: E-government and e-governance: definitions/domain framework and status around the world. In: International Conference on E-governance, pp. 1–12 (2007)
20. Parasuraman, A., Zeithaml, V.A., Malhotra, A.: ES-QUAL: a multiple-item scale for assessing electronic service quality. J. Serv. Res. **7**(3), 213–233 (2005)
21. PWC: Total Retail 2017. https://www.pwc.com/gx/en/industries/assets/total-retail-2017.pdf (2017). Accessed 15 Aug 2017
22. Rahayu, R., Day, J.: E-commerce adoption by SMEs in developing countries: evidence from Indonesia. Eurasian Bus. Rev. **7**(1), 25–41 (2017)
23. Randheer, K., Al-Motawa, A.A.: Measuring commuters' perception on service quality using SERVQUAL in public transportation. Int. J. Mark. Stud. **3**(1), 21 (2011)
24. Rillo, A.D., dela Cruz, V.: The Development Dimension of E-Commerce in Asia: Opportunities and Challenges (2016)

25. Srinivasan, S.S., Anderson, R., Ponnavolu, K.: Customer loyalty in e-commerce: an exploration of its antecedents and consequences. J. Retail. **78**(1), 41–50 (2002)
26. Teoli, et al.: Mobile Web Development. Separate sites for mobile and desktop. https://developer.mozilla.org/en-US/docs/Web/Guide/Mobile/Separate_sites (2013). Accessed 15 Aug 2017
27. The World Bank: Enterprise surveys. What business experience. Lao PDR 2016. http://www.enterprisesurveys.org/data/exploreeconomies/2016/lao-pdr#innovation-and-technology (2016). Accessed 17 Aug 2017
28. Warf, B.: E-Government in Asia: Origins, Politics, Impacts, Geographies. Chandos Publishing (2016). Wells, J.D., Valacich, J.S., Hess, T.J.: What signal are you sending? How website quality influences perceptions of product quality and purchase intentions. MIS Quarterly, 373–396 (2011)

A GUI-Based Automated Test System for Android Applications

Tianxiang Chen, Tao Song, Shusheng He and Alei Liang

Abstract Android application testing has always been a serious problem for mobile developers. To support developers, this paper presents GATS, a GUI-based automated test system for Android apps. This tool uses finite-state machine to learn a model of the app during testing, then uses the learned model to generate user inputs or system event to visit the rest states of the app, and then uses the result of the input to refine the model. The goal of the tool is to trigger crashes. When a crash is happened, GATS will generate a crash report containing screenshot, logcat info with stack trace crash, reproduction steps, and so on. We evaluate GATS on ten Android applications from the top list of several app markets with Monkey, a fuzzing tool from Android platform, and Dynodroid, a previous research. Our result shows that our system has less running time and more bugs found.

Keywords Android testing · Finite-state machine · GUI testing · Test automation

1 Introduction

In recent years, smartphones are becoming increasingly popular. Hundreds of thousands of applications are available for Android. Until the first quarter of 2016, the number of Android apps in Google Play, which is available to download, had reached 1,900,000 and the number of apps, which downloaded from the Google Play, was more than 50 billion [1]. Since the number of apps is continuously growing, it

T. Chen (✉) · T. Song · S. He · A. Liang
School of Software, Shanghai Jiao Tong University, Shanghai, China
e-mail: jordan13game@sjtu.edu.cn

T. Song
e-mail: songt333@sjtu.edu.cn

S. He
e-mail: eternalone@sjtu.edu.cn

A. Liang
e-mail: liangalei@sjtu.edu.cn

© Springer Nature Singapore Pte Ltd. 2019
S. K. Bhatia et al. (eds.), *Advances in Computer Communication
and Computational Sciences*, Advances in Intelligent Systems
and Computing 760, https://doi.org/10.1007/978-981-13-0344-9_44

becomes more and more important to automatically test Android apps instead of manual testing.

There are several testing tools, such as Monkeyrunner [2], Hierarchy Viewer [3], and Robotium [4], proposed to facilitate automatic testing. However, these tools need developers to create test scripts manually, so as to execute the tests automatically. This is a huge workload to developers on developing test scripts for applications. And manual testing is usually error-prone. Prior approaches for automated GUI exploration [5–9] have one or more limitations that some of them need instruction on platform or app, some of them only generate UI operation, and some of them need source code of the app.

To reduce testers' workload on developing test scripts for Android applications and tackle these challenges, this paper presents GATS, a GUI-based automated test system for Android apps. GATS tests an app using an input generation algorithm and produces several crash reports with the steps for reproduction which is in a human-readable format. Our system has different testing strategies which include sending different system events, such as battery low, wireless turned off, phone call coming in and so on, and generating text based on the context of character allow lists. Our system is completely automated, other than reading reproduction reports.

The rest of the paper is organized as follows. Section 2 overviews existing work which related to our approach. Section 3 overviews the design of our system. Section 4 presents our testing algorithm. Section 5 describes how the model updated. Section 6 presents our evaluation of GATS. Section 7 presents the conclusion.

2 Related Work

In this section, we briefly describe the existing automatic testing approaches for Android. These automatic tools can be grouped into two parts: random exploration and model-based testing.

2.1 Random Exploration

Random exploration approach is choosing arbitrary UI events. The advantage is that it can efficiently generate events and suitable for stress testing. The main drawback is that a random strategy cannot be able to generate specific inputs.

Dynodroid [5] presents an approach in a context-sensitive way by maintaining the history of events' execution times that finds the relevant events throughout the testing.

Intent Fuzzer [6] uses intents to test apps in an offline static analysis by taking into consideration the extracted information. However, it cannot easily scale for large applications.

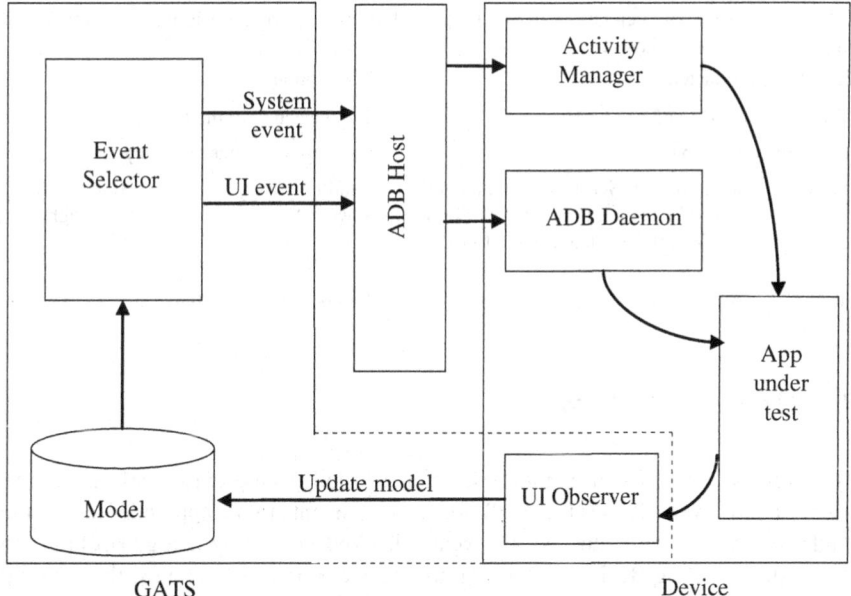

Fig. 1 Dataflow diagram of GATS

2.2 Model-Based Testing

Model-based testing approach tries to build several models of apps. These models are usually finite-state machines.

GUIRipper [7] is the upgrade of AndroidRipper and models the states of an app's UI. It only generates UI events and cannot expose system event.

Swifthand [8] uses a dynamic finite-state machine of the app and generates inputs by using the learned model that leads the execution of the app to the unexplored states. It aims to minimize the restarts of the app, and it also cannot generate system events.

PUMA [9] exposes high-level events for which users can define handlers. Developers can implement their own exploration strategies.

3 Design of GATS

This section overviews the system design of GATS. Figure 1 illustrates a dataflow of GATS that presents more details about the mechanisms it used.

Table 1 UI event parameters: l, r, t, and b denote left position, right position, top position, and bottom position of the layout

Event	Parameters	Description
Tap	Tap $((l+r)/2, (t+b)/2)$	Tap the center of the layout
Longtap	LongTap $((l+r)/2, (t+b)/2)$	Longtap the center of the layout
Scroll	random one of: Scroll $((l+r)/2, t, (l+r)/2, b)$, Scroll $((l+r)/2, b, (l+r)/2, t)$, Scroll $(l, (t+b)/2, r, (t+b)/2)$, Scroll $(r, (t+b)/2, l, (t+b)/2)$	Randomly trigger one of gestures: scroll down, scroll up, scroll left, scroll right
Text	Arbitrary fixed string	Trigger arbitrary string

3.1 The Event Selector

The event selector first uses the model that learned from previous information to trigger a new event. It uses the ADB host to send events to the app under test on the Android device or emulator. For UI events, the Android Debug Bridge Host talks to the Android Debug Bridge daemon on the device. For system events, the Android Debug Bridge host talks to the AM tool on the device, which sends system events as intents to apps.

There are several input mechanisms in Android. GATS supports two input mechanisms: touch screen and navigation buttons (specifically, "back" button and "menu" button). We found these sufficient in practice, because they are the most common input mechanisms.

Once the event selector obtains the view hierarchy from the model, it only considers the view objects which can interact with. It extracts the bound and the attributes (clickable, scrollable, and long-clickable) of the UI element on the touch screen. Finally, the event selector computes the parameters of UI events, as dictated by Table 1.

3.2 The UI Observer

The UI observer collects the layouts about the app's GUI elements currently displayed on the device screen. When an operation finished, UI observer gets current layout sending to the model, and then, model uses the information to update self.

4 Testing Algorithm

The testing algorithm combines active learning with testing. Informally, the algorithm works as follows.

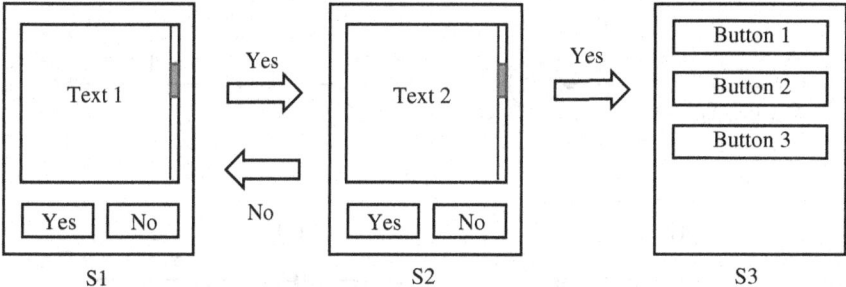

Fig. 2 First three screens of test app

Step 1: GATS installs and launches the app to be tested and waits to reach an idle state. This is the first state of the app. For each state, we compute all of the elements that enabled user inputs in the layout. Initially, the model only contains one state, the initial app state. And the current app state is the initial app state.

Step 2: If the current app state has at least one unexplored outgoing transition, the event selector will choose one of these events using appropriate mechanism and execute the transition.

Otherwise, GATS will find transitions to an app state which has at least one unexplored outgoing transition and execute the transitions to change the state of the app under test. Then, the event selector will choose one of these events using appropriate mechanism and execute the transition.

All of the executions will be written into operation log that is used to reproduce when app crash occurred.

Step 3: GATS waits for the app to reach an idle state. UI observer gets the layout of the app currently displayed and sends it to the model to update the model. After updated the model, the current app state changes. If UI observer detects app crash occurred, it will take a screenshot and log the error information to the log file and restart the app.

If the app restart times reaches the max app restart times and the number of steps reaches the max number of steps, the algorithm finishes. Otherwise turns to Step 2.

5 Model Update Algorithm

We use Fig. 2, the first three screens of test app, to explain how the model update algorithm works.

The first two screens have four input choices: (a) No button, (b) Yes button, (c) Scroll Down the text, and (d) Scroll Up the text. These two screens will be considered

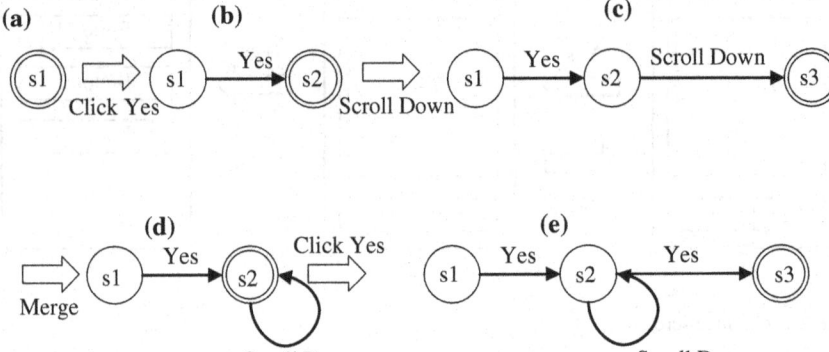

Fig. 3 Process of learning the model on test app. A circle with one line denoted a state in the model. A circle with double line denotes the current app state. The line with arrow denotes a transition

as two different app states because the position or the content of the elements is different.

There are two scenarios that can arise when the system gets a state to update the model.

Case 1. The app state reached has not been occurred previously. GATS adds a new app state to the model and adds a transition.

Case 2. If the app state reached is equal to the app state occurred before, GATS merges the two app states together and adds a transition to the previous app state.

Figure 3 shows how our system works.

Initialization: After launching the app, app reaches the first screen. The model only has one state s1, and the current app state is s1; it is shown in Fig. 3a.

First operation: The app is now at state s1, as shown in Fig. 3a. GATS now chooses to tap the button Yes. The result state has not occurred before. According to the first case of the algorithm, GATS adds a new app state s2 to the model. Figure 3b shows the model.

Second operation: The app now is at state s2, as shown in Fig. 3b. GATS now chooses to scroll down the screen and reaches a new state s3, as shown in Fig. 3c. However, the new state is as same as the app state s2. Therefore, according to Case 2 of the algorithm, GATS merges s3 with s2. The model is shown in Fig. 3d.

Then, GATS will do as the above until reaches the terminal state.

6 Evaluation

In this section, we evaluate GATS on several real-world Android apps from the top list of some app market. We compare GATS with Monkey [10] and Dynodroid.

Table 2 Comparison of GATS with other techniques

App	GATS		Monkey		Dynodroid	
	Bug (s)	Time (s)	Bug (s)	Time (s)	Bug (s)	Time (s)
com.tencent.reading	1	1219	0	751	0	10152
com.athoc.panic	2	1568	0	635	1	14387
com.call.plus	0	1085	0	315	0	9848
app.safeSchool	1	980	1	421	1	10487
com.sphene.sphene	1	1248	1	754	1	13687
es.tid.gconnect	0	1157	0	456	0	12485
com.fvm.who	0	1374	0	731	0	11358
com.lvstudio.emotionchat	1	1579	1	948	1	18451
org.mozilla.fennec_root	1	1487	1	780	1	13578
com.hippi	0	1207	0	507	1	11489

We randomly choose ten apps from the top list of several open app markets, such as Baidu, Google Play. But the apps with a native entry are all excluded.

We performed the experiment on a four-core Intel(R) Core(TM) i5-3450 CPU @ 3.10 GHz Windows machine with 16 GB RAM using one emulator. The emulator has 1536 MB RAM and 64 MB VM heap with Android 5.1 API using x86 system image.

In this study, we evaluated the following three approaches: GATS using random strategy; the Monkey testing tool provided in Android platform; Dynodroid using BiasedRandom strategy.

We ran GATS for 1000 events, ran Monkey for 5000 events, and ran Dynodroid for 1000 events. For each app, we ran ten times on each approach. We used different number of events for Monkey and others that the three tools have the events coverage because of Monkey having many redundant events.

The total bug number found on ten times running and average execution time for all these approaches are summarized in Table 2.

Compared to Monkey, GATS has higher bug found in all cases, although run 2 times slower. And compared to Dynodroid, GATS has almost the same bug found in most cases, but run about ten times faster.

7 Conclusion

We present GATS, a GUI-based automated test system for Android apps. We proposed a novel learning-based testing algorithm. Our experiments show the effectiveness of the technique in finding bugs in real-world apps. Moreover, our algorithm achieves less execution time than Dynodroid.

References

1. Google Play Wiki. https://en.wikipedia.org/wiki/Google_Play
2. Monkeyrunner. http://developer.android.com/tools/help/monkeyrunner_concepts.html
3. Hierarchy Viewer. http://developer.android.com/tools/help/hierarchy-viewer.html
4. Robotium. http://code.google.com/p/robotium/
5. Machiry, A., Tahiliani, R., Naik, M.: Dynodroid: An input generation system for android apps. In: Proceedings of the 2013 9thJoint Meeting on Foundations of Software Engineering, ser. ESEC/FSE2013. New York, NY, USA, pp. 224–234. ACM (2013)
6. Sasnauskas, R., Regehr, J.: Intent Fuzzer: crafting intents of death. In: Proceedings of the 2014 Joint International Workshop on Dynamic Analysis (WODA) and Software and System PerformanceTesting, Debugging, and Analytics (PERTEA), ser. WODA + PERTEA2014. New York, NY, USA, pp. 1–5. ACM (2014)
7. Amalfitano, D., Fasolino, A.R., Tramontana, P., De Carmine, S., Memon, A.M.: Using GUI ripping for automated testing of android applications. In: Proceedings of the 27th IEEE/ACM InternationalConference on Automated Software Engineering, ser. ASE 2012.New York, NY, USA, pp. 258–261. ACM (2012)
8. Choi, W., Necula, G., Sen, K.: Guided GUI testing of android appswith minimal restart and approximate learning. In: Proceedings of the 2013 ACM SIGPLAN International Conference on Object OrientedProgramming Systems Languages & Applications, ser. OOPSLA'13, pp. 623–640. ACM, New York, NY, USA (2013)
9. Hao, S., Liu, B., Nath, S., Halfond, W.G., Govindan, R.: PUMA: programmable UI-automation for large-scale dynamic analysis of mobileapps. In: Proceedings of the 12th Annual International Conference on Mobile Systems, Applications, and Services, ser. MobiSys '14, pp. 204–217. ACM, New York, NY, USA (2014)
10. Android monkey. http://developer.android.com/guide/developing/tools/monkey.html

Fine-Grained Access Control on Android Through Behavior Monitoring

Ximing Tang, Tao Song, Kun Wang and Alei Liang

Abstract Google's Android platform includes a permission system that protects privileged resources from applications' abuse, such as Internet, Location, and Telephony. The permission system has great importance to the privacy security for users, but it is coarse grained that applications will have broader access than they actually require. In our paper, we design and implement a system with fine-grained access control of the privileged resources on Android. The system will contain two main modules: Behavior Monitor Module and Behavior Decision Module. The former will closely monitor the applications' behavior such as attempting to access a user's private data, send SMS, and access network. The latter will notify the user to make the decision for the application's behavior and store the user's decision. Experiments show that the system has a fine-grained access control through behavior monitoring.

Keywords Android platform · Fine-grained access control · Behavior monitoring

1 Introduction

In the current market of mobile platform, Google's Android OS is undoubtedly the fastest growing mobile operating system in the world. Famous market research company Kantar's statistics shows that Google's Android's market share has reached 76% among EU5 until April 2016 [1]. To help address security concerns, Android uses a permission system [2] to restrict operations applications can perform on privileged

X. Tang (✉) · T. Song · K. Wang · A. Liang
School of Software Engineering, Shanghai Jiao Tong University, Shanghai, China
e-mail: tangximing@sjtu.edu.cn; 1040606555@99.com

T. Song
e-mail: songt333@sjtu.edu.cn

K. Wang
e-mail: wk476855@sjtu.edu.cn

A. Liang
e-mail: liangalei@sjtu.edu.cn

© Springer Nature Singapore Pte Ltd. 2019
S. K. Bhatia et al. (eds.), *Advances in Computer Communication and Computational Sciences*, Advances in Intelligent Systems and Computing 760, https://doi.org/10.1007/978-981-13-0344-9_45

resources. Permissions correspondent to the privileged resources should be defined in the application's AndroidManifest.xml file. During application installation, users are notified to grant or deny the permissions the application required.

The current Android permission system is coarse grained, which causes that applications have broader access than they actually require. For example, when an application creates a socket to connect to a domain, the INTERNET permission is required. However, because INTERNET permission is coarse grained which does not restrict which domain the application can access, the application is provided with more privilege than it truly needs.

Several methods have been proposed to enforce fine-grained access control of the privileged resources on Android. The first method is the bytecode rewriting technique [3–6]. By instrumenting applications' Java bytecode or Dalvik bytecode, security policies could be enforced upon security-sensitive Android API methods. But the existing work [4, 7] have pointed out that the bytecode rewriting technique can be attacked. The second method is native library rewriting, which enforces fine-grained access control by intercepting the native invocations to Bionic library [8], which needs to rewrite every application. A third way to implement fine-grained access control can be achieved by modifying the Android Operating System [9, 10] which has significant usability and hinders any efforts for widespread, called "fragmentation problem."

Our Approach: We address these problems by intercepting the native functions in the Bionic library and retrieving the parameters of the functions to analysis the behavior of the applications and then notify the user to make a decision for the behavior. There is no need to modify the apk file of the application, either to modify the Android OS. In our system, there are two main modules: one for monitoring the applications' behavior, another for the users to make a decision for the behavior.

2 Android

2.1 Zygote

Android at its core has a process named *Zygote*, which starts up at init time. *Zygote* is a daemon and its only mission is to launch applications. This means that *Zygote* is the parent of all application processes. Once *Zygote* starts, it loads all necessary libraries and opens a socket named */dev/socket/zygote* to listen to requests for starting applications. When a request for starting an application comes in, the function *fork()* triggered, *Zygote* will create an exact and clean new Dalvik VM, preloaded with all necessary libraries that an application will need.

In our system, we inject to *Zygote* to achieve the goal of monitoring the behavior of all launched applications.

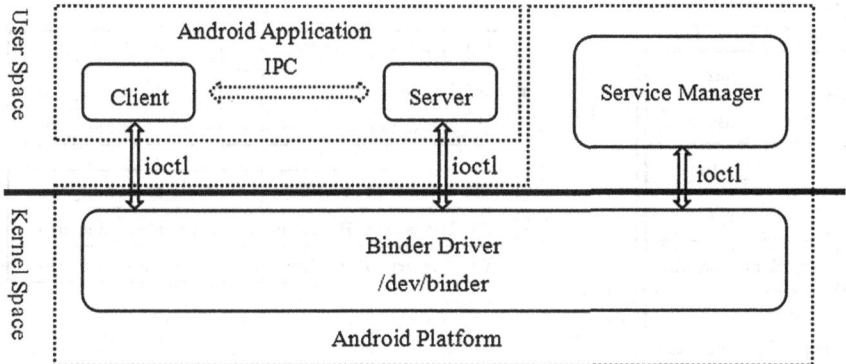

Fig. 1 Android binder mechanism

2.2 Android Binder Mechanism

In the Android system, each application is composed of a number of activity and service, while activity and service are possible to run in the same process or in different processes. Android builds a new kind of IPC mechanism called Binder for different processes to exchange data.

In the Binder Mechanism, there are four components: Client, Server, Service Manage, and Binder. Client, Server, and Service Manager are running in the user space while Binder is running in the kernel space. Client is the process requesting services while Server is the process providing the relevant service. Service Manager manages all registered services, such as Location and Telephony. Figure 1 presents the Android Binder Mechanism.

Client and Server exchange data through Binder driver with the virtual device */dev/driver* mounted in the kernel. Client and Server interact with binder driver through the function *ioctl* in Bionic library *libc.so*. The function *ioctl*'s prototype is *int ioctl(int fd, int request, void* arg);*

The first parameter *fd* is the file descriptor of the virtual device */dev/driver*. The second parameter *request* is the control command for the device, such as BINDER_WRITE_READ and BINDER_VERSION. The third parameter *arg* is the data transferred to the device. If the request is BINDER_WRITE_READ, it means that the client or server is to write data to the binder driver or read data from the binder driver, while the *arg* is a point to the *structure binder_write_read*. During one transaction, multiple instructions can be written to the binder driver or multiple results can be read from the binder driver. Figure 2 presents the structure in binder.

The data in the *write_buffer* is stored sequentially as format of *command code* and *command content*, namely an *instruction*. The behavior of applications can be identified exactly by analyzing the command content of every instruction. For example, when an application is trying to access the IMEI or Phone Number, the command content of the instruction will contain a string

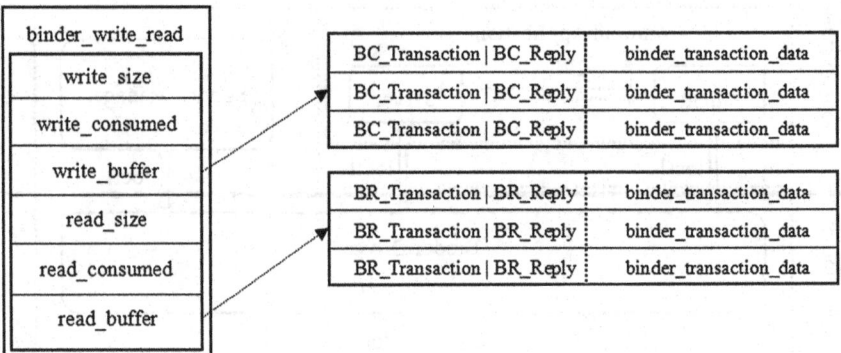

Fig. 2 Data structure in binder

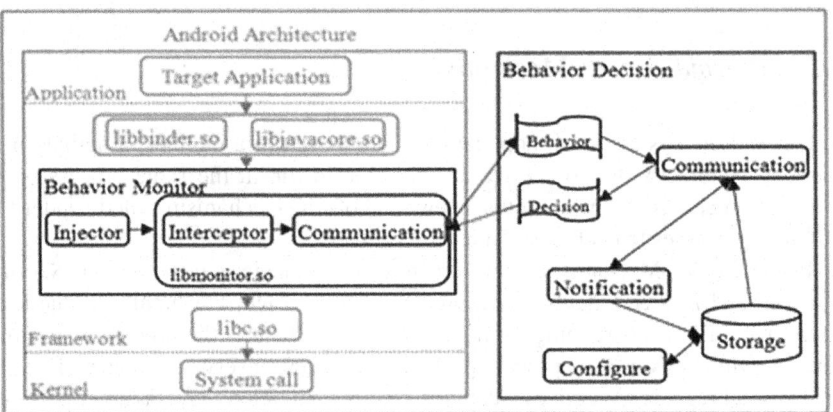

Fig. 3 Framework of the system

"com.android.internal.telephony.IPhoneSubInfo". Therefore, in our system, we will hook the function *ioctl* and analyze the data to monitor the behavior of the applications.

3 System Design and Implementation

3.1 Framework

The system mainly contains two modules: Behavior Monitor Module, called *BMM* and Behavior Decision Module, called *BDM*. Figure 3 presents the framework of our system.

Three components are included in BMM:

- **Injector** Inject the shared library *libmonitor.so* to the target process which in our system is *Zygote*.
- **Interceptor** Hook the native functions in shared libraries to monitor the behavior of the applications and redirect the functions to our functions.
- **Communication** This module is the client part of the Communication Module, which is responsible for transfer the data representing the behavior of the applications to *BDM* and get the decision from *BDM*.

Four components are included in *BDM*:

- **Communication** This module is the server part of the Communication Module, which is responsible for get the behavior of the applications from *BMM* and transfer user's decision for the behavior to *BMM*.
- **Notification** Notify the user of the behavior of the applications and wait for the user to make the decision.
- **Storage** Store the user's decision for every behavior of every application which can be used later.
- **Configure** Display all decisions to the user, and user can update the decision for the behavior.

3.2 System Implementation

Inject Shared Library

The injection of the shared library, which means inject the shared library to the target process mainly includes four steps:

- Attach to the target process with the ptrace function;
- Invoke the dlopen function to load the shared library to the target process;
- Executing code in the shared library;
- Detach from target process.

Follow the steps, Injector is implemented to inject our libmonitor.so to Zygote process. Since the Zygote process is the parent of all application process, the memory of all application processes will contain the shared library in Zygote. So when an application starts to run, it will contain libmonitor.so, which means the Injector module will only run once.

Redirection

Global Offset Table (GOT) and Procedure Linkage Table (PLT) are used in the dynamic linking mechanism in Linux. Shared Libraries are ELF [11] files, which contain many sections, including *rel.got* and *rel.plt*.

Figure 4 presents the functions will be hooked in our system. We can retrieve the address of the shared library libbinder.so and libjavacore.so by reading the file */proc/< pid >/maps* (pid is the process id of *Zygote*). Then, we can get the address

Shared library	Functions in libc.so	Functions in libmonitor.so
/system/lib/libjavacore.so	open, connect	hook_open, hook_connect
/system/lib/libbinder.so	ioctl	hook_ioctl

Fig. 4 The functions to be hooked in the native library

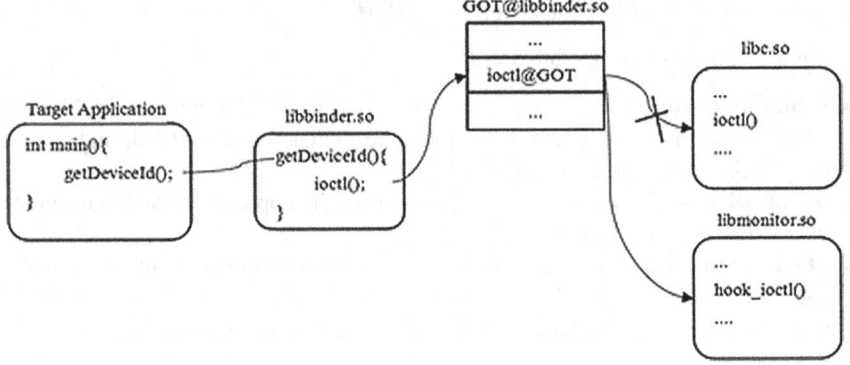

Fig. 5 The procedure of redirection

of the function in GOT of libbinder.*so* and *libjavacore.so* and then replace them with our new function address. Finally, when original functions are invoked, it will redirect to our functions. Figure 5 presents the procedure of the Redirection.

Communication

Since *libmonitor*.so is injected to *Zygote* running at Linux Level, and the BDM is running as a service at Android Application Level. Our system uses Linux Socket and Java Socket to achieve the communication function.

Behavior Decision

BDM is implemented as a service in Android Application Level, listening for the behavior of the applications. When a behavior comes in, *BDM* will first get the decision for the behavior from *Storage*. If the decision is *Allow*, our function in *BMM* will redirect to the original function in Bionic Library, named *libc.so*. If the decision is *Deny*, our function will return. If the decision is *Query*, *Notification* will be triggered, and user will make a decision *Allow* or *Deny* which will be transferred to BMM and also be stored in Storage.

The decision for all behaviors of all applications will be displayed in *Configure*, and user can update the decision for later use.

Fig. 6 **a** Read IMEI; **b** access network; **c** send SMS

4 Experimental Results

From the Android Development Web site, we can figure out that Android 4.4 is the most widely used Android version [2] among all Android versions. So we test our system mainly on the Android 4.4.

4.1 Privacy Monitor

Our system monitors the behavior of the applications which will access users' private data, presented in Fig. 6a. The users' private data include IMEI, IMSI, Phone Number, Location, Phone Call Logs, and SMS.

4.2 Network Monitor

Our system monitors the behavior of the applications which will interact with the Internet, presented in Fig. 6b. Since the application has the Internet permission which does not restrict which domain the application can access, we enforce fine-grained control that restrict the application from connecting to a malicious IP address or let users make the decision.

4.3 SMS Monitor

In this paper, we design and implement a system to enforce fine-grained access control of the privileged resources on Android. The system achieves the goal through behavior monitoring with injection of the shared library and redirection of the native function. The users can make the decision for the behavior with the socket communication. The experimental results tested on the real Android device indicate that the system has a fine-grained access control of the privileged resources through Behavior monitoring.

5 Conclusion

In this paper, we design and implement a system to enforce fine-grained access control of the privileged resources on Android. The system achieves the goal through behavior monitoring with injection of the shared library and redirection of the native function. The users can make the decision for the behavior with the socket communication. The experimental results tested on the real Android device indicate that the system has a fine-grained access control of the privileged resources through Behavior monitoring.

References

1. Google's Android's market share in the Kantar's statistics. http://www.kantarworldpanel.com/global/smartphone-os-market-share/
2. Android Developer. https://developer.android.com
3. Bartel, A., Klein, J., Monperrus, M., et al.: Improving privacy on android smartphones through in-vivo bytecode instrumentation. uni. lu (2012)
4. Davis, B., Sanders, B., Khodaverdian, A., et al.: I-arm-droid: A rewriting framework for in-app reference monitors for android applications. Mob. Secur. Technol. **2012**(2), 17 (2012)
5. Jeon, J., Micinski, K.K., Vaughan, J.A., et al.: Dr. Android and Mr. Hide: Fine-grained security policies on unmodified Android (2011)
6. Reddy, N., Jeon, J, Vaughan, J., et al.: Application-centric security policies on unmodified Android. UCLA Comput. Sci. Dep. Tech. Rep. 110017 (2011)
7. Hao, H., Singh, V., Du, W.: On the effectiveness of API-level access control using bytecode rewriting in Android. In: Proceedings of the 8th ACM SIGSAC Symposium on Information, Computer and Communications Security, 25–36. ACM (2013)
8. Xu, R., Saïdi, H., Anderson, R.: Aurasium: practical policy enforcement for android applications. In: Presented as Part of the 21st USENIX Security Symposium (USENIX Security 12), pp. 539–552 (2012)
9. Hornyack, P., Han, S., Jung, J., et al.: These aren't the droids you're looking for: retrofitting android to protect data from imperious applications. In: Proceedings of the 18th ACM Conference on Computer and Communications Security, pp. 639–652. ACM (2011)
10. Nauman, M., Khan, S., Zhang, X.: Apex: extending android permission model and enforcement with user-defined runtime constraints. In: Proceedings of the 5th ACM Symposium on Information, Computer and Communications Security, pp. 328–332. ACM (2010)
11. Standard, T.I.: TIS Committee. Executable and linkable format (elf). Tool Interface Stand. (TIS) **1**, 1 (2008)

Research on Rendezvous Algorithm Based on Asymmetric Model in Cognitive Radio Networks

Xiu Wang, Li Yu and Fangjian Han

Abstract Cognitive radio is an important communication technology, which can relieve the situation of shortage and waste for spectrum source, and the rendezvous of cognitive users is the prerequisite of communication. This paper focuses on the algorithms for blind rendezvous, i.e., rendezvous without using any centralized controller and common control channel (CCC). Channel hopping sequence is an effective method to solve the problem of blind intersection. In this paper, we describe three kinds of rendezvous algorithm in detail, which based on the symmetric network, and apply these algorithms to asymmetric model with synchronous and asynchronous ways. We derive the maximum time-to-rendezvous (MTTR) and the expected time-to-rendezvous (ETTR) of algorithms for two users. Extensive simulations are conducted to evaluate the performance of these algorithms.

Keywords Cognitive radios · Rendezvous · Asymmetric · TTR
Synchronous and asynchronous

1 Introduction

In the cognitive radio (CR) networks, rendezvous is a necessary condition for communication among cognitive users [1], i.e., find each other on a common channel (rendezvous channel). The traditional approach is to set up one or more unlicensed channels as dedicated common control channels, and all of the cognitive users can exchange information on CCC. However, with the proliferation of wireless devices

X. Wang (✉) · L. Yu · F. Han
School of Electronic Science and Engineering, National University of Defense Technology, Changsha, China
e-mail: 15116296384@163.com

© Springer Nature Singapore Pte Ltd. 2019
S. K. Bhatia et al. (eds.), *Advances in Computer Communication and Computational Sciences*, Advances in Intelligent Systems and Computing 760, https://doi.org/10.1007/978-981-13-0344-9_46

and services, the unlicensed spectrums become more and more crowded, and this is contrary to the principles of not interfering with the licensed users if set up CCC on the licensed channel. To solve this problem, the blind rendezvous algorithms are proposed [2, 3], in which all vacant channels are potentially available for the exchange of control and data. There are two types of blind rendezvous: synchronous and asynchronous. Synchronous approaches require that cognitive users attempt to rendezvous at the same global time; this is easy to success. But in certain types of networks, such as ad hoc networks [4], precise clock synchronization among nodes may not be feasible, while asynchronous approaches can be used without global time. For asynchronous, there are many approaches to achieve rendezvous, and the popular is channel hopping (CH) algorithm [1], i.e., each cognitive user hops on channels according to the sequence which is generated for rendezvous. According to the different construction strategies, CH sequence can be divided into random, permutation-based, and Quorum-based. For random algorithm, N. Theis and R. Thomas proposed Modular Clock (MC) algorithm [5], which can guarantee the performance of channel hopping by using the arithmetic of prime numbers. But in symmetric model, if two users have the same indexes and steps, rendezvous cannot be achieved. To solve this problem, the authors presented Modified Modular Clock (MMC) algorithm further, but the situation of infinite for MTTR is still not be avoided. L. DaSilva proposed Generated Orthogonal Sequence-based algorithm (GOS) [6, 7], which can guarantee two users rendezvous in m $(m + 1)$ time lots under symmetric, m is the available channels. K. Bian and J. Park proposed optimal Asynchronous Quorum-based Channel Hopping System (optimal AQCH) [8–10], which can have good performance on time-to-rendezvous and utilization of channels. Above several algorithms will be described in the second chapter.

In this paper, we investigate the performance of these algorithms in the asymmetric network and make comparison between synchronous and asynchronous ways. The performance is commonly evaluated by the following two metrics [11, 12].

- Expected Time-to-Rendezvous (ETTR): Do not consider the activities of the primary users; the expected time for two cognitive users achieve rendezvous.
- Maximum Time-to-Rendezvous (MTTR): The maximum time spent in all of the two cognitive users of the rendezvous success number of times.

The rest of this paper is organized as follows: We describe above several algorithms in detail in Sect. 2. In Sect. 3, we simulate and compare these algorithms in the asymmetric model under two ways of synchronous and asynchronous. We conclude this paper in Sect. 4.

2 Rendezvous Algorithm Based on Channel Hopping

2.1 Modular Clock and Modified Modular Clock

Modular Clock algorithm uses prime number to guarantee TTR. The parameters involved in the algorithm are as follows [13]:

- m is the number of available channels in the current network.
- j is the current channel index.
- r represents the step of cognitive user hops between channels. On a time lot, the user forwards hops r channels, and if the channel index exceeds the max index, it employs modulo arithmetic.
- t is the minimum time unit for hopping, i.e., time slot [14].
- p indicates the prime number of the cognitive user, which must be lowest prime greater than or equal to m

The channel hopping sequence is determined by the above parameters. When $t=0$, the cognitive user selects randomly one channel in $[0, m)$ to work, after, it forwards steps r each time slot. If the resulting index is greater than $m-1$, then the index is remapped by the mod function between 0 and $m-1$ and the user hops to that channel. In addition, this algorithm makes $2p$ as the period of time. When t exceeds $2p$, this algorithm works again from $t=0$. For example, when $m=4$, $p=5$, the index is 1 at $t=0$, and assume r is 3, the resulting channel hopping sequence is

$$1, 0, 3, 2, 1, 0, 3, 2, 1, \ldots$$

MC algorithm can complete rendezvous at most p time slots, but in the symmetric network, if two users have the same indexes and steps at $t=0$, rendezvous cannot be achieved. To avoid this, authors modified MC to MMC; the comparison of the two algorithms are as follows (Table 1).

MMC reduces the possibility of not rendezvous, but MTTR is still may infinite, and performance of ETTR becomes worse.

Table 1 Comparison of MC algorithm and MMC algorithm

	r	Period of rendezvous	p	J
MC	$[0, p)$	$2p$	A fixed value	$j=j \bmod (m)$
MMC	$[0, m)$	$2p^2$	$[m, 2m]$	$j=\mathrm{rand}\,([0, m))$

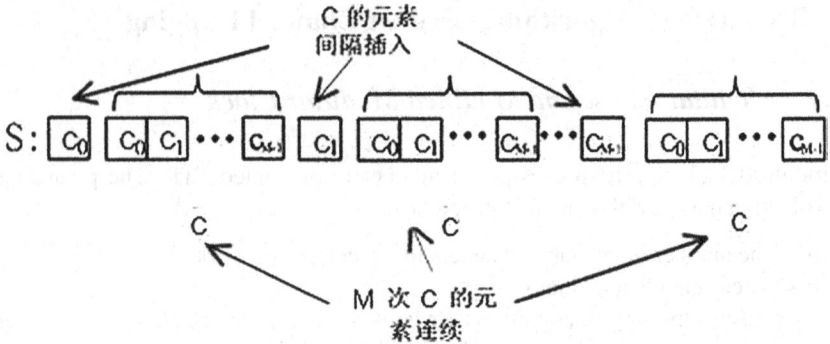

Fig. 1 Construction of S

2.2 Permutation-Based

The idea of designing channel hopping sequence by using permutation and combination theory was discussed by DaSilvas L [8], and proposed Generated Orthogonal Sequence algorithm (GOS). We index m available channels as $\{0, 1, 2, \ldots, m-1\}$ and combine them into an array called C, according to the principle of non repetition. Then repeat $C(m+1)$ times to derive the channel hopping sequence S with length of m $(m+1)$. Among them, there are m times elements of C, and they are continuous, the other elements of C are inserted intervally. The following figure shows the construction of S (Fig. 1).

Assume $m=5$, the random sequence $C = \{3, 2, 0, 1, 4\}$, the channel hopping sequence S is

3, 3, 2, 0, 1, 4, 2, 3, 2, 0, 1, 4, 0, 3, 2, 0, 1, 4, 1, 3, 2, 0, 1, 4, 4, 3, 2, 0, 1, 4.

Observe this sequence, the ETTR is

$$\text{ETTR} = \frac{m^4 + 2m^2 + 6m - 3}{3m(m+1)}$$

$$\text{MTTR} = m(m+1)$$

Under the symmetric network, GOS algorithm can guarantee the users achieve to rendezvous within $m(m+1)$ time slots.

Fig. 2 MMC algorithm

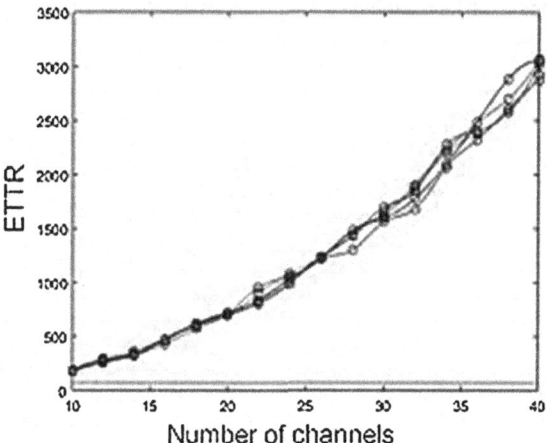

Number of channels

2.3　Quorum-Based [15]

The channel hopping sequence method based on Quorum system was proposed by Bian K. and Park J. For asynchronous network, the authors proposed the optimal AQCH system. When the cognitive user has no data to send, the channel indexes $\{0, 1, 2, \ldots, m - 1\}$ are constructed randomly to an array x. Then repeat x m times and combine them into a receiver sequence u. When the users have data to send, make x cyclic shift successively and combine them to the sender sequence v.

For example, $m = 5$, the random sequence $x = \{3, 1, 2, 0, 4\}$. If there are no data to send, the hopping sequence of sender and receiver is

$$3, 1, 2, 0, 4, 3, 1, 2, 0, 4, 3, \ldots, 3, 1, 2, 0, 4$$

In contrast, the sequence of sender is

$$3, 1, 2, 0, 4, 1, 2, 0, 4, 3, 2, 0, 4, 3, 1, \ldots, 4, 3, 1, 2, 0$$

Optimal AQCH algorithm makes the overlap degree of two channel hopping sequence is m, and the period of sequence is minimum, i.e., m^2.

3　Simulation and Analysis

In this section, we simulate and analyze the above several algorithms in the ways of synchronous and asynchronous under non-symmetric network. We consider that two users are in the range of one hop for each other and ignore the channel competition or the situation of the primary users suddenly occupies channels. Set the change range

Fig. 3 GOS algorithm

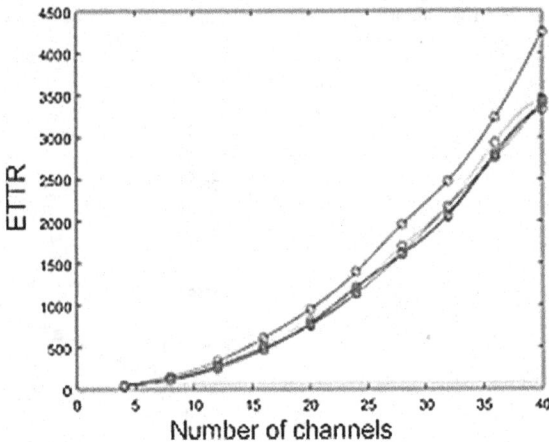

Fig. 4 Optimal AQCH algorithm

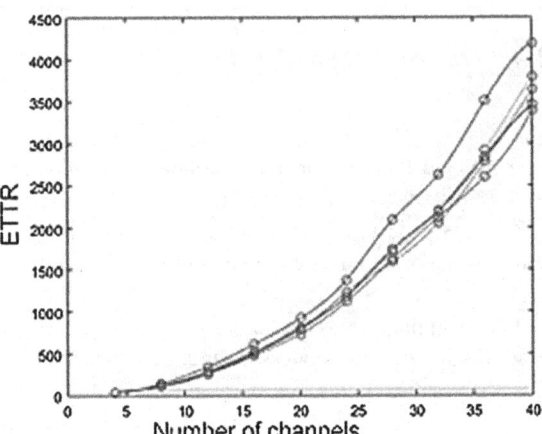

of the m available channels is from 10 to 40, $r = 2$, and the period of rendezvous is $2p_1p_2 \cdot p_1$ is the random prime generated by sender, and p_2 is generated by receiver. In the asymmetric network, because the available channels are different for sender and receiver, we set the common channels successively as 1, 3, 5, 7, 9. The simulation results of synchronous and asymmetric as shown in Figs. 2, 3, and 4 are as follows.

As shown in the figures, with the increase of the available channels, ETTR is also increasing, and the GOSs and AQCHs have the increasing trend of index. The five lines in the graph represent successively the common channels 1, 3, 5, 7, 9. Because MMC selects channels randomly, the performance of different common channels has no significant difference. In addition to the common channel of number 1, the ETTR of GOS and optimal AQCH do not have significant difference.

Fig. 5 MMC algorithm

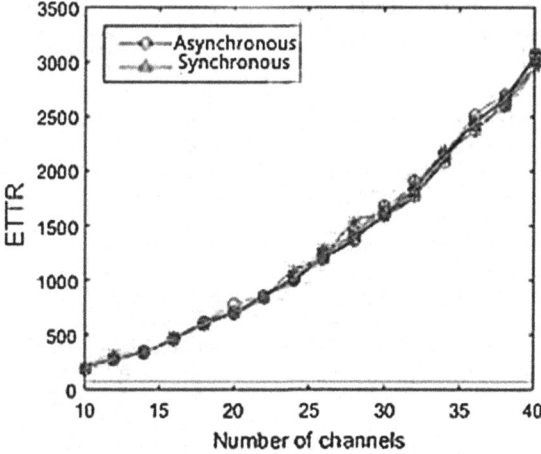

Fig. 6 GOS algorithm

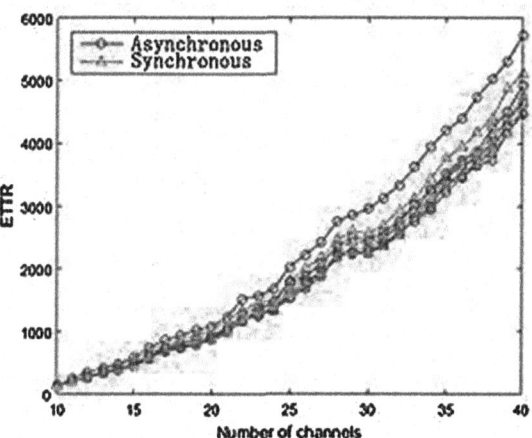

Further, we simulate these algorithms in synchronous and asynchronous networks, and compare the ETTR and MTTR. The results are as follows (Figs. 5–10).

Under the asymmetric network, synchronous and asynchronous have the closed ETTR, because sender and receiver select channels in random way. For GOS and optimal AQCH, the two nodes hop according to the sequence strictly; the ETTR shows exponential growth, but the asynchronous way has higher ETTR than synchronous.

In terms of MTTR, synchronous way and synchronous way have closed MTTR for MMC algorithm. Similarly, asynchronous way has higher MTTR than synchronous for GOS and optimal AQCH overall.

Fig. 7 Optimal AQCH
algorithm

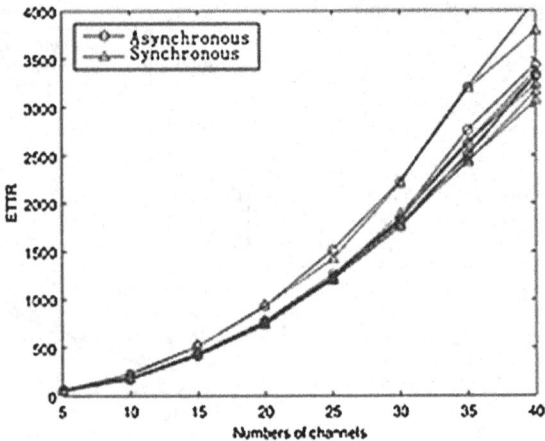

Fig. 8 MMC algorithm

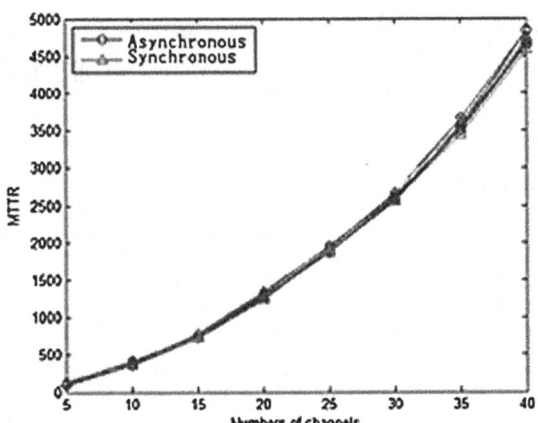

4   Conclusions

In this paper, three kinds of asynchronous rendezvous algorithms based on symmetric model are reviewed. We apply these algorithms to the asymmetric network and simulate them in asynchronous and synchronous ways. The resulting simulation shows that whether it is synchronous or asynchronous, the ETTR or MTTR of MMC is similar. But for GO and optimal AQCH, synchronous way can rendezvous more faster and stable than asynchronous.

Fig. 9 GOS algorithm

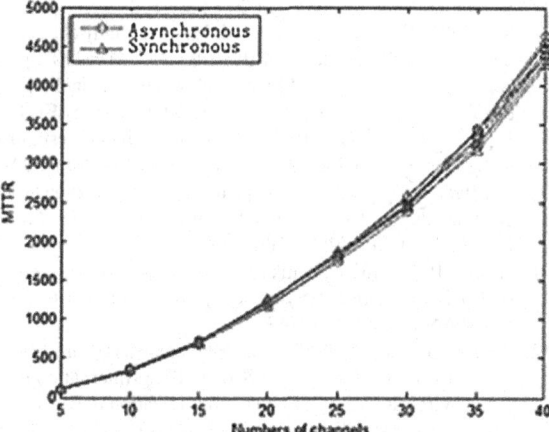

Fig. 10 Optimal AQCH algorithm

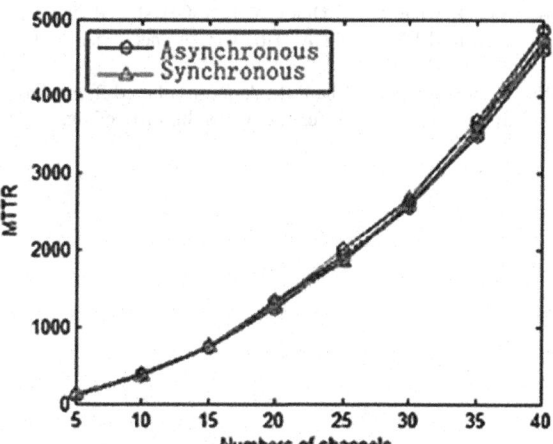

References

1. Xiao, Y., Hu, F.: Cognitive Radio Networks. CRC Press (2008)
2. Lin, Z.Y., Liu, H., Chu, X.W., Leung, Y.W.: Jump-Stay based channel-hopping algorithm with guaranteed rendezvous for cognitive radio networks. In: Proceedings of the IEEE INFOCOM, 2011, pp. 2444–2452. https://doi.org/10.1109/infcom.2011.5935066
3. Liu, H., Lin, Z.Y., Chu, X.W., Leung, Y.W.: Jump-Stay rendezvous algorithm for cognitive radio networks. IEEE Trans. Parallel Distrib. Syst. (99), 1 (2012). https://doi.org/10.1109/tpds.2012.22
4. So, H.W., Walrand, J., Mo, J.: McMAC: a multi-channel MAC proposal for Ad Hoc wireless networks. In: Proceedings of IEEE WCNC, 2007, pp. 334–339
5. Theis, N., Thomas, R., DaSilva, L.: Rendezvous for cognitive radios. IEEE Trans. Mob. Comput. **10**(2), 216–227 (2011)
6. DaSilva, L., Guerreiro, I.: Sequence based rendezvous for dynamic spectrum access. In: Proceedings of IEEE International Symposium on New Frontiers in Dynamic Spectrum Access

Networks (DySPAN'08), Oct 2008, pp. 1–7

7. So, H.W., Walrand, J., Mo, J.: McMAC: a multi-channel MAC proposal for ad hoc wireless networks. In: Proceedings of IEEE WCNC, Mar 2007, pp. 334–339

8. Bian, K., Jerry Park, J.-M.: Asynchronous channel hopping for establishing rendezvous in cognitive radio networks. In: Mini-Conference at IEEE INFOCOM 2011

9. Bian, K., Park, J.M.J., Chen, R.: Control channel establishment in cognitive radio networks using channel hopping. IEEE J. Sel. Areas Commun. JSAC (2011)

10. DaSilva, L., Guerreiro, I.: Sequence based rendezvous for dynamic spectrum access. In: Proceedings of IEEE International Symposium on New Frontiers in Dynamic Spectrum Access Networks (DySPAN'08), 2008, pp. 1–7

11. Horine, B., Turgut, D.: Link rendezvous protocol for cognitive radio networks. In: Proceedings of IEEE International Symposium New Frontiers in Dynamic Spectrum Access Networks (DySPAN'07), pp. 444–447

12. Horine, B., Turgut, D.: Performance analysis of link rendezvous protocol for cognitive radio networks. In: Proceedings of Second IEEE International Conference on Cognitive Radio Oriented Wireless Networks and Communication (CROWNCOM'07), Aug 2007, pp. 503–507

13. Li, X., Zhang H., Guo C., et al.: Asynchronous channel hopping algorithm for cognitive radio networks. J. Xi'an Jiaotong Univ. **46**(12), 1–6 (2012) (in Chinese)

14. Bahl, P., Chandra, R., Dunagan, J.: SSCH: slotted seeded channel hopping for capacity improvement in IEEE 802. 11 Ad Hoc wireless networks. In: Proceedings of ACM MobiCom, 2004, pp. 216–230

15. Bian, K., Park, J., Chen, R.: A quorum-based framework for establishing control channels in dynamic spectrum access networks. In: Proceedings of ACM MobiCom, 2009, pp. 25–36

Secure Localization Algorithms in Wireless Sensor Networks: A Review

Hua Wang, Yingyou Wen, Yao Lu, Dazhe Zhao and Chunlin Ji

Abstract Localization plays a vital role in understanding the application context for wireless sensor networks (WSNs). However, it is vulnerable to various threats due to the open arrangement area, the nature of radio broadcasting, and resource constraint. Two kinds of attacks on localization process need to be investigated. On the one hand, adversaries may capture, impersonate, replicate, or fabricate nodes to mislead the target to an incorrect position. On the other hand, adversaries may tamper, interfere, replay, or modify jointly localization information to disturb the localization. This paper describes two kinds of localization attacks and gives a systematic survey of existing secure localization solutions, including the design ideas, application scopes, and limitations. Next, several novel secure localization schemes for special WSNs applications, including Underwater Wireless Sensor Networks (UWSNs), Wireless Body Area Network (WBAN) and three-dimensional (3D), are analyzed. Finally, many open problems for secure localization are presented.

H. Wang · Y. Wen · D. Zhao (✉)
School of Computer Science & Engineering, Northeastern University, Shenyang 110819, China
e-mail: zhaodazhe@mail.neu.edu.cn

H. Wang
e-mail: wanghua@neusoft.com

Y. Wen
e-mail: wenyy@neusoft.com

Y. Wen
State Key Laboratory of Software Architecture, Neusoft Corporation, Shenyang 110179, China

Y. Lu
Beijing Jiaotong University, Beijing 100044, China
e-mail: luyao12@163.com

D. Zhao
Key Laboratory of Medical Image Computing, Ministry of Education, Northeastern University, Shenyang 110819, China

C. Ji
Kuang-Chi Institute of Advanced Technology, Shenzhen 518000, China
e-mail: chunlin.ji@kuang-chi.org

© Springer Nature Singapore Pte Ltd. 2019
S. K. Bhatia et al. (eds.), *Advances in Computer Communication and Computational Sciences*, Advances in Intelligent Systems and Computing 760, https://doi.org/10.1007/978-981-13-0344-9_47

543

Keywords Wireless Sensor Networks · Localization · Threats · Security

1 Introduction

Location determining is one of the most significant research subjects in WSNs due to many fundamental techniques, e.g., geographical routing [1], geographical key distribution [2], and location-based authentication [3] which require the positions of sensors [4]. The essential feature of a WSN is to monitor and report events which can be meaningfully assimilated and responded to, only if the position information of the event is known [5]. For instance, the positions of intruders and enemy tanks in a war are vital to deploy rescue troops. Hence, localization is very significant for enabling many applications such as wildfire detection, battlefield surveillance, and traffic regulation, which makes localization in WSNs very challenging.

As location-based applications become increasingly prevalent, localization, especially, is vulnerable to attacks. When a WSN was deployed in untrusted or hostile environments, adversaries might attack the localization process to make the position estimates wrong. For some applications such as military applications and environmental applications [6], wrong locations may result in severe consequences, e.g., incorrect military decisions on the war and false alarms [7]. In these applications, it is required for location-based technology to ensure the accuracy and validity of location information. Therefore, this situation poses the demand for secure localization scheme.

In this paper, we study the problem of the secure localization from two aspects. To begin with, adversaries may disturb the localization to make the estimated locations wrong. Hence, we need secure location determine (SLD) to estimate sensor locations. Moreover, as sensors may be compromised, the position claimed by sensors may be unreliable. Thus, we need to location verification (LV) to verify reported locations.

The remained of paper is organized as follows: Sect. 2 states attack model. Sections 3 and 4 describe the state of research in SLD and LV, respectively. Section 5 provides a detailed comparison between schemes. Section 6 gives several novel localization schemes for special applications. Section 7 gives the conclusions and open problems.

2 Attack Model [4]

According to the attack goal, attacks on localization can be divided into two categories: attacks on sensors and attacks on location information.

2.1 Attacks on Sensors

Malicious nodes may contain adversaries and compromised sensors. An adversary is an external sensor intruded into the WSNs, and then a compromised sensor is an internal sensor, which may be a localization or a beacon sensor, controlled by the adversaries. Attacks on sensors are listed as follows:

Compromising. Compromised sensor nodes can be formed by: (1) capturing common sensors and reprogramming them; (2) deploying a larger number of powerful sensors to attack common sensors [8].

Sybil Attack. The adversary can fabricate a new identity or steal a certification from a legitimate sensor node to send messages with different identities [9].

Replication Attack. The adversary can clone sensor from the captured sensors. The replicas can be regarded as normal nodes of the network [10].

Wormhole Attack. The adversary can hijack a packet or individual bits of a packet at one position, and then tunnel the packet to another position to replay it [11].

2.2 Attacks on Location Information

During the localization process, localizing sensors always use the locations from beacon sensors to locate themselves. Adversaries aim to make the location estimations wrong. Attacks on location information are listed as follows:

Tampering. The malicious sensors tamper with the sensing data, alter the coordinates, time, or hop count to inject to the wrong location information and mislead to the localization [12].

Interference. The adversary can obstruct signals between a transmitter and a receiver to delay transmission time, modify the angle of arrival, or weaken the received signal strength (RSS) [13].

Replay. The adversary can congest the location information transmission between a transmitter and a receiver, and then replay the outdated message.

Position-Reference Attack. Position-reference attack is against positioning systems in which each normal node obtains the position references from the reference nodes to localization, and the attack is to modify partial position references [8]. Based on the smart level, the attack can be divided into three categories: non-collusion attack, collusion attack, and pollution attack. For non-collusion attack, wrong position references aim to mislead the normal sensor to different wrong positions, while for collusion attack, to a random, but the same wrong position [10], and for pollution attack, to a specially chosen incorrect position [14].

3   Secure Location Determination

According to the sensor classification in WSNs, the SLD can be classified into two types: beacon-oriented and sensor-oriented.

3.1   Secure Localization for Beacons

Since most current sensor localization algorithms rely on the location information provided from beacon sensors, then the localization accuracy depends on the accuracy of position information from the beacon sensors. Therefore, the security and reliability of the beacon sensors become significant for determining the position of the localizing node.

According to the examine ways, existing solutions can be classified into two types: (1) Detect and revoke the beacon sensors that producing erroneous location information (the detection method), and (2) tolerate false or misleading location information from beacon sensors (the robustness method).

The Detection Method. In [15], Yang et al. presented a cluster-based MMSE (CMMSE) which utilizes MMSE to inspect and construct a consistent location reference set for determining the unknown sensors position. However, the random selection of initial location references makes CMMSE get different results in different runs and may lead to more rounds of execution failure.

In [16], Wu et al. presented a label-based secure localization approach, which is resistant to wormhole attack for the DV-Hop localization process. The main idea of this approach is to generate a pseudo-neighbor list for each beacon sensor, utilize neighbor lists received from neighboring beacon sensors to classify compromised sensors into different groups. Based on the labels of neighboring sensors, each sensor prohibits the communications with its pseudo-neighbors, which are compromised by the wormhole attack.

In [17], Garg et al. presented a gradient-based secure localization method, which set the target function as the distance error between the known distances and the estimated distances on the basis of the estimated locations of sensors. However, prunes 50% of anchor nodes may help to increase the localization error. Only one sensor location is updated in each iteration.

The Robustness Method. Many recent approaches tolerate the presence of abnormal beacon sensors instead of removing them to minimize their effect on the localization accuracy.

In [18], Jadliwala et al. theoretically analyzed the necessary and sufficient conditions to guarantee a bounded localization error for the distance-based position determination, with the presence of cheating beacons. If n represents the total number of beacons, the upper bound $k_{max} = (n - 3)/2$ on the number of malicious beacons k. To be located with a bounded error, an unknown sensor needs at least k + 3 beacons in which there are at least three genuine beacons. This condition called the "$k + 3$"

condition. Moreover, they outlined three distance-based localization algorithms to ensure the localization accuracy, when the number of untrusted beacons is below a given threshold.

In [19], Wang et al. presented a cooperative attack-resistant semi-definite programming (CARSDP). The Cooperation method provides more precise position estimate and better coverage with same deployment. It improves the capability of monitoring abnormal beacons. The reliability of the localization process is also improved by employing the semi-definite programming (SDP) to relax the non-convex process into a convex one. Nonetheless, CARSDP poses demanding computational and communication challenges on the hardware of the sensors due to their high complexity.

3.2 Secure Localization for Unknown Sensors

Several methods have been proposed to securely determining the locations of localizing sensors, which are called secure localization for unknown sensors [20, 21]. According to the dependency of beacons, secure localization for unknown sensors can be divided into two categories: trusted beacon-based (TBB) and beacon-free (BF).

The Trusted Beacon-Based Method. A number of localization methods rely primarily on trusted beacon sensors, which already know their exact physical locations, to estimate the locations of unknown sensors by exchanging collected information.

In [20, 21], Capkun et al. presented a secure cooperative positioning mechanism (SPINE) based on the verifiable multilateration (VM) technique. With a central authority and several trusted beacon sensors which are also named verifiers, VM enables a secure computation and verification of the positions for unknown sensors with the presence of adversaries. In VM, if a sensor is inside the triangle formed by three beacons, its position can be uniquely calculated by distance bounding protocol [22]. However, SPINE has many weaknesses. For instance, to perform VM, a large number of trusted verifiers are needed.

In [23], Lazos et al. presented a distributed range-free positioning scheme called SeRLoc, in which trusted sensors equipped with a set of higher-power sectored antennas called locators to instead beacons. When a sensor hears multiple locators, it computes the center of gravity of the sectors corresponding to locators as its locations. Moreover, the authors presented an improved approach HiRLoc [24], which obtained higher accuracy by employing rotatable antennas and variable transmission powers.

In [25], He et al. presented an enhanced secure localization scheme (ESLS) according to an attack-driven model, which extends SLS [26] and defends against distance reduction attacks and enlargement attacks. ESLS is the new one, which utilize the Petri net to validate a method for WSNs. However, it relies on trusted beacons and MMSEE for positioning.

The Beacon-Free Method. And yet, very little research has addressed beacon-free secure localization.

In [27], Garg et al. first proposed an iterative-based technique with low computational complexity to determine the position of the mobile sensors without beacon sensors. The algorithm combines gradient descent with selective pruning of inconsistent measurements to achieve a high localization accuracy, with the help of the relative location map preserves pairwise distances. The set of relative positions is only a rotation and translation of the absolute positions.

4 Location Verification

Based on the goals of verification, the existing sensor location verification methods can be classified into two types: in-region and on-spot. The on-spot verification intends to verify whether the locations claimed by sensors are far from their true spots beyond a certain distance. The in-region verification verifies whether a sensor is inside an application-specific verification region [28].

4.1 On-Spot Verification

In [28], Wei et al. proposed two algorithms for on-spot verification, namely the Greedy Filtering by Matrix (GFM) algorithm and the Greedy Filtering by Trustability indicator (GFT) algorithm. Both algorithms exploit the inconsistency between sensors' geographical relationships according to their claimed locations and those implied by their neighborhood observations.

In [29], Du et al. proposed a Localization Anomaly Detection (LAD) algorithm to examine whether the location estimates of the localizing sensors are consistent with the pre-prepared deployment information, based on three metrics: the Difference metric, Add-All metric, and Probability metric. LAD can achieve a low false alarm rate and a high detection rate, but did not handle abnormal sensors.

In [30, 31], Capkun et al. presented a sensor location verification by utilizing the Covert Base Station (CBS). The method calculates the difference in between the calculated distance and the distance between the claimed position of the sensor and the position of CBS, and then removes the position if the difference is beyond a threshold.

In [32, 33], Ekici et al. proposed a probabilistic location verification (PLV) algorithm to verify a sensor position with trusted verifiers. PLV leverages the statistical relationship between the hop counts in the network and the covered distances. The central sensor collects the probability slack and maximum probability values obtained from verifiers and computes the common plausibility to accept or reject the position.

In [34], Mandal et al. presented a secure position verification (SPV) scheme in noisy channels for WSNs, without trusted entities. SPV compares two different distance estimates. One is calculated from a sensor position, and the other is computed utilizing the RSSI localization. SPV is conceptually quite simple, and easy to implement.

4.2 In-Region Verification

In [35], Sastry et al. presented the Echo method to inspect whether a verifying sensor is actually inside the particular region. Echo is similar to the distance bounding [22]; however, it does not need any complicated hardware. The adversary can listen the trusted verifier's response and fabricate its own identity without any authentication.

In [36], Li et al. presented a secure location verification (SVLE) algorithm. The base station continues to perform the VerSec algorithm to verify the RSS value of the verifying sensor. A sensor can be regarded as an adversary, if its signal strength is inconsistent with that of other sensors in region.

In [28], Wei et al. also presented a lightweight algorithm to perform in-region verification by employing the verification center (VC). A probabilistic method is utilized to calculate the confidence that a sensor is inside the verification region.

5 Discussion

In Table 1, we compare each of the studied schemes, showing which type of security they belong to, and provide observations about them and security attacks addressed.

As we can see, most localization schemes do not require the additional hardware. All SLD schemes are beacon-based, except for iterative-based [27] scheme depending on the relative map. However, many LV schemes are beacon-free, in which each sensor collects information from other sensors.

6 Novel Secure Localization Schemes for Applications

A number of security schemes have been presented to ensure the localization security. Trade-offs between energy efficiency and power consumption are investigated in all sorts of literature, a few concentrated on the special attacks and density of beacons, some on the network structure and operational environment, and several on additional hardware requirements, but beacon deployment remained a priority for a few [37].

Additionally, several novel secure localization methods for WSNs applications, including UWSNs, WBAN, and 3D schemes are gradually appeared. Accurate and reliable sensor locations are needed for these applications. Some system functions such as network topology and communication protocols also require the sensor locations to achieve. So, secure localization in these applications has already become a challenge [38].

Table 1 Secure localization schemes comparison

Algorithms	Objective	Beacons-based/free	Range-based/free	Hardware requirement	Defense attacks
CMMSE [15]	SLB	Based	Based	NO	R, W
Label-based [16]	SLB	Based	Free	NO	W
Gradient-based [17]	SLB	Based	Based	NO	Sy
Three distance-based [18]	SLB	Based	Free	NO	Sy
CARSDP [19]	SLB	Based	Based	NO	Sy
SPINE [20, 21]	SLUS	Based	Based	Nanosec clocks	R
SeRLoc [23]	SLUS	Based	Free	Sectored antennas	R, W, Sy
HiRLoc [24]	SLUS	Based	Free	Directional antennas	R, W, Sy
ESLS [25]	SLUS	Based	Based	Nanosec clocks	W, Sy
Iterative-based [27]	SLUS	Free	Based	NO	Sy
GFM and GFT [28]	OSLV	Free	Based	NO	P, I
LAD [29]	OSLV	Based	Based	NO	P, I
CBS [30, 31]	OSLV	Free	Based	NO	T, P, I
PLV [32, 33]	OSLV	Based	Based	NO	T, P, I
SPV [34]	OSLV	Based	Based	NO	P, T, I
Echo [35]	IRLV	Free	Free	NO	T, R, I
SVLE [36]	IRLV	Free	Free	NO	I
VC [28]	IRLV	Free	Based	NO	I

SLB secure localization for beacons; *SLUS* secure localization for unknown sensors; *OSLV* on-spot location verification; *IRLV* in-region location verification; *C* compromising; *Sy* sybil; *R* replication; *W* wormhole; *T* tampering; *I* interference; *R* replay; *P* position-reference

In [38], Han et al. presented a trust model-based collaborative secure localization for UWSNs to determine the sensor location. It first utilizes trust model to mitigate the influence of abnormal sensors and ensure sensor safety, which decreases localization errors and improves location precision. Moreover, localization accuracy can be further enhanced on the basis of the collaboration of sensors.

In [39], Lo et al. presented a new scheme to recognize the positions of wearable sensors in a WBAN automatically, enabling continuously monitor unassisted sensors' locations without beacons. It experimentally demonstrated an enhancement scheme aiming to reduce false-positive (Type I) errors in conventional accelerometer-based on-body fall detection. It also explored on-body energy management mechanisms in the context of emerging WBAN.

In [40], Chen et al. presented a 3D secure localization algorithm. It picks up the excellence beacons by utilizing the steepest descent method and the malicious sensor detection, and then combining with the Newton iterative method to achieve a high localization accuracy in three-dimensional space. It can resist the non-coordinated attack and have a better advantage in the localization accuracy and rate.

7 Conclusion

Without effective security mechanisms to prevent or minimize the effect from incorrect location information, localization would a wrong estimate lead to serious consequences. An overview of current secure localization schemes is presented, classifying them into two categories on the basis of two issues about secure localization, to highlight their strengths and weaknesses. Then we further classify them according to the targets and provide a comparison between different localization schemes, showing the attacks addressed by each. Finally, we propose some novel secure localization schemes for special WSNs applications.

Future trends of secure positioning schemes possibly are: (1) research new lightweight secure localization schemes to reduce energy consumption and computation overhead; (2) evaluate the security and performance of secure positioning schemes with a series of standards; (3) extend new challenges for more applications.

Acknowledgements This work was supported by the National High Technology Research and Development Program of China (Grant No. 2015AA016005), the National Natural Science Foundation of China (Grant Nos. 61402096 and 61173153), Shenzhen Key Laboratory of Ultrahigh Refractive Structural Material (Grant No. CXB201105100093A), and Shenzhen Key Laboratory of Data Science and Modeling (Grant No. CXB201109210103A).

References

1. Karp, B., Kung, H.T.: GPSR: greedy perimeter stateless routing for wireless networks. In: 6th annual international conference on Mobile computing and networking, pp. 243–254. Massachusetts, Boston (2000)
2. Liu, D., Ning, P.: Location-based Pairwise key establishments for static sensor networks. In: 1st ACM workshop on Security of ad hoc and sensor networks, pp. 72–82. ACM (2003)
3. Sastry, N., Shankar, U., Wagner, D.: Secure verification of location claims. In: 2nd ACM workshop on Wireless security, pp. 1–10. ACM (2003)
4. Jiang, J., Han, G., Zhu, C., et al.: Secure localization in wireless sensor networks: a survey. J. Commun. **6**(6), 460–470 (2011)
5. Srinivasan, A., Wu, J.: A survey on secure localization in wireless sensor networks. In: Encyclopedia of Wireless and Mobile Communications. Taylor and Francis Group (2007)
6. Akyildiz, F., Su, W., Sankarasubramaniam, Y., Cayirci, E.: Wireless sensor networks: a survey. Comput. Netw. **38**(4), 393–422 (2002)
7. Zeng, Y., Cao, J., Hong, J., et al.: Secure localization and location verification in wireless sensor networks: a survey. J. Supercomput. **64**(3), 685–701 (2013)
8. Chen, X.: Defense against node compromise in sensor network security. FIU Electronic Theses and Dissertations (2007)
9. Newsome, J., Shi, E., Song, D., et al.: The Sybil attack in sensor networks: analysis and defenses. In: 3rd international symposium on Information processing in sensor networks, pp. 259–268. ACM (2004)
10. Yu, C.M., Lu, C.S., Kuo, S.Y.: Efficient and distributed detection of node replication attacks in mobile sensor networks. In: VTC 2009, pp. 1–5. IEEE (2009)
11. Jain, M., Kandwal, H.: A survey on complex wormhole attack in wireless ad hoc networks. In: 2009 International Conference on Advances in Computing, Control, and Telecommunication Technologies, pp. 555–558. IEEE (2009)
12. Wang, Y., Attebury, G., Ramamurthy, B.: A survey of security issues in wireless sensor networks. IEEE Commun. Surv. Tutor. **8**(2), 2–23 (2006)
13. Cao, X., Yu, B., Chen, G., et al.: Security analysis on node localization systems of wireless sensor networks. China J. Softw. **19**(4), 879–887 (2008)
14. Xing, K., Liu, F., Cheng, X., et al.: Real-time detection of clone attacks in wireless sensor networks. In: ICDCS 2008, pp. 3–10. IEEE (2008)
15. Yang, W., Zhu, W.T.: Voting-on-grid clustering for secure localization in wireless sensor networks. In: ICC 2010, pp. 1–5. IEEE (2010)
16. Wu, J.F., Chen, H.L., Lou, W., Wang, Z., Wang, Z.: Label-based DV-hop localization against wormhole attacks in wireless sensor networks. In: 5th International Conference on Networking, Architecture and Storage, pp. 79–88. Macau, China (2010)
17. Garg, R., Varna, A.L., Wu, M.: Gradient descent approach for secure localization in resource constrained wireless sensor networks. In: ICASSP 2010, pp. 1854–1857. IEEE (2010)
18. Jadliwala, M., Zhong, S., Upadhyaya, S., Qiao, C., Hubaux, J.: Secure distance-based localization in the presence of cheating beacon nodes. IEEE Trans. Mob. Comput. **9**(6), 810–823 (2010)
19. Wang, D., Yang, L.: Cooperative robust localization against malicious anchors based on semidefinite programming. In: MILCOM 2012, pp. 1–6. IEEE (2012)
20. Čapkun, S., Hubaux, J.P.: Secure positioning of wireless devices with application to sensor networks. In: INFOCOM 2005, pp. 1917–1928. IEEE (2005)
21. Čapkun, S., Hubaux, J.P.: Secure positioning in wireless networks. IEEE J. Sel. Areas Commun. **24**(2), 221–232 (2006)
22. Brands S., Chaum D.: Distance-bounding protocols. In: EUROCRYPT '93, pp. 344–359. Springer (1993)
23. Lazos, L., Poovendran, R.: SeRLoc: secure range-independent localization for wireless sensor networks. In: WiSe 2004, pp. 21–30. Philadelphia (2004)

24. Lazos, L., Poovendran, R.: HiRLoc: high-resolution robust localization for wireless sensor networks. IEEE J. Sel. Areas Commun. **24**(2), 233–246 (2006)
25. He, D., Cui, L., Huang, H., Ma, M.: Design and verification of enhanced secure localization scheme in wireless sensor networks. IEEE Trans. Parallel Distrib. **20**(7), 1050–1058 (2009)
26. Zhang, Y., Liu, W., Fang, Y., Wu, D.: Secure localization and authentication in ultra-wideband sensor networks. IEEE J. Sel. Areas Commun. **24**(4), 829–835 (2006)
27. Garg, R., Varna, A.L., Wu, M.: A gradient descent based approach to secure localization in mobile sensor networks. In: ICASSP 2012, pp. 1869–1872. IEEE (2012)
28. Wei, Y., Guan, Y.: Lightweight Location verification algorithms for wireless sensor networks. IEEE Trans. Parallel Distrib. Syst. **24**(5), 938–950 (2013)
29. Du, W., Fang, L., Ning, P.: LAD: location anomaly detection for wireless sensor networks. J. Parallel Distrib. Comput. **66**(7), 874–886 (2006)
30. Čapkun, S., Čagalj, M., Srivastava, M.: Secure localization with hidden and mobile base stations. In: INFOCOM 2006, pp. 23–29. Barcelona (2006)
31. Čapkun, S., Čagalj, M., Srivastava, M.: Secure location verification with hidden and mobile base stations. IEEE Trans. Mob. Comput. **7**(4), 470–483 (2008)
32. Ekici, E., Mcnair, J., Al-Abri, D.: A probabilistic approach to location verification in wireless sensor networks. In: ICC 2006, pp. 3485–3490. IEEE (2006)
33. Ekici, E., Vural, S., McNair, J., Al-Abri, D.: Secure probabilistic location verification in randomly deployed wireless sensor networks. Ad Hoc Netw. **6**(2), 195–209 (2008)
34. Mandal, P.S., Ghosh, A.K.: Secure position verification for wireless sensor networks in noisy channels. In: Ad-hoc, Mobile, and Wireless Networks, pp. 150–163. Springer, Berlin, Heidelberg (2011)
35. Sastry, N., Shankar, U., Wagner, D.: Secure verification of location claims. In: ACM WiSe 2003, pp. 1–10. ACM (2003)
36. Li, C., Chen, F., Zhan, Y., et al.: Security verification of location estimate in wireless sensor networks. In: WiCOM 2010, pp. 1–4. IEEE (2010)
37. Arisar, S.H., Kemp, A.H.: A comprehensive investigation of secure location estimation techniques for WSN applications. Secur. Commun. Netw. **4**(4), 447–459 (2011)
38. Han, G., Liu, L., Jiang, J., et al.: A collaborative secure localization algorithm based on trust model in underwater wireless sensor networks. Sensors **16**(2), 229 (2016)
39. Lo, G.Q., González-Valenzuela, S., Leung, V.: Wireless body area network node localization using small-scale spatial information. IEEE J. Biomed. Health Inform. **17**(3), 715–726 (2013)
40. Chen, Y., Jiang, C.H., Guo, C.: A 3D secure localization algorithm resistant to non-coordinated attack in wireless sensor network. Chin. J. Sensors Actuators **28**(11), 1702–1707 (2015)

Author Index

© Springer Nature Singapore Pte Ltd. 2019
S. K. Bhatia et al. (eds.), *Advances in Computer Communication
and Computational Sciences*, Advances in Intelligent Systems
and Computing 760, https://doi.org/10.1007/978-981-13-0344-9

Printed in the United States
By Bookmasters